CIVIL ENGINEERING AND ENERGY-ENVIRONMENT
VOLUME 2

Civil Engineering and Energy-Environment focuses on the research of civil engineering, environment resources and energy materials. This proceedings gathers the most cutting-edge research and achievements, aiming to provide scholars and engineers with preferable research direction and engineering solution as reference. Subjects in this proceedings include:

- Engineering Structure
- Environmental Protection Materials
- Architectural Environment
- Environment Resources
- Energy Storage
- Building Electrical Engineering

The works of this proceedings will promote development of civil engineering and environment engineering. Thereby, promote scientific information interchange between scholars from top universities, research centers and high-tech enterprises working all around the world.

PROCEEDINGS OF THE 4TH INTERNATIONAL CONFERENCE ON CIVIL ENGINEERING, ENVIRONMENT RESOURCES AND ENERGY MATERIALS (CCESEM 2022), SANYA, CHINA, 21–23 OCTOBER 2022

Civil Engineering and Energy-Environment Volume 2

Edited by

Qingfei Gao
Department of Bridge and Tunnel Engineering, School of Transportation Science and Engineering, Harbin Institute of Technology, China

Zhenhua Duan
Department of Structural Engineering, Tongji University, China

CRC Press is an imprint of the
Taylor & Francis Group, an **informa** business

A BALKEMA BOOK

First published 2023
by CRC Press/Balkema
4 Park Square, Milton Park, Abingdon, Oxon, OX14 4RN
e-mail: enquiries@taylorandfrancis.com
www.routledge.com – www.taylorandfrancis.com

CRC Press/Balkema is an imprint of the Taylor & Francis Group, an informa business

© 2023 selection and editorial matter, Qingfei Gao & Zhenhua Duan; individual chapters, the contributors

The right of Qingfei Gao & Zhenhua Duan to be identified as the authors of the editorial material, and of the authors for their individual chapters, has been asserted in accordance with sections 77 and 78 of the Copyright, Designs and Patents Act 1988.

All rights reserved. No part of this book may be reprinted or reproduced or utilised in any form or by any electronic, mechanical, or other means, now known or hereafter invented, including photocopying and recording, or in any information storage or retrieval system, without permission in writing from the publishers.

Although all care is taken to ensure integrity and the quality of this publication and the information herein, no responsibility is assumed by the publishers nor the author for any damage to the property or persons as a result of operation or use of this publication and/or the information contained herein.

Library of Congress Cataloging-in-Publication Data
A catalog record has been requested for this book

SET
ISBN: 978-1-032-44029-3 (hbk)
ISBN: 978-1-032-44033-0 (pbk)

Volume 1
ISBN: 978-1-032-56057-1 (hbk)
ISBN: 978-1-032-56058-8 (pbk)
ISBN: 978-1-003-43364-4 (ebk)
DOI: 10.1201/9781003433644

Volume 2
ISBN: 978-1-032-56059-5 (hbk)
ISBN: 978-1-032-56061-8 (pbk)
ISBN: 978-1-003-43365-1 (ebk)
DOI: 10.1201/9781003433651

Typeset in Times New Roman
by MPS Limited, Chennai, India

Civil Engineering and Energy-Environment – Gao & Duan (Eds)
© 2023 the Editor(s), ISBN 978-1-032-56059-5

Table of Contents

Preface	xi
Committee Members	xiii

VOLUME 2

Coal mine utilization and energy management research

Study on the key technologies of U.S. smart mines
Xiao Wang & Nan Zhang
3

Application and development suggestions of domestic drilling rigs in
high-temperature geothermal drilling in Kenya
Wenzheng Chen, Rutao Ma & Wen Shi
9

Consideration of the key technologies of exploration and development
based on the standard system of shale gas
Xiaolan Lv, Zhiyong Zhang, Beibei Yang & Lin Zhang
14

Comprehensive evaluation of China's mineral exploration technology
Xianying Huang & Chunfang Wang
23

Research on China's mineral resources exploration strategy
Xianying Huang, Wei Li & Chunfang Wang
29

Thermal and mechanical numerical simulation of a multipoint elastic
restraint system that considers thermal effects
Jibin Wang & Huan He
39

The road to coal transformation under the "carbon neutrality" goal
GuoChen Zhao
48

Numerical simulation of wave action on sand topping section
Xiaodong An, Bin Zhou, Xianliang Yan, Siyuan Dong,
Chengye Liu & Tingting Cui
58

Study on evaluation method of slope toughness of muck deposit
Shaojie Feng, Renzhuo Zhang & Wangqian Deng
64

A research on environmental accounting statements for sustainable
development in China—A case study on coal industry
You Wu, Qini Deng, Qi Shen, Yiran Fang & Yufei Hong
72

Analysis of carbon emission accounting and carbon neutral capacity in
Gansu Province
Rong Huang, Zhicheng Ma, Tian Liang, Jiafu Xi, Yongli Wang,
Lixiang Gong & Bo Yuan
78

Research progress of carbon capture technology in combustion
Guochen Zhao
85

Experimental study on shear properties of modern rammed earth
Xiang Zhao, Hui Tao, Tiegang Zhou, Zengfei Liang & Wei Tan
91

Fuzzy matching of historical conditions-based prediction of carbon
emission intensity of coal-fired power plants
Fanliang Meng, Qinpeng Zhang & Huahai Qiu
100

Influences of fly ash on water and salt transportation of alkali-saline soil
Yue Ma, Guoli Xie, Qiguang Cao, Hongmei Chen & Peng Lu
112

A new natural gas prediction model based on polynomials and its application
Huanying Liu, Changhao Wang, Yulin Liu, Jiaxiang Yang & Bingyuan Hong
119

Research on inversion of surface temperature in Hefei City
Kunmeng Zhou, Xiaobing Zheng, Lan Xiao & Qing Kang
126

The engineering geological suitability assessment of Xiaolongtan
lignite deposit in Yunnan Province, China
Shuran Yang & Qianrui Huang
139

Toward a sustainable Wyndham: An agriculture-led action for the
food security of Wyndham, Australia
Jialing Xie
149

A review of application of deep eutectic solvents as lubricants and
lubricant additives
Ting Li, Zhipeng Zhang, Rui Wang, Junmiao Wu, Qianqian Zou & Yulan Tang
156

Effect of electric field on combustion characteristics of ethanol-air mixture
Zihao Wang, Boyun Liu & Shuai Zhao
162

Occurrence characteristics and mining technology of coal seam in
Dananhu No.2 Mine
Xiaoqian Yuchi
168

A framework of carbon emission comprehensive evaluation standard for
operation and maintenance of parks
Anshan Zhang, Jian Yang, Feiliang Wang, Xihong Ma, Chaofang Zhuang,
Xiwei Qian & Yan Jiang
174

Numerical simulation of the swirling effect of an elastic vertical bulkhead in
a swirling liquid chamber
Liang Chen & Wenfeng Wu
186

Research on transmission line design under electromagnetic environment
impact assessment based on life cycle concept
Yin Du
192

Research on injection, production, and workover engineering technology
for fire flooding development pilot test in Menggulin oilfield
Zhonghai Qin, Xuanqi Yan, Wenjie Wu, Jiancheng Qi, Kejia Wang, Zheyong Sun,
Ying Liu, Lili Wei, Suzhen Guo, Zhi Ma, Zhonghua Shao & Fengqun Li
199

Study on the sensitivity of mechanical properties of two-dimensional random
granules matter to hole depth
Zhongshan Lu, Yuan Liu & Yun Lei
207

Multifractal characteristics of Cu element grade distribution in Jiama porphyry copper deposit 216
Hui Liu & Li Wan

Analysis of the experience and enlightenment of EU carbon market construction 222
Wenjing Ruan

Statistics and restoration governance research on gradation geological environment problems in limestone mines 226
Lei Cheng & Dongdong Li

Study on pretreatment and leaching technology of a fine disseminated refractory gold mine 234
Zhongbo Lu, Guangsheng Li, Xingfu Zhu, Mingming Cai & Yanbo Chen

Research progress of electrochemical oxidation and its coupling technology to remove algal in water 241
Di Jia, Li Lin, Yueqi Cao, Sheng Zhang, Xiong Pan, Lei Dong & Yuting Zhang

Advances in treatment technology of microcystins in water 249
Jun Wang, Dong Liang, Fei Wang, Jing Li, Tao Wang & Zhongwu Zhang

Remote sensing ecological environment assessment based on GF image data – Taking Anshan open-pit mine as an example 254
Zhiwen Hu, Shunbao Liao & Yuna Qi

Ecological data modeling and environmental technology

Maintenance and treatment technology of oil-based drilling fluid polluted by high-pressure brine 263
Shuanggui Li, Sheng Fan, Guohe Xu, Zhong He, Shafei Shu & Cheng Zhai

Research on physical quantity pricing method for new mechanical excavation of transmission line foundation under rock geology 271
Xuemei Zhu, Fangshun Xiao, Cong Zeng, Ye Ke & Ying Wang

Construction of carbon budget management mechanism for power generation enterprises amid national strategy of carbon peaking and carbon neutrality 277
Hongji Li & Lin Hu

Summary of energy management technology for electromechanical system of more electric aircraft 285
Pengyu Wang, Wei Li & Wei Liang

Layout and construction organization design of river ecological control project — A case study of urban section of Shichuan River 293
Kairen Yang, Jiwei Zhu, Liang Li, Wangyu Luo & Chao Zou

Saihanba ecological protection construction based on GIS and its impact on the environment 303
Bing Xia & YuLong Lei

Research progress of rural wastewater treatment technology 309
Cong Xiao, Ning Tian, Hongtao Wang, Xiaoning Qu, Lizhen Ma, Bin Shuai, Mengshan Yu & Weichong Xu

Research progress on the use of wastewater to cultivate oil-producing microalgae 316
Yuqi Huang & Wenhao Jian

Effective mitigation strategies for improving the outdoor thermal environment of university campuses using the ENVI-met Program 326
Lina Yang & Jiying Liu

Problems and countermeasures found in the safety supervision of water projects based on river basin comprehensive management 335
Qian Fu, Meng Ting Huang, Hui Tan & Le Kang

Research on operation mechanism of distributed electro-hydrogen coupling system 342
He Wang, Caixia Tan, Yida Du, Leiqi Zhang, Qiliang Wu & Zhongfu Tan

Changing trend and influencing factors of ambient air quality in Gansu Province: Based on grey correlation analysis 350
Jia Liang, Shan Wang, Longlong Wang, Guohua Chang, Panliang Liu & Jinxiang Wang

Progress of electrocatalytic technology in treating organic chemical wastewater 361
Jianguang Wang, Haifeng Fang, Shiyi Li, Shengjie Fu & Xiaohu Lin

An evolutionary transfer optimization framework on well placement optimization via kernelized autoencoding 369
Ji Qi, Kai Zhang & Xingyu Zhou

Review of research progress on pharmaceutical wastewater treatment technology 375
Jianguang Wang, Shengjie Fu, Haifeng Fang, Shiyi Li & Xiaohu Lin

Research on the impact factors of urban terminal industrial carbon emissions based on the Kaya-LMDI method 382
Hao Chen, Yewei Tao, Peng Wang, Yahui Ma & Wenbo Shi

Study on conventional process optimization after water source switching in the drinking water plant 390
Shuo Zhang, Jiajiong Xu, Yan Wang & Ruhua Wang

Design of on-site wastewater treatment facilities in highway service areas 396
Bo Fan, Jiawei Wang & Zhong Yan

Experimental study on influencing factors of grouting ability and fluidity of cement slurry 402
Weiquan Zhao, Jianhua Zhou, Wei Lu & Zengzeng Ren

Research on the risk measurement and management methods of power purchase by power grid agents 408
Nannan Xia, Bo Dong, Zhanyuan Feng, Jian Zhang & Xilin Xu

Prospect of typical post-combustion carbon capture technology in fossil-fuel power station 415
Guochen Zhao

Growth conditions and growth kinetics of Chlorella Vulgaris cultured in domestic sewage 423
Xingguan Ma & Wenhao Jian

Coordinated control of Ozone and $PM_{2.5}$ under the perspective of air pollution treatment

Dong Li

436

The regional difference and convergence of tourism eco-efficiency in the Yellow River basin

Yu Jie & Wen Ya

443

Environmental evaluation using the analytic hierarchy process

Yihan Jiang

451

Research on the digital twin intelligent management platform in the communication industry existing data center

Wei Wei Kou, Xue Mei Zhang & Yun Long Fan

457

Simulation method of flood routing based on mathematical morphology

Shaobo Wang, Yulong Xiong & Wanhua Yuan

467

Comparative analysis of structural changes and rationality of water conservancy investment in the five Northwestern Provinces of China

Jia He, Jiwei Zhu, Hong Zhao & Jianmei Zhang

474

A numerical simulation of the impact of hydropower development on regional air temperature in Canyon district

Hailong Wang, Bei Zhu, Chang Liu, Shiyan Wang, Shilin Zhao, Xu Ma, Yiqian Tan, Xing Yang & Huazhang Sun

485

An empirical research on the ecological and economic effects of CCER trading in Chinese rural areas: Using synthetic control method

Qianhui Ma, Yixin Lv, Guangyang Yu, Yuqi Liu & Yuhui Fan

495

Analysis of the siltation trend of beach on the south bank of Hangzhou Bay based on remote sensing images in China

Bohu Zhang & Taoxiao Chen

503

Intelligent inspection of enterprise environment based on AR helmet

Yongliang Peng, Hongyu Zhao & Ming Li

509

Study on mechanical properties of soil modified by ash in power equipment foundation under freeze-thaw cycle

Keyu Yue, Zhigang Wang, Yu Zheng & Hongdan Zhao

516

Improvement in the extraction and classification accuracy of vegetation

Xuefei Zhang

523

Retail package design in the context of full-scale commercial and industrial entry

QingChun Li, Nan Liu, Ye Zhang, Yang Qi & QianQiao Zhao

534

Author index

541

Civil Engineering and Energy-Environment – Gao & Duan (Eds)
© 2023 the Editor(s), ISBN 978-1-032-56059-5

Preface

The International Conference on Civil Engineering, Environment Resources and Energy Materials is a leading conference held annually. It aims at building an international platform for the communication and academic exchange among participants from various fields related to civil engineering, environment resources and energy materials. Here, scholars, experts, and researchers are welcomed and encouraged to share their research progress and inspirations. It is a great opportunity to promote academic communication and collaboration worldwide.

This volume contains the papers presented at the 2022 4th International Conference on Civil Engineering, Environment Resources and Energy Materials (CCESEM 2022), held during October 21st–23rd, 2022 in Sanya, China (virtual form). Under the influence of COVID-19 and for the safety concern of all participants, we decided to hold it as a virtual conference which is also effective and convenient for academic exchange and communication. Everyone interested in this field were welcomed to join the online conference and to give comments and raise questions to the speeches and presentations.

The online conference was composed of keynote speeches, oral presentations, and online Q&A discussion, attracting 200 individuals from all over the world. We have invited three sophisticated professors to perform keynote speeches. Among them, Prof. Zhenhua Duan from Tongji University deliver a keynote speech on Early-age Properties of 3D Printed Concrete with Recycled Coarse Aggregate. In his research, a preparation method of 3D printed concrete (3DPC) is proposed based on the secondary mixing of fine aggregate commercial concrete at the construction site. This study investigated the effect of different materials, including cement, metakaolin (MK), and cellulose ether (HPMC) on the workability, rheological properties and density of fine aggregate commercial concrete and the early-age properties of 3DPC, including the extrudability and fresh compressive strength, were studied. All keynote speakers made brilliant speeches and share their unique experience and insights, and here we are hoping the pandemic soon come to an end and we could see each other face to face next year.

CCESEM 2022 received a great many submissions in the areas of civil engineering, environment resources and energy materials. Each submission was reviewed by at least two review experts and the committee picked out some excellent papers that are included in the proceedings, including but not limited to the following topics: Wind Power, Diagnosis and Sensing Systems, Structural Engineering, Plant Protection, Semiconductor Technology, New Energy Materials, etc.

On behalf of the Conference Organizing Committee, we would like to thank the Program Committee members and external reviewers for their hard work in reviewing and selecting papers. And we would like to acknowledge all of those who have supported CCESEM 2022. In particular, our special thanks go to the CRC Press. Hopefully, all participants and other interested readers can benefit scientifically from the proceedings and find it rewarding in the process.

The Committee of CCESEM 2022

Civil Engineering and Energy-Environment – Gao & Duan (Eds)
© 2023 the Editor(s), ISBN 978-1-032-56059-5

Committee Members

Co-Chairman
Prof. Yingbo Ji, *North China University of Technology (NCUT), China*

Program Committee Chairman
A. Prof. Qingfei Gao, *Harbin Institute of Technology, China*

Organizing Committee Chairman
Prof. Jingzhou Lu, *Yantai University, China*

Publication Chair
A. Prof. Zhenhua Duan, *Tongji University, China*

Program Committees
Prof. Songlin Zhang, *Northwest Normal University, China*
Prof. Hui-Mi Hsu, *National Dong Hwa University, Taiwan, China*
Prof. Guangliang Feng, *Chinese Academy of Sciences, China*
A. Prof. Jui-Pin Wang, *National Central University, Taiwan, China*
A. Prof. Zawawi Bin Daud, *University Tun Hussein Onn Malaysia, Malaysia*
A. Prof. Leszek Chybowski, *Maritime University of Szczecin, Poland*
Prof. Haluk Akgün, *Middle East Technical University, Turkey*
Prof. Salvatore Grasso, *University of Catania, Italy*
Prof. Víctor Yepes, *Universitat Politècnica de València, Spain*

Organizing Committees
Prof. Chunzhen Qiao, *North China University of Technology, China*
Prof. Wei Song, *North China University of Technology, China*
Prof. Kamel Khlaef Jaber Al Zboon, *Al-Balqa Applied University, Jordanian*
A. Prof. Md. Naimul Haque, *East West University, Bangladesh*
A. Prof. Damian Pietrusiak, *Wroclaw University of Science and Technolog, Poland*
A. Prof. Supriya Mohanty, *Indian Institute of Technology, India*
A. Prof. Luigi Palizzolo, *University of Palermo, Italy*
Dr. Samia Hachemi, *University of Biskra, Algeria*

*Coal mine utilization and energy
management research*

Civil Engineering and Energy-Environment – Gao & Duan (Eds)
© 2023 the Author(s), ISBN 978-1-032-56059-5

Study on the key technologies of U.S. smart mines

Xiao Wang & Nan Zhang*
Army Engineering University of PLA, Xuzhou, China

ABSTRACT: Smart mine is an anti-maneuvering obstacle weapon that can effectively shape the terrain, damage armored vehicles such as tanks, and delay the enemy's attacks on the battlefield. With their excellent ability to block attacks, smart mines based on advanced technologies such as sensor technology, battery technology, etc., have become a research hot spot in military-related fields, and developed countries such as the United States have begun to strengthen their research on new advanced mine-weapons. Considering the characteristics of smart mines, the article focused on the application status of smart mines and summarized the key technologies and research achievements. In order to evaluate the development trends of smart mines, an analysis model for evaluating technologies was proposed. The result shows that sensor technology and remote control are critical technologies to renovate smart mines and eliminate humanitarian issues. The development trends of smart mines were summarized and prospected.

1 INTRODUCTION

The battlefield's forms have changed dramatically due to the rapid development of military technology. Future wars will therefore feature greater hostility and a more dynamic process, posing significant challenges for weapons and equipment. The western developed countries, such as the United States, began to strengthen their research on new advanced weapons. The Smart Mine is an anti-maneuvering obstacle weapon that plays a crucial role on the battlefield by shaping the terrain, damaging armored vehicles like tanks, and delaying enemy attacks. Traditional mines are passive defense weapons that are victim-triggered because they use target-triggered detonation mode and are typically buried underground. Mines can be controlled in real-time and actively attack targets to quickly block the terrain by incorporating cutting-edge technologies such as communication technology, sensor technology, locating technology, and battery technology. This new kind of mine, known as the Smart Mine, eliminates the drawbacks of conventional mines, which can only be passively triggered by targets, and switches the attack mode from passive defense to active attack. The AHP model summarizes and analyzes the application status, key technologies, and research accomplishments of smart mines in order to assess the development trends of smart mines.

2 DEVELOPMENTS OF U.S. SMART MINES

Mines are a powerful defensive weapon for blocking terrain. Mines gradually became more prevalent on the battlefield during World War I as various fuses were developed.

---

*Corresponding Author: kelsi03@163.com

DOI: 10.1201/9781003433651-1

Mines' shells began to be made of plastic and other modern materials in the 1950s, which made it more difficult to detect and remove them from the battlefield. This significantly increased mines' ability to survive and perform well in combat. However, there is also the issue of mines injuring civilians, including children, which made mines notorious after the war.

Since 1970, new technologies such as microelectronics technology, information technology, and sensor technology have been continuously developed. The United States took the lead in applying these advanced technologies to mines and invested a lot of money to develop smart mines that can detect, identify, track, and attack in a complex battlefield environment. In the early 1990s, the United States successfully developed the M93 wide area mine (WAM), also known as the "Hornet," an anti-tank mine that attacked the top of the tank. M93 WAM captures the vibration and sound of the tank through the various sensors. When the target enters the attacking range, the submunitions will be launched. The M93 WAM can attack any target within a radius of 100 meters. At this time, the smart mines developed by the US military have shown the ability to launch independent attacks. At the beginning of this century, the US army raised the concept of self-healing smart minefield systems, which applied wireless communication and ad-hoc networking technology to the mines. While setting the mines, an ad-hoc network is quickly formed. If one of the mines in the field is damaged, mines from other locations can be controlled to bounce to the damaged area by changing the network topology to repair the network and realize the self-healing of the minefield. The self-healing smart minefield system uses the multi-connections within the minefield to attack and maintain the minefield completely to achieve an effective area blockade. Later, based on the M93 wide-area smart mine, the US military developed the "Raptor Intelligent Combat Outpost" system by combining the wireless sensor network and ad-hoc networking technology of a self-healing smart minefield system. The system consists of a warhead, a communication and networking system, a detection and identification system, and a control system. After being settled in the target area, automatic networking starts to function. Sensors detect the environment, collect information, and search for targets. Data is transmitted to the remote-control station using the remote communication system. In this way, two-way communication has been established, which not only realizes the control of mines and minefields but also broadens the operational use of mines on the battlefield. It can be used to support various operations, provide tactical information, monitor and perform reconnaissance, and maneuver and attack missions in both offensive and defensive operations.

3 KEY TECHNOLOGIES OF SMART MINES

In recent years, the U.S. military has been dedicated to developing new types of mines. On January 31, 2020, the White House officially announced that it would abolish the Obama administration's ban on using anti-personnel mines outside the Korean Peninsula, meaning the United States would again "freely" deploy mines everywhere. Meanwhile, on April 1st, the arsenal of the US Army in New Jersey (Picainny) released a contract. It solicited a prototype described as a "terrain shaping obstacle (TSO) top attack prototype" and required it to be able to conduct a two-way connection through the remote-control unit to confirm the status of a single mine, control the effectiveness of the minefield, selectively switch the minefield, and achieve the ability of friend or foe identification and civilian identification. As a result, mines will continue to be an important counter-mobility weapon capable of temporarily blocking terrain on future battlefields. The research and development of new smart mines outfitted with cutting-edge wireless communication technology, sensor network technology, positioning technology, and battery-powered technology under the framework of compliance, as shown in Figure 1, is an inevitable trend due to the humanitarian issues mines cause, the restrictions placed on conventional weapons.

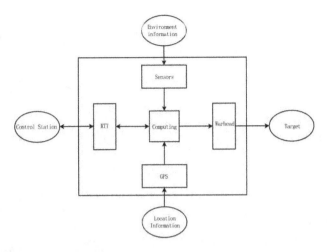

Figure 1. Typical smart landmines system structure.

3.1 *Intelligent technology*

Intelligent technology is constantly changing as a result of ongoing advancements in sensors, information processing, and other critical technologies. Using intelligent technology, landmines can think, see, and hear, enabling sensor units to recognize and track the targets. For the purpose of target identification and precise attack, other information processing characteristics, such as movement and radiation, are used in addition to an acoustic wave, vibration, infrared, and magnetic induction. Mines can now detect the state of the battlefield thanks to composite sensors and information processing technology, allowing the development of smart mines that work with other systems (Li & Xuan 2018).

According to Army Times, the US Army hopes to replace all traditional mines with a new generation of smart mines, which are expected to pose an effective threat to high-tech troops. For example, the M7 Spider smart mine can provide early warning, detection, and identification of intruders by connecting an extended-range trigger line. The XM1100 Scorpion smart mine has perfect network sensors that can detect, track, classify, report, and attack enemy vehicles to shape the terrain effectively and realize force protection.

Applying intelligent technology to mines is equivalent to giving them the ability to think, so they can independently analyze and report. In order to determine the target accurately and to give precise instructions for detonating the target, data from multiple sensors are acquired and analyzed. The wireless sensor network-based smart mine system can both increase the power of the mine itself and attack the target more successfully. The smart mine of the future will be able to "detect, track, and attack" armored vehicles and personnel, transform the minefield into a sensor, realize identification, prevent unintentional harm to civilians, and alert friendly forces in a timely manner. When the headquarters receives word that the enemy is trapped in a minefield, they can dispatch ground troops, aircraft, and even suppress fire at the minefield's location. To expand the use of mines, intelligent technology can turn every mine into a "sentinel" that gathers and transmits complex battlefield data.

3.2 *Communication technology*

Mines are able to perceive and think due to intelligent technology, and communication technology allows mines to "speak" and keeps them under human control. The majority of mines, with the exception of those that are manually detonated, are in an "off-control" state after settling and can only be activated passively by the enemy. Mines typically take the shape of a wireless network based on the characteristics of random deployment in the field in order to adapt to the varying and unpredictable structure of minefields. Given the fact that systems nowadays usually cooperate with each other through networks to support missions, communication technology enables the connection between mines and minefields, minefields and minefields, and minefields and remote-control stations. To control mines and lessen unintentional harm to civilians, humans are programmed into the initiation circuit of mines, which can switch according to remote instructions and remotely perform self-destruction. Cooperative detection, information sharing, and receiving instructions from remote control terminals ensure the greatest combat effectiveness and achieve two-way communication and control (Guo & Zhai 2014; Yuan *et al.* 2003).

The M7 mines developed by the U.S. Army, also known as spider mines, are an excellent example of the application of communication technology in mines. This system is the only anti-infantry mine that meets the requirements of the convention. It has the characteristics of rapid manual setting, recovery, and reuse with good terrain adaptability. Spider Mine consists of up to 84 munitions control units (MCUS), a remote-control station (RCS), and a repeater for extending the communication range. Each MCU can control up to six independent submunitions, each covering an attack range of 60 degrees of arc. Through manual setting, six tripwires are deployed around the mine, and a repeater is placed one mile away from the mine to extend the communication distance up to 3 kilometers. When a tripwire is touched, the mine sends a wireless signal to the remote-control station, allowing the operator to choose whether to attack or not and the opportunity to correct a second command after the first input, thus significantly reducing the chance of friendly fire (Zhang 2017).

Communication technology enables tactical information sharing between mines and between mines and other military subsystems in a communication-networked mine system. Mines can optimize target allocation among themselves, make coordinated attack decisions, improve the success rate of attacks, and cooperate with other military subsystems to complete more complex and flexible tactical tasks.

3.3 *Position technology*

Traditional mines are highly uncontrollable after emplacement. The incomplete mine-laying documents that are missing and the topographic changes after a long time led to uncertainty about mine location, which brought great challenges to mine clearance after the war. The remaining minefields are difficult to remove, which also restricts local economic development and poses a threat to the lives of residents, thus causing a series of humanitarian problems. With the continuous advancement of mine-laying technology, the majority of mines are now laid by remote distribution. To control the location of the minefield, the U.S. military has proposed a networked minefield that can be located through satellite communication for the future battlefield. Positioning technology has the potential to radically lower the risk of demining and the cost of demining after a conflict. Additionally, it can assist in enhancing the capacity to place obstacles over a long distance.

3.4 *Battery technology*

The primary distinction between smart mines and conventional mines is that the former is outfitted with electronic technology. In the 1970s, electronic technology, integrated circuits, and a variety of electronic components were used in mines. With the advancement of electric fuse technology, the firing of mines has evolved from the traditional mechanical way to electric detonators. These electronic components are highly susceptible to power supply issues. The US Army's request for a "terrain shaping obstacle (TSO) top attack prototype" calls for an anti-tank mine to have a standby time of at least six months, which puts the dependability of mine battery technology to the test. Because of their size, mines typically don't use large-scale power supplies, but by using intelligent power management, the designated working time can be met. The electronic components inside the mines are designed for as low power consumption as possible, and the surrounding environment can be monitored with very little power consumption before entering sleep mode to reduce power consumption. When the target is detected, the mines will "wake up" from sleep mode and switch to combat mode to attack the target (Li *et al.* 2018).

4 ANALYSIS OF KEY TECHNOLOGIES

The weighted analysis of the factors affecting the consistency of key technologies in smart mines was carried out by experts. The established judgment matrix was then subjected to a consistency test to determine the weight values of the influential consistency factors under the assumption of a single expert. The analytic hierarchy process (AHP) is a type of decision theory. A comparison matrix was established by comparing the two multi-objectives. The effectiveness of decision-making is significantly increased and decreased when the experience of the expert, the rational analysis, and the uncertain factors are combined. Multi-objective programming can greatly benefit from this in terms of applications. If the results showed that the consistency was good, the weighted arithmetic mean method was then applied to combine the weight of the experts with the weight of the consistency influencing factors index. The weighted value obtained was the comprehensive weighted value of the consistency influencing factors of key technologies in smart mines. The results are shown in Tables 1 and 2.

Table 1. Factors affecting the consistency of key technologies of smart mines.

No	Factors
F1	Microelectronic technology
F2	Sensor technology
F3	Communication technology
F4	Remote-control
F5	Locating technology
F6	Computing technology
F7	Powering technology

Table 2. Comprehensive weight Ψ for influencing factors of key technologies of smart mines.

No	Ψ
F1	0.0666
F2	0.3319
F3	0.0880
F4	0.2641
F5	0.0878
F6	0.1394
F7	0.0222

5 CONCLUSION

In this paper, the key technologies of smart mines are summarized, and the AHP method is adopted to study the key technologies of smart mines, which provides a theoretical reference for future development and has practical guiding significance for engineering practice.

The key technologies that significantly influence smart mines are sensor technology, remote control technology, computing technology, etc. Applying sensor technology to smart mines makes it possible for the mines to attack independently. The dependability of various sensor technology will improve the efficacy of smart mine operations and will gradually be applied to the front lines of the battlefield. Meanwhile, based on remote-control technology and command decision-making, the purpose of "human in the loop" is achieved to keep the mines under human control to reduce citizen harm and other humanitarian issues. Based on these technologies, smart mines can accomplish integrated detection, recognition, tracking, and attacking; they can also realize the connection between weapons and between weapons and remote-control stations.

REFERENCES

Guo Ying, Zhai Yanlong. Research on Channel Detection Algorithm of Intelligent Mine Field Based on WSN [J]. *Computer Science*, 2014(S1):5.

Li Hanming, Xuan Zhaolong, et al. Application Situation and Development Direction of Battery n Ammunition. *Chinese Journal of Power Sources*, 2018, 42(11):1761–1763.

Li Hanming, Xuan Zhaolong. Development Status and Key Technology of Intelligent Mine. *Modern Defence Technology*, 2018, 46(02):6–11.

Yuan Ping, Chen Xiang, Zhang He. The Research on Wireless Communication Network of Smart Minefield. *Journal of Projectiles, Rockets, Missiles and Guidance*, 2003, 23(002):43–45.

Zhang Lin. Development Status of "spider" Mine System in the U.S. Army. *Foreign Tank*, 2017(2):42–43.

Civil Engineering and Energy-Environment – Gao & Duan (Eds)
© 2023 the Author(s), ISBN 978-1-032-56059-5

Application and development suggestions of domestic drilling rigs in high-temperature geothermal drilling in Kenya

Wenzheng Chen
China Petroleum Technology & Development Corporation, Beijing, China

Rutao Ma*
CNPC Engineering Technology R&D Company Limited, Beijing, China

Wen Shi
China Petroleum Technology & Development Corporation, Beijing, China

ABSTRACT: This paper examined the difficulties of geothermal drilling in Kenya, the development status of Chinese-funded enterprises in Kenya's geothermal drilling business, the application of domestic drilling rigs in Kenya's high-temperature geothermal drilling, and the development prospects of Kenya's geothermal market, as well as problems in the development of geothermal projects. It raised suggestions for relevant Chinese-funded enterprises to carry out research on enhanced geothermal drilling technology in Kenya from the aspects of high-temperature geothermal drilling technology study, drilling rig assembly optimization, after-sales service, market development, etc.

1 INTRODUCTION

Kenya is located in eastern Africa, spanning the north-south equator, and is one of the most active countries in terms of economic and social development in East Africa. Increasing power supply has become a big issue facing the Kenyan government and the people's livelihood. Kenya's primary energy sources are hydropower, nuclear power, and geothermal power. Geothermal resources are currently the most local potential source for Kenya to generate power. According to the plans of the Kenyan government, by 2030, the total amount of geothermal power generation will increase from the current 165,000 kW to 900,000 kW. With the development of geothermal power generation, there will be a large market demand for high-temperature geothermal rigs and high-temperature geothermal drilling services in the next few years. It also provides huge business opportunities for Chinese drilling contractors and drilling machine exporters (Bu *et al.* 2011; Li & Zhu 2007; Wang *et al.* 2010). Since 2006, China Petroleum Technology Development Corporation and Great Wall Drilling have carried out geothermal drilling operations in Kenya, developed air foam drilling and high-temperature directional drilling technology, and successfully completed geothermal wells with a temperature of 350°C. By analyzing the difficulties of geothermal drilling in Kenya, the author found the future application of domestic drilling rigs in Kenya's geothermal drilling and evaluated the development prospects of Kenya's geothermal market, while also pointing out problems in the development of geothermal projects. The paper raised suggestions for relevant Chinese-funded enterprises to carry out research on

---

*Corresponding Author: marutao8251@163.com

DOI: 10.1201/9781003433651-2

9

enhanced geothermal drilling technology in Kenya from the aspects of high-temperature geothermal drilling technology study, drilling rig assembly optimization, after-sales service, market development, etc.

2 THE DIFFICULTY OF HIGH-TEMPERATURE GEOTHERMAL DRILLING IN KENYA

2.1 *Complicated geological structure, severe drilling fluid leakage, and frequent stuck pipe*

Kenya's geothermal development area is mostly located on the open structure of volcanic eruption deposits, and its geological structure shows different fault structures. There are large empty holes in some well sections, and the space between faults is rough and over-grown. During spud operations, drilling fluid is lost from circulation after about 20 meters. Most of the drilling cuttings exist in the annulus and bottom of the well. Torque and drag are extremely high on the BHA (bottom hole assembly), and the service rig is generally equipped with air drilling equipment.

2.2 *High temperature and corrosive gases*

During the drilling process, 150°C and higher-temperature steam appeared at a depth of about 60 m, and the temperature at about 1200 m rose to above 250°C. The temperature at the well's bottom could reach 300°C, and the steam contains a high concentration of H2S.

2.3 *High environmental protection requirements and strict restrictions on drilling fluids*

The geothermal drilling operating area is environmentally sensitive. Strict requirements for the protection of wild animals and the environment and the ban on conventional fluids used in oil drilling increase the difficulty of removing drilling cuttings (Dai *et al.* 2011; Kohlt *et al.* 1991; Lu *et al.* 2009).

3 THE CURRENT SITUATION OF DOMESTIC RIGS IN KENYA HIGH -TEMPERATURE GEOTHERMAL DRILLING

3.1 *Status of chinese-funded enterprises that carry out geothermal drilling*

The geothermal drilling industry is currently active in Kenya, largely owing to Chinese-funded businesses. Drilling contractors and suppliers of rig equipment both exist. China Petroleum Technology Development Corporation, Shandong Dongying Haixin Petroleum Equipment Co., Ltd., and Sichuan Honghua Petroleum Equipment Co., Ltd. are the main equipment suppliers. In the meantime, Sinopec Jianghan Oilfield Drilling Company and Great Wall Drilling Company are the main drilling contractors.

In terms of equipment suppliers, China Petroleum Technology Development Corporation has won the bidding project for the Kenya Geothermal Drilling Project twice since 2009. There are four sets of geothermal rigs, providing services such as air drilling, directional drilling, and high-performance drilling fluids. For the first time, Chinese-funded enterprises sold high-temperature geothermal drilling rigs to the international market. The geothermal rig exported by the company met the harsh working conditions during the geothermal dril-ling process. Shandong Dongying Haixin Petroleum Equipment Co., Ltd. signed two 2,000 HP rig supply contracts with Kenya National Electric Power Corporation (Kengen) in 2011. It was the first domestic private enterprise to provide a full set of high-temperature geothermal rigs to Kenya. In addition, Sichuan Honghua Petroleum Equipment Co., Ltd. exported three rigs to Kenya in 2013 (Mu 2000; Shen *et al.* 2009).

At the end of 2006, a new drilling company, PetroChina Great Wall Drilling Company, entered the Kenyan geothermal drilling market. Since then, four drilling rigs have served the Kenyan market. They are large capacity rigs capable of drilling to depths of over 5,000 meters. The main operation areas are in Alkaria. After several years of hard work, Great Wall Drilling Company has achieved perfect performance in geothermal drilling technology (Chen & Ma 2005; Sun & Su 2006; Xiang 2007). Sinopec Jianghan Oilfield No. 2 Drilling Company successfully won the Kenyan petroleum block exploration service after entering the Kenyan market. Later, they entered the geothermal drilling field. From November 2010 to March 2011, two geothermal wells were constructed for the Kenya National Electric Power Company.

3.2 *Configuration and application of kenya high-temperature rig*

The Chinese drilling rig that is servicing the Kenya geothermal well is a large-capacity electric drilling rig with 1,500 to 2,000 horsepower. The typical difference between this rig and conventional rigs is that this rig is equipped with air drilling equipment, utilizing an SCR room and an independent generator room. The promotion system is equipped with two 800 kW draw works and a 450-ton hook. It also contained an air foam drilling system with five air pressure machines, two pressure compressors, and a top drive system.

Domestic drilling rigs have been successfully applied in Kenya for high-temperature geothermal drilling, which meets the environmental requirements of high-temperature drilling. Great Wall Drilling Company and China Petroleum Technology Development Corporation conduct drilling operations in the Menngai area of Kenya. Several techniques are performed and well tested, such as foam drilling, high-temperature directional drilling, drill bit selection, anti-stick drilling, etc., which were used to successfully complete geothermal wells with a temperature of 350°C. The application of foam drilling technology solves the problem of severe drilling fluid leakage, effectively purifies boreholes, and reduces the occurrence of stuck pipe accidents. In terms of high-temperature directional drilling, the temperature is as high as 300°C, and the only choice of a downhole measurement instrument is a single-point orientation device. In Kenya, the borehole track control is adapted to the single-point diagonal measurement. By utilizing a turbo drill without rubber seals, it can be adapted to temperatures up to 300°C. ROP was greatly improved by measures such as optimizing drills to match the strata and using reasonable drilling combinations. The duration is reduced from the original 75-day 2200-meter straight well to 50–55 days. Through measures such as using protective fluid agents, controlling the pH value of the water, and purifying and cooling the drilling fluid, the working life of the drilling tools has been effectively guaranteed.

4 DEVELOPMENT SUGGESTIONS

4.1 *Reinforcement of geothermal drilling process research*

Compared with traditional drilling technologies, high-temperature geothermal drilling may encounter complicated strata during drilling, as well as frequent stagnation, high-temperature corrosion, and stuck pipe accidents. The ability of the drillers directly affects the life and performance of the equipment. Chinese engineers recognized the importance of drilling technology in the protection operation and training services. In order to improve the benefits of the project, the drill contractor has invested a certain amount of manpower and material resources in the high-temperature geothermal drilling process. Improper drilling operations or drilling process design can also cause equipment damage. However, there was a lack of research on crafts and weak information support, causing difficulties in project implementation. It is recommended that equipment suppliers in the future strengthen the research of high-temperature geothermal drilling technology and research on related tools and equipment.

4.2 *Improve the complete set of capabilities and methods of the product*

The bidding for Kenya's high-temperature rigs requires fully equipped rigs. At present, most domestic companies are relatively familiar with the derrick, substructure, hoisting system, drilling pump, power system, etc. The technical details, such as cementing systems, directional drilling tools, fishing tools, and drilling bits, are not understood. The comprehensive set of complete capacity sets is relatively weak. Imported equipment needs to be deducted from domestic exports.

4.3 *Rig export enterprises should improve after-sales service capabilities*

Kenyan owners have limited knowledge of rig operation and maintenance. They lack high-level technical personnel. Purchasing bidding has made detailed requirements for personnel training, installation, and trial operation. The language abilities of the after-sales technical service personnel are poor, and their understanding of local culture is limited; meanwhile, there are situations that cause friction due to cultural differences. Standard management procedures and English-translated traceable documents are hard to come by in rig transfer, training records, and after-sales service. Because of a lack of capabilities in the emergency treatment department, even if the equipment is damaged due to improper user operation, it will be strongly claimed due to quality issues. It is recommended that rig export enterprises create a professional and high-quality after-sales service team and establish a scientific service team management system to improve the level of after-sales service, providing customers with nanny-like services so that after-sales service will become the income of enterprises.

4.4 *Market development recommendations*

Related enterprises should pay close attention to the development plan for Kenya's geothermal power, study and analyze the development trend of important projects, actively participate in the preliminary work of the project, and make full use of the valuable experience accumulated in the geothermal drilling market.

Although the Kenyan geothermal market has broad prospects, the Kenyan government lacks corresponding support funds. Chinese companies should participate more flexibly in the Kenyan geothermal projects and make full use of the World Bank and China-Africa Development Bank to carry out projects.

Judging from the previous Kenyan high-temperature geothermal rigs, the competitive strength is the Chinese rig supplier. The price is becoming more and more transparent, and the profit window is getting smaller and smaller. They should focus on analyzing and cultivating their own comparative advantages, respond rapidly to market changes, and improve project execution capabilities and other aspects.

Due to the high cost of hiring foreign drilling contractors, it is not conducive to promoting Kenya's national employment and economic development. The Kenyan government has the willingness to support the development of state-owned drilling contractors and put it into action.

Geothermal resource development has been extremely successful in Kenya, where it has also sparked a lot of interest on a national level. Chinese-funded businesses should think about actively expanding in neighboring potential markets like Rwanda and Uganda.

5 CONCLUSION

Chinese companies have repeatedly won bids in the international tendering projects for Kenya's high-temperature geothermal drilling projects in recent years, highlighting the competitive strength and development level of Chinese companies.

Improving comprehensive sets of capacity, reducing complete sets of costs, strengthening the construction of after-sales service teams, and customer-oriented procedural management with an emphasis on high-temperature geothermal drilling process research and cognition are the efforts of China's high-temperature geothermal rig suppliers in the future.

High-temperature geothermal drilling service projects, relatively few competitors, and perfect project economic benefits are an effective way to improve the quality and efficiency of drilling contractors.

REFERENCES

Bu Pingping, Zhao Jinchang, Zhao Yangsheng et al. (2011). Key Technical Research on High-Temperature Rock Body Geothermal Drilling Construction, *Journal of Rock mechanics and engineering*. 30 (11):2234–2242.

Chen Liren, Ma Guang snake (2005). The Current Status and Thinking of the Development of Petroleum Drilling Equipment at Home and Abroad, *Petroleum Machinery*. 33 (4): 193–196.

Dai Yiqin, Tang Qiyu, Zhou Jinjin (2011). Kenya OW-41 well Drilling Technology, *Jianghan Petroleum Technology*. 21 (2): 44–46.

KOHLT, Evans K.F., Hopkir R.J. et al. (1991). Recent Geothermal Technology in Kakkonda and Matsukawa, *Geothermal Resources Council Bulletin*. 5 (7): 166–174.

Li Zhimao, Zhu Tong (2007). The Current Status of Geothermal Power Generation in the World, *Solar Forum*. 16 (8): 10–14.

Lu Liqiang, Shan Zhengming, Deng Faibing et al. (2009). The Application of Air Bubble Drilling Technology in Kenya High-Temperature Geothermal Drilling. *Drilling Process*. 32 (6): 5–7.

Mu Degang (2000). ZK201 High-Temperature Geothermal Rhombus Technology, *Petroleum Drilling Technology*. 28 (4): 11–13.

Shen Yan, Liu Jun and Cheng Xiaonian (2009). Application of Foam Fluid Drilling Technology in Kenya OW904 Ultra-high Temperature Geothermal Wells. *Journal of Chongqing University of Science and Technology (Natural Science Edition)*. 11(4): 16–18.

Sun Ning, Su Yi brain (2006). *Progress of Drilling Engineering Technology*. Beijing: Petroleum Industry Press. 9: 60–92.

Wang Long, Zhang Ziqiao, Zhang Zhiwei (2010). Waiting for the Application of High-Temperature Geothermal Drilling Technology in Kenya Olkaria, *Foreign Oil Field Project*. 26 (10): 32–34.

Xiang Wenjun (2007). Direct Drilling Technology and its Applications, *Minerals*. 28 (5):28–32.

Civil Engineering and Energy-Environment – Gao & Duan (Eds)
© 2023 the Author(s), ISBN 978-1-032-56059-5

Consideration of the key technologies of exploration and development based on the standard system of shale gas

Xiaolan Lv
Development Research Center of China Geological Survey, Beijing, China

Zhiyong Zhang*
Beijing Stimlab Oil & Gas Technology Co., Ltd, Beijing, China

Beibei Yang
Development Research Center of China Geological Survey, Beijing, China

Lin Zhang
Haikou Marine Geological Survey Center of China Geological Survey, Haikou, Hainan Province, China
Oil and Gas Resources Survey Center of China Geological Survey, Beijing, China

ABSTRACT: The high-quality standard system plays a leading and guiding role in the development of the shale gas industry. China has established a technical standard system for the exploration and development of shale gas by which it has been regulated. In this paper, 55 characteristic technical standards of 6 specialties, including geological evaluation, earthquake and logging, drilling and completion technology, reservoir transformation, gas reservoir development, and safe and clean production, have been sorted out and published. This paper analyzes the technical system of exploration and development of shale gas in line with China's reality, including 27 characteristic technologies of 6 main technologies, namely geological comprehensive evaluation technology, shale gas development optimization technology, horizontal well optimal drilling technology, horizontal well volume fracturing technology, factory operation technology as well as shale gas characteristic efficient and clean extraction technology. It introduces the application effect of six major technologies in the exploration and development of shale gas in the Sichuan Basin. According to the four majors with more standards in the technical standard system of shale gas, 15 technical difficulties and key technologies in the corresponding four key stages, geological evaluation, drilling and completion, reservoir stimulation and development, are sorted out. Finally, two suggestions are drawn: one is to build a standard system for high-quality development, and the other is to look forward to the key technology research direction of the exploration and development of shale gas.

1 INTRODUCTION

On October 10, 2021, the National Outline for Standardization Development (hereinafter referred to as "the Outline") was issued, pointing out that standards are the technical support for economic activities and social development. The Outline proposes to speed up the establishment of a standard system for promoting high-quality development. It requires that the proportion of standard research results of generic key technologies and applied science and technology planning projects during the 14th Five-Year Plan period should reach more

---

*Corresponding Author: lffyzzy@163.com

than 50% ("National standardization development outline"). During the 13th Five-Year Plan period, shale gas production increased by 4.5 times in China (Lu 2021). Shale gas production was 23 billion cubic meters in 2021, and natural gas production was 205.3 billion cubic meters (Wang 2022). The proportion of shale gas in natural gas production has stabilized at about 10%, and shale gas will be the main force to increase reserves and production of unconventional natural gas during the 14th Five-Year Plan period. Considering the key technologies of shale gas exploration and development from the standard system is of great significance to implement the spirit of the Outline and to promote the high-quality development of the shale gas industry.

2 SHALE GAS INDUSTRY USHERS IN A PERIOD OF HIGH-QUALITY DEVELOPMENT IN CHINA

2.1 *The promotion of oil and gas reserves and production in China is a strategic choice to ensure the national energy security*

Oil and natural gas security is the core of energy security. China's external dependence on oil and natural gas is as high as 70% and 40%, respectively. Only by continuously enhancing the ability of oil and gas self-sufficiency can we firmly put the energy rice bowl in our hands (Wang 2022). On October 24, 2021, to effectively ensure the supply of oil and gas, the State Council issued an action plan for a carbon peak by 2030. On October 24, 2021, China proposed to accelerate the large-scale development of unconventional oil and gas resources such as shale gas, coalbed methane, tight oil (gas), etc. On December 18, 2021, SASAC held a meeting of heads of central enterprises, emphasizing the exploration of domestic resources, promoting the reserves and production, and better playing a supporting role with self-sufficiency in the production of important energy resources.

2.2 *China's shale gas industry has entered the stage of industrial development*

In China, the favorable exploration area of shale gas is 43×10^4 km^2, and recoverable resources reach 31.6×10^{12} m^3, ranking first in the world (Wang 2017). Shale gas production has gone through three stages: the cooperation and reference stage (2007–2009), the independent exploration stage (2010–2013), and the industrial development stage (since 2014). Important events (Zou et al. 2021), (Yue et al. 2020) and annual production (Wang 2022), (Yue et al. 2020), (Zhao et al. 2021) are shown in Figure 1. The marine shale gas in Sichuan Basin, with a depth of less than 3500 m, has been effectively developed on a large scale, and a breakthrough has been made over 3500 m.

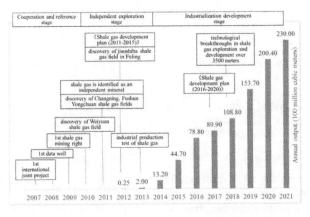

Figure 1. Development history and the annual output of the shale gas industry in China.

2.3 *Standardization plays a leading and guiding role in industrial development*

As the means of management standardization, the thrust of industrial technological progress and the starting point of the healthy development of industry have been paid more and more attention by all walks of life. The establishment of the standard system is the basic and strategic standardization work. Guided by the needs of shale gas development in the energy industry, a standardized top-level design and a shale gas standard system can standardize and drive the healthy and rapid development of China's shale gas industry and promote scientific and technological innovation.

3 CONSTRUCTION OF SHALE GAS STANDARD SYSTEM IN CHINA

In order to meet the needs of standardization for the development of China's shale gas industry, the National Energy Administration approved the establishment of the Energy Industry Shale Gas Standardization Technical Committee in the energy industry (hereinafter called "Shale Gas Standardization Committee") in August 2013. A standard system covering the business development of the whole industry chain of shale gas has been established to guide the orderly development of shale gas and standardize production operations for shale gas. According to the special characteristics, the standard shale gas technology system comprises a general basis and six majors, including geological evaluation, earthquake and logging, drilling and completion technology, reservoir reconstruction, gas reservoir reconstruction, and safe and clean production, as shown in Figure 2. A total of 1773 standards were formulated in the standard system planning, including 1600 conventional oil and gas standards related to shale gas exploration and development formulated by the National Natural Gas Standardization Technical Committee and the National Oil and Gas Standardization Technical Committee, and 173 shale gas characteristic standards were formulated by the Shale Gas Standardization Committee (Yue *et al.* 2020).

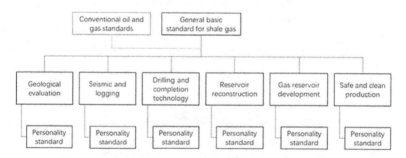

Figure 2. The standard system structure for shale gas.

So far, 55 national and industrial standards for shale gas have been issued in China, as shown in Table 1.

3.1 *General basic standard*

Technical requirements and test methods for shale gas (GB/T 33296-2016) mainly aim at the standardization requirements of the product technology involving the whole industry in shale gas exploration, development, transmission, and utilization.

Table 1. The standard level and name for geothermal energy.

| Major | Published standards (items) | | | Sub majors |
	Subtotal	National standard	Industry standard	
General basis	1	1	0	
Geological evaluation	10	4	6	Including geological evaluation, reserve evaluation, geological analysis, and logging
Earthquake and logging	2	0	2	Including geophysical exploration and logging
Drilling and completion technology	15	0	15	Including pre-drilling engineering, drilling engineering, cementing engineering, well completion engineering, and drilling equipment
Reservoir reconstruction	14	0	14	Including fracturing technology, oilfield chemistry, and fracturing tools
Gas reservoir development	11	1	10	Including development plan, reserve calibration, development modeling, conceptual design, development experiment, dynamic reserve calculation, numerical simulation, and economic benefit evaluation
Safe and clean production	2	0	2	Including ground construction, safety, and environmental protection
Total	55	6	49	

Data source: The website of the Shale Gas Standardization Technical Committee of the energy industry, http://www.cscsg.cn/, enter on July 24, 2022

3.2 Geological evaluation standard

Geological evaluation standards include geological evaluation, reserve evaluation, geological analysis, and logging. The standards have been issued, including the identification of shale and mudstone rock slices, specification for resource evaluation, and exploration of the targeted evaluation technology.

3.3 Earthquake and logging standards

Earthquake and logging standards include geophysical exploration and logging. Two standards have been issued, including processing seismic exploration data and monitoring microseismic fracturing.

3.4 Drilling and completion technology standard

Drilling and completion technology standards include pre-drilling engineering, drilling engineering, cementing engineering, well completion engineering, and drilling equipment. Fifteen standards have been issued, including drilling design, drilling fluid, well control, HSE management, etc.

3.5 Reservoir reconstruction standard

Reservoir reconstruction standards include fracturing technology, oilfield chemistry, and fracturing tools. Fourteen standards have been issued, including fracturing design and chemical reagents for fracturing.

3.6 *Gas reservoir development standard*

Gas reservoir development standards include development scheme, reserve calibration, development modeling, conceptual design, development experiment, dynamic reserve calculation, numerical simulation, and economic benefit evaluation. Eleven standards have been issued, including the feasibility demonstration for shale gas development, recommended practices for the development of gas reservoirs, and economic benefit evaluation.

3.7 *Safety and clean production standards*

Safety and clean production standards include ground construction, safety, and environmental protection. Two standards have been issued to reduce the impact of fracturing on the ground and to design the shale gas field gathering and transmission engineering.

4 SHALE GAS EXPLORATION AND DEVELOPMENT IN CHINA

4.1 *The technology system for shale gas exploration and development in China*

Combined with the characteristics of exploration and development of shale gas in China, the technology system for shale gas exploration and development in line with China's actual situation is established, including 27 characteristic technologies of 6 main technologies, including geological comprehensive evaluation technology, shale gas development optimization technology, horizontal well optimization, and fast drilling technology, horizontal well volume fracturing technology, industrialized operation technology and shale gas characteristic high-efficiency and clean production technology (Yue *et al.* 2020), as shown in Table 2.

Table 2. The technology system of Shale gas exploration and development in China.

Comprehensive geological evaluation technology	1) Shale gas analysis and test technology 2) Seismic reservoir prediction technology 3) Logging evaluation technology of horizontal wells 4) Evaluation index system and optimization technology of favorable areas
Shale gas development optimization technology	1) Integrated modeling technology of geological engineering 2) Develop optimized deployment technology 3) Optimization design technology of horizontal wells 4) Dynamic tracking and quantitative analysis and prediction technology
Optimized and fast drilling technology for horizontal wells	1) Drilling optimization design technology of platform horizontal well group 2) Optimized and fast drilling technology for horizontal wells 3) Integrated guiding technology of geological engineering 4) Cementing technology under the condition of oil-based drilling fluid
Volume fracturing technology of horizontal well	1) Integrated fracturing design technology of Geological Engineering 2) Low viscosity slick water + low-density proppant fracturing technology 3) Dense cutting segmented clustering + high strength sanding technology 4) Real-time adjustment technology for fracturing construction 5) Temporary plugging and steering (multiple) fracturing technologies
Industrialized operation technology	1) Factory layout technology 2) Industrialized drilling technology 3) Industrialized fracturing technology

(continued)

Table 2. Continued

Shale gas characteristic efficient and clean exploitation technology	1) Standardized design technology 2) Data acquisition and data integration technology 3) Real-time monitoring and remote-control technology 4) Fracturing fluid reuse technology 5) Water and soil protection technology 6) Combined skid-mounted technology 7) Collaborative analysis and assistant decision technology

In the practice of shale gas exploration and development in the Sichuan Basin, the six main technologies have achieved good results (Yue *et al.* 2020). The comprehensive geological evaluation technology has optimized six favorable areas in Sichuan Basin, such as Changning, Weiyuan, and Fuling, and solved the problems of the distribution scope and resource potential of the favorable areas. Shale gas development optimization technology solves the problems of utilization for shale gas reservoir resources and prediction for production performance and realizes the efficient development of shale gas wells. The coincidence rate of geological modeling and dynamic prediction exceeds 90%. The optimized and fast drilling technology of horizontal wells solves the problems of complex accidents and low ROP and realizes that the average drilling cycle of a single well within 5000 meters is less than 70 days, which independently researches and develops key tools and liquid systems, forms the integrated fine fracturing technology of geological engineering, and realizes the volume transformation of shale reservoir with high-level stress differences. The industrialized operation technology solves the problems of the long well construction cycle, low construction efficiency, and low utilization of personnel and equipment, which realizes "resource sharing, reuse, speed-up, and efficiency increase." Efficient and clean production technology has realized the rapid construction and investment of production capacity, reduced labor intensity and production cost, completed the harmless treatment of drilling and fracturing waste, realized the "standardized design, skid mounted equipment, industrialized prefabrication and integrated device" for surface gathering and transportation, and realized the "automatic production, digital office, and intelligent management" for digital gas fields.

4.2 *Key technologies of shale gas exploration and development in China*

There are 50 characteristic standards for the four majors of geological evaluation, drilling and completion technology, reservoir reconstruction, and gas reservoir development. Therefore, their corresponding four stages are key. The technical difficulties and key technologies of the four stages are sorted out (Chen 2018; Guo & Zhao 2018; He *et al.* 2021; Liu *et al.* 2018; Liu *et al.* 2019; Liang *et al.* 2017; Ma *et al.* 2020; Wang *et al.* 2018; Wang 2019, 2022; Xie *et al.* 2017; Zou *et al.* 2017, 2021; Zhu *et al.* 2021), as shown in Table 3.

Table 3. Technical difficulties and key technologies for exploration and development of shale gas.

Stage	Technical difficulties	Key technologies
Geological evaluation	1) The formation and enrichment mechanism of shale gas is not clear, and the uncertainty of shale gas resources is large.	1) Evaluation for shale reservoir classification and optimized targets

(continued)

Table 3. Continued

Stage	Technical difficulties	Key technologies
	2) There is no classification evaluation for the shale gas adaptability and favorable targeted optimization method in China, which makes it difficult to locate gas-rich areas. 3) The rock cuttings are fine or powdery, which increases the difficulty of logging lithology identification.	2) Evaluation of shale gas 3) Cutting logging and sample identification description
Drilling and completion	1) The adaptability of conventional rotary guide tools is not strong. 2) The precise geological guidance technology of "needle type" horizontal wells is not mature. 3) The drilling encounters rate of the class I reservoir is not high. 4) "One trip drilling" technology has not yet been formed. 5) The drilling fluid system cannot meet the needs of prevention from formation collapse and leakage.	1) Horizontal well drilling technology 2) Rotary guidance system 3) Underbalanced drilling technology 4) Logging and drilling technology 5) Foam cement cementing technology
Reservoir re-construction	1) High-level stress differences make it difficult for the deep shale reservoir to form a complex fracture network. 2) High fracture pressure and closure stress lead to difficulty in crack initiation and extension. 3) The control mechanism of bedding on fracture height and the influence of the mechanism on the natural fracture zone for fracture propagation are unclear. 4) It is difficult to inject proppant under the condition of high closure stress and elastic modulus, and it is difficult to obtain high fracture conductivity.	1) Subdivision dense cutting volume fracturing technology 2) Well factory fracturing technology 3) Horizontal well-refracturing technology 4) CO_2 fracturing technology 5) Research and development of low-cost fracturing fluid system 6) Research and development of key fracturing tools
development	1) The phase behavior of CH_4 in the pores of deep shale reservoirs is still unclear. 2) The multi-scale flow law of gas in deep shale reservoirs is not clear, and the microflow mechanism of gas-water two phases is still under exploration. 3) The technical countermeasures for deep shale gas development are not clear.	1) "Factory" operation of a well pad 2) Pressure limiting control and the rapid development of production decline

5 CONCLUSIONS

We can draw two conclusions as follow:

Since the establishment of the Shale Gas Standardization Committee in 2013, it has established a technical standard system suitable for China's shale gas exploration and development and has formulated and issued 55 characteristic technical standards to regulate exploration and development, meeting the needs of China for supervision of shale gas development. The formulated technical standards have provided operation specifications for shale gas exploration and development, effectively improving operation efficiency, promoting the substantial improvement of relevant technical levels, and strongly supporting the healthy development of China's shale gas industry.

After more than ten years of scientific exploration and technology introduction, China has gradually formed a technology system for shale gas exploration and development in line with China's actual situation, including six technology series and 27 characteristic technologies. In this paper, 15 technical difficulties and key technologies in 4 key stages of shale gas exploration and development are sorted out according to 4 majors with more standards in the technical standard system of shale gas.

6 SUGGESTIONS

6.1 *Establishment of a high-quality technical standard system*

According to the GB/T 13016-2018 standard system construction principles and requirements, a systematic analysis of international standards, national standards, industry standards, community standards, local standards and enterprise standards, etc., in combination with the geothermal energy development planning, consultation, builds the quality "difference." Even the longer planning phase of the shale gas standard system of the energy industry standardizes and leads the shale gas industry to achieve scientific development.

The standard system is to be improved through demonstration projects. The shale gas standard system is studied and established on the basis of relevant national and industrial standards and needs to be adjusted and supplemented by practical results. The operation experience for the demonstration project of shale gas exploration and development is summarized to provide the basis for further improvement of the standard system.

6.2 *Clarification of the key technology research direction of shale gas exploration and development*

In the coming period, the key technologies of shale gas exploration and development in China will focus on drilling engineering for deep horizontal wells and later fracturing and stimulation technology.

In terms of drilling, foreign countries are still in a monopoly position in the core technologies of shale gas drilling, such as rotary steering, geological steering, high-performance water-based drilling fluid, efficient PDC bits, automated drilling rigs, etc., which are the technical difficulties that need to be overcome in the subsequent research of shale gas development in China.

A shale gas reservoir belongs to an "artificial gas reservoir," and fracturing is the core of its development. Given the complex surface conditions of China's shale gas fields and the need for large-scale fracturing technology, the next step will mainly study and practice deep ultra-deep marine shale gas fracturing, continental marine continental transitional shale gas fracturing, little or no water fracturing, research and development of key fracturing tools, geological engineering economic integrated fracturing, and the fracturing technology for efficient exploitation of adsorbed gas.

ACKNOWLEDGMENT

This paper is supported by the following funding projects: Special Project of Geological Mineral Resources and Environment Investigation "geological survey standardization and standard formulation and revision (2019–2021)" (No. DD20190470) and "geological survey standardization and standard formulation and revision (2022–2025)" (No. DD20221826).

REFERENCES

Chen Yufeng, "Analysis and Research on Key Technologies in Shale Gas Exploration and Development," *Contemporary Chemical Research*, pp. 112–113, July 2018.

Guo Yuemiao and Zhao Longmei, "Research on the Current Situation and Countermeasures of Shale Gas Exploration and Development Technology," *China Petroleum and Chemical Industry Standards and Quality*, pp. 149–150, July 2018.

He Xiao, Li Wuguang and Dang Lurui et al., "Key Technological Challenges and Research Directions of Deep Shale Gas Development," *Natural Gas Industry*, vol. 41, pp. 118–124, January 2021.

Liang Xing, Wang Gaocheng and Zhang Jiehui et al., "High-efficiency Integrated Shale Gas Development Model of Zhaotong National Demonstration Zone and its Practical Enlightenment," *China Petroleum Exploration*, vol. 22, pp. 29–37, January 2017.

Liu He, Meng Siwei and Su Jian et al., "Reflections and Suggestions on the Development and Engineering Management of Shale Gas Fracturing Technology in China," *Natural Gas Industry*, vol. 39, pp. 1–7, April 2019.

Liu Naizhen, Wang Guoyong and Xiong Xiaolin, "Practice and Prospect of Geology-engineering Integration Technology in the Efficient Development of Shale Gas in Weiyuan block," *China Petroleum Exploration*, vol. 23, pp. 59–68, February 2018.

Lu Hongqiao, *"In 2020, China's Shale Gas Production Increased by more than 30%, Becoming the Main Force of Natural Gas Production [on Increase,"* China Youth Network, February 10, 2021.

Ma Xinhua, Li Xizhe and Liang Feng et al., "Dominating Factors on well Productivity and Development Strategies Optimization in Weiyuan Shale Gas Play, Sichuan Basin, SW China," *Petroleum Exploration and Development*, vol. 47, pp. 555–563, March 2020.

"National standardization development outline."

Wang Junfeng, "Discussion on the Application of Geological Logging Technology in Chale Gas Exploration and Development," *Western Exploration Engineering*, pp. 77–78, June 2022.

Wang Ruoxi, *"Benefit Development and Deep Potential Tapping – China's Oil and Gas Exploration and Development Achieved Gratifying Results in 2021,"* China Power News Network, May 12, 2022.

Wang Shiqian, "Shale Gas Exploitation: Status, Issues, and Prospects," *Natural Gas Industry*, vol.37, pp. 115–130, June 2017.

Wang Xiaochuan, Wu gen and Yan Jinding, "Current Situation and Trend of World Shale Gas Development and Technology Development," *Science and Technology China*. pp. 17–21, December 2018.

Wang Yike, "Overview of Shale Gas Exploration and Development Technology," *Petroleum Knowledge*, pp. 42–43, February 2019.

Wang Yichen, "We Should not Relax our Efforts to Increase Oil and Gas Reserves and Production," *Economic Daily*, May 12, 2022.

Xie Jun, Zhang Haomiao and She Chaoyi et al., "Practice of Geology-Engineering Integration in Changning State Shale Gas Demonstration Area," *China Petroleum Exploration*, vol. 22, pp. 21–28, January 2017.

Yue Hong, Chang Honggang and Fan Yu et al., "Construction and Prospect of China's Shale Gas Technical Standard System," *Natural Gas Industry*, vol. 40, pp. 1–8, April 2020.

Zhao Jinzhou, Ren Lan and Jiang Tingxue et al. "Ten Years of Gas Shale Fracturing in China: Review and prospect," *Natural Gas Industry*, vol. 41, pp. 121–142, August 2021.

Zhu Weiyao, Chen Zhen and Song Zhiyong et al., "Research Progress in Theories and Technologies of Shale Gas Development in China," *Journal of Engineering Science*, vol. 43, pp. 1397–1412, October 2021.

Zou Caineng, Ding Yunhong and Lu Rongjun et al., "Concept, Technology, and Practice of 'Man-made Reservoirs' Development," *Petroleum Exploration and Development*, vol. 44, pp. 144–154, January 2017.

Zou Caineng, Zhao Qun and Cong Lianzhu et al., "Development Progress, Potential and Prospect of Shale Gas in China," *Natural Gas Industry*, vol. 41, pp. 1–14, January 2021.

Civil Engineering and Energy-Environment – Gao & Duan (Eds)
© 2023 the Author(s), ISBN 978-1-032-56059-5

Comprehensive evaluation of China's mineral exploration technology

Xianying Huang* & Chunfang Wang*
Chinese Academy of Natural Resources Economics, Beijing, China

ABSTRACT: On the basis of the data about human resources, capital, and assets of geological exploration units, this paper makes a comprehensive evaluation of mineral exploration technology in 31 provinces in China and confirms the unequal relationship between the level of mineral exploration technology and resource situation. The regions with the highest level of mineral exploration technology include Beijing, Shaanxi, Hebei, Shandong, Sichuan, and Jiangxi. The mineral exploration activities carried out by practitioners in other places are related to the lack of local resources. The level of mineral exploration technology in Central China and Northeast China is second, and it maintains a relatively balanced relationship with the resource situation of the region. Xinjiang, as a province with large resources, needs to further improve the level of mineral exploration technology. The economically underdeveloped Gansu, Guizhou, Guangxi, and other regions have a low level of mineral exploration technology, and it is necessary to formulate policies such as talent incentives and capital allocation to support the development of the mineral resources industry in the region. The mineral exploration level in Chongqing, Shanghai, Hainan, and other regions is the lowest, but it is suitable for the local resource situation and does not need to formulate encouraging or restrictive policies.

1 INTRODUCTION

Some studies believe that the competitiveness of mineral resources in a region is inversely proportional to resource reserves (Zhang *et al.* 2012), so the weak economic agglomeration of the mineral resources industry in western China has led to the lack of motivation for industrial competitiveness (Chen & Yan 2011). Environmental regulation will stimulate mining enterprises to innovate ecological compensation and enhance their competitiveness (He *et al.* 2018), reflecting the delicate relationship between resources and economic development. The relationship between resources and the economy provides a new perspective for the comprehensive evaluation of China's mineral exploration technology and analysis of the resource situation. There are over 2,000 geological exploration units in China, including institutions and companies. The regional distribution is extremely unbalanced, and it plays an irreplaceable role in resource discovery and exploration technology (Huang & Li 2022). What are the differences in the level of mineral exploration technology in different regions? How do we improve the overall level of mineral exploration requires in-depth research?

---

*Corresponding Authors: xyhuang@canre.org.cn and cfwang@canre.org.cn

DOI: 10.1201/9781003433651-4

2 CONSTRUCTION OF A COMPREHENSIVE EVALUATION MODEL

The comprehensive evaluation is a global evaluation of the objects of the multi-attribute structure system, and the evaluation objects are ranked and compared through the method of assignment. The comprehensive evaluation theory must meet the three standards of universality, accuracy, and indirectness, which is very practical in the multidisciplinary field. Among many evaluation methods, quantitative evaluation methods such as Grey Incidence Analysis (GIA), Entropy Analysis (EA), and Artificial Neural Networks (ANN) are more commonly used. The development of theories and methods in comprehensive evaluation has its unique historical background and specific applicability, which can be continuously improved in empirical research (Peng *et al.* 2015). The comprehensive evaluation reflects the comprehensive measurement of information, and its "representativeness" and "comprehension" should be fully considered when constructing the index system rather than simply deleting the index with overlapping information (Su 2012).

2.1 *Evaluation index*

In China's national economic industry classification, the Geological Prospecting Industry does not belong to the Mining Industry but is a sub-category of scientific research and experimental research and development. Geological exploration units are the main body of the geological exploration industry, and their comprehensive capabilities represent the level of China's mineral exploration technology. There are various indexes to measure the technical level of mineral exploration, not only the progressiveness of exploration equipment but also the technical ability of personnel engaged in mineral exploration and the scale of capital investment. A comprehensive evaluation of China's mineral exploration technology can overcome the problem of one-sided information caused by a single index and can also guide the balanced development of geological exploration units.

To evaluate China's mineral exploration technology, it is necessary to investigate three aspects. The first is human resources. Mineral exploration is an exploratory work that requires practitioners with basic theoretical knowledge of earth science and field practice. The number of employees determines the overall level of the geological exploration units, and the senior technicians determine the leading level of the exploration technology. The second is the capital. Mineral exploration requires a large amount of long-term financial support. The total income of geological exploration units and geological exploration projects determines the breadth and continuity of the implementation of exploration technology. The financial subsidy is a special source of funds for China's mineral exploration. It is mainly used to support the basic development of public institutions. It also shows that there is external dependence on the operation and development of geological exploration units. The financial subsidy is the only negative index among the indexes for evaluating mineral exploration technology. The third is assets. Net assets represent the economic interests of the owner of the geological exploration unit and affect the investment in the exploration technology. Special instruments and equipment are the main means and tools for implementing mineral exploration projects. The progressiveness of instruments and equipment determines the efficiency of mineral exploration.

2.2 *Evaluation method*

Entropy is a measure of information uncertainty, and entropy evaluation can obtain the order degree and effect of information. The entropy weight method is the main method of multi-attribute decision-making, with high accuracy, strong objectivity, and good information interpretation effect (Liu *et al.* 2022). The primary task of the comprehensive evaluation is to construct a judgment matrix $R = \left(x_{ij}\right)_{mn}$ and uses the entropy weight method to determine the index weight, which can eliminate the interference of human factors and make the evaluation result objective. The calculation steps of the weight coefficient are as follows:

(1) Index normalization processing

De-quantify and temper each index to get a normalized judgment matrix $B = (b_{ij})_{mn}$

Positive Index

$$b_{ij} = \frac{x_{ij} - x_{\min}}{x_{\max} - x_{\min}} \tag{1}$$

Negative Index

$$b_{ij} = \frac{x_{\max} - x_{ij}}{x_{\max} - x_{\min}} \tag{2}$$

(2) Calculate the proportion f_{ij} of each index

$$f_{ij} = \frac{b_{ij}}{\sum_{j=1}^{m} b_{ij}} + \gamma \tag{3}$$

In Formula (3), γ is the translation coefficient, which takes a small value, such as 1/10,000 of the proportion of the index to ensure that the proportion of all indexes can be logarithmic.

(3) Calculate the information entropy H_i

$$H_i = \frac{-1}{\ln m} \sum_{j=1}^{m} f_{ij} \ln f_{ij} \tag{4}$$

In Formula (4), $i = 1, 2, \cdots n; j = 1, 2, \cdots m$

(4) Calculate the weight matrix W of the index system

$$W = (w_i)_{1 \times n} = \frac{1 - H_i}{n - \sum_{i=1}^{n} w_i} \tag{5}$$

2.3 *Determination of index weights*

The analysis data comes from statistical yearbooks related to China's geological exploration industry. The cross-sectional data is the relevant index data of 31 provinces in China and the time series are from 2012 to 2021. The analysis data constitutes a judgment matrix $R = (x_{ij})_{mn}$, where n = 7, m = 320. The weight values of different indexes are calculated, as seen in Table 1.

Table 1. Index weights of the comprehensive evaluation of China's mineral exploration technology.

Indexes	Practitioners	Senior technicians	Total income	Geological exploration income	Financial subsidies	Net assets	The net value of geological exploration instruments and equipment exploration instruments and equipment
Information entropy	0.956922	0.964563	0.928353	0.949702	0.993781	0.929272	0.94563
Index weight	0.12984	0.106811	0.215948	0.151602	0.018746	0.213179	0.163874

3 EVALUATION RESULTS

The results of the comprehensive evaluation of the mineral exploration technology can be obtained by the product of the index weight and the index normalization matrix, as shown in Table 2.

Table 2. Comprehensive evaluation values of mineral exploration technology.

Years	2012	2013	2014	2015	2016	2017	2018	2019	2020	2021
Beijing	0.62610	0.67104	0.73807	0.63225	0.64632	0.63682	0.50335	0.58273	0.55766	0.61471
Tianjin	0.06855	0.06251	0.07504	0.06787	0.07205	0.06141	0.07636	0.08228	0.08358	0.09000
Hebei	0.44210	0.49384	0.43882	0.42349	0.44232	0.36132	0.39031	0.43561	0.50385	0.56674
Shaanxi	0.38648	0.25778	0.25844	0.24897	0.22719	0.20082	0.29880	0.19941	0.19288	0.21741
Inner Mongolia	0.24165	0.23190	0.20042	0.20441	0.17281	0.11384	0.20775	0.28168	0.22571	0.26302
Liaoning	0.22692	0.23125	0.22758	0.23973	0.24466	0.18461	0.17829	0.18470	0.21670	0.21545
Jilin	0.28702	0.26465	0.22888	0.18695	0.19211	0.15889	0.14891	0.21659	0.23929	0.28634
Heilongjiang	0.25371	0.26180	0.24815	0.23979	0.22944	0.20686	0.16771	0.12181	0.12985	0.14544
Shanghai	0.08043	0.08881	0.08230	0.12109	0.11320	0.11650	0.04332	0.14923	0.08099	0.08676
Jiangsu	0.15199	0.14883	0.16482	0.19213	0.14975	0.16506	0.19023	0.19659	0.19135	0.32515
Zhejiang	0.14065	0.15761	0.14896	0.10340	0.06506	0.11543	0.13684	0.53109	0.49725	0.56899
Anhui	0.26944	0.26777	0.23762	0.21909	0.20767	0.20019	0.19432	0.26250	0.24128	0.26337
Fujian	0.28147	0.33618	0.32270	0.29584	0.27336	0.30100	0.32080	0.10849	0.11479	0.09279
Jiangxi	0.26198	0.26156	0.27003	0.26491	0.27020	0.26085	0.26503	0.43494	0.54953	0.55964
Shandong	0.50784	0.46239	0.38210	0.44444	0.41601	0.39633	0.52792	0.39305	0.36067	0.46423
Henan	0.46202	0.48149	0.26334	0.19910	0.23754	0.21456	0.19998	0.20366	0.20586	0.23626
Hubei	0.21843	0.21721	0.21828	0.24707	0.25842	0.25123	0.27884	0.21562	0.26065	0.32600
Hunan	0.33334	0.32025	0.30316	0.29612	0.28662	0.27584	0.24428	0.16988	0.18326	0.18140
Guangdong	0.26416	0.25143	0.22724	0.22131	0.24029	0.26217	0.24340	0.27490	0.27664	0.26744
Guangxi	0.16835	0.17128	0.15944	0.15946	0.14853	0.13530	0.14512	0.14579	0.15688	0.15821
Hainan	0.03897	0.03920	0.03758	0.03412	0.03511	0.03507	0.03052	0.02615	0.02734	0.02830
Chongqing	0.08912	0.09990	0.10141	0.12009	0.12700	0.07556	0.09834	0.11392	0.11096	0.12590
Sichuan	0.31547	0.37165	0.36280	0.35989	0.36174	0.33870	0.30389	0.33088	0.34400	0.37768
Guizhou	0.15072	0.18247	0.17107	0.19913	0.18125	0.18117	0.16013	0.15796	0.17492	0.19622
Yunnan	0.29472	0.30494	0.28384	0.28079	0.28265	0.22540	0.16708	0.27283	0.28072	0.32119
Tibet	0.06109	0.06048	0.05895	0.06196	0.05325	0.06653	0.07125	0.05406	0.05066	0.07387
Shaanxi	0.43832	0.41018	0.38330	0.37270	0.42708	0.48198	0.53068	0.51408	0.61425	0.62408
Gansu	0.17253	0.20964	0.18681	0.17291	0.17425	0.15967	0.17771	0.18372	0.18606	0.18700
Qinghai	0.15430	0.13892	0.14190	0.11861	0.11574	0.11412	0.11777	0.14860	0.15531	0.16895
Ningxia	0.06281	0.06735	0.06532	0.06147	0.05455	0.05044	0.04813	0.03348	0.03835	0.03850
Xinjiang	0.24114	0.25162	0.26760	0.27530	0.29299	0.21788	0.16618	0.20929	0.18593	0.22040

Judging from the comprehensive evaluation results, the level of exploration technology in the same areas in the time series remains generally stable. From the average value of the data during the ten years of the evaluation results, there are obvious differences between regions, and five intervals can be divided according to the size of the value, as shown in Table 3. Beijing is in a leading position, mainly due to the dense practitioners, sufficient funds, and extensive international exchange platforms in the region. Shaanxi, Hebei, Shandong, Sichuan, and Jiangxi, which are in the second interval, have a large number of mineral exploration practitioners, who often migrate to other areas for exploration operations with strong mineral exploration technology. The level of exploration technology in areas rich in mineral resources, such as Central China, Northeast China, and Xinjiang is higher than that of some underdeveloped western regions, such as Gansu, Guizhou, and Guangxi. Chongqing, Shanghai, Tianjin, Tibet, Ningxia, and Hainan, are at the bottom, mainly due to the limited exploration space and few practitioners in this area.

Table 3. Comparison of mean values of a comprehensive evaluation for mineral exploration technology.

Interval 1	Beijing 0.620905					
Interval 2	Shaanxi 0.479666	Hebei 0.449841	Shandong 0.435496	Sichuan 0.346667	Jiangxi 0.339867	
Interval 3	Yunnan 0.271417	Henan 0.270381	Hunan 0.259415	Guangdong 0.252898	Hubei 0.249175	Shaanxi 0.248819	Zhejiang 0.246528
	Fujian 0.244742	Anhui 0.236325	Xinjiang 0.232833	Jilin 0.220964	Liaoning 0.214989	Inner Mongolia 0.214319	Heilongjiang 0.200456
Interval 4	Jiangsu 0.18759	Gansu 0.181032	Guizhou 0.175504	Guangxi 0.154838	Qinghai 0.137422	Chongqing 0.10622
Interval 5	Shanghai 0.096263	Tianjin 0.073966	Tibet 0.061212	Ningxia 0.052041	Hainan 0.033235	

4 CONCLUSION

There are obvious regional differences in the level of exploration technology for China's mineral resources. There are not only historical reasons for regional economic development and changes in the management system of mineral exploration but also the adjustment of industrial structure caused by changes in the economic situation. The resource endowments of Beijing, Shaanxi, Hebei, and Shandong are not prominent in China. The existence of some scientific research institutions related to mineral exploration in this region and the export-oriented development model have jointly promoted the level of mineral exploration technology in the regions. Shanghai, Ningxia, Hainan, and other regions restrict the exploration and development of mineral resources, and the orientation of regional development policies is the investigation, monitoring, and restoration of the ecological environment. The level of exploration technology in Sichuan, Jiangxi, Yunnan, Hubei, and Hunan is in the upper middle position in China, which is rich in mineral resources and has become a good test base for improving exploration technology. The relative backwardness of exploration technology in Northeast China, Gansu, Guizhou, and Guangxi is directly related to the comprehensive economic strength, investment environment, and geographical conditions of the region.

Mineral exploration is highly exploratory work, and practitioners need years or even decades of practical experience to discover major mineral resources. There is a certain degree of mismatch between China's mineral exploration technology level and resource situation, which leads to the cross-regional mobility of practitioners. Unfamiliarity with the geological environment is a direct obstacle for practitioners to make major discoveries. China needs to adjust the current mismatch at the national level and guide the permanent transfer of practitioners and capital investment in Beijing, Hebei, Shandong, and other regions to regions rich in mineral resources. Economically poor but resource-rich western regions need to increase financial subsidies or enhance the financing capacity of mineral exploration and establish talent training, introduction, and incentive mechanisms to improve the technical level of mineral exploration.

REFERENCES

Chen Lianfang and Yan Liang. Research on the Relevance Between the Degree of Mineral Industrial Agglomeration and Industrial Competitiveness in China's Western Region. *China Population, Resources and Environment*, 2011, 21(05):31–37.

He Yumei, Luo Qiao and Zhu Xiaowei. Environmental Regulation, Ecological Innovation, and Enterprise Competitiveness: An Analysis Based on Data of Mineral Resources Enterprises. *Commercial Research*, 2018(03):132–137.

Huang Xianying and Li Wei. New Formation and Guarantee Mechanism of the New Development Pattern of the National Geological Prospecting Development. *Natural Resources Information*, 2022(03):40–44 + 32.

Liu Hongyu, Liu Youcun and Meng Lihong et al. Research Progress of Entropy Weight Method in water Resources and Water Environment. *Journal of Glaciology and Geocryology*, 2022, 44(01):299–306.

Peng Zhanglin, Zhang Qiang and Yang Shanlin. Overview of Comprehensive Evaluation Theory and Methodology. *Chinese Journal of Management Science*, 2015, 23(S1):245–256.

Su Weihua. Review and Recognition on the Research of Multi-indicators Comprehensive Evaluation in China. *Statistical Research*, 2012, 29(08):98–107.

Zhang Baoyou, Xiao Wen and Zhu Weiping. Evaluation of Competitiveness of Regional Mineral Resources and Correlation with Regional Economy. *Journal of Natural Resources*, 2012, 27(10):1623–1634.

Civil Engineering and Energy-Environment – Gao & Duan (Eds)
© 2023 the Author(s), ISBN 978-1-032-56059-5

Research on China's mineral resources exploration strategy

Xianying Huang*, Wei Li & Chunfang Wang*
Chinese Academy of Natural Resources Economics, Beijing, China

ABSTRACT: China has a large number of mineral resources depending on imports and faces an unstable market supply. The Chinese government has put forward new requirements for ensuring the supply of primary products. The implementation of the mineral resource exploration strategy will help achieve the goal. The relationship between the exploration and development of mineral resources in China is weak, but ferrous metal mineral resources exploration investment and mining industry profits have a statistically linear relationship. On the basis of considering the advantages and disadvantages of mineral resources exploration and facing the opportunities and challenges of future development, the strategic direction of China's mineral resources exploration includes: cultivating outstanding leading talents in mineral resources exploration, optimizing the echelon structure of professional and technical personnel, learning from overseas exploration technology and management experience; providing more high-quality basic geological survey information to protect the legitimate rights and interests of mining rights holders; building an environmental, social responsibility and corporate governance system to optimize the investment environment; encouraging the research and development of mineral resources exploration technology; increasing investment in mineral resources exploration of scarce minerals.

1 INTRODUCTION

China's economic development has shifted from high-speed growth to medium-low-speed growth and the demand for mineral resources has begun to be weak, which does not change the situation of China's shortage of bulk mineral products. Oil, natural gas, iron ore, copper ore, bauxite, nickel ore, potash, and other minerals are still reliant on large imports (Chen *et al.* 2022). At the same time, in the face of complex and volatile geopolitical situations, uncertainties in the international energy resource market, and blocked mineral supply chains, the Chinese government raised the issue of a safe supply of primary products at the 2021 Economic Conference. Reducing the dependence on external mineral products requires increasing domestic supply. Resource exploration is the only way to change from mineral resources to mineral products (Ju *et al.* 2019).

The situation of China's mineral resource exploration is not optimistic. In 2021, the investment in mineral resource exploration was 8.6 billion yuan, a decrease of 79.3% compared with 2012; the number of newly discovered mineral deposits decreased by 62.2% during the same period. China's financial investment in mineral resource exploration will cause some projects to focus on design plans, physical workload, and technical evaluation while despising the prospecting effect and economic evaluation. Without obvious incentive policies, practitioners' enthusiasm for the exploration of mineral resources is not high.

---

*Corresponding Authors: xyhuang@canre.org.cn and cfwang@canre.org.cn

DOI: 10.1201/9781003433651-5

Uncertainty in the external market environment makes mining investment biased toward mining exploration, and the scale of investment in grassroots exploration is gradually shrinking, which is not conducive to the long-term goal of a safe supply of mineral resources (Yang *et al.* 2019).

In recent years, the changes in investment in China's mineral resources exploration have not matched the trend of rising prices of international mineral products. The deserted market for mineral resources exploration is mainly caused by various domestic factors. Only by objectively treating the relationship between mineral resources exploration and national economic development and analyzing the current advantages, disadvantages, opportunities, and challenges of mineral resources exploration can we formulate strategic goals for mineral resources exploration that are in line with the long-term interests of China's economic development.

2 REGRESSION ANALYSIS OF RESOURCE EXPLORATION AND DEVELOPMENT

Obtaining income is an important goal of investment for mineral resources exploration, and its investment decision depends on: first, the expected certain mineral resources exist; second, mineral resources can be exploited under the legal and technical framework; third, mineral product prices remain at a certain level. The legal and technical requirements for mining specific minerals in specific areas are clear, and investors do not have many choices. When other conditions are stable, the profit of the mining industry depends on the number of mineral resources and the price of mineral products, which indirectly affects the scale of investment in mineral resource exploration.

2.1 *Evaluation indicators and methods*

The total profit of mining enterprises can comprehensively reflect the number of mineral resources and the price of mineral products. Using statistical methods to quantitatively analyze the relationship between the investment of mineral resources exploration and mining development profits can correct people's subjective cognition of the relationship between mineral resources input and output and even find some cognition misunderstandings. There are huge differences in the technical mining methods and market environment changes of different minerals, so three different mineral categories of coal, ferrous metals, and non-ferrous metals are selected for analysis.

Regression analysis is a statistical analysis method to determine the interdependent quantitative relationship between two or more variables. In general understanding, the investment scale of mineral resources exploration depends on the growth of mining industry profits, so greater profits mean increasingly conducive investment. In the regression analysis, the investment in mineral resources exploration is used as the dependent variable, and the profit of the mining industry is used as the independent variable. The time series prediction model is established to discover the mathematical relationship between the variables.

2.2 *Data sources*

The data on investment for mineral resources exploration come from the statistical yearbook released by the Ministry of Natural Resources of China, and the total profit data of mining enterprises above the designated size come from the data publicly released by the National Bureau of Statistics of China. After data collation, the statistical data are divided into three categories: coal, non-ferrous metals, and ferrous metals, as seen in Table 1.

Table 1. Changes in the profits and exploration investment of mining enterprises (data unit: billion yuan).

Years	Investment in coal exploration	Profits of coal mining enterprises	Investment in ferrous metal exploration	Profits of ferrous metal mining enterprises	Investment in non-ferrous metals exploration	Profits of non-ferrous metal mining enterprises
2008	7.9	234.8	2.3	70.0	10.1	40.7
2009	9.1	220.8	2.8	43.9	11.6	33.9
2010	10.5	344.6	3.8	89.3	14.6	57.2
2011	11.7	456.1	4.3	121.0	15.3	81.5
2012	12.2	380.8	5.4	113.3	18.5	78.7
2013	8.5	268.0	3.9	113.7	19.0	66.6
2014	5.9	142.4	2.9	85.1	16.5	58.2
2015	3.2	40.5	2.1	51.9	14.1	45.1
2016	1.7	115.9	1.3	41.1	9.8	45.9
2017	1.6	295.2	0.6	16.0	6.4	53.3
2018	1.3	300.9	0.4	12.1	4.6	50.2
2019	1.0	283.7	0.3	23.5	4.4	32.1
2020	1.2	222.1	0.4	37.3	3.5	38.7

2.3 Data analysis

Through the statistical analysis of the data, a series of statistical parameters are obtained. From the statistical results, the goodness of fit of coal, ferrous metals, and non-ferrous metals are 0.287502, 0.880173, and 0.49137, respectively. The goodness of fit for coal and non-ferrous metals is too low to account for a linear relationship between the variables.

Table 2. Regression statistics of different ore types.

Regression statistics	Coal	Ferrous metals	Non-ferrous metals
Multiple R	0.536192	0.938176	0.700978
R square	0.287502	0.880173	0.49137
Adjusted R square	0.216252	0.868191	0.440507
Standard error	39.73251	6.392261	42.323
Observations	12	12	12

The regression curve of ferrous metals can explain the change in mineral resources exploration with the profit of the mining industry, and the P-value in the t-test is less than 0.0001, with a confidence level of more than 99.99%.

2.4 Analysis of results

From the analysis results, the investment of mineral resources exploration and development profits of ferrous metals have a statistical correlation, and the future investment for mineral resources exploration can be predicted according to the profit data of the mining industry in the simulation curve. Investment in mineral resources exploration of coal and non-ferrous metals has a very poor statistical correlation with development profits, and it is impossible to predict through regression analysis. Unlike the investment return relationship of general commodity production, the high risk and high return characteristics of mineral resource exploration and the influence of uncertain environmental factors lead to the change of its investment return rate from zero to dozens of times. Even though the regression model can predict the investment in the exploration of ferrous metal mineral resources, there is still a gap between the simulated curve and the actual point of investment in the exploration of mineral resources and the profit of the mining industry.

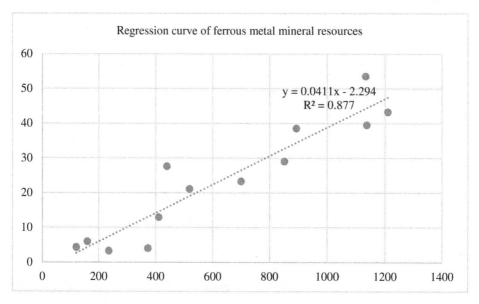

Figure 1. Regression curve of investment for ferrous metal mineral resources exploration and mining enterprise profit.

3 STRATEGIC ANALYSIS OF MINERAL RESOURCES EXPLORATION

Discovering and identifying industrial deposits is the basic task of mineral resource exploration, and it is an inevitable process for people to obtain production and living materials. Compared with the age of copper and iron in history, under the background of industrialization and informatization demand, contemporary people have a greater demand for bulk mineral products and key minerals. Considering factors such as global geopolitical risks and supply chain crises, the country needs to have an objective understanding of the distribution of internal resources. It is necessary to fully respect the status and influence of mineral resource exploration in the development of the national economy.

3.1 *Enhancements to analysis*

3.1.1 *Large targeted area*

China's vast land and seas contain rich mineral resources, and 173 minerals have been discovered. By the end of 2020, China has completed 1:50,000 regional geological mapping covering an area of 4.34 million square kilometers accounting for 45% of the country's land area, providing prerequisites for further exploration of mineral resources (Ministry of Natural Resources 2021). China has completed the exploration and development of the Fuling shale gas field and the 7,018-meter-deep Songke No. 2 Well, indicating that the exploration scope of mineral resources will continue to expand with the advancement of horizontal drilling and deep exploration technology.

3.1.2 *High ROI*

Many cities in China have arisen because of resources, including Daqing, Karamay, Ordos, and Panzhihua. The return on investment ratio of discovering a large deposit can be more than ten times, and a resource-based city can meet the production and living needs of millions of people. According to statistics, from 2011 to 2020, the total profit of the mining industry above the designated size in the country accounted for 8.76% of the total profit of industrial enterprises. The mining industry plays a pivotal role in the national economy, and the inevitable way to realize the profit of the mining industry includes the exploration of mineral resources.

3.1.3 *Strong human resources*

In China, nearly 1,000 geological exploration units have been operating for more than 30 years. As of the end of 2021, there is 170,000 geological exploration personnel, of which 27.3% have senior technical titles. The main personnel engaging in mineral resources exploration have received higher education and professional level, and the regional distribution is balanced. They are very familiar with the local geological situation and play an irreplaceable role in discovering mineral resources.

3.2 *Weaknesses for analysis*

3.2.1 *Long exploration period*

According to the survey data (Shi *et al.* 2008), the average age of discovering a large deposit is 19 years, of which 34% of the deposits require exploration within five years, and 33% of the deposits require intermittent exploration for more than 20 years. From the discovery of the deposit to the start of production, it takes an average of 7 years. The long exploration cycle of mineral resources has reduced the interest of most investors and also put more pressure on practitioners.

3.2.2 *Poor resource endowment*

China's total mineral resources are large, but the per capita possession is low. There are a few large deposits and many small deposits. For example, in mining areas of different scales (Zhou *et al.* 2015): copper and lead-zinc mines account for more than 90% of the small-scale mining areas; tin and molybdenum mines account for more than 80% of the small-scale mining areas; only rare earth and potash belong to large and medium-sized mining areas, accounting for than 40%. The poor resource endowment makes prospecting difficult and reduces the return on investment.

3.2.3 *Low influence of geologists*

Compared with the role of professional technical equipment, people's theoretical knowledge level and fieldwork experience are more important to whether positive progress in mineral resources exploration can be achieved. In 2022, the China Mining Federation released the list of "geologists" for the first time, showing that China is striving to highlight the value of people in the evaluation of mineral resources. However, the biggest gap between China's geologist system and the "Competent Person" signature reporting system, which is implemented by developed mining countries such as Canada, Australia, and South Africa, is that China's mineral resource exploration reports are less credible, and do not have financing function.

3.3 *Analysis opportunities*

Continuously improved and updated mineral resource exploration technologies and methods have significantly reduced the cost of resource discovery and increased the success rate of prospecting. Chinese geological exploration units actively introduced the V8 array magnetotelluric system launched by Phoenix Canada and the EH 4 electromagnetic system launched by American Geometrics and EMI, which greatly improved the accuracy of mineral resources exploration information (Wang *et al.* 2008). The development of geophysical technology in China is slow, but the pace of independent research and development of scientific instruments and equipment is accelerating (Lv *et al.* 2019). Geochemical analysis with greatly improved sensitivity and accuracy is widely used in the delineation of mineral resource exploration targets in covered or concealed areas (Jiang *et al.* 2010). Satellite remote sensing technology has outstanding performance in regional geological surveys and evaluation of mineral resources potential. Advances in shallow coring drilling technology and surface bedrock horizontal drilling technology and equipment have made it possible for small-angle drilling to replace trenching and pit drilling, reducing disturbance to the ecological environment and facilitating the smooth progress of mineral resource exploration (Wang *et al.* 2019).

3.3.1 *Data integration analysis*
The rapid development of contemporary information technology has led to the generation of Geographic Information Systems (GIS). Based on the GIS platform, the collection, storage, and analysis of data generated by technologies such as geophysics, geochemistry, and satellite remote sensing are converted into practical comprehensive geological information and visualized three-dimensional images, which have become the main working mode of mineral resources exploration in China.

3.3.2 *Large resource consumption*
China is the world's largest consumer of mineral resources, and its consumption will continue to grow in the future. The growth rate of mineral resource output will not keep up with that of consumption. While maintaining the balance between supply and demand of mineral products by means of trade, China must increase the self-sufficiency of mineral resources and pay more attention to the exploration of mineral resources in order to resolve the problem of structural resource contradictions (Zhi *et al.* 2010). In particular, it is important to emphasize that oil, natural gas, rich iron ore, copper ore, lithium ore, and other scarce minerals are the focus of mineral resource exploration.

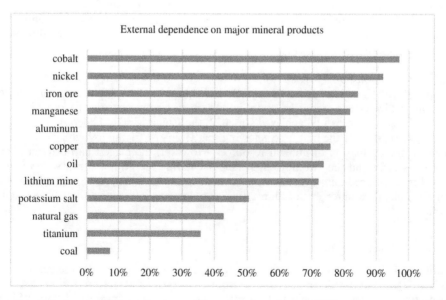

Figure 2. China's external dependence on major mineral products in 2020.

3.4 *Threats of analysis*

3.4.1 *Increase of difficulty in prospecting*
Judging from the newly discovered important deposits, the surface mines and shallow mines are decreasing day by day, and the direction of mineral resource exploration has changed to concealed mines and deep mines. In recent years, the development cost of new resource reserves has increased, and the proportion of "low permeability oil, tight oil, and heavy oil" in the new oil reserves is high. Some metal deposits with large geological reserves face difficulties in mineral processing and smelting.

3.4.2 *Disturbance to the ecological environment*
In the process of mineral resources exploration, the movement of personnel, the use of vehicles and equipment, and prospection for projects such as trenches and inclined shafts will cause disturbance to the surface, accelerating soil erosion and degradation of the primary environment. The drilling mud waste liquid is produced in geological drilling, and the waste oil is produced in equipment operation and maintenance, damaging surface vegetation and soil microorganisms.

3.4.3 *Difficulties in overseas exploration*
Affected by the investment resistance of developed countries, China's mineral resources exploration mainly invests in developing countries such as Africa, South America, and Southeast Asia and faces some prominent problems: high hidden costs, investment mainly in the grassroots exploration stage, insufficient cooperation with local enterprises, and backward infrastructure, etc.

4 THE STRATEGIC CHOICE OF CHINA'S MINERAL RESOURCES EXPLORATION

Based on the current situation of land resources and the mining market environment, China needs to grasp opportunities and overcome challenges on the basis of recognizing its advantages and disadvantages and formulate a mineral resources exploration strategy that meets economic development and social needs.

Table 3. SWOT Analysis of China's mineral resources exploration.

Internal factor analysis / External factor analysis	Strengths	Weakness
	1. Large target area; 2. High ROI; 3. Strong human resources.	1. Long exploration period; 2. Poor resource endowment; 3. Weak technical experts.
Opportunities	Opportunity-Strength Strategies	Opportunity-Weakness Strategies
1. Advances in technology; 2. Integrated analysis of data; 3. High resource consumption.	1. Basic geological survey information services are provided; 2. Investment in exploration technology research and development is encouraged; 3. Leaders in the field of mineral resources exploration are cultivated.	1. Investment in a national shortage of minerals is increased; 2. The rights and interests of mining rights holders are protected; 3. A credit evaluation system for technicians is established.

(continued)

Table 3. Continued

Opportunities	Opportunity-Strength Strategies	Opportunity-Weakness Strategies
Threats	Threat-Strength Strategies	Threat-Weakness Strategies
1. Difficulty in exploration;	1. The Structure of the Technical Personnel Echelon is optimized;	1. The cost of the ecological environment is considered;
2. Disturbance of the ecological environment;	2. An environmental and social responsibility and corporate governance system are built;	2. Inefficient projects are reduced;
3. Risks of overseas exploration.	3. Overseas companies join the cooperation.	3. The management experience and technology of overseas exploration are learned.

4.1 *Opportunity-strength strategies*

The huge consumption of mineral resources is the biggest opportunity for China's mineral resources exploration, and the vast territory provides the possibility of resource supply. Advances in mineral resource exploration technology have reduced investment risks, and high returns have further stimulated venture capital. Huge personnel input and integrated analysis of comprehensive data ensure the necessary work intensity for mineral resource exploration. The advantage-opportunity strategy of China's mineral resources exploration includes: first, it should provide the public with more convenient basic geological survey information services; second, the private sector to invest in research and development of mineral resources exploration technology should be encouraged; third, it should continue to export technologies related to geology applications personnel, especially the training of leaders in the field of mineral resources exploration.

4.2 *Opportunity-weakness strategies*

The long mineral exploration cycle determines that investors need strategic investment capabilities, and technical service entities need to have a high professional level and good reputation. The difference in resources determines the investment in mineral exploration of different minerals. The disadvantage-opportunity strategy of China's mineral resources exploration includes: first, social capital to invest in mineral exploration for minerals that are in short supply in the country should be encouraged, and government subsidies are provided when necessary; second, the rights to a protection system for mining rights holders should be provided; third, the credit evaluation system of geological experts should be established, and risk financing through exploration reports should be encouraged.

4.3 *Threat-strength strategies*

When technological progress cannot keep up with the speed of resource consumption, the difficulty of geological prospecting begins to increase, so increasing overseas exploration efforts has become an inevitable route for China's economic development. In the process of exploration and development of advantageous resources, China should pay more attention to the voice of the mining market and its impact on the social environment. The advantage-challenge strategy of China's mineral resources exploration includes: first, the structure of technical personnel should be optimized, and a small and sophisticated technical team should be established; second, the investment strategy of mining enterprises should be adjusted, fully considering the environment and social responsibility and corporate governance; third, the long-term investment concept of mineral resources exploration should be established, and local enterprises in overseas exploration should join the cooperation.

4.4 *Threat-weakness strategies*

The mining market dominated by small deposits is fiercely competitive, and the development of China's mining industry needs to regulate the order of mineral resource exploration further. In the process of China's economic development and transformation, the emphasis on the ecological environment also constrains the mode of mineral resource exploration. The threat-weakness strategy of China's mineral resources exploration includes: first, the cost of ecological and environmental protection into the investment plan for mineral resources exploration should be incorporated; second, mineral resources exploration projects with low resource utilization should be reduced; third, mineral resources exploration technology and management experience in the cooperation with overseas enterprises should be learned.

5 CONCLUSION

China's economic development relies heavily on the import of important mineral resources such as oil, iron ore, and copper ore. China's pricing power in the global mining market is relatively weak even with advantageous minerals such as tungsten ore and rare earth ore. At a time of high global uncertainty, China needs to formulate a mineral resource exploration strategy to ensure long-term and stable economic prosperity. In terms of investment and income, there is no linear relationship between the investment of coal and non-ferrous metal mineral resources exploration and mining industry profits. The regression curve of investment for ferrous metal mineral resources exploration and mining industry profits has reference significance for predicting future investment scale.

On the basis of analyzing the advantages, disadvantages, opportunities, and challenges of China's mineral resources exploration and the SWOT model, four selection paths of China's mineral resources exploration strategy are obtained. First, advantages, opportunities, and government departments provide more high-quality basic geological survey information services to encourage social capital to participate in the research and development of mineral exploration technology and cultivate excellent management and technical talents in mineral resources exploration. Second, regarding disadvantages and opportunities, China needs to make key investments in minerals that are in short supply in the country to protect the legitimate rights and interests of mining rights and establish a credit evaluation system for technicians. Third, in terms of advantages and challenges, China should optimize the echelon structure of professional and technical personnel in mineral resources exploration and encourage mining enterprises to formulate investment strategies for the environment, social responsibility and corporate governance, conducting cooperative exploration with overseas enterprises. Fourth, in terms of disadvantages and challenges, China should list the cost of ecological and environmental protection as a rigid cost of mineral resource exploration, reduce inefficient projects in mineral resource exploration, and learn from the technology and management experience in overseas mineral resource exploration.

REFERENCES

Chen Jiabin, Huo Wenmin, Fang Dandan, et al. Analysis of the Supply Situation of Mineral Products in China—Based on the Basic Situation in 2021. *Natural Resource Economics of China*, 2022, 35(05):42–48.

Jiang Yongjian, Wei Junhao, Zhou Jingren, et al. The Application of New Geochemical Exploration Methods to Mineral Exploration and its Geological Effect. *Geophysical & Geochemical Exploration*, 2010, 34 (02):134–138.

Ju Jianhua, Wang Qiang, Chen Jiabin. Study on the High-Quality Development of China Mining Industry in the New Era. *China Mining Magazine*, 2019, 28(01):1–7.

LV Qingtian, Zhang Xiaopei, Tang Jingtian, et al. Review on Advancement in Technology and Equipment of Geophysical Exploration for Metallic Deposits in China. *Chinese Journal of Geophysics*, 2019, 62 (10):3629–3664.

Ministry of Natural Resources, PRC. *China Mineral Resources 2021[M]*. Geological Publishing House: Beijing, 2021.

Shi Junfa, Tang Jinrong, Zhou Ping, et al. Experience in Exploration for Buried Deposits and its Implications—discussed from Information-based Mineral Exploration Strategy and One Hundred Mineral Exploration Case Histories. *Geological Bulletin of China*, 2008(04):433–450.

Wang Da, Zhao Guolong, Zuo Ruqiang, et al. The Development and Outlook of Geological Drilling Engineering-To Review the 70th Anniversary of Exploring Engineering. *Exploration Engineering Rock & Soil Drilling and Tunneling*, 2019, 46(09):1–31.

Wang Rruijiang, Wang Yitian, Wang Gaoshang, et al. Analysis of the State of the Worldwide Mineral Exploration. *Geological Bulletin of China*, 2008, 27(1):154–162

Yang Jianfeng, Ma Teng, Zhang Cuiguang, et al. Global Economic Changes and Mineral Exploration Trends in the Last 20 Years. *China Mining Magazine*,2019, 62(10):3629–3664.

Zhi Yingbiao, Wang Zailan, Ma Zhong, et al. China's Mineral Resources Endowment with the Pattern of International Reserve. *China Population, Resources, and Environment*, 2010, 20(S1):321–324.

Zhou Pu, Liu Tianke, Hou Huali. Study on Mining Structure and Adjustment Effect in China. *China Mining Magazine*, 2015, 24(11):28–32.

Civil Engineering and Energy-Environment – Gao & Duan (Eds)
© 2023 the Author(s), ISBN 978-1-032-56059-5

Thermal and mechanical numerical simulation of a multipoint elastic restraint system that considers thermal effects

Jibin Wang* & Huan He*

College of Aeronautics, Nanjing University of Aeronautics and Astronautics, Nanjing, China

ABSTRACT: In engineering, some structures inevitably need to operate under high-temperature conditions, which causes a non-uniform distribution of temperature within the structure. Under the condition of a non-uniform temperature field, the dynamic characteristics of the structure will be affected by uncertain factors such as contact state, friction, and material properties. The ground test can only simulate various uncertain conditions one by one, which will greatly increase the design cost and prolong the test period. As an important tool for the analysis of various working conditions, numerical simulation has great advantages in simulating the physical properties of complex structures in a multi-physics environment. In this paper, a statically indeterminate multipoint elastic restraint system is designed for vertical containers, and its main function is to support the containers and transfer the load. Based on the heat conduction theory, the steady-state thermal analysis of the multipoint elastic restraint system is carried out to obtain the temperature field distribution, and then the stress state is analyzed under the uneven temperature field to verify the rationality and reliability of the structural design.

1 INTRODUCTION

In recent years, with the continuous development of the industrial level, the working environment of all kinds of mechanical equipment has become more and more complex. For example, in aerospace, energy and power engineering, mechanical engineering, and many other fields, many mechanical systems need to meet the demands of service under a variety of harsh conditions, such as high temperature and vibration, which puts forward high requirements for the safety, practicality, and durability of mechanical systems. After the structure has been heated, temperature changes have primarily two effects on it: firstly, the change in material thermal physical parameters such as elastic modulus and thermal expansion coefficient will cause the structure to be deformed; secondly, due to the constraints of the structure, the material cannot freely expand or compress in all directions under a temperature field, resulting in uncoordinated structural deformation and mechanical stress. The root cause of this mechanical stress is the temperature change, so the structure generates thermal stress in this case. The thermal expansion and contraction of objects are the most common physical phenomena in nature, and the thermal stress and thermal deformation of objects have become major problems in practical engineering. Cui (2007) studied the modeling theory of rigid-soft coupled dynamics of flexible multibody systems, considering the influence of geometric nonlinearity and thermal effects. Based on structural dynamics theory and heat transfer theory, He (2014) researched complex structural dynamics modeling methods in a thermal environment by combining theoretical analysis and simulation-based

---

*Corresponding Authors: 1257011049@qq.com and hehuan@nuaa.edu.cn

DOI: 10.1201/9781003433651-6

experimental verification. Snyder & Kehoe (1991) proposed that the structural reliability of high-speed aircraft is affected by temperature gradients. The test results showed that with an increase in temperature, the frequency would decrease and the damping would increase. The prediction results have a good correlation with the experimental results.

Some of the structures that work in high temperatures must be compact, and the gap between the various parts is small, even if the initial state does not collide. But as the temperature gradually increases, the parts are heated and expanded, and the two parts of the structure may come into contact. Therefore, as the temperature continues to change, the boundary conditions of the structure and material properties are also constantly changing. The use of structural dynamics models in the thermal environment can not only reduce the cost of experiments and shorten the test cycle but also provide a theoretical basis for solving thermo-structural dynamics problems in engineering. Therefore, it is of practical significance to analyze complex structures with gaps and multiple constraints in the thermal environment (Feng 2011).

2 BASIC THEORY OF THERMAL ANALYSIS AND THERMO-STRUCTURAL MECHANICS

2.1 Theories related to thermal analysis

All physical problems in nature need to meet the physical laws of conservation, including fluid problems. When we establish the basic equations of fluid motion, we must follow the laws of conservation of mass, conservation of momentum, and conservation of energy. Based on these laws, we can obtain the control equation of fluid heat transfer (Shao *et al.* 2003).

(1) Law of conservation of mass

Any flow problem must satisfy the law of conservation of mass. This law can be formulated as follows: the increase in mass in a fluid microelement per unit of time is equal to the net mass flowing into the microelement in the same time interval. According to this law, the equation of conservation of mass (1) can be derived as:

$$\frac{\partial \rho}{\partial t} + \frac{\partial}{\partial x}(\rho u) + \frac{\partial}{\partial y}(\rho v) + \frac{\partial}{\partial z}(\rho w) = 0 \tag{1}$$

This equation is a general form of the conserved mass equation, which applies to both compressible and non-compressible flows. Where ρ is the density, u, v, and w are the components of the velocity vector in the x, y, and z directions.

(2) Law of conservation of momentum

The law of conservation of momentum is also a fundamental law that any flow system must meet. This law can be formulated as follows: the rate of change of the fluid momentum versus in a microbody is equal to the sum of the various forces acting on this microbody. According to this law, taking the x direction as an example, the equation of conservation of momentum (2) can be derived as:

$$\frac{\partial}{\partial t}(\rho u_i) + \frac{\partial}{\partial x_j}(\rho u_i u_j) = -\frac{\partial p}{\partial x_i} + \frac{\partial \tau_{ij}}{\partial x_j} + \rho g_i \tag{2}$$

where p is the static pressure; τ_{ij} is the stress tensor; g_i is the volume force in the i direction

(3) Law of conservation of energy

The law of conservation of energy (3) can be expressed as follows: the energy increase rate in the micro is equal to the net heat flow into the micro plus the work done by the volume force and the surface force on the micro body. This law is actually known as the first law of thermodynamics, expressed as:

$$\frac{\partial(\rho T)}{\partial t} + \text{div}(\rho u T) = \text{div}\left(\frac{k}{c_p} gradT\right) + S_T \tag{3}$$

where, T is the temperature; k is the heat transfer coefficient; c_p is the specific heat capacity; S_T is the viscous dissipative term.

(4) Equation of state

For the solution of the fluid problems above, five partial differential equations are listed, namely the mass conservation equation, the momentum conservation equation in three directions, and the energy conservation equation. They contain six variables: ρ, u, v, w, p, and t. Then an equation between ρ and p is needed to ensure the closure of the system of equations. The equation is called the equation of state (4) and is used to describe the change in the fluid state. For the ideal gas, the equation of state can be expressed as:

$$p = \rho RT \tag{4}$$

where, R is the ideal gas constant.

2.2 *Principles of thermal structure mechanics*

Using the geometric relation, derive the equation (5) for representing the strain of the element with the node displacement (Mahi *et al.* 2010).

$$\{\varepsilon\} = [B]\{\delta\}^e \tag{5}$$

where, $\{\varepsilon\}$ is any point of the cell strain array and $[B]$ is the element strain matrix.

Using physical relations, Equation (6) expressing element stresses in terms of node displacement is derived.

$$\{\sigma\} = [D](\{\varepsilon\} - \{\varepsilon_0\}) + \{\sigma_0\} = [D][B]\{\delta\}^e - [D]\{\varepsilon_0\} + \{\sigma_0\} \tag{6}$$

where, $\{\sigma\}$ is a stress matrix at any point within an element, $[D]$ is an elastic matrix related to the element material, $\{\varepsilon_0\}$ is the element initial strain array, and $\{\sigma_0\}$ is the initial stress array of units.

Using the principle of virtual work, the equation between the junction force and the node displacement on the element is derived, that is, the element stiffness equation (7).

$$\{R\}^e = [K]^e\{\delta\}^e \tag{7}$$

where, $[K]^e$ is the element stiffness matrix, which is represented as:

$$[K]^e = \int\int\int [B]^T[D][B]dxdydz \tag{8}$$

The process of assembling the stiffness equations of all elements to establish a structural equilibrium equation collection process consists of assembling the element stiffness matrix into a structural stiffness matrix and collecting the equivalent nodal arrays acting on each element into a total load array, getting Equation (9).

$$\{R\} = [K]\{\delta\} \tag{9}$$

It is possible to combine the thermal load with other loads to obtain combined stress, including thermal stress. The calculation of stresses should include the outgoing strain terms.

$$\{\sigma\} = [D](\{\varepsilon\} - \{\varepsilon_0\}) \tag{10}$$

where, $\{\varepsilon_0\}$ is the temperature strain due to temperature changes.

3 INTRODUCTION TO MODELS OF MULTIPOINT ELASTIC RESTRAINT SYSTEM

For some large structural components, when supporting them, it is necessary to constrain the six degrees of freedom of the component space to achieve fixation. The main idea of multipoint elastic restraint is to disperse the overall support surface of the traditional support method into multiple support points, and each elastic support device is used as a basic unit through the mutual coordination among multiple elastic restraint devices to ensure the safety and reliability of the supporting parts. The multipoint elastic restraint system in this paper is designed as shown in Figure 1.

Figure 1. Multipoint elastic restraint system and its support devices.

The protected parts are designed as a combination of three closed tank containers; each tank is connected by a connecting tube. Figure 2 shows the combined container structure and its lugs' number.

The combination container is located based on the support. To make the system have enough stiffness to ensure its safety and reliability, there are ten elastic support devices, each group of which includes a support lug and support. Supports are located under the corresponding lugs, which become the bearing structure between the combined container and the supporting foundation, and the support device can limit the movement of the combined container in the axial direction of the supporting lugs, release the thermal displacement of the system superimposed, and generate thermal expansion from the heating of the container and pipe itself.

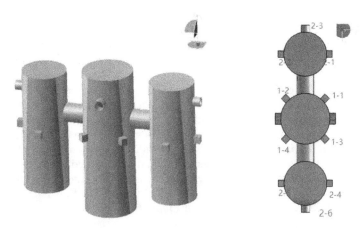

Figure 2. Combined container structure and its lugs' number.

4 TEMPERATURE FIELD SIMULATION OF A MULTIPOINT ELASTIC RESTRAINT SYSTEM

In this paper, the combined container stores thermal conduction oil with its surface in direct contact with the fluid, generating convective heat transfer. Then, through the thermal conductivity of the metal material, increasing the temperature of the outer wall surface of the combined container (including the surface of the supporting bond). Finally, transferring energy is between the supporting bond and the support and supporting foundation, so that the overall system is in an uneven temperature field.

In the numerical simulation of convective heat transfer in the container, the specific aerodynamic boundary conditions are set as follows: the inlet temperature is set to 573K, the inlet flow rate is set to 2 m/s, and the average static pressure at the outlet is set to natural atmospheric pressure. The third type of boundary condition is selected to determine the temperature of the container's surface based on the temperature of the medium around the container and the convective heat transfer coefficient of its surface. The convective heat transfer coefficient is 30 W/(m^{2}K).

The oil enters from two sides of the containers, flowing out of the middle container. The calculated liquid flow lines in the container are shown in Figure 3. The streamlined diagram shows that the oil exchange rate in the middle container is relatively low, especially in the lower part of the container and its bottom area, with almost no participation in the oil exchange, resulting in the middle container's bottom temperature being too low. Figure 4

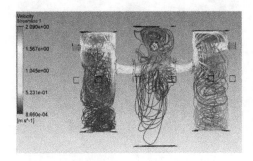

Figure 3. Streamline diagram inside the container.

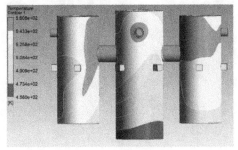

Figure 4. Container surface temperature field.

shows a temperature distribution plot that corroborates the effect of the streamline. This creates unnecessary thermal stress due to excessive temperature gradients. The flow path of the oil can be improved by adding a deflector to the container so that the flow of the liquid is more uniform and the temperature distribution on the outer surface of the natural container is more uniform.

The deflector plate design of the inlet and outlet containers is shown in Figure 5. A hollow round tube with a ring in the inlet container is added for the deflector plate. An arc plate is added to the upper part of the ring. Then, at the bottom of the tank, symmetrically open the deflector plate in two grooves so that oil flows through the guide of this arc plate to the bottom of the tank. A pair of vertical plates are added to the outlet container for the deflector plate. The upper and lower edges of the plate are slotted separately, so that the deflector plate divides the middle container into four areas, and the oil can flow from one partition to another through the bottom slot and the top slot. The flow path of the oil after replacing the outlet deflector plate is shown in Figure 6.

Figure 5. Inlet and outlet container deflector plate.

Figure 6. The oil path inside the container.

The combined container model with a deflector plate is imported into CFX for heat transfer calculations, and the results of the streamlined calculation in the container after the deflector plate and the temperature field of the container surface are obtained, as shown in Figures 7 and 8.

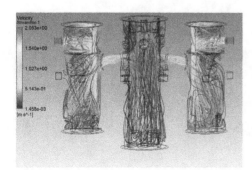

Figure 7. The streamlined diagram with the deflector.

Figure 8. Container surface temperature field.

Compared to tanks without deflectors, the temperature distribution of the middle tank is significantly more uniform; the main temperature is about 515K, and the floating temperature up and down is 10K. Since the floating value is small, there is no separate high-temperature zone and low-temperature zone. The minimum temperature on the surface of the middle tank (except the supporting bond part) was raised from 456K to 490.3K, which increased the temperature of the cryogenic zone at the bottom of the tank by about 34K.

As a result, the design of the deflector does serve to allow the oil to be fully exchanged in all areas of the combined tank, resulting in a more uniform temperature distribution on the container surface.

5 SIMULATION ANALYSIS OF SYSTEM MODELS IN THERMAL ENVIRONMENTS

The temperature distribution results obtained by heat transfer calculation in the previous section are loaded as the overall system temperature load and analyzed for simulation. The gap between the support and the lug is 10 mm.

The contact force analysis results for cold and hot environments are shown in Table 1.

Table 1. The force of the lugs in cold and hot states.

lugs' number	Component direction	Contact force/N	
		Cold state	Hot state
1-1	Lateral force	0.74	40.2
	Axial force	1.55	−506
	Vertical force	871.51	1791
1-2	Lateral force	0.78	101
	Axial force	1.49	−610
	Vertical force	869.93	1723
1-3	Lateral force	0.75	1.65
	Axial force	−1.55	−419
	Vertical force	871.05	1501
1-4	Lateral force	0.8	139
	Axial force	1.48	−635
	Vertical force	871.28	2069
2-1	Lateral force	2.65	86.4
	Axial force	−2.4	−13
	Vertical force	792.65	188
2-2	Lateral force	2.61	0
	Axial force	2.4	0
	Vertical force	792.64	0
2-3	Lateral force	0	0
	Axial force	1.1	0
	Vertical force	389.19	0
2-4	Lateral force	−2.64	−74
	Axial force	−2.48	−16
	Vertical force	792.07	161
2-5	Lateral force	−2.6	0
	Axial force	2.47	0
	Vertical force	791.87	0
2-6	Lateral force	0	0
	Axial force	1.1	0
	Vertical force	390.11	0

When modeled, the mass of the combined container is 0.7584 t. It can be seen from the table data that the resultant force of the vertical support and reaction force is 7432.32 N, which is equal to the gravity of the equipment, proving that the calculation result is accurate. The vertical support reaction force of each support lug changes compared with the cold state; only the lugs of the middle container are subjected to the vertical force. And compared with the self-weight in the cold state, the forces in the three directions are increased. So the lugs other than lugs 1-1, 1-2, 1-3, and 1-4 are lifted. Under the non-uniform temperature field, the temperature distribution of the lug is uneven. The uneven temperature will lead to the problem of inconsistent deformation of the lug itself. This situation will cause the support and reaction forces to be redistributed in the thermal environment of the entire equipment system, and even some lugs will not be loaded. Since the middle four lugs bear almost all the loads, it is necessary to check whether they meet the strength requirements. The stress state of lug 1-1 in the hot state is shown in Figure 9.

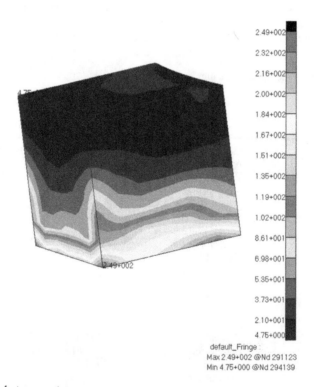

Figure 9. Lug 1-1 stress contour.

In the hot state, the friction between the lug and the support increases, resulting in an increase in stress. The maximum stress of the lug is located at the lower corner and is 249 MPa. The allowable stress is set to 80% of the yield strength, which means that the maximum stress of the system cannot exceed 310 MPa, preventing the support structure from exceeding the allowable stress. Figure 10 is the stress cloud diagram of the overall support system.

The maximum stress of the overall support system occurs at the contact between the lugs and supports. The maximum stress is 287 MPa, which meets the strength requirements, and the structure design is reasonable.

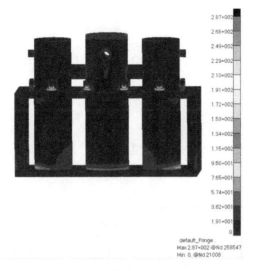

Figure 10. Overall support system stress contour.

6 CONCLUSION

1. The steady-state thermodynamics simulation was carried out, the flow field distribution under the actual heating condition was obtained, and the temperature field distribution of the container was simulated. To improve the uniformity of the temperature distribution on the surface of the container, a design scheme for the deflector was proposed.
2. Under the conditions of thermal and self-weight, the strength of the designed multipoint elastic restraint system is checked to ensure that the structure will not be damaged. The statically indeterminate structure is designed to improve the safety and reliability of the vessel.

ACKNOWLEDGMENTS

I would like to thank Professor He for his careful guidance in my thesis writing and the students for their help. The last is to thank the conference organizer for providing a platform.

REFERENCES

Cui Lin. *Dynamics of Flexible Multibody Systems Considering Geometric Nonlinearity and Thermal Effects*. Shanghai Jiao Tong University, 2007.
Feng Min. *Mechanical Stress and Deformation Analysis of Large-scale Ports Based on Temperature Field Changes*. Wuhan University of Technology, 2011.
He Cheng. Research on Key Technology of Structural Dynamics Modeling in High-Temperature Environment. *Nanjing University of Aeronautics and Astronautics*, 2014.
Mahi A., Adda Bedia E A., Tounsi A et al. An Analytical Method for Temperature-dependent Free Vibration Analysis of Functionally Graded Beams with General Boundary Conditions. *Composite Structures*, 2010, 92(8):1877–1887.
Shao Hongyan, Zhu Runxiang, Ren Chaxian. Finite Element Analysis of Structure Temperature Field and Temperature Stress Field. *Journal of Ningbo University* 2003(01):57–60.
Snyder H.T., Kehoe M.W. *Determination of the Effects of Heating on Modal Characteristics of an Aluminum Plate with Application to Hypersonic Vehicles*. NASA Technical Memorandum 4274, Washington D.C.: NASA, 1991.

Civil Engineering and Energy-Environment – Gao & Duan (Eds)
© 2023 the Author(s), ISBN 978-1-032-56059-5

The road to coal transformation under the "carbon neutrality" goal

GuoChen Zhao*

Datang Northeast Electric Power Test & Research Institute, China Datang Corporation Science and Technology Research Institute, Changchun, P.R. China

ABSTRACT: Under the "carbon neutrality" goal, the traditional energy system is facing fundamental changes. As a cornerstone of national energy security, determining how to lead green development and achieve the dual-carbon goal has become an unavoidable responsibility of industry development. To this end, starting from the objective requirements of the carbon neutrality goal, it is crucial to explore the feasibility of coal technological transformation and innovative development under different technological paths. In order to achieve green and low-carbon development, the coal industry has adopted a variety of clean utilization technologies, which have a significant effect on reducing carbon emissions from coal production and utilization, but there is still a huge gap between achieving the goal of carbon neutrality. Therefore, we further seek a more effective zero-carbon utilization path and propose a future technological path for subversive transformation of the coal industry that meets the requirements of the carbon neutrality vision. The advantages and disadvantages of the two technical paths are discussed in this paper, which provides scientific, theoretical, and technical support for the innovative development and technological transformation of the coal industry under the carbon neutrality goal.

1 INTRODUCTION

Under the "carbon neutrality" goal, vigorously developing carbon dioxide capture, utilization, and storage (CCUS) technology is not only a strategic choice for China to reduce carbon dioxide emissions and ensure energy security in the future, but also an important means to build an ecological civilization and achieve sustainable development. The clean technology of CO_2 separation in integrated coal gasification combined cycle power generation systems (IGCC) is one of the key technologies in the future scenario of coal's transition from main energy to basic energy. From the perspective of clean coal utilization, it provides technical options for realizing the dual-carbon goal (Wang *et al.* 2022).

From the perspective of Chinese emission reduction practice, the primary task of carbon reduction is to improve energy efficiency, followed by vigorously increasing the proportion of renewable energy. The coal industry must achieve disruptive technological breakthroughs as a traditional fossil energy source. It is necessary to completely abandon the idea of "patching" and upgrading the current technical system, even breaking the traditional four links of mining-processing-transportation-utilization, to find a suitable living space for coal in the future carbon-neutral energy system. In July 2021, the Central Committee of the Communist Party of China pointed out that it is necessary to correct the campaign-style "carbon reduction," insist on establishing it first and then breaking it, and have a

---

*Corresponding Author: 1257593514@qq.com

comprehensive plan. The gradual withdrawal of traditional fossil energy must be based on the safety and reliability of new energy sources. Under the carbon neutrality goal of the future, coal must achieve near-zero emissions and have a price advantage to survive. Therefore, subversive coal technological change is of equal importance to the development of renewable energy in terms of ensuring national energy security and achieving carbon neutrality goals (Chen *et al.* 2022).

The coal industry should make full use of the limited time before the carbon peak, carry out a new round of profound self-reform and top-level design, and explore how to achieve the near-zero emission goal with the smallest economic and social cost under the carbon neutrality goal. This will profoundly promote feasible technological changes in the coal industry and provide scientific, theoretical, and technical support to ensure that the country achieves carbon peaking and carbon neutrality goals as scheduled.

2 CURRENT SITUATION OF TRADITIONAL TECHNOLOGY

2.1 *Current status of foreign research*

In 2003, the US Department of Energy proposed the "Future Gen" project (FutureGen), which planned to build a prototype of a 275 MW coal-fired power generation and hydrogen production near-zero emission demonstration power plant. The power plant utilizes state-of-the-art IGCC technology combined with carbon capture and storage (CCS) technology. The US Department of Energy is responsible for 74% of the initial budget investment of $950 million, and the rest of the investment is shared by the Future Generation Alliance. On January 30, 2008, the US Department of Energy announced that it would reorganize the "Power Generation of the Future" project plan, planning to demonstrate the cutting-edge CCS technology in a number of commercial-scale IGCC clean coal power plants. Under the restructuring plan, each commercial-scale power plant should have at least 300 MW of installed capacity. Each power plant should be able to capture and safely store at least twice the amount of CO2 originally planned. In the demonstration phase, at least 1 million tons of CO_2 will be stored in salt-bearing reservoirs each year, and another 1 million tons will be used for oil and gas stimulation or other ways to achieve permanent storage. Emissions of sulfur dioxide, nitrogen oxides, particulate matter, and mercury do not exceed the original regulatory levels (Li *et al.* 2009).

In 2004, the European Union launched the "Hydrogen and Electricity Cogeneration" (HypoGen) program in its "Sixth Framework Program" (FP6), intending to develop a 400 MW IGCC for coal gasification-based power generation, hydrogen production, and CO_2 capture and storage demonstration power plant. Combining the strengths of 12 major EU member states and 32 energy industry organizations, it is hoped that the commercial operation of CCS in Europe will be officially launched in 2012. Its first phase, called the Dynamis program, hopes to develop a large-scale commercial power generation demonstration project based on carbon-free coal or hydrogen by 2010 and achieve a carbon capture rate of 90%. The project started in 2004 and completed construction and demonstration operations in 2015, with a total investment of 1.3 billion euros (Ma 2002).

The Japanese "EAGLE" (Application of Coal Gas, Liquid, and Electricity for Energy) project is a coal gasification technology for fuel cells being developed by Japan's "New Energy Industrial Technology Development Organization" (NEDO). A pilot power plant with a daily coal consumption of 150 tons has been built to develop a coal gasification system (IGFC) suitable for integrated coal gasification fuel cells. The project started in 1998 and ended in 2006, coupling the research and development of IGCC and fuel cells, and the energy efficiency of the system reached at least 53–55%. Japan is expected to deploy an IGCC-fuel cell program starting in 2010, starting with a 50,000-kilowatt distributed generation unit and then commercializing a 600,000-kilowatt system by 2020 (Li *et al.* 2009).

Australia launched the "COA21" plan in 2004, taking coal gasification-based power generation, hydrogen production, synthesis gas production, and CO_2 separation and treatment systems as the future development direction for near-zero emissions (Zhang et al. 2013).

The "2020 Clean Coal Technology Roadmap," formulated by Canada, has identified four R&D priorities and implementation goals: coal optimization, combustion, oxy-fuel combustion, gasification, and chemical synthesis. And through the "ZECA" program to develop advanced coal-to-hydrogen and CO_2 capture, separation, and storage technology (Li et al. 2012).

2.2 Current status of chinese research

In order to further improve the efficiency of coal-fired power generation and reduce pollutant emissions, the Huaneng Group proposed the "green coal power" plan in 2004. The overall goals of the program are: (1) We expect to research, develop, demonstrate, and promote a coal-based energy system based on coal gasification for hydrogen production, hydrogen turbine combined cycle power generation, fuel cell power generation, and CO_2 separation and treatment; (2) We expect to significantly improve the efficiency of coal power generation so that coal power generation achieves near-zero emissions of CO_2 and pollutants; (3) We expect to master its core technology, supporting technology, and system integration technology to form a "green coal power" technology with independent intellectual property rights; (4) We expect to make it economically acceptable and gradually popularize the application to realize the sustainable development of coal power generation.

In August 2009, the State Council officially issued the "Jiangsu Coastal Development Plan," which requires "supporting the cooperation between Jiangsu Province and the Chinese Academy of Sciences in energy power research, promoting the transformation of technological achievements, and building a clean energy innovation industrial park." The construction content of the Demonstration Zone of the Clean Energy Innovation Industrial Park includes 1200 MW of IGCC power generation, co-generation of steam, and co-production of oil, SNG, polyethylene, polypropylene, etc. The demonstration project conducts experiments and demonstrations of capturing 1 million tons of CO_2 per year; 500,000 tons of CO_2 are used for the production of urea and soda ash, and 500,000 tons of CO_2 are stored in saline aquifers (Wu 2007; Wang 2010).

The national "Twelfth Five-Year Plan" 863 major projects "Research and Demonstration of IGCC-based CO_2 Capture, Utilization, and Storage Technology" has achieved a zero breakthrough in my country's coal-based power generation pre-combustion CO_2 capture project. In 2016, China Huaneng launched a project to build the world's first 100,000-ton-grade pre-combustion carbon dioxide capture demonstration device. The unit energy consumption of the capture system is 1.907 GJ per ton of CO_2, the dry basis concentration of CO_2 after capture is 98.11% (Vol.), the CO_2 recovery rate is 91.61%, and the capture cost is 164.3 yuan per ton of CO_2. The index has reached the international leading level. The process design of the 400 MW-based full-flow CO_2 capture, utilization and storage system based on IGCC has been completed, and a process package has been formed. After the CO_2 capture system is integrated, the power supply efficiency reduction value of IGCC can be less than 8%, and the capture energy consumption is 1.5 GJ per ton of CO_2 (2021).

3 TRADITIONAL TECHNOLOGY

3.1 Technical introduction

IGCC is an advanced power system that combines coal gasification technology with an efficient combined cycle. IGCC consists of two parts: the coal gasification and purification part and the gas-steam combined-cycle power generation part. The main equipment of the first part includes a gasifier, an air separation unit, and gas purification equipment (including a sulfur recovery unit). The main equipment of the second part includes a gas turbine power generation system, a waste heat boiler, and a steam turbine power generation system (Wang et al. 2011).

The coal enters the gasifier, is driven by nitrogen, and reacts with the pure oxygen sent by the air separation system in the gasifier. After the reaction, syngas is generated, which is purified to remove pollutants such as sulfides, nitrides, dust, and others, and becomes a clean gaseous fuel, which is then sent to the combustion chamber of the gas turbine for combustion. The gas working medium is heated to drive the gas turbine to do work, and the exhaust gas of the gas turbine enters the waste heat boiler to heat the feed water, generating superheated steam to drive the steam turbine to do work (Shi et al. 2009).

The IGCC carbon capture process begins with the fuel entering the gasifier for gasification to produce syngas. The main components of the syngas are CO and H_2. The syngas is then subjected to a water-gas shift reaction, which turns the gas into CO_2 and H_2, transfers the chemical energy of the fuel to H_2, and then separates the CO_2 and H_2. All the energy of the fuel is carried by H_2, which is burned in the combustion chamber of the gas turbine to form a quasi-zero emission system (Figure 1) (Kanniche & Bouallou 2007).

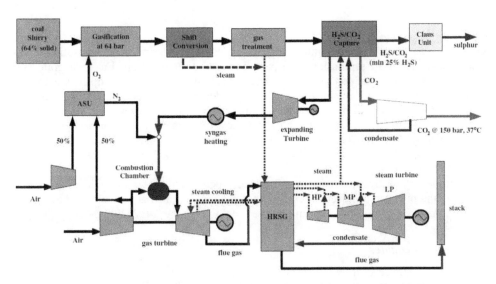

Figure 1. Advanced coal IGCC system with CO_2 capture (Kanniche & Bouallou 2007).

3.2 Key technology

3.2.1 Coal gasification technology selection

Coal gasification technology is the core technology of IGCC, which affects the investment, availability, and efficiency of IGCC power plants. Presently, foreign IGCC power plants use six entrained bed gasification technologies, including GE-Texaco, E-Gas, Shell, Prenflo, MHI (air gasification), and TPRI. Tianjin IGCC in China uses two-stage dry pulverized coal pressurized gasification technology.

3.2.2 Gas turbine selection

The Cool Water and Wabash River IGCC power stations in the United States use 7E and 7FA gas turbines from GE. The two demonstration power stations in Buggenum, the

Netherlands, and Puertollano, Spain, use German Siemens gas turbines V94.2 and V94.3. Tianjin IGCC adopts the SGT5-2000E (LC) type syngas gas turbine jointly provided by Shanghai Electric Group and Siemens of Germany.

3.2.3 *Syngas purification system*

Syngas purification methods are divided into normal-temperature wet purification and high-temperature dry purification. The normal temperature wet purification technology is mature. The equipment is simple, including a cyclone separator (a medium-temperature ceramic filter), a Venturi scrubber, desulfurization equipment, and sulfur recovery. High-temperature (500–600°C) dry purification makes full use of the sensible heat of crude gas, improves the power supply efficiency of IGCC, and reduces equipment investment, but has low operating reliability and a high operating cost.

3.2.4 *System heat recovery method*

Medium-pressure or high-pressure saturated steam is generated by absorbing high-grade sensible heat from crude syngas through a waste boiler. Low-grade heat recovery method: (1) hot water is produced through a saturator to heat and humidify the syngas; (2) low-pressure steam is produced to provide steam for other processes in the system; (3) heating the syngas raises the temperature of the syngas entering the gas turbine; and (4) boiler feed water is preheated.

3.2.5 *Air separation integration*

IGCC air separation is divided into three categories: fully integrated air separation, completely independent air separation, and partial overall air separation. Fully integrated air separation means that the compressed air required for air separation is completely extracted from the gas turbine compressor, and completely independent air separation means that the compressed air is provided by an independent air compressor. The overall air separation has the advantages of high-power supply efficiency, low investment, and improved gas turbine flow problems. Many IGCC plants use integrated air separation systems.

3.2.6 *NO_x removal*

The methods of nitrogen oxide removal include fuel-heated water humidification, steam injection, and nitrogen re-injection. Fuel humidification and nitrogen re-injection in foreign power plants are the two most commonly used methods (Yu *et al.* 2020).

3.3 *Technical features*

3.3.1 *High power generation efficiency*

The carbon conversion rate of the gasifier can reach 96%–99%. Due to the rapid development of gas-steam combined-cycle power generation technology, its thermal efficiency has reached 60% on the power island. The associated IGCC power generation efficiency is likely to increase from the current 43%–45% to more than 50%.

3.3.2 *Environmentally friendly*

Because the gas is efficiently purified under pressure before it is sent to the gas turbine for combustion, the emission of pollutants from IGCC power plants is only 10% of that of conventional coal-fired power plants. Its desulfurization efficiency can reach 99%, and the sulfur dioxide emission concentration is about 25 mg/Nm^3 (the national sulfur dioxide emission concentration is 1200 mg/Nm^3). The nitrogen oxide emission concentration is 15%–20% of that of conventional coal-fired power plants. Water usage is only 30%–50% of that of conventional power plants. It is of great significance for environmental protection.

3.3.3 *Good load applicability and strong peak shaving ability*

IGCC power plants can operate smoothly at 35%–100% load, whereas conventional coal-fired power plants can operate at 50%–100%. The load change rate can reach 7%–15%/min, and the conventional coal-fired power station is 2%–5%, which has a good peak shaving effect.

3.3.4 *Wide range of fuel applicability*

From general high-sulfur coals to low-grade inferior coals and even biological waste, the performance of IGCC gasifiers has little effect. It has good adaptability to different coal types, and the feed price is much lower than the price of natural gas.

3.3.5 *Multi-generation, high economic efficiency*

The crude gas produced by the IGCC gasifier is rich in H_2, which can produce clean H_2 energy. If a conversion module is installed in the system, part of the crude gas can be converted to produce pure H_2. H_2 can be used as a clean and cheap urban transportation fuel with good social, environmental, and economic benefits (Xu 2005).

4 TRANSFORMATIVE TECHNOLOGY

4.1 *UCG-IGCC-CCUS*

The fluidized coal mining (UCG) + IGCC + CCUS technology system was created to address the issue of difficult-to-mine coal seams. The basic principle is to introduce oxygen-enriched gas to make coal chemically react in the gasifier, thereby forming combustible gases such as CO, CH_4, H_2, etc., and collect these mixed combustible gases for power generation and hydrogen production to achieve the utilization of coal energy that is not easy to mine. At present, the technology has been gradually mastered, and many underground coal gasification projects have been successful around the world, such as the Angren project in the former Soviet Union, the Chinchilla project in Queensland, Australia, and the underground coal gasification (power generation) project near Forth Bay in Edinburgh, UK.

UCG not only solves the problem of the utilization of difficult-to-mine coal seams but also reduces land occupation and damage caused by coal mining, ecological environment damage, and coalbed methane gas emissions. It can be said that it is a green and low-carbon mining technology. In addition, the technical system has several advantages, such as solving the safety problem of coal mine production personnel without the need for underground operations and meeting the demand for hydrogen used in the coal chemical industry. UCG-IGCC combined power generation technology can not only directly use the mixed gas to achieve clean power generation but also save carbon emissions in coal mining, washing, transportation, and other links and solve environmental problems such as CH_4 escape, polluted groundwater, and land disturbance (Figure 2) (Bu 2014; Feng *et al.* 2021; Imran *et al.* 2014).

4.2 *UCG-H$_2$-CCUS*

The UCG + hydrogen production + CCUS technical system is similar to the technical principle of UCG-IGCC-CCUS. After UCG produces mixed gas, H_2 is further separated, and other gases containing CH_4, CO, etc., are further reacted with coal to produce H_2 and capture CO_2. It can realize the transformation from coal-to-ash hydrogen (CG) plus CCUS negative carbon technology to coal-to-blue hydrogen (CG-CCUS).

Affected by scale, raw materials, and transportation costs, the cost of hydrogen production varies greatly. It is necessary to comprehensively consider capital cost, operation, and maintenance cost, and raw material and power cost, which is generally calculated on a

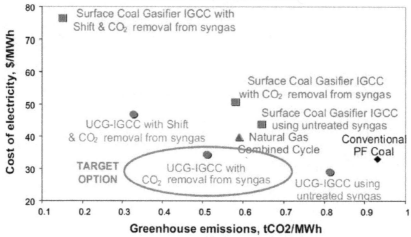

Figure 2. Greenhouse emission (Imran *et al.* 2014).

levelized basis. Compared with wind energy electrolyzed water (WELE), light energy electrolyzed water (P-ELE), and biomass gasification to hydrogen (BG), coal-to-ash hydrogen (CG) is quite cheap, but the carbon emission of CG can reach 20.0–30.0 kg eCO$_2$/kg H$_2$. Therefore, CCUS technology must be introduced under the carbon neutrality goal to achieve zero-carbon CG-CCUS. According to relevant research, the cost of CG-CCUS is about 440 yuan/t higher than that of CG (UCG does not consider the cost of CO$_2$ transportation), plus the carbon emissions caused by the energy consumption of CCUS. Compared with W-ELE, P-ELE, and BG hydrogen production methods, CG-CCUS still has significant cost advantages. It can be considered that UCG-CCUS technology is one of the innovative technology paths for zero-carbon utilization of coal that meets the requirements of carbon neutrality goals (Figure 3) (Kempka *et al.* 2011; Nakaten *et al.* 2014; Wang *et al.* 2021).

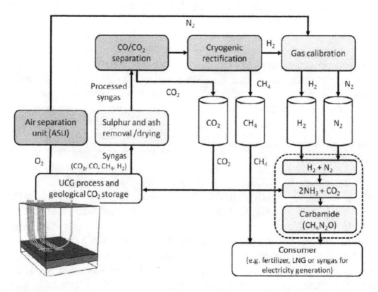

Figure 3. Schematic view of the coupled UCG-CCS process (Kempka *et al.* 2011).

5 FUTURE OUTLOOK

Zero carbonization and economic feasibility are the basic requirements of the carbon neutrality goal for the future energy system. The coal industry must meet these two requirements in order to occupy a space. However, the current coal utilization technology system is far from meeting the above requirements. Therefore, it is necessary to fundamentally achieve subversive technological breakthroughs, break the traditional mining-processing-transportation-utilization links, and even sacrifice a part of resource utilization in order to achieve zero emissions. Judging from the currently reserved technologies, one is the UCG-IGCC-CCUS technical system. It not only greatly reduces the energy consumption of mining, washing, and transportation in the traditional technical path but also greatly reduces the emission of CH4 from coal-measure gas. The synthesis gas after underground gasification can be separated to produce hydrogen, which can meet the production demand of the coal chemical industry using traditional technology. IGCC can also be used directly, and CO2 utilization and storage can be realized with the help of CCUS negative carbon technology to achieve near-zero emissions in the entire life cycle of coal and achieve the carbon neutrality goal. This is particularly concerned with UCG technology, which mainly considers the whereabouts of CO2 after large-scale capture and the risk of shallow groundwater contamination. A small amount of CO2 can be mineralized and utilized and can also be used to drive oil and gas to achieve storage and economic benefits. However, in the future, only geological storage can be implemented after the capture of tens of millions of tons or even 100 million tons of CO_2. The tightness and stability of the caprock of the shallow ore body are insufficient, and the pressure is also insufficient. It is difficult for CO2 to have a miscible reaction with the slag after underground gasification to form a relatively fixed storage state. The second is the UCG-H2-CCUS technology system, which is similar to the hydrogen production process in the coal chemical industry. It is just that the hydrogen produced by the separation of synthesis gas after underground gasification is no longer used as a raw material for the coal chemical industry but is directly used as secondary energy for power generation or energy storage, and CO_2 is captured, utilized, and stored in underground mines. In this way, the transition from gray hydrogen to green hydrogen in traditional technology is realized. In the foreseeable future, coal-to-green hydrogen will still be one of the most cost-effective technology paths. In addition, coal can also be widely used in the synthesis of coal-based high-energy fuels, the production of advanced coal-based carbon materials, and the synergistic utilization of biomass and waste. The carbon-neutral technology system alters the traditional coal utilization path, and significant amounts of carbon emissions are saved in mining and washing, transportation, the coal chemical industry, power generation, and heating supply. At the same time, to a certain extent, the coordinated treatment of pollutants has been realized. It is not necessary to rely on promoting and using new energy technologies such as solar and wind energy. This will be based not only on China's actual energy resource endowment, but also on a feasible carbon-neutral path.

6 CONCLUSIONS

The role of coal in the future energy system needs to be judged carefully, and it should be considered from the perspective of energy elasticity and the safety zone. As a guaranteed safe energy source, coal will still play an important role in a carbon-neutral energy system. At present, the carbon footprint of coal in its entire life cycle includes six links: mining and washing, transportation, power generation and heating, iron and steel smelting, the coal chemical industry, and other links. Therefore, it is determined that the implementation of zero-carbon high-efficiency power generation, terminal electrification, and coal-to-green hydrogen under the current carbon neutrality goal in the future is the first choice for coal technological transformation and to realize the integration of coal-fired power or coal-to-green hydrogen energy storage. UCG-IGCC-CCUS technology and UCG-H$_2$-CCUS technology have significant economic and carbon reduction advantages and will become the only way for coal to be safe, efficient, green, and zero-carbon. Compared with ultra-supercritical coal-fired power generation and IGCC power generation technologies, UCG-IGCC-CCUS power generation consumes more coal but can achieve zero emissions. Compared with blue hydrogen production from new energy sources, UCG-H$_2$-CCUS green hydrogen production technology has significant cost advantages and achieves near-zero emissions.

ACKNOWLEDGMENTS

This work was financially supported by the "Biomass Hydrothermal Carbonization Coupled Carbon Capture Technology for Coal-Fired Power Plants" project of the Datang Northeast Electric Power Test & Research Institute.

REFERENCES

Bu X.P. (2014) Carbon Dioxide Capture Technology and Application Analysis. *Clean Coal Technology*, (5):9–13.

Chen F., Wang S.Y., Yu H.C., et al. (2022) On the Technical Path of Coal Reform Under the Carbon Neutrality Goal. *Journal of China Coal Society*.

Feng Y., Yang B., Hou Y., et al. (2021) Comparative Environmental Benefits of Power Generation from Underground and Surface Coal Gasification with Carbon Capture and Storage. *Journal of Cleaner Production*, 310:127383.

Imran M., Kumar D., Kumar N., et al. (2014) Environmental Concerns of Underground Coal Gasification. *Renewable and Sustainable Energy Reviews*, 31, 600–610.

Kanniche M. and Bouallou C. (2007) CO$_2$ Capture Study in Advanced Integrated Gasification Combined Cycle. *Applied Thermal Engineering*, 27(16), 2693–2702.

Kempka T., Plötz M.-L., Schlüter R., et al. (2011) Carbon Dioxide Utilisation for Carbamide Production by Application of the Coupled UCG-urea Process. *Energy Procedia*, 4, 2200–2205.

Li G.J., Zhang J. and Ji L.C. (2009) Update on Future Zero-emission Coal-fired Power Generation Projects in the United States. *Sino-Global Energy*, 14(5):96–100.

Li Q., Hou Y.L. and Wang B.H. (2009) NEDO: Technology Knows No Borders – Interview with Yuzo Goto, Chief Technical Representative of NEDO Beijing Office. *China Construction News: Sunshine Energy*, (4):28–29.

Li G.J., Zhang J., Li X.C., et al. (2012) Analysis and Enlightenment of Carbon Dioxide Capture and Storage Technology Roadmap. *Science and Technology Management Research*, 7(5):17–19.

Ma J.Z. (2002) The EU Launches the Sixth Framework Programme. *Science News*, 22.

Nakaten N., Schlüter R., Azzam R., et al. (2014) Development of a Techno-Economic Model for Dynamic Calculation of Cost of Electricity, Energy Demand and CO$_2$ Emissions of an Integrated UCG-CCS Process. *Energy*, 66:779–790.

Shi Q., Wu X.J., Xu X.Y., et al. (2009) Integrated Coal Gasification Combined Cycle (IGCC) Power Generation Technology and Energy Saving and Emission Reduction. *Energy Conservation Technology*, 27(1):18–20.

Wang X.J., Liu P., Li R.C., et al. (2022) Research Progress and Prospect of Advanced Power Generation Technology Under the "Dual Carbon" Goal. *Thermal Power Generation*, 51(1):52–59.

Wu R.S. (2007) Coal-fired Power Plants of the Future – China Green Coal Power Plan. *Electric Power*, 40(3):6–8.

Wang Z.L. (2010) "Green Coal Power" Set Sail in China. *Resources and Human Settlements*, (7):46–48.

Wang X.L., Wu J.H., Zhao, Q. (2011) Research Status and Prospect of CO$_2$ Capture Technology in Coal-fired Power Plants. *Dongfang Electric Review*, 25(2):1–9.

Wang Y.Z., Zhou S., Zhou X.W., et al. (2021) Cost Analysis of Different Hydrogen Production Methods in China. *Energy of China*, 43(5):29–37.

Xu L.B. (2005) Development Status and Prospects of Integrated Coal Gasification Combined Cycle Power Generation Technology. *Electric Power Survey & Design*, (6):8–11.

Yu L.H., Li T., Li C.Y., et al. (2020) Development Status of Integrated Coal Gasification Combined Cycle Power Generation System. *Shandong Chemical Industry*.

Zhang J., Li G.J., Chen W., et al. (2013) The Clean and Efficient Comprehensive Utilization of Coal Resources will Form an Emerging Industry. *Bulletin of Chinese Academy of Sciences Bull Chin Acad Sci*, 28 (5):625–627.

(2021) IGCC-based Pre-combustion CO_2 Capture Technology and Engineering Demonstration. *High-Technology and Industrialization High-Technol Ind*, 27(06).

Civil Engineering and Energy-Environment – Gao & Duan (Eds)
© 2023 the Author(s), ISBN 978-1-032-56059-5

Numerical simulation of wave action on sand topping section

Xiaodong An
Shandong Ocean Culture Tourism Development Group Co. LTD, Rizhao, China

Bin Zhou
Guocheng Group Co. LTD, China

Xianliang Yan & Siyuan Dong
Shandong Provincial Communications Planning and Design Institute Group Co. LTD, Jinan, China

Chengye Liu & Tingting Cui*
College of Energy and Electrical Engineering, Hohai University, Nanjing, China

ABSTRACT: In order to build a continuous and complete multi-level disaster prevention and reduction system and improve the safety and stability of the regional ecosystem, this paper, based on Delft3D-FLOW, achieves the solution of water body dynamics and sediment dynamics in the sand paving area through the coupling simulation of wave and tide effects and sediment erosion and deposition calculation under three sand paving adjustment schemes. The influence of the hydrodynamic field on sediment transport is explored, and the hydrodynamic characteristics of water and sediment under normal sea wave conditions and wave action with 2, 10, and 50-year return periods are predicted. It is found that the hydrodynamic field reduction effect of Scheme III is the best as far as the index of wave reduction rate is concerned. Under the action of ENE directional wave, the attenuation rate of effective wave height induced by Scheme III is 41.1%~52.8%; The maximum attenuation rate of wave height is 37.8%~55.2%. The attenuation effect of wave and tide coupling in Scheme II is the worst. However, the three sand laying adjustment schemes also have the effect of resisting the impact and erosion of tidal waves.

1 INTRODUCTION

The coastal zone is located at the junction of land and sea and is affected by multiple natural and human processes of land, sea, and air. It is not only a complex and dynamic natural system on the earth's surface but also a space unit under the influence of high-intensity human activities. It has unique land and sea attributes in the dynamic and complex natural system (Shen & Zhu 1999; Yu *et al.* 2010). The disorderly and extensive large-scale development of coastal zones by human beings has also caused many problems. Excessive use of the coast, the curved natural coastline is artificially straightened, and the disorderly reclamation of the sea leads to the deterioration of the ecological environment of the coastal waters (Johnson *et al.* 2015; Liu *et al.* 2015; Zhang & Hou 2020; Zhang *et al.* 2016). The hydrodynamic and numerical models are established to reflect the interaction among wind, wave, and flow factors. It is helpful to reasonably simulate sediment movement, and

---

*Corresponding Author: cuitt061@163.com

accurately predict sediment transport process and silting problems, and it is of great significance to the coastal restoration project and the resistance to beach erosion.

Based on the two-dimensional shallow water equation, Yu Yingxia *et al.* (2022) constructed a two-way coupling between the storm surge model and the numerical model of shallow water waves, established a wave-tide coupling model and described the complex nonlinear interaction between storm surge and waves. They found that due to the special topography of the Shanghai coastal area, the tidal current field of Hangzhou Bay was greatly influenced by the wave at the highest tide level of the storm surge, and the water-increasing effect of the Yangtze estuary was obvious. Zhou *et al.* (2022) used an improved two-dimensional morphological dynamic model to simulate and study the sediment transport process and morphological adjustment process of the three central sandbars in the river reach and studied the impact of different factors on the channel regulation engineering of the natural evolution of the river channel. It is considered that riverbed erosion is unlikely to occur in the protected area due to the protection of river regulation works unless there is a newly formed sediment layer in the previous simulation period. Zuo *et al.* (2022) obtained a new synthetic expression for the depth-averaged suspended sediment concentration in the vortex bed by integrating the process-based suspended sediment concentration profiles. They discussed the change of sediment content under the condition of enhanced wave dynamics and believed that the proposed formula could better describe the effect of the riverbed shape during enhanced wave dynamics.

Studying the temporal and spatial variation characteristics of coastlines is of great significance for understanding the ecological environment process, evolution mechanism, comprehensive management and sustainable development of coastal zones. This paper focuses on the coastal protection zone of Sun Bay in Rizhao City, through the coupling simulation of wave and tidal action under three sand-laying adjustment schemes, as well as the calculation of sediment erosion and silting, to explore the ability of artificial beaches to resist the impact and erosion of tidal waves in coastal restoration and renovation projects. The purpose is to build a continuous and complete multi-level disaster prevention and mitigation system, improve the safety and stability of the regional ecosystem, and achieve the comprehensive improvement of coastal disaster prevention and mitigation capacity.

2 COMPUTATIONAL MODELS AND METHODS

2.1 *Computational model and meshing*

The area of this numerical simulation is the near-shore sea area covering the project area, with an east-west distance of about 6328 m and a north-south distance of about 7980 m. The FLOW model uses variable cells to partially densify the project area. The minimum grid size is about 5 m × 7 m, and the maximum grid size is 23 m × 45 m. Since the calculation results of the WAVE module are insensitive to the spatial accuracy of the results of the FLOW module, in the same region, the WAVE cells are twice as many as FLOW cells, and cell nesting mode is adopted.

2.2 *Numerical calculation method*

The numerical simulation in this paper is based on the establishment of Delft3D-FLOW, which achieves the coupling solution of hydrodynamics, sediment dynamics and other issues in the sand-laying area, explores the influence of the hydrodynamic field on the sediment transport, and then predicts the hydrodynamic characteristics of water and sediment under the condition of constant sea waves and the action of waves with 2, 10 and 50-year return periods in this area. Among them, the damping characteristics of the seabed surface of the

WAVE module are calculated by using the module Johnson equation, and the damping coefficient is selected as 0.067. The FLOW module adopts the Manning equation, and the Manning coefficients of the plane coordinates in the U and V directions which are both taken as 0.0263. And the Van Rijn formula is selected to solve the shearing action induced by waves. In the numerical simulation of the sediment dynamic module, the porosity ratio of sediment is set to 0.4, the median particle size of the sediment under natural terrain is set to 0.1 mm, the gradation coefficient is selected to be 1.1, and the median particle size of the sand laid on the beach is set to 0.3 mm, and the gradation coefficient is selected as 1.1. The zero-flux and zero-gradient boundaries are adopted.

2.3 Research scheme

In this paper, three-wave directions (E, ENE, ESE) and different return periods (2 years, 10 years, and 50 years) are combined to calculate the wave boundary conditions. It should be pointed out that there are three sand-laying adjustment schemes involved here, the specific order is as follows: Scheme 1: the initial shoreline seaward elevation + 1.5 m is used for sand laying, and the lateral span of the beach shoulder is 60 m. Option 2: the initial shoreline seaward elevation + 1.0 m is used for sand laying, the lateral span of the beach shoulder is 60 m, and the slope is 1:34 to the initial mud line; Option 3: the initial seaward seaward elevation + 2.0 m is adopted for sand laying, the lateral span of the shoulder of the beach is 40 m, and it is graded to the initial mud line according to 1:34.

3 RESULTS AND ANALYSIS

Based on the above conditions, the coupling simulations of wave and tidal action, as well as sediment flushing and siltation calculations were carried out for the project area under the current shoreline. The simulation results are shown in Figures 1 to 3. According to the results, the wave height of the wave load acting on the coastal area is roughly in the range of 0.6~1.8 m. The wave height in the area adjacent to the beach decreases sharply to less than 0.1 m due to the decrease of water depth. The wave situation at the south side of Xiaohai estuary in the south of the calculation project area is complex, with a wave height of about 1.0 m, while the wave height entering the estuary and the north side of the estuary is 0.6~0.8 m. The specific wave heights in the area south of the Xiaohai estuary are shown in Table 1.

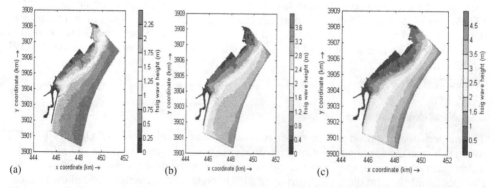

Figure 1. Contours of significant wave height of E wave direction in different years. (a) Effective wave height out of one in 2 years (b) Effective wave height out of one in 10 years (c) Effective wave height out of one in 50 years.

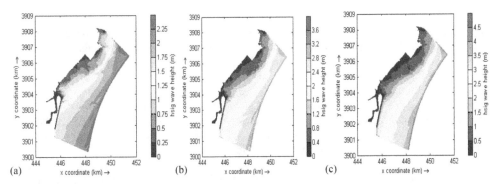

Figure 2. Contours of significant wave height of ENE wave direction in different years. (a) Effective wave height out of one in 2 years (b) Effective wave height out of one in 10 years (c) Effective wave height out of one in 50 years.

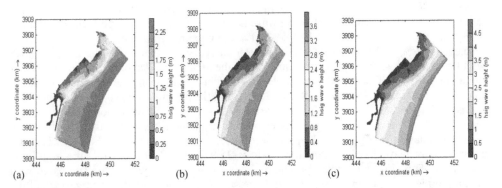

Figure 3. Contours of significant wave height of ESE wave direction in different years. (a) Effective wave height out of one in 2 years (b) Effective wave height out of one in 10 years (c) Effective wave height out of one in 50 years.

Table 1. Wave Height and period in the area south of Xiaohai estuary.

Wave Height/ Wave Direction	H1%	H4%	H13%	T
E	1.454	1.277	1.115	9.0
ENE	1.116	0.971	0.811	9.0
ESE	1.504	1.301	1.151	9.0

On the basis of the above-mentioned coupled simulation of wave and tidal action under the current coastline, the numerical module is further carried out to solve the coupled hydrodynamic field of waves and tides in the engineering area under the conditions of three sand-laying adjustment schemes.

There are certain differences in the wave height attenuation rates before and after the project under the action of different wave directions, which are as follows:

Under the action of the E wave, the effective wave height attenuation rate caused by the sand laying project (+1.5 m elevation) in Scheme 1 is in the range of 31.7%~47.7%. Since the effective wave height is in the wavelet category, the higher the incident wave height of the load, the higher the wave height. The larger the attenuation range is, the general rate accounts for 20.8% of the overall protection effect; the maximum wave height attenuation rate is 29.6%~48.3%, and the good rate of large waves in the protection attenuation of E waves at different sections to the maximum wave height accounts for 100%.

Under the action of the ESE wave, the effective wave height attenuation rate caused by the sand laying project (+1.5 m elevation) of Scheme 1 is in the range of 32.6%~46.7%, and the attenuation characteristics of the effective wave height of the ESE wave load are the same as those of the E wave. The attenuation rate is 30.7%~49.6%, and the good attenuation rate of ESE waves in the case of large waves also accounts for 100%.

Under the action of the ENE wave, the effective wave height attenuation rate caused by the sand laying project is 29.4%~42.4%. The maximum wave height attenuation rate is 24.3%~45.2%, which is compared with the first two wave directions (E-direction waves and ESE waves). Although the effective wave height, maximum wave height and decay rate of ENE directional waves have decreased slightly, the overall effect compliance rate is still guaranteed.

Similarly, three module solutions of wave-direction load decay rates were carried out for the sand-laying adjustment scheme (+1.0 m elevation) of scheme 2. It is found that under the action of E direction waves, the effective wave height attenuation rate generated by the sand-laying scheme 2 is in the range of 17.1%~31.1%, the maximum wave height attenuation rate is 16.0%~35.6%, and the load is in the wavelet category. So the higher the incident wave height of the load is, the higher the value is, and the greater the wave attenuation is. Under the action of the ESE wave, the effective wave height attenuation rate generated by sand-laying scheme 2 is in the range of 18.9%~30.8%, and the maximum wave height attenuation rate is 17.3%~33.8%. The attenuation characteristics of the ESE wave load effective wave height are the same as the E wave. Under the action of the ENE wave, the effective wave height attenuation rate generated by the second sand-laying scheme is 17.6%~32.1%, and the maximum wave height attenuation rate is 18.9%~33.2%. In general, the amplitude attenuation caused by sand dressing scheme 2 has a relatively large decrease compared with the amplitude attenuation caused by sand dressing scheme 1, and the average value fluctuates around 25.0%.

Further, three module solutions for the attenuation rate of wave loads were carried out for the sand laying adjustment scheme (+2.0 m elevation) of scheme three. Comparing the above results, it can be seen that under the action of the E-wave, the effective wave height attenuation rate induced by sand-laying scheme 3 is in the range of 36.6%–53.2%, and the maximum wave-height decay rate is 37.0%–53.6%. The effective wave height attenuation rate induced by scheme three is in the range of 37.8%~52.3%, and the maximum wave height attenuation rate is 37.3%~54.5%. Under the action of the ENE wave, the effective wave height attenuation rate induced by sand laying scheme three is 41.1%~52.8%. The maximum wave height attenuation rate is 37.8%~55.2%. In general, the amplitude attenuation induced by sand-laying scheme 3 increases significantly compared with the amplitude attenuation caused by sand-laying scheme 1, and the average value fluctuates around 46.8%.

4 CONCLUSION

Based on the coupling simulation of wave and tide and the calculation of sediment erosion and deposition under three sand paving adjustment schemes, this paper solves the problems of water and sediment dynamics in sand paving areas and the influence of hydrodynamic fields on sediment transport. For the three sand spreading adjustment schemes proposed for Rizhao Sun Bay coastal protection, in terms of the wave attenuation rate, Scheme 3 has the best attenuation effect on the hydrodynamic field, Scheme 2 has the worst attenuation effect on the coupled wave and tidal effects, and Scheme 1 has the middle function. However, the three sand-laying adjustment schemes are also effective in resisting the impact of tides and waves and can achieve the objectives of building a continuous and complete multi-level disaster prevention and mitigation system, improving the safety and stability of the regional ecosystem, and promoting biodiversity, thus achieving a comprehensive enhancement of the coastal zone's disaster prevention and mitigation capacity and promoting the synergy of ecology and disaster mitigation in the coastal zone. In the future, it is of great interest to optimize the wave-tidal interaction model and the sand paving adjustment scheme.

REFERENCES

Johnson J.M., Moore L.J., Ells K., et al. Recent Shifts in Coastline Change and Shoreline Stabilization Linked to Storm Climate Change. *Earth Surface Processes and Landforms*, 2015, 40(5): 569–585.

Liu B.Q., Meng W.Q., Zhao J.H., Hu B.B., Liu L.D., Zhang F.S. Changes in the Characteristics of Development and Utilization of Coastline Resources in Mainland China from 1990 to 2013. *Journal of Natural Resources*, 2015, 30(12): 2033–2044.

Shen H.T., Zhu J.R. On the Study of Land-sea Interaction in China's Coastal Zone. *Ocean Bulletin*, 1999(06): 11–17.

Yu L., Hou X., Gao M., et al. Assessment of Coastal Zone Sustainable Development: A Case Study of Yantai, China. *Ecological Indicators*, 2010, 10(6): 1218–1225.

Yu Y.X., Zhang X.J. and Song H.H. Research on Storm Surge Simulation Based on Two-dimensional Shallow Water Equation and SWAN Model. *Journal of Henan University of Science and Technology (Natural Science Edition)*, 2022,43(06):67–73 + 9.

Zhang X., Zhang Y., Ji Y., et al. Shoreline Change of the Northern Yellow River (Huanghe) Delta After the Latest Deltaic Course Shift in 1976 and its Influence Factors. *Journal of Coastal Research*, 2016, 74 (10074): 48–58.

Zhang Y.X. and Hou X.Y. A Review of International Research Progress on Coastline Change—based on Bibliometric Methods. *Chinese Journal of Applied Oceanography*, 2020, 39(02): 289–301.

Zhou M.R., Xia J.Q., Deng S.S., Li Z.W. Two-dimensional Modeling of Channel Evolution Under the Influence of Large-scale River Regulation Works. *International Journal of Sediment Research*, 2022, 37 (04): 424–434.

Zuo L., Roelvink D., Lu Y., et al. Process-based Suspended Sediment Carrying Capacity of Silt-sand Sediment in Wave Conditions. *International Journal of Sediment Research*, 2022, 37(2): 229–237.

Study on evaluation method of slope toughness of muck deposit

Shaojie Feng, Renzhuo Zhang* & Wangqian Deng
School of Civil Engineering, North China University of Technology, Beijing, China

ABSTRACT: With the development of urban underground space, a large number of engineering slag or construction waste slag disposal sites have emerged, which has brought great potential safety hazards to urban development. Therefore, it is of great theoretical value to study the toughness evaluation method of slag accumulation slope for building a resilient city. This paper selects 12 indexes under the four influencing factors of material characteristics, slope characteristics, environment and infrastructure, and organizational mechanism of muck disposal site as toughness evaluation indexes. Secondly, based on the fuzzy analytic hierarchy process, the evaluation model of the toughness of the slag soil slope is established. The membership function of the slag soil slope is constructed by the fuzzy mathematics method, and the weight of the influencing factors is determined by the analytic hierarchy process. Finally, the toughness evaluation is carried out with an engineering example of the muck field. The results show that the slope toughness value of the muck yard is grade II (strong toughness); the toughness evaluation method of the muck yard slope can provide theoretical support for the toughness evaluation of similar projects.

1 INTRODUCTION

With the acceleration of urbanization, super high-rise buildings have sprung up, and the depth of excavation is deeper and deeper, which inevitably produces a large amount of muck. There are more and more accumulation fields of slag soil, especially under the condition of rainstorms. Slope instability is very likely to occur in the slag soil field, and the slag soil field becomes one of the hidden dangers of urban safety development (Li *et al.* 2021).

The slope safety problem has attracted much attention. Its structure is complex, and there are many influencing factors (Huang & Liu 2020). Many scholars have carried out relevant research on the landslide problem. For example, Wang Yanxia (Wang 2010) believes that the slope of a large number of data statistics, by using the fuzzy mathematics approach, deals with a comprehensive evaluation of slope stability, and more realistic results are available. By using the maximum entropy principle and engineering fuzzy set theory, Ruan Hang (Ruan *et al.* 2015) *et al.* determined the index weight and put forward an improved highway slope stability evaluation theory.

Due to the characteristics of large discreteness and low strength of slag soil (Hu 2021), its particularity makes the mechanism of muck field landslide different from the traditional slope. At present, there are few studies on the slope problem of the muck yard at home and abroad. In addition, a "resilient city" is the main means of future urban disaster response. For the first time, the author combines fuzzy theory with urban resilience safety evaluation. Based on the fuzzy analytic hierarchy process, the evaluation method of slope resilience of

---

*Corresponding Author: 2471491228@qq.com

the muck field is established, which provides an exploration for urban safety resilience evaluation.

2 FUZZY ANALYTIC HIERARCHY PROCESS

The basic principle of the fuzzy analytic hierarchy process is to combine the principle of fuzzy transformation with the principle of maximum membership degree and make a strict and detailed analysis and research on the target layer and the related factors affecting it. Then the factor set is selected, and the evaluation set is determined according to the characteristics of fuzzy transformation, and the membership relation between the factor set and evaluation set is established. Then fuzzy relation matrix is constructed. Finally, the comprehensive evaluation results are obtained by combining each factor (Xue *et al.* 2021). Therefore, the combination of fuzzy mathematics and analytic hierarchy process overcomes the shortcomings of traditional analytic hierarchy process and fuzzy evaluation method in theory and practice and can better solve the problems encountered in reality. The evaluation results of this method are more convincing. This method can fully consider the influencing factors of the toughness of the slag soil slope, so the author uses this method to evaluate the toughness of the slag soil slope.

3 DREG SITE SLOPE TOUGHNESS INDEX AND EVALUATION MODEL

3.1 *Selection of toughness evaluation index and establishment of model*

According to engineering address conditions and landslide characteristics in the survey area, factors affecting residue soil slope toughness are divided into internal factors and external factors, so as to construct the target layer and index layer of the resilience evaluation model.

According to the characteristics of slag field materials, slag field slope characteristics, environment and infrastructure, and organizational mechanisms that affect the toughness of slag slope, the slag toughness evaluation model is divided into the following three layers. (1) the goal layer is the toughness evaluation of the residue field; (2) the normative level, the intermediate steps involved in influencing the ultimate achievement of objectives; (3) the indicator layer, this layer consists of 12 indicators. That is, the material characteristics of the slag yard are composed of four influencing factors. The slope characteristics of the slag yard are composed of two influencing factors, the environment and infrastructure are composed of three influencing factors, and the organizational mechanism is composed of three influencing factors. The slope toughness evaluation model of the slag field is shown in Figure 1.

On the basis of establishing the evaluation model of muck slope, the classification standard of each index is determined according to the evaluation target. The evaluation set is the language expression of landslide toughness evaluation. Four grades (grade I: strong toughness, grade II: strong toughness, grade III: general toughness, grade IV: weak toughness) were selected to establish the fuzzy rating of the slope toughness of the slag field. Combined with the example of a slag field, the evaluation indexes are quantitatively and qualitatively expressed. The grading standards and scoring values of each evaluation index are shown in Table 1.

3.2 *Determination of membership function of evaluation index*

The first-level matrix A-B of muck slope toughness evaluation consists of the membership degree of the first-level evaluation index relative to the four evaluation levels. Therefore, the determination of the first-grade evaluation matrix can be transformed into the determination of the membership degree of the first-grade evaluation index for each grade.

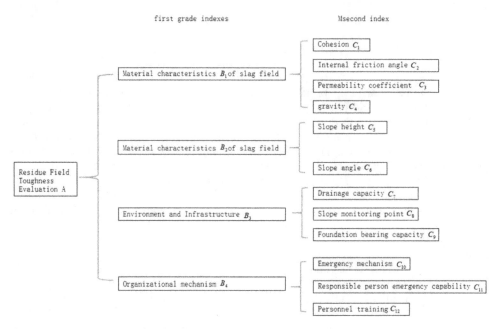

Figure 1. Evaluation model of slag slope toughness.

Table 1. Slag yard slope toughness evaluation standard table.

Evaluating Indicator		I (Strong Toughness) ≥ 90	II (Strong Toughness) 75~90	III (General Toughness) 60~75	VI (Weak Toughness) < 60
Characteristics of Slag Material B_1	Cohesion C_1/kPa	>60	40~60	20~40	<20
	Angle of Internal Friction C_2/(°)	>40	30~40	20~30	<20
	Permeability Coefficient C_3/m·s^{-1}	<0.05	0.05~0.75	0.75~1	>1
	Gravity C_4/kN·m^{-3}	>18	17~18	16~17	<16
Characteristics of Muck Slope B_2	Slope Height C_5/m	<6	6~12	12~18	>18
	Slope Angle C_6/(°)	<20	20~5	35~50	>50
Environment and Infrastructure B_3	Drainage Capacity C_7	strong	relatively strong	normal	simple
	Slope Monitoring Points C_8	>11	7~10	1~6	2
	Bearing Capacity of Foundation C_9/ kPa	800	600–800	300–600	<300
Organization Mechanism B_4	Emergency Mechanism C_{10}	robust	relatively robust	general	defect
	The Emergency Capability of the Person in Charge (years of working) C_{11}	>10 a	6~10 a	1~5 a	1 a
	Personnel Training C_{12}/ (times · a^{-1})	12	6~11	2~6	2

Toughness comprehensive evaluation set B represents the membership degree of landslide toughness grade, which can be determined according to the principle of maximum membership degree. According to the membership function to calculate the value, the common membership function types are triangular, normal, ring (down) type, and drop (rise) half trapezoid. The above secondary indicators are continuous; the larger the quantitative score index, the better. Therefore, the membership function is established by "rising half trapezoid" distribution (Wang *et al.* 2012). The calculation formula of membership degree of 4 evaluation grades can be obtained: Formula (1)~Formula (4), where x is the measured data score (score) of the evaluation index.

$$\mu_I(x) = \begin{cases} 0 & x < 82.5 \\ \dfrac{x - 82.5}{7.5} & 82.5 \leq x < 90 \\ 1 & x \geq 90 \end{cases} \quad (1)$$

$$\mu_{II}(x) = \begin{cases} \dfrac{x - 67.5}{15} & 67.5 \leq x < 82.5 \\ 1 - \mu_I(x) & 82.5 \leq x < 90 \\ 0 & x < 67.5 \ or \ x \geq 90 \end{cases} \quad (2)$$

$$\mu_{III}(x) = \begin{cases} \dfrac{x - 60}{7.5} & 60 \leq x < 67.5 \\ 1 - \mu_{II}(x) & 67.5 \leq x < 82.5 \\ 0 & x < 60 \ or \ x \geq 82.5 \end{cases} \quad (3)$$

$$\mu_{IV}(x) = \begin{cases} 1 & x < 60 \\ 1 - \mu_{III}(x) & 60 \leq x < 67.5 \\ 0 & x \geq 67.5 \end{cases} \quad (4)$$

4 ANALYSIS OF SLAG FIELD ENGINEERING EXAMPLE

4.1 *Overview of the slag field project*

The author takes a muck field in Beijing as an example, the waste dump is located in Shijingshan District, and the scene is shown in Figure 2.

(a) Top view of the slag field　　　　　　　(b) Site map of the slag field

Figure 2.　Site map of the slag field.

The muck yard is mainly a dump site for engineering excavation trench soil. The stacking time is eight months, and the project site and staff living area is below the dump site. The site muck deposit surface is relatively loose and deep due to the weight of the soil, and the dense degree is larger. The main component of the slag soil is silty clay, which is doped with a small number of broken bricks and concrete slag, and the foundation soil at the bottom of the slag field is cohesive. There are spatial differences in the upper and lower toughness parameters of the slag accumulation part, so the average value is taken as the measured index. According to the site conditions of the muck field slope, combined with the evaluation criteria of the muck field slope toughness in table 1, each secondary index is scored and quantified, and the membership degree vector R of each secondary index is calculated according to Formula (1)∼Formula (4). The quantitative results of the secondary index and its membership degree vector R are shown in Table 2.

Table 2. Evaluation index value of a slag field slope.

Secondary Evaluation Index	Index Value/ Description	Index Evaluation Score	Subordinated Degree Vector R
Cohesion C_1/kPa	45	79	(0, 0.77, 0.23, 0)
Angle of Internal Friction C_2/(°)	35	83	(0.07, 0.93, 0, 0)
Permeability Coefficient C_3(m/d)	0.05	75	(0, 0.5, 0.5, 0)
Gravity C_4/kN·m^{-3}	18	90	(1, 0, 0, 0)
Slope Height C_5/m	30	50	(0, 0, 0, 1)
Slope Angle C_6/(°)	33	88	(0.73, 0.27, 0, 0)
Drainage Capacity C_7	relatively strong	83	(0.07, 0.93, 0, 0)
Slope Monitoring Points C_8	9	85	(0.33, 0.67, 0, 0)
Bearing Capacity of Foundation C_9/kPa	750	86	(0.47, 0.53, 0, 0)
Emergency Mechanism C_{10}	relatively robust	85	(0.33, 0.67, 0, 0)
The Emergency Capability of the Person in Charge (years of working) C_{11}	10	79	(0, 0.77, 0.23, 0)
Personnel TrainingC_{12}/(times·a^{-1})	10	86	(0.47, 0.53, 0, 0)

4.2 Engineering example toughness evaluation

The toughness evaluation of the muck yard slope is carried out by using the established toughness evaluation method of the muck yard slope. The analytic hierarchy process is used to determine the weight value W of each index. The maximum eigenvalue λ_{max} of the judgment matrix is used to calculate the consistency index of the matrix CI ($CI = (\lambda_{max} - n)/n - 1$), where n is the matrix order. RI is a random consistency index to measure the size of CI, and its value is shown in Table 3.

Table 3. Corresponding value of consistency index RI.

Order of Matrix	1	2	3	4	5	6	7	8	9
RI	0	0	0.58	0.90	1.12	1.24	1.32	1.41	1.45

CR ($CR = CI/RI$) is the test coefficient. If $CR < 0.1$, it is considered that the judgment matrix is a consistency test; otherwise, you can not pass the test, and you need to re-adjust the matrix.

A first-level and second-level indicator judgment matrix are established, as shown in Table 4 and Table 5. W is the index weight vector.

Table 4. First-level index weight value (A-B).

A-B	B_1	B_2	B_3	B_4	W	Remark
B_1	1	1	4	5	0.40	λmax = 4.082
B_2	1	1	4	6	0.41	CI = 0.027
B_3	1/4	1/4	1	3	0.13	CR = 0.031 < 0.1
B_4	1/5	1/6	1/3	1	0.06	Through the consistency test

Table 5. Secondary indicator weight value (B1-C).

B_1-C	C_1	C_2	C_3	C_4	W_1	Remark
C_1	1	1	2	2	0.33	λmax = 4.061
C_2	1	1	2	2	0.33	CI = 0.020
C_3	1/2	1/2	1	1/2	0.14	CR = 0.023 < 0.1
C_4	1/2	1/2	2	1	0.20	Through the consistency test

Similarly, a judgment matrix is constructed for the second-level indicators B_2, B_3, and B_4. The weight vectors are respectively W_2 = (0.25,0.750), W_3 = (0.55,0.21,0.24), W_4 = (0.26,0.33,0.41).

In order to obtain the influence weight value of the second-class index (C) to the target layer (A), it is necessary to sort the weight vectors from high to low to determine the order of the influencing factors. The results are shown in Table 6.

It can be seen from Table 6 that the influence factors such as slope angle, slope height, cohesion, internal friction angle, unit weight, and drainage capacity have a great influence on the slope toughness of the muck yard. Therefore, these six factors should be taken into account when evaluating the toughness of the muck slope.

Table 6. Hierarchical total sorting matrix (A-C).

Hierarchy C_j	Hierarchy B_i				Hierarchy C Combined Sort Weight Value W_i	Level C Total Sort
	B_1 0.40	B_2 0.41	B_3 0.13	B_4 0.06		
C_1	0.33				0.132	2
C_2	0.33				0.132	2
C_3	0.14				0.056	7
C_4	0.20				0.080	5
C_5		0.25			0.103	4
C_6		0.75			0.308	1
C_7			0.55		0.072	6
C_8			0.21		0.027	9
C_9			0.24		0.031	8
C_{10}				0.26	0.016	12
C_{11}				0.33	0.020	11
C_{12}				0.41	0.025	10

According to the second-level index evaluation weight vector and Table 2, the fuzzy comprehensive evaluation model can be used to calculate the membership vector of slag material characteristics, such as Equation (5):

$$E_1 = W_1 \cdot R_1 = (\,0.223 \quad 0.631 \quad 0.146 \quad 0\,) \tag{5}$$

Similarly, the membership vector of the remaining first-level evaluation indicators can be calculated, as shown in Table 7.

According to the first-class index evaluation weight vector and Table 7, the fuzzy comprehensive evaluation results can be obtained by applying the fuzzy comprehensive evaluation theory $B = W \cdot E$, as shown in Formula (6):

$$B = W \cdot E = (\,0.359 \quad 0.476 \quad 0.063 \quad 0.103\,) \tag{6}$$

According to the principle of maximum membership degree, the slope toughness evaluation result of the muck yard is grade II, and the toughness is strong.

Table 7. Membership vector of first-class evaluation index.

First-level Evaluation Index	Subordinated Degree Vector E
Characteristics of Slag Material B_1	(0.223, 0.631, 0.146, 0)
Characteristics of Muck Slope B_2	(0.548, 0.203, 0, 0.250)
Environment and Infrastructure B_3	(0.221, 0.779, 0, 0)
Organization Mechanism B_4	(0.279, 0.646, 0.076, 0)

5 CONCLUSION

Based on the investigation of the present situation of the waste dump in Beijing, the author summarizes the present situation and regional distribution characteristics of the waste dump in Beijing. Based on the fuzzy analytic hierarchy process, the toughness index and evaluation model of the slag field slope are constructed, and the weight and membership degree of each evaluation index is determined. The analysis and verification are carried out in combination with engineering examples, which provides an exploration of the toughness evaluation method of the slag field slope.

ACKNOWLEDGMENTS

This work was supported by the National Key R&D Program of China (Grant No. 2018YFC0809900).

REFERENCES

Hu Hang. *Study on Disaster Mechanism of Large Accumulated Soil Landslide Induced by Rainstorm.* Beijing: North China University of Technology, 2021:1–2.

Huang Ming, Liu Jun. Slope Safety Monitoring and Early Warning System Based on Fault Prediction and Health Management. *Industrial Construction*, 2020, 50(5):66–70.

Li Cheng, Cai Liming, Zhang Weifeng et al. Influence of Initial Seepage Field on Rainfall Infiltration Characteristics and Stability of Muck Slope. *Journal of Civil and Environmental Engineering*, 2021,43 (2):1–9.

Ruan Hang, Zhang Yonghui, Zhu Zeqi et al. Research on an Improved Fuzzy Evaluation Method of Highway Slope Stability. *Rock and Soil Mechanics*,2015,36(11):3337–3344.

Wang Yanxia. Application of Fuzzy Mathematics in Slope Stability Analysis. *Rock and Soil Mechanics*,2010,31(9):3000–3004.

Wang Zhe, Yi Facheng, Chen Tingfang. Susceptibility Evaluation of Geological Disasters in Mianyang City Based on Fuzzy Comprehensive Evaluation. *Science & Technology Review*,2012,30(31):53–60.

Xue Xiaohui, Zhou Ling, Qin Aihong. Reservoir Bank Wading Landslide Risk Status Analysis and Prediction Evaluation. *Journal of Safety Science and Technology*, 2021,17(5):169–175.

Civil Engineering and Energy-Environment – Gao & Duan (Eds)
© 2023 the Author(s), ISBN 978-1-032-56059-5

A research on environmental accounting statements for sustainable development in China—A case study on coal industry

You Wu*, Qini Deng*, Qi Shen*, Yiran Fang* & Yufei Hong*
Shanghai Lixin University of Accounting and Finance, Shanghai, China

ABSTRACT: In recent years, China's environmental pollution has gradually worsened, which requires us to prepare updated environmental accounting statements for China's sustainable development. Based on the data of Shanxi Coal Industry and Peabody Energy Company, this paper compared the environmental accounting statements of China and developed countries, pointed out the gap and related reasons, and put forward specific solutions.

1 INTRODUCTION

Since Industrial Revolution in the 1760s, the increasingly advanced modern industry has provided ample material goods for human society, but simultaneously, it also causes serious environmental pollution (Liu & Wang 2016). The development of science and technology, the surge in population, and the growth of social needs have led to the depletion of natural resources, which fundamentally limits economic development and affects daily life (Zhang & Su 2018).

In order to solve these problems, economists, environmentalists, sociologists, and ecologists began to study the coordinated development of the economy and environment in the 1970s, and then the research and development of green accounting gradually came into people's view. However, the research on environmental accounting was lagging and didn't attract enough attention (Liu & Wang 2016). With further research, people realized that green accounting could guide and supervise enterprises to protect resources and maintain ecological balance through certain social and economic activities, constantly maintaining their competitive advantages (Bai *et al.* 2014).

On the one hand, externally, proper environmental accounting information disclosure can not only render government departments a service to recognize the status quo of the companies' environment better so as to make optimal decisions to protect the environment effectively but also raise public awareness of environmental protection activities, which is conducive to improving the external environment of the company. On the other hand, internally, relevant disclosure enables enterprise managers to understand the business conditions of the enterprise more accurately, evaluate the business results of the enterprise more objectively, and help make corresponding decisions to improve the internal management of the enterprise as well. And the shareholders can infer the company's economic benefits and sustainable development capabilities (Zhang 2019).

In China, environmental accounting is still a new research field. Its structural system needs to be refined and improved since theory and practice have not been combined for advanced

---

*Corresponding Authors: 15156592282@163.com; 2631687286@qq.com; 1026200349@qq.com; 2218593481@qq.com and 2411174383@qq.com

72

DOI: 10.1201/9781003433651-10

and systematic research. The purpose of our study is to compare Chinese environmental accounting with that of developed countries, explore problems, and propose solutions to promote the sustainable development of environmental accounting in China.

2 RESEARCH METHOD

This study adopted a case study approach. We compared the environmental accounting data of the Shanxi Coal industry and Peabody Energy, including disclosure, environmental protection measures, and so on.

3 ENVIRONMENTAL ACCOUNTING REPORTING SYSTEM FOR DOMESTIC AND FOREIGN ENTERPRISES

3.1 *Shanxi coal industry*

Shanxi Province produces most of the raw coal, more than a quarter of the total output in China, which even achieved approximately 1.063 billion tons in 2020. Coal has become the label of Shanxi with the title of "coal sea." Taking the Shanxi coal industry as an example, in addition to resource tax, urban maintenance, construction tax, and other contents that must be disclosed, Datong coal industry annual report only disclosed the incentive payment for the protection and comprehensive utilization of mineral resources. Other enterprises in environmental litigation, reward and punishment, environmental protection investment, and other contents with uncertainty and discontinuity are almost not mentioned in the annual report.

According to Table 1, firstly, it is noteworthy that the 11 quoted coal companies above in Shanxi have disclosed environmental accounting information in their annual reports, but there are differences in the ways and degrees of disclosure. Besides, all enterprises disclose environmental accounting information in the financial accounting statements and notes to the accounting statements from 2015 to 2017. In addition, most companies chose to disclose environmental accounting information in the board of directors report, while few were mentioned in the corporate governance structure of enterprise environment accounting information. At the same time, some significant problems with environmental accounting information which occur annually in these enterprises were disclosed. For example, Shanxi

Table 1. Disclosure methods and quantity statistics of the following coal companies in Shanxi from 2015 to 2017.

	2015		2016		2017	
	Number of Enterprises	Proportion	Number of Enterprises	Proportion	Number of Enterprises	Proportion
Corporate Governance Structure	3	27%	3	27%	5	45%
Director's Report	6	55%	3	45%	7	64%
Important Items	5	45%	4	36%	4	36%
Audit Report	0	0%	0	0%	0	0%
Social Responsibility Report	4	36%	4	36%	4	36%
Financial Reports	11	100%	11	100%	11	100%
Notes to the Financial Statements	11	100%	11	100%	11	100%

Coking, in the 2017 annual report, disclosed the important items of sewage processing integrated services, such as the related party transactions. However, among the coal enterprises above, only four issued separate social responsibility reports, which specifically elaborated on the theme of environmental protection, and reflected the content of environmental accounting information from multiple perspectives, such as environmental measures and environmental impact. Unfortunately, there is no authentication institution to assess the environmental accounting information issued by quoted coal companies.

3.2 *Inadequate environmental accounting in China*

As for the counterpart abroad for comparison, we take Peabody Energy as an example since it is the largest privately traded coal company in the United States and the world, primarily producing coal in the United States and Australia.

According to Table 2, compared with foreign countries, we can analyze the major deficiencies of China's environmental accounting system.

1. Environmental accounting information disclosure lacks significance and clarity. At present, China has not established a unified environment accounting information disclosure standard. Enterprises almost always disclose environmental accounting information too dispersed and disperse specific environmental items in their financial indicators (Xie & Tang 2016). For example, the environmental liabilities and environmental costs of environmental accounting elements are rarely disclosed in terms of sewage discharge fees and green fees in the project of enterprise management expenses. Similarly, environmental rights and interests, environmental assets, and environmental income are also rarely disclosed. Enterprises fail to set up independent environmental projects to reflect environmental accounting information, and the subject disclosed is incomplete and not fixed. Environmental accounting items of different enterprises have the same nature, but the name of disclosure is different, which increases the difficulty of environmental accounting information collection and enterprise environmental performance evaluation (Wang 2005).

2. The scope of environmental accounting information disclosure is narrow. In terms of annual reports in China, enterprises' environmental accounting information disclosure focuses on environmental costs, environmental policies and government environmental subsidies, but rarely involves other aspects of information. For some major environmental investment projects, the corresponding cost-benefit analysis is rarely carried out, involving the expenditure and liabilities related to environmental matters. Moreover, the enterprise's compliance with environmental laws and regulations, environmental capital expenditure, expense expenditure and other environmental information disclosure is less. Though these Shanxi coal quoted companies in their prospectuses have different degrees of environmental accounting information disclosure, they mainly concentrated on the

Table 2. Basic information of peabody environmental accounting disclosure.

Disclosure Method		Social responsibility report, annual report, investor report
Disclosure Content	Quantitative	Annual environmental expenditure
	Qualitative	All kinds of American environmental protection laws, environmental management methods, environmental protection advanced figures
	Qualitative and quantitative	Advanced technology, pollution waste discharge and treatment, social environmental protection actions, social environmental protection measures, environmental protection achievements and awards

consideration and countermeasures of environmental risks. In addition, they reflect the information of the past, but rarely forecast the amount of pollution discharged in the future, the number of coal resources developed by the reporting enterprises, and the life of coal in the development areas (Cai *et al.* 2021).

3. The disclosure of environmental accounting information is casual. Due to the lack of unified environmental accounting information disclosure regulations, quoted companies of coal enterprises disclose environmental accounting information at will and lack comparability of information. Some companies decide whether to disclose environmental information and how to disclose environmental information according to whether the disclosure benefits the company and their own reporting mode. In the study of the annual report, it was found that the contents of environmental control reported by some boards of directors were the same for two consecutive years (Wang 2019).

4. The results disclosed in the report may not be true, with serious "formalism." As the environmental accounting information disclosure regulations are not unified, the environmental accounting information disclosure of quoted coal enterprises lacks authenticity. On the one hand, most of the environmental accounting information published by coal enterprises in the annual report is "good news but not bad news." To maintain a good impression, enterprises deliberately delete negative environmental information, and as a result, the environmental information disclosed in the annual report is not comprehensive. On the other hand, enterprises only treat the issuance of annual reports as superficial work, and the information related to environmental accounting in financial statements is not comprehensive or even a falsified data phenomenon, which seriously affects the authenticity of information between enterprises and stakeholders (Shen & Chen 2020).

4 THE CAUSES OF THESE DEFICIENCIES

1. There is a lack of sound laws and regulations as the basis. China has issued the clean Production Promotion Law of the People's Republic of China and the Environmental Impact Assessment Method of the People's Republic of China, but the provisions are not detailed enough, and the content is not clear enough. Therefore, enterprises escape the punishment of laws and regulations and develop at the expense of the environment.

2. The relevant audit supervision system is not perfect. At present, China has not established an audit supervision system for environmental accounting information disclosure. The government and supervision departments have not performed their responsibilities for environmental accounting information disclosure. Because of their own protection, enterprises will not fully disclose environmental accounting information.

5 SPECIFIC MEASURES TO PROMOTE ENVIRONMENTAL ACCOUNTING REPORTING FROM THE PERSPECTIVE OF SUSTAINABLE DEVELOPMENT

1. Constructing the framework of enterprise environmental accounting in China.

    Important non-financial information is connected with the company's reputation and image. More information related to major events, such as those relating to social responsibility and the environment, is seldom disclosed by Chinese enterprises. Therefore, enterprises should pay attention to the integrity and authenticity of accounting information disclosure and attach importance to non-financial information.

    In the existing accounting subjects, we can plan some rules on environmental accounting, such as adding "environmental assets course" to "asset subject," "environmental liabilities" to "liabilities subject," "environment income" and "environmental cost" to "profit and loss," and "environmental activities of cash inflows and outflows caused by the project" to "the cash flow statement." The corresponding environmental accounting policies and detailed explanations shall be disclosed in the notes to the financial statements.

2. Countermeasures to improve environmental accounting information disclosure.

Above all, we should improve laws and regulations related to environmental accounting. Observing the status quo of environmental accounting information disclosure in western developed countries, we can find that strong laws and regulations play an important guiding role in the work of environmental accounting information disclosure. With the increase of relevant environmental legislation in China in recent years, the information disclosed by enterprises has gradually become standardized.

Moreover, we have the responsibility to strengthen the supervision of government departments. If perfect laws and regulations want to maximize their effectiveness, strong law enforcement supervision is indispensable. Several departments have jointly implemented law enforcement to build a scientific assessment, reward, and punishment mechanism, striving to do regular inspections. In addition, the audit department and industry associations should establish special auditing standards for environmental accounting as soon as possible, and verify the reliability and authenticity of the environmental accounting information disclosed by quoted companies, so as to provide a factual basis for the public to supervise the implementation of corporate environmental responsibility.

The government should perfect the management system of environmental information. Based on the survey, we have found that there are fewer enterprises with a sophisticated management system of environmental information among the investigation objects, which is of particular importance to the highly polluted industry. Therefore, companies should set up technical environmental management departments and cultivate excellent environmental knowledge-based talents. Environmental Performance Indicators should be regarded as one important indicator of judging the performance of enterprises. Social responsibility should be shouldered by each department or even each employee. Enterprises may adopt the measure of fining to develop an environmental protection-oriented industry. Enterprises should be encouraged to build a complete environmental management system and then meet the certification standards of IS environmental management system.

6 CONCLUSION

This paper compares the environmental accounting statements of China and developed countries, points out the gap between them and their reasons and puts forward concrete solutions. This provides a scientific reference for the future development of environmental accounting statements in China.

REFERENCES

Bai L.Z., Gui L., Wang J.Z. et al. (2014) Comparison and Enlightenment of Rural Tourism Development at Home and Abroad. *Ence Technology and Industry*, 5–10.

Cai R., Lv T. and Deng X. (2021) Evaluation of Environmental Information Disclosure of Listed Companies in China's Heavy Pollution Industries: A Text Mining-Based Methodology. *Sustainability*, 13.

Liu Y.Z. and Wang Y. (2016) The Empirical Study on Environmental Accounting Information Disclosure of Coal Industry. *Technology and Innovation Management*, 7–10.

Liu Y.Z. and Wang Y. (2016) The Empirical Study on Environmental Accounting Information Disclosure of Coal Industry. *Technology and Innovation Management*, 7–12.

Shen J., Chen Y. (2020) *A Comparative Study of Environmental Accounting Information Disclosure between China and Developed Countries*. IOP Publishing Ltd. IOP Publishing Ltd, 012011.

Wang H. (2019) Research on the Present Situation and Countermeasures of Environmental Accounting Information Disclosure of Listed Companies in China. *Modern Management Forum*, 3(2):19.

Wang J. (2005) Research on the Construction of Integrated System of Environmental-Economic Accounting in China. *World Sci-tech R & D*, 27(2):83–88.

Xie M. and Tang W. (2016) The Research on Environmental Accounting Information Disclosure of Chinese Listed Companies in the Steel Industry. *International Conference on Service Systems & Service Management*. IEEE.

Zhang L. (2019) Research on Enterprise Environmental Accounting Information Disclosure from the Perspective of Environmental Protection-taking Sinopec as an example. *IOP Conference Series: Earth and Environmental Science*, 267(2):22008.

Zhang S., Su-Bo X.U. (2018) Environmental Accounting: Theoretical Review and Research Prospects. *Ecological Economy*, 14(2):69–82.

Civil Engineering and Energy-Environment – Gao & Duan (Eds)
© 2023 the Author(s), ISBN 978-1-032-56059-5

Analysis of carbon emission accounting and carbon neutral capacity in Gansu Province

Rong Huang, Zhicheng Ma, Tian Liang & Jiafu Xi
State Grid Gansu Electric Power Company, Ltd, Lanzhou, Gansu, China

Yongli Wang, Lixiang Gong* & Bo Yuan
School of Economics and Management, North China Electric Power University, Changping, Beijing, China

ABSTRACT: Carbon neutrality is an important component of China's high-quality economic development, and carbon accounting plays a fundamental role in supporting carbon neutrality pathway research. Based on the primary energy and electricity consumption of Gansu province during the 10-year period from 2010 to 2019, the total carbon emissions were calculated and the trends were analyzed. Based on the assessment of the total carbon sequestration of arable land, forest land, garden land and pasture land in Gansu province, the carbon neutrality capacity of Gansu province was analyzed. The results show that the carbon emissions from primary energy sources in Gansu province fluctuate from 2011 to 2011, and the variation range is within ±5%, while the carbon emissions from power sources increase rapidly after 2010. Carbon emissions from electricity sources are about twice as high as primary energy emissions, and the proportion is increasing year by year, while the total amount of carbon sequestration on various types of land in Gansu province is relatively stable. In 2019, carbon sequestration accounts for 51.57% of primary energy carbon emissions and 24.26% of power source carbon emissions.

1 INTRODUCTION

Practicing Xi Jinping's thought of ecological civilization, building a green economic system, and enhancing green development are important measures for Gansu Province to build an ecological security barrier in the west (Jin 2021), as well as an important means to achieve peak carbon and carbon neutrality. Achieving carbon peaking and carbon neutrality by 2030 and 2060, respectively, is a solemn commitment made by China to the world (Xi 2020). As the largest developing country in the world, China's total carbon emissions are among the highest in the world, and General Secretary Xi Jinping's commitment sets an important time point for China's carbon dioxide emission reduction and also specifies important basic research areas such as carbon emission pathways, carbon emission effects, and low-carbon economy. The goal is to reduce carbon dioxide emissions in China.

Gansu Province has a diverse topography and rich mineral resources, with two oil regions in the region, Yumen and Changqing, and a highly concentrated distribution of resource reserves, making it an important energy base in the northwest and even in China. At the same time, Gansu Province is influenced by its geographical location, and its economic development has been in a relatively slow growth state for a long time, and the contradiction

---

*Corresponding Author: 120212206288@ncepu.edu.cn

between economic and social development and ecological and environmental protection is very prominent. This paper accounts for the total carbon emissions in Gansu Province in the past 10 years in the context of carbon neutrality and analyzes the carbon emission trends, with a view to providing a basis for decision-making on the control of carbon neutrality paths and the formulation of measures for the province.

2 RESEARCH METHODOLOGY AND DATA SOURCES

2.1 Carbon emission accounting method for primary energy consumption

Based on the IPCC carbon emission calculation model (Li 2020), the total primary energy carbon emissions in Gansu Province are accounted for with the following formula (Chen 2020):

$$EC = \sum ec_i = \sum En_i \times S_{ei} \times f_i \tag{1}$$

where EC is the total carbon emission of primary energy; ec_i is the carbon emission of the ith energy source; En_i is the consumption of the ith energy source; S_{ei} is the standard coal conversion factor of the ith energy source (Jing 2019); f_i is the carbon emission factor of the ith energy source. As shown in Table 1, the primary energy sources selected for the study include raw coal, coke, crude oil, gasoline, kerosene, diesel, and natural gas, and their standard coal conversion factors and carbon emission factors are referred to the values given by IPCC (Sun 2019).

Table 1. Main energy standard coal conversion factor and carbon emission factor table.

Type of Energy	Raw Coal	Coke	Crude Oil	Gasoline	Kerosene	Diesel	Gas
The Natural Gas Standard Coal Conversion Factor	0.71	0.97	1.43	1.47	1.47	1.46	1.33
Carbon Emission Factor	0.76	0.86	0.59	0.55	0.57	0.59	0.45

2.2 Accounting method for carbon emissions from power sources

Based on the method of accounting for carbon dioxide emissions from power sources in the "Guidelines for the Preparation of Provincial Greenhouse Gas Inventories (for Trial Implementation)", the total carbon emissions from power sources (Bai 2013) in Gansu Province are calculated with the following formula:

$$EC = En \times f_{co_2} \tag{2}$$

where EC is the total carbon emission from power sources; En is the total power consumption; f_{co_2} is the carbon dioxide emission factor.

The carbon dioxide emission coefficients of different regional power grids in China are shown in Table 2. Gansu Province belongs to the northwest region, and the carbon dioxide emission coefficient is selected as 0.977 kg/kW-h.

2.3 Carbon sequestration accounting method

Based on the net ecosystem production (NEP) coefficient (Xie 2008), the direct carbon emission coefficient method (Fang 2007) is used to calculate the amount of carbon

Table 2. Average CO_2 emissions per unit power supply of regional power grids in China [kg/(kW-h)].

Name of the Grid	Covering Provinces, Districts and Cities	Carbon Dioxide Emissions
North China Region	Beijing, Tianjin, Hebei Province, Shanxi Province, Shandong Province, the western region of Inner Mongolia Autonomous Region	1.246
Northeast Region	Liaoning Province, Jilin Province, Heilongjiang Province, and the eastern part of the Inner Mongolia Autonomous Region	1.096
East China Region	Shanghai, Jiangsu Province, Zhejiang Province, Anhui Province, Fujian Province	0.928
Central China Region	Henan Province, Hubei Province, Hunan Province, Jiangxi Province, Sichuan Province, Chongqing Municipality	0.801
Northwest Region	Shaanxi Province, Gansu Province, Qinghai Province, Ningxia Hui Autonomous Region, Xinjiang Uygur Autonomous Region	0.977
Southern Region	Guangdong Province, Guangxi Zhuang Autonomous Region, Yunnan Province, Guizhou Province	0.714
Hainan	Hainan Province	0.917

sequestered on various types of land in Gansu Province, such as arable land, forest land, garden land and pasture land (Wu 2016), with the following formula:

$$E = \sum e_i = \sum S_i \times \delta_i \tag{3}$$

where E is the total amount of carbon sequestered in each type of land; e_i is the amount of carbon sequestered in type i land; S_i is the area of type i land; δ_i is the coefficient of net ecosystem production in type i land. According to the climate zone and distribution characteristics of forests and grasslands in Gansu Province, the NEP coefficient of forest land is taken as 4.50×104 t/(hm^2-a) regarding temperate forests, the NEP coefficient of garden land is taken as 0.60×104 t/(hm2-a) concerning boreal forests, the NEP coefficient of pasture land is taken as the average value of world grasslands, i.e. 0.90×104 t/(hm^2-a), and the NEP coefficient of cropland was taken as -0.497×104 t/(hm^2-a).

2.4 Data sources

The basic data of primary energy consumption, electricity consumption and various types of land area in the Gansu Province Statistical Yearbook for 10 years from 2010 to 2019 were selected.

3 RESULTS AND ANALYSIS

3.1 Analysis of primary energy and electricity source carbon emissions and change trends in Gansu Province

The total primary energy carbon emissions and growth rate of Gansu Province from 2010 to 2019 are shown in Figure 1. The total primary energy carbon emissions in Gansu Province had a high growth rate of 13.76% in 2011, after which the growth rate slowed down and stayed within ±5%, and showed negative growth in 2015 and 2016, -3.06% and -3.55%

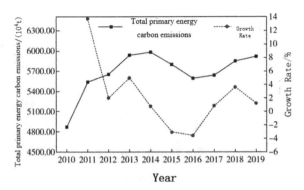

Figure 1. Total primary energy carbon emissions and growth rate curve in Gansu Province.

respectively. And the curve of total primary energy carbon emissions showed a wave-like up and down in general. The trend of total carbon emissions from electricity consumption in Gansu Province from 2010 to 2019 is shown in Figure 2. The overall trend of carbon emissions from electricity consumption in Gansu Province is increasing, with only a slight decrease in 2016, and the emissions remain basically unchanged in 2019 relative to 2018.

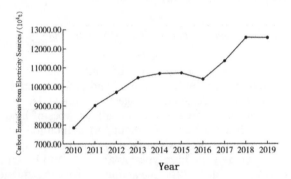

Figure 2. Total carbon emission curve of electricity sources in Gansu Province.

3.2 *Analysis of the total carbon emissions in Gansu Province and its share*

The total carbon emissions in Gansu Province from 2010 to 2019 and the proportion of the two types of emission sources are shown in Figures 3 and 4. As can be seen from Figure 3,

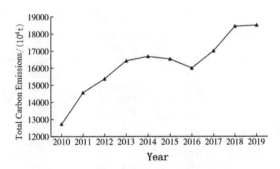

Figure 3. Total carbon emission curve of Gansu Province.

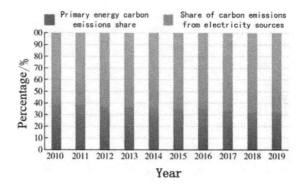

Figure 4. Carbon emissions from primary energy and electricity sources in Gansu Province.

the total carbon emissions in Gansu Province have been growing continuously since 2010, with only a small decrease of 0.91% and 3.23% in 2015 and 2016, respectively, and then growing all the way to reach a peak of 18505.59×104 t in 2019. In the carbon emission share, electricity sources have always been in the lead and have continued to grow from 61.74% in 2010 to 68.00% in 2019. In contrast, the carbon emission share of primary energy has decreased from 38.26% in 2010 all the way to 32.00% in 2019.

3.3 *Analysis of the carbon sequestration capacity of various types of land in Gansu Province*

The carbon sequestration capacity of four types of land in Gansu Province from 2010 to 2019, namely, arable land, garden land, forest land and pasture land, is shown in Table 3. Since the land area of Gansu Province has changed from about 45.4 million hm² to about 42.58 million hm² since 2012, the area of the above four types of land has also changed, especially the pasture land has changed from about 14.1 million hm² to about 5.92 million hm². Therefore, the data of the four types of land after 2012 were taken for analysis. The amount of carbon sequestered by the four types of land in Gansu Province is maintained at 3053×104 t~3057×104 t per year. In 2019, the amount of carbon sequestered accounted for 51.57% of carbon emissions from primary energy and 24.26% of carbon emissions from electricity sources, respectively.

Table 3. Summary of carbon sequestration on various types of land in Gansu Province, 2010-2019.

Year	Arable Land	Garden	Woodland	Grazing Land	Total
2010	-229.698	12.336	2333.070	1340.004	3455.711
2011	-229.331	12.336	2332.215	1340.004	3455.224
2012	-231.567	12.258	2332.215	1340.441	3453.346
2013	-267.664	15.600	2746.575	562.809	3057.319
2014	-267.326	15.516	2745.135	562.666	3055.991
2015	-267.282	15.474	2744.640	562.562	3055.394
2016	-267.008	15.384	2744.280	562.372	3055.027
2017	-267.272	15.360	2743.245	562.267	3053.600
2018	-267.098	15.312	2743.110	562.163	3053.487
2019	-267.222	15.312	2743.110	562.153	3053.353

Column header: Amount of Solid Carbon

4 DISCUSSION AND CONCLUSION

1. Primary energy carbon emissions in Gansu Province show up-and-down changes since 2011, but the changes are not large, and carbon emissions basically remain at the same level, while carbon emissions from electricity sources maintain a faster growth rate since 2010, indicating that primary energy carbon emissions in Gansu Province tend to saturate, and carbon emissions from electricity sources will become the main growth direction in the future. The continuous growth of carbon emissions from electricity sources, to some extent, reflects the increasing urbanization rate of Gansu Province, which has increased by 12.37% from 2010-2019. And the reasonable layout of urban space and optimization of the electricity supply structure will provide the possibility to reduce carbon emissions from electricity sources.
2. The total carbon emissions from primary energy and power sources in Gansu Province show an overall growth trend, and carbon emissions from power sources reach about twice the carbon emissions from primary energy, and the proportion shows a trend of year-on-year expansion. Improving the clean ratio of power production will effectively reduce the intensity of carbon dioxide emissions in Gansu Province. In 2019, Gansu Province's hydroelectric power, photovoltaic, wind power and other clean energy generation exceeded thermal power, reaching the province's total power generation by 50.48%, but the ratio to the grid is still low. Therefore, further improving the ratio of clean energy to the grid and increasing the utilization rate of power generation will effectively reduce carbon dioxide emissions.
3. The total amount of carbon sequestered on arable land, forest land, garden land and pasture land in Gansu Province is relatively stable, but it is not enough to completely neutralize carbon emissions from primary energy sources. While looking for new carbon sequestration paths, reducing primary energy consumption is still the main measure that Gansu Province needs to take to achieve the carbon neutrality target. Gansu Province has vast Gobi and deserts, and it is an important measure to enhance the total carbon sequestration in the province by continuously strengthening sandy land management, improving the regional ecological environment, and gradually increasing the greening area of desert land.
4. Gansu Province's Hexi Corridor is a vast and sparsely populated area with a large area of the Gobi desert, which has obvious regional advantages for the development of photovoltaic and wind power and other clean energy. At the same time, Gansu Province is a major channel and an important node for the transmission of electricity from the west to the east, and clean energy power output also has an inherent advantage, gradually eliminating thermal power generation and other high energy consumption. High emissions of electricity production methods improve clean energy production, storage and Utilization efficiency, and is an effective way to achieve carbon peak and carbon neutrality in Gansu Province.

ACKNOWLEDGMENTS

This paper is funded by the "Research on the construction path of near-zero carbon parks and the response strategies of power grid enterprises" (SGGSKY00BGJS2200248)

REFERENCES

Bai Weiguo, Zhuang Guiyang, Zhu Shouxian. Research Progress and the Prospect of Urban Greenhouse Gas Inventory in China. *China, Population, Resources and Environment*, 2013, 23(1): 63–68.

Chen Qizhen, Wang Wentao, Wei Xinfeng *et al.* Establishment, Mechanism, Impact and Controversy of IPCC, China's Population, *Resources And Environment*, 2020, 30(5): 1–9.

Fang Jingyun, Guo Zhaodi, Park Sai Lung *et al.* Estimation of Terrestrial Vegetation Carbon Sinks in China, 1981-2000. Chinese *Science: Part D*, 2007, 37(6): 804–812.

Jin Wangqiang, Yang Bin, Liu Peng. Study on the Construction Countermeasures of "Two Mountains" Practice and Innovation Base of Babosha Forest Farm in Gulang County. *Environmental Ecology*, 2021, 3 (5): 98–101.

Jing Zhaorui, Wang Jinman An Analysis of the Decoupling Relationship Between Land Use Carbon Footprint and Economic Growth in Shandong province. *Jiangsu Agricultural Sciences*, 2019, 27(6): 310–314.

Lee Yuenting, Sung Namchul, Shen Yingcai. Study on the Application of Mathematical Models in the Calculation and Prediction of Carbon Emissions. *Environmental Science and Management*, 2020, 45(12): 41–45.

Sun Ying, Liu Ning., Industrialization Urbanization and Regional Carbon Emission Intensity: Analysis of Provincial Panel Data Based on STIRPAT Model. *Reform and Opening Up*, 2019(17): 16–20.

Wu Lihua. *Carbon Emission Effects of Land Use Change in Lanzhou, China, 2002-2012*. Lanzhou: Northwest Normal University, 2016.

Xi Jinping. Speech at the General Debate of the 75th United Nations General Assembly. *Bulletin of the State Council of the People's Republic of China*, 2020 (28): 5–7.

Xie Hongyu, Chen Xiansheng, Lin Kairong *et al.* Ecological Footprint of Fossil Energy and Electricity Based on Carbon Cycling. *RJournalrnalEcologylogy*, 2008, 28(4)1729–1735.

Civil Engineering and Energy-Environment – Gao & Duan (Eds)
© 2023 the Author(s), ISBN 978-1-032-56059-5

Research progress of carbon capture technology in combustion

Guochen Zhao*

Datang Northeast Electric Power Test & Research Institute, China Datang Corporation Science and Technology Research Institute, Changchun, P.R. China

ABSTRACT: Achieving carbon peaking and carbon neutrality has become the main goal of the present international society. In order to achieve the goal of carbon peaking in 2030 and carbon neutrality in 2060, not only changes in the energy structure but also break-throughs in existing traditional fossil energy utilization technologies are required. To realize the goal of low-carbon and high-quality utilization of fossil energy, China should vigorously promote the energy technology revolution. Carbon capture technology obtains high-concentration CO_2 by enriching, compressing, and purifying CO_2. The capture technologies in combustion mainly include oxy-fuel combustion technology and chemical looping combustion technology. Oxygen-enriched combustion carbon capture technology has many advantages, such as relatively low cost, easy scale, and suitability for retrofitting existing units. The chemical looping combustion technology realizes the efficient and cascade utilization of fuel chemical energy through the orderly decoupling of the chemical reaction process and can realize the low-cost source capture of CO_2. In this work, we introduced the research progress of oxy-fuel combustion technology and chemical looping combustion technology in detail, aiming to provide a comprehensive introduction for people to understand and learn carbon capture technology in combustion.

1   INTRODUCTION

Energy security, the ecological environment, climate change, and other issues facing global economic and social development are increasingly prominent and intertwined. In 2015, the "Paris Agreement" on global climate change clearly stated that by the end of this century, compared with the pre-industrial level, the global warming temperature should be controlled within 2°C, and efforts should be made to achieve the goal of 1.5°C. The Chinese government has actively responded to climate change, played an irreplaceable role in the adoption of the agreement, and made efforts to fulfill relevant climate commitments. According to the assessment report of the United Nations Intergovernmental Panel on Climate Change (IPCC), in order to achieve the temperature rise target proposed in the agreement, it is necessary to improve energy efficiency, develop renewable energy on a large scale, and develop carbon capture, utilization and storage (CCUS) (Xie 2021; Yang 2022).

Currently, there are 22 carbon capture and storage demonstration projects based on electricity generation worldwide. Among them, pre-combustion capture and post-combustion capture projects account for the majority, with 10 and 9 projects, respectively, while only three projects are based on oxy-fuel combustion. In terms of carbon capture and storage investment countries, the United States leads the world with seven projects, and China ranks second with five demonstration projects. In July 2019, the trading price of carbon emission allowances within the EUETS was set at 35.4 USD/t CO_2, and it is expected

---

*Corresponding Author: 1257593514@qq.com

DOI: 10.1201/9781003433651-12

to reach 47.25 USD/t CO_2 by 2023. Pricing of carbon is considered one of the most effective ways to encourage the deployment of carbon capture and storage technologies.

Oxygen-enriched combustion technology and chemical looping combustion technology are large-scale carbon capture technologies for coal-fired power plants with great potential. The development of the two carbon capture technologies is becoming more and more mature, and the main bottlenecks are high cost and energy consumption, and lack of extensive large-scale demonstration project experience. But there is no denying the future development potential of the two carbon capture technologies (Lu *et al.* 2021; Zhang *et al.* 2022).

In this work, we give a brief overview of the oxy-fuel combustion technology and the chemical looping combustion technology, combine the research status of the two technologies at home and abroad, and propose the future development direction of the two technologies. We hope this work can provide some inspiration and ideas for related researchers and those who are interested in it.

2 OXYGEN-ENRICHED COMBUSTION TECHNOLOGY

2.1 *Technical introduction*

Oxygen-enriched combustion technology was first proposed by Horn, Steinberg, and Abraham in the early 1980s. It was proposed earlier that the purpose of oxy-fuel combustion had nothing to do with pulverized coal combustion and carbon dioxide capture. Abraham proposed the concept of oxy-fuel combustion mainly to produce large amounts of carbon dioxide in enhanced oil recovery technologies, while Horn and Steinberg aimed to reduce the influence of environmental factors on the reaction process in the process of generating energy from fossil fuels. In the mid-1990s, due to the global warming problem caused by the rising CO_2 level in the atmosphere, oxy-fuel combustion technology, as an important carbon dioxide capture technology approach, once again attracted widespread attention from the scientific community all over the world. The technology uses a mixture of high-purity oxygen obtained by air separation and partially recycled flue gas to replace air and fuel for combustion, thereby increasing the concentration of CO_2 in the exhaust. The combustion temperature is adjusted by circulating flue gas, and the circulating flue gas also replaces N_2 in the air to carry heat to ensure the heat transfer and thermal efficiency of the boiler. Oxygen-enriched combustion technology has a significant energy-saving effect, can effectively prolong furnace life, is conducive to improving product output and quality, and has outstanding environmental protection effects (Figure 1) (Toftegaard *et al.* 2010).

2.2 *Research status*

2.2.1 *Current status of foreign research*

At present, oxygen-enriched combustion technology has been valued and developed in the United States, Japan, Canada, Australia, the United Kingdom, Spain, France, the Netherlands, and other countries. Since 2005, the industrial demonstration of oxy-fuel combustion has made outstanding progress. In 2008, the Swedish Waterfall Power Company built the world's first full-process 30 MWth oxy-fuel combustion test device in Black Pump, Germany. In 2011, CS Energy of Australia built the world's first and largest 30 MWe oxy-fuel power generation demonstration power plant in Calide. The CIUDEN Technology R&D Center in Spain has built a set of 20 MWth oxygen-enriched pulverized coal boilers and the world's first 30 MWth oxygen-enriched fluidized bed test device (Huang *et al.* 2018; Zheng *et al.* 2014).

2.2.2 *Current status of Chinese research*

The basic research on oxy-fuel combustion in China began as early as the mid-1990s. Huazhong University of Science and Technology, Southeast University, North China

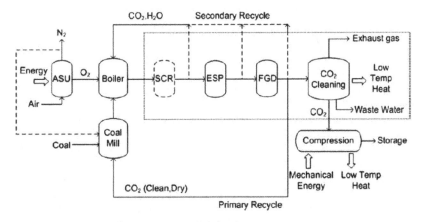

Figure 1. Possible configuration ofanoxy-fuelpowerplant (Toftegaard *et al.* 2010).

Electric Power University, Zhejiang University, etc., were the first to pay attention to the combustion characteristics, pollutant emissions and removal of oxy-fuel combustion in China. In 2006, Huazhong University of Science and Technology undertook the first national 863 plan project and national 973 plan project for CO2 emission reduction and started the research and development, and testing of oxy-fuel combustion technology in China. A 300 kW oxygen-enriched combustion bench has been built, and more systematic basic research results and small-scale test bench operation experience have been obtained. It is proved that it is a near "zero" emission combustion method that can easily obtain a CO2 concentration in the flue gas of more than 90%, and has the ability of desulfurization, denitration, and mercury removal at the same time. In recent years, the research and platform construction around the two main types of coal combustion: pulverized coal furnace and fluidized bed oxy-fuel combustion, have been very active (Huang *et al.* 2011; Sturgeon *et al.* 2009).

2.3 *Future outlook*

Given that Chinese greenhouse gas emissions have risen sharply due to accelerated economic development, China will inevitably become the focus of global CO_2 reduction and will be under increasing pressure. As a responsible big country, the Chinese government is fulfilling its due obligations. Oxygen-enriched combustion technology is one of the easiest technologies to achieve large-scale CO_2 emission reduction in China, which is dominated by coal-fired thermal power generation due to its own unique technical advantages. The basic theoretical and experimental research in the laboratory has been completed in China, a 3 MWth full-process test platform has been built and put into operation, and a 35 MWth oxygen-enriched combustion demonstration device will be built soon. Commercial-scale large-scale units are undergoing feasibility studies, and the current stage is the golden age of rapid development of oxy-fuel combustion technology, which is subject to funding, government policy support, carbon sequestration conditions and regulations, the establishment of transmission networks, permits and other factors. Although no commercial-scale projects have been implemented yet, Chinese research institutions and large enterprises have clarified the technical development route of oxy-fuel combustion, and clarified the overall planning and arrangement of basic research, pilot and demonstration projects, and strive to take the lead in building commercial projects. With the strong support of government funds and policies, China is expected to complete the reserve and industrial demonstration of oxy-fuel combustion technology with its own characteristics in the next five years or so, and promote and lead the development of international carbon capture technology.

3 CHEMICAL LOOPING COMBUSTION CARBON CAPTURE TECHNOLOGY

3.1 Technical introduction

Chemical Looping Combustion (CLC) is a clean and efficient new flameless combustion technology. Compared with traditional combustion, the biggest advantage of this technology is that it avoids direct contact between fuel and air by means of the circulation of intermediate carriers between oxidation-reduction reactors. The separation of CO_2 is realized at the same time as chemical conversion, which has the essential characteristics of CO_2 separation and has advantages in reducing the energy consumption of CO_2 capture. The total heat released by the chemical looping combustion process is the same as the heat released by the direct contact combustion of fuel and oxygen. This technology can separate CO_2 from combustion products without additional energy consumption to achieve a high concentration of CO_2 capture. The chemical looping combustion is based on a two-step chemical reaction, which realizes the cascade utilization of the chemical energy of the fuel, reduces the exergy loss during the combustion process, and thus improves the energy utilization efficiency of the system (Figure 2) (Adanez et al. 2012; Ishida & Jin 1997).

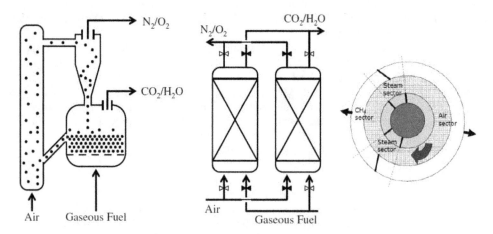

Figure 2. Possible reactor concepts for chemical-looping combustion (Adanez et al. 2012).

3.2 Research status

3.2.1 Current status of foreign research

The concept of chemical looping first appeared in Lane's 1913 invention patent for a hydrogen production process: a steam-metal hydrogen production fixed bed reactor. Then in 1954, Lewis et al. proposed the idea of using similar technology to produce pure CO_2. In 1994, Japanese scholar Ishida and Chinese scholar Jin Hongguang first organically combined chemical looping combustion with a thermodynamic cycle and proposed a new wet air turbine system for chemical looping combustion to control CO_2 emissions, which achieved high-efficiency conversion of energy and low-consumption separation of CO_2. In 2004, the Swedish scholar Lyngfelt et al. realized the pilot-scale experiment of the serial fluidized bed chemical looping combustion, which proved that the chemical looping combustion could realize the separation of CO_2, thus setting off a worldwide upsurge in the research of chemical looping combustion (Ishida & Jin 1994; Jin & Ishida 2000).

3.2.2 *Current status of Chinese research*

Since the chemical looping combustion was proposed in 1983, it has experienced more than 30 years of development. China Southeast University established a solid-fuel 10 kWth chemical looping combustion reactor in 2009, followed by a pilot-scale pressurized chemical looping combustion unit in 2012. Jin Hongguang from the Institute of Engineering Thermophysics of the Chinese Academy of Sciences analyzed the chemical looping combustion system and based on the concept of energy cascade utilization, a polygeneration system coupled with solar energy and chemical looping combustion was proposed. And the analysis has verified the feasibility of the new system. Compared with the conventional combined cycle with carbon capture, the exergy efficiency is increased by 14.2%, and the carbon emission reduction is 18.2%. Tsinghua University, Huazhong University of Science and Technology, etc., have conducted related research on the development and testing of oxygen carriers (Shen *et al.* 2009a, 2009b).

3.3 *Future outlook*

As a new low-carbon emission technology developed in the past ten years, chemical looping technology has shown advantages in high-efficiency and low-consumption carbon capture, air separation oxygen production, and hydrogen production in the conversion and utilization of fossil resources. It has great application potential in the development of clean production technology for the coal chemical industry and coal-fired power generation. Coal chemical looping combustion technology can realize high-efficiency and low-consumption carbon capture while generating coal-fired power and can control the production of NO_x at the same time. At present, there are kilowatt-level test devices, and there are experimental records of continuous operation for more than 100 hours. As a separate decarbonization unit, the calcium chemical looping decarburization technology has a carbon capture rate higher than 90% and reduces energy consumption by more than 60% compared with the traditional process, which will play an important role in tail gas decarbonization. The chemical looping oxygen production technology can greatly reduce the energy consumption of oxygen production. Compared with cryogenic air separation oxygen production, energy consumption can be reduced by more than 40%.

4 CONCLUSION

The power system must ensure the flexibility, stability, and reliability of the power supply while reducing carbon emissions to achieve the carbon peaking goal. CCUS is the main technical means to maintain the flexibility of the power system under the carbon neutrality goal. Oxygen-enriched combustion technology is considered a relatively secure technology that can be used in energy or industrial production processes closely related to fossil fuel combustion processes, such as power plants and cement plants. And it's easy to retrofit legacy projects, and it's an efficient and near-zero-emission technology. Compared with the traditional amine-based absorbent capture technology, the chemical looping combustion technology reduces the complexity of the system, and its greatest benefit is that the energy consumption in the capture process is greatly reduced. Compared with the conventional oxygen-enriched combustion technology, its energy consumption will be competitive to a certain extent because it eliminates the high energy consumption of air separation and oxygen production.

ACKNOWLEDGMENTS

This work was financially supported by Datang Northeast Electric Power Test & Research Institute.

REFERENCES

Adanez J., Abad A., Garcia-Labiano F. et al. (2012). Progress in Chemical-Looping Combustion and Reforming Technologies. *Progress in Energy and Combustion Science*, 38(2), 215–282.

Huang Q., Zhang L.Q., Zhou D. et al. (2018) Research Progress of Compression and Purification of Oxygen-enriched Combustion Flue Gas. *Chemical Industry and Engineering Progress*, 37(03): 1152–1160.

Huang X., Li J., Liu Z., et al. (2011) *Ignition Behavior of Pulverized Coals in Lower Oxygen Content O2/CO2 Atmosphere.*

Ishida M. and Jin H. (1994) A New Advanced Power-generation System Using Chemical-looping Combustion. *Energy*, 19(4): 415–422.

Ishida M. and Jin H. (1997) CO_2 Recovery in a Power Plant with Chemical Looping Combustion. *Energy Conversion and Management*, 38: S187–S192.

Jin H. and Ishida M. (2000) A Novel Gas Turbine Cycle with Hydrogen-fueled Chemical-looping Combustion. *International Journal of Hydrogen Energy*, 25(12): 1209–1215.

Lu B.W., Zhang L.Q., Xu Y.Q. et al. (2021) Carbon Capture, Utilization and Storage (CCUS) Technology Contributes to Carbon Neutrality. *Industrial Safety and Environmental Protection*, 47(S1): 30–34.

Shen L., Wu J., Gao Z., et al. (2009a) Reactivity Deterioration of NiO/Al_2O_3 Oxygen Carrier for Chemical Looping Combustion of Coal in a 10 kW_{th} Reactor. *Combustion and Flame*, 156(7): 1377–1385.

Shen L., Wu J., Xiao J. et al. (2009b) Chemical-looping Combustion of Biomass in a 10 kW_{th} Reactor with Iron Oxide as an Oxygen Carrier. *Energy & Fuels*, 23(5): 2498–2505.

Sturgeon D.W., Cameron E.D. and Fitzgerald F.D. (2009) Demonstration of an Oxyfuel Combustion System. *Energy Procedia*, 1(1): 471–478.

Toftegaard M.B., Brix J., Jensen P.A., et al. (2010) Oxy-fuel Combustion of Solid Fuels. *Progress in Energy and Combustion Science*, 36(5): 581–625.

Xie H. (2021) Application Status and Research Progress of Carbon Dioxide Capture *Technology. Chemical Fertilizer Design*, 59(6): 1–9.

Yang K.Y. (2022) *Research Progress of Carbon Dioxide Capture Technology in Coal-fired Power Plants. Energy and Energy Conservation.*

Zhang F., Fu A.Y., Liu K. et al. (2022) *Research Progress in Carbon Capture, Storage and Utilization Technologies.* Leather Manufacture and Environmental Technology.

Zheng C.G., Zhao Y.C., Guo X. (2014) Research and Development Progress of Oxygen-enriched Combustion Technology in china. *Proceedings of the CSEE*, 34(23): 3856–3864.

Civil Engineering and Energy-Environment – Gao & Duan (Eds)
© 2023 the Author(s), ISBN 978-1-032-56059-5

Experimental study on shear properties of modern rammed earth

Xiang Zhao, Hui Tao* & Tiegang Zhou
School of Civil Engineering, Xi'an University of Architecture and Technology, Xi'an, China

Zengfei Liang
School of Human Settlements and Civil Engineering, Xi'an Eurasia University, Xi'an, China

Wei Tan
School of Civil Engineering, Xi'an University of Architecture and Technology, Xi'an, China

ABSTRACT: To study the influence of modification ratio on the shear properties of modern rammed earth materials, 25 cubic and 15 cylindrical specimens, both in five groups, were designed and made for large-scale direct shear and triaxial tests, respectively. Through the observation of the failure modes of rammed earth specimens with different modification ratios, the shear strength, cohesion, and internal friction angle of rammed earth specimens were obtained to analyze the patterns of variation of the data obtained in the two test methods. The study results show that no matter which test method is adopted, the shear strength of the rammed earth specimen mixed with 15% cement is the greatest, while that of the rammed earth specimen containing no curing agent is the smallest. The curing agent can significantly improve the cohesion of rammed earth specimens, but incorporating it into the rammed earth specimens can slightly reduce their internal friction angles. Considering the shear properties, the usage of curing agents, and the economic benefits of rammed earth specimens with different modification ratios, it is concluded that the shear properties of rammed earth specimens mixed with 10% cement are superior.

1 INTRODUCTION

Characterized by their energy-conservation and environment-friendly, low-cost, excellent thermal stability, recyclability, and other advantages, rammed earth buildings are widely used in Western China (Lu 2015; Zhang 2013). However, as such buildings are composed of multiple rammed layers that allow layer-by-layer material filling and compaction, they often undergo brittle failures caused by external forces. Therefore, rammed earth buildings have poor safety properties, low material strength, and other defects.

Over recent years, scholars at home and abroad have carried out a large amount of research. Walker *et al.* (Walker 1991; Zhao *et al.* 2011) have expressly defined the ramming process, the earthiness of rammed materials, and their property indicators. Through experimental studies on the properties of rammed earth walls with different ratios, Yang (2010) explored the main factors in the shear capacity of rammed earth walls. Additionally, by experimentally studying the compressive strength of rammed earth materials with different modification ratios, Burroughs (2008, 2010) conducted a quantitative analysis of the differences in mechanical properties of such materials. Moreover, Zhou *et al.* (2015, 2016) thoroughly analyzed the dynamic characteristics and dynamic response of two types of houses through shaking table experiments on rammed earth houses and found that the seismic property of buildings constructed with modern ramming technology was improved

---

*Corresponding Author: 1074620779@qq.com

DOI: 10.1201/9781003433651-13

substantially. Nevertheless, most previous studies focus on studying traditional rammed earth houses, so more attention remains to be paid to the shear characteristics of modern rammed earth materials.

Although the direct shear test and triaxial test are two classic test methods for determining the shear properties of rammed earth materials, they are slightly different. Specifically, the former better represents the failure features of horizontal shear, while the latter best reflects the actual shear properties of materials. Therefore, for materials of the same type, there are fine distinctions (Sun *et al.* 2013) between the two test methods. Likewise, the shear properties of rammed earth materials with different modification ratios vary.

This study has performed large-scale direct shear and triaxial tests to observe the failure modes of rammed specimens with different modification ratios. During the process, the shear strength, cohesion, and internal friction angle of rammed earth specimens were obtained, thereby comparatively analyzing the patterns of variation of the data obtained therein. Based on the experiment results, a relatively reasonable modification ratio of modern rammed earth materials was concluded to provide a reference for further study on modern rammed earth materials and practical projects.

2 SPECIMEN DESIGN AND MAKING

2.1 *Apparatus and raw materials for the tests*

The apparatus for large-scale direct shear and triaxial tests was the Shear Trac III fully-automated direct shear apparatus produced by Geocomp in the US and the large fully-automated triaxial testing system of GDS in Britain.

Raw materials for the tests: The soil was that used for the Erlitou Site Museum of the Xia Capital project and was tested indoors following the Standard for Soil Test Method (GB/50123-2019). The optimal ratio of soil, sand, and stone (mass ratio: 4:3:3) was adopted in the tests. According to the tests, the physical indicators of the soil are as follows. Numerically, the liquid and plastic limits are 26% and 16%, respectively. The plasticity index is 10, and the natural water content is 3.01%. Besides, the sand used for the tests was natural sand (medium sand) that meets standards in the Sand for Construction (GB/T14684-2011), and the rubble used was broken stones (size: 10 mm to 20 mm). The modification materials included ordinary Portland cement and slaked lime.

2.2 *Specimens of the large-scale direct shear and triaxial tests*

The dimensions of the specimens of the large-scale direct shear test are 200 mm × 200 mm × 200 mm (Zhou & Peng 2005). Additionally, a total of 25 cubic specimens with different modification ratios in five groups were designed and made. The specimen number (No.) and dimensions are presented in Table 1.

Table 1. Numbers and dimensions of specimens of the large-scale direct shear test.

Curing Agent	Soil: Sand: Stone	Dimensions/mm	No.
No	4:3:3	200 × 200 × 200	Z01-Z05
10% cement			Z11-Z15
15% cement			Z21-Z25
20% lime			Z31-Z35
30% lime			Z41-Z45

A metal mold with the dimensions of 300 mm (diameter) × 630 mm (height) was used for making the specimens of the large-scale triaxial test, and the ramming tool was a rammer used for the specimen making of the test. The ramming process throughout the test was conducted six times. For each time, the height of the materials used for filling was 150 mm, and the thickness of the rammed layer was 100 mm. The rammed specimen was a cylindrical one composed of six rammed layers. By doing so, the study simulated the construction process of the modern rammed earth walls. In this test, a total of 15 rammed earth specimens with different modification ratios were designed in five groups, and specimen No. and dimensions are presented in Table 2.

Table 2. Numbers and dimensions of specimens of the large-scale triaxial test.

Curing Agent	Soil: Sand: Stone	Dimensions	No.
No	4:3:3	ϕ300mm × 600mm	SZ01-SZ03
10% cement			SZ11-SZ13
15% cement			SZ21-SZ23
20% lime			SZ31-SZ33
30% lime			SZ41-SZ43

3 LOADING AND PHENOMENA OF THE TESTS

3.1 Large-scale direct shear test

The continuous loading mode was used in the large-scale direct shear test. The loading continued until the overall failures of the specimens, and then the data on the specimens during the shear process were extracted. The loading status of the experimental test is presented in Table 3.

Table 3. Loading status of the large-scale direct shear test.

Curing Agent	Specimen No.	Shear Rate	Normal Pressure (kPa)
No	Z01-Z05	2.0 mm/min	50, 100, 200, 300, and 400
10% cement	Z11-Z15		
15% cement	Z21-Z25		
20% lime	Z31-Z35		
30% lime	Z41-Z45		

The failure modes of specimens with the same modification ratio are broadly the same. The shear planes of Specimens No. Z03, No. Z13, No. Z23, No. Z33 and No. Z43 is presented in Figure 1.

As shown in Figure 1, the shear failure surfaces of specimens mixed with curing agents are bumpy soil-and-stone contact ones. Such a failure mode with a convex upper shear plane and a hollow lower shear plane indicates that the shear failure process within rammed layers always progresses toward the loose area in rammed layers, representing the non-homogeneous characteristic of rammed earth materials. Unlike the failure modes of rammed earth specimens mixed with no curing agents, modification materials can largely enhance the constitutive property of rammed earth materials and reduce the spalling of soil, sand, and

Figure 1. Failure phenomena of specimens with different modification ratios.

stone. The comparative observation of the failure surfaces of Specimens No. Z13 and No. Z33 reveals that Specimen No. Z13 has a better structure, while Specimen No. Z33 has a large amount of lime soil on its failure surface, which is caused by the differences in the curing mechanism of cement and lime. Cement is a kind of hydraulic cementing material (Wang 2004). Its components can produce various hydrates through a chemical reaction with water molecules, thereby hardening rammed earth materials. By comparison, slaked lime is a kind of air-hardening inorganic cementing material (Liang 1984), and the $Ca(OH)_2$ contained therein needs to react with water molecules and carbon dioxide in the air to produce a carbonization effect on rammed earth materials. However, considering the thin carbon dioxide contained in the rammed layers, the carbonization effect is inhibited. Therefore, the curing effect of lime is not as good as that of cement.

3.2 Large-scale triaxial test

Given the exceptional nature of rammed earth materials, this study referred to the test methods in the field of rammed earth and adopted the large-scale unconsolidated undrained triaxial test for the current stage of study (Jiang *et al.* 2014; Yuan 2003). The test parameters, such as the pre-applied confining pressure and loading rate, were defined by running the matching GDS-Lab software of the triaxial test apparatus. The entire shear process consisted of two parts: a) applying confining pressure. b) loading at a constant shear rate. The loading status of the experimental test is presented in Table 4.

The failure modes of some cylindrical rammed earth specimens in the current stage of study are presented in Figure 2.

Table 4. Loading status of the large-scale triaxial test.

Curing Agent	Specimen No.	Shear Rate	Confining Pressure (kPa)
No	SZ01-SZ03	2.0 mm/min	0, 50, and 100
10% cement	SZ11-SZ13		
15% cement	SZ21-SZ23		
20% lime	SZ31-SZ33		
30% lime	SZ41-SZ43		

As shown in Figure 2, the shear failure modes of cylindrical rammed earth specimens with different modification ratios are broadly similar. Specifically, all of them present conical failure modes, and the failure surfaces are soil-and-stone contact ones. During the loading process, the specimens underwent vertical cracks, and then the middle parts of the specimens swelled out and cracked. As the loading progressed, the cracks constantly extended over the weak surfaces and gradually formed oblique shear planes.

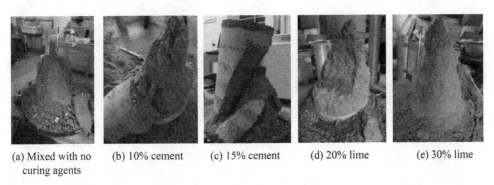

(a) Mixed with no curing agents   (b) 10% cement   (c) 15% cement   (d) 20% lime   (e) 30% lime

Figure 2. Failure modes of specimens with different modification ratios.

4 TEST DATA AND ANALYSIS

4.1 Relation curve between the shear stress and shear displacement in the large-scale direct shear test

As shown in Figure 3, each relation curve between the shear stress and shear displacement is peak-type. When the shear stress reaches the shear strength, it falls drastically as the shear displacement rises, obviously presenting the characteristics of brittle failure. Additionally, the residual friction phases in the curves present almost straight lines, and the residual shear stress increases with the normal pressure. The increase in the normal pressure strengthens the interlocking force among particulates, thereby resisting the shear stress. Therefore, macroscopically, the shear strength appears to increase.

4.2 Mohr's circles and shear strength envelopes of rammed earth specimens in the large-scale triaxial test

As shown in Figure 4, the shear strength envelopes of rammed earth specimens with different modification ratios are nearly straight lines. Besides, the shear strength parameters are in line with the Mohr-Coulomb equation. Modification materials can significantly enhance the

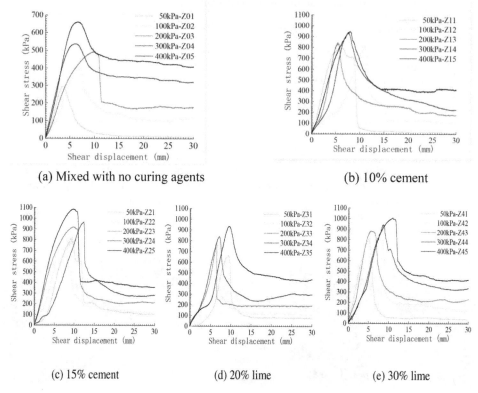

Figure 3. Relation curves between shear stress and shear displacement.

cohesion of rammed earth specimens. Compared with those mixed with no curing agents, the cohesion of rammed earth specimens containing 10% cement, 15% cement, 20% lime, and 30% lime has risen by 360.0 kPa (173.8%), 495.3 kPa (239.2%), 299.2 kPa (144.5%), and 301.4 kPa (145.5%), respectively. Meanwhile, the incorporation of modification materials will slightly reduce the internal friction angles of rammed earth specimens by approximately 4.7° to 7.2°.

5 COMPARATIVE ANALYSIS OF DIFFERENT TEST RESULTS

5.1 *Comparative analysis of parameters of shear strength*

Statistics for the cohesion and internal friction angles of rammed earth specimens with different modification ratios in the two tests are presented in Table 5.

As can be seen in Table 5, for rammed earth specimens with the same modification ratio, the cohesion tested through the triaxial test is smaller than that tested through the direct shear test, but the internal friction angle tested through the former is slightly larger than that through the latter, because the latter limits the failure surfaces of the specimens. Moreover, the contact surface between the upper and lower shear boxes is not necessary for the weak surface of the rammed layers. By comparison, the shear failure surfaces of rammed earth specimens in the triaxial test are caused by the development, extension, and transfixion of cracks in the areas where the bonding strength of particulates is relatively

Table 5. Statistics for the parameters of shear strength.

Curing Agent	Test Method	Cohesion (kPa)	Internal Friction Angle
No	Direct shear	297.3	39.3 °
	Triaxial shear	207.1	45.6 °
10% cement	Direct shear	761.1	28.5 °
	Triaxial shear	567.1	40.4 °
15% cement	Direct shear	795.2	35.7 °
	Triaxial shear	702.4	39.6 °
20% lime	Direct shear	631.9	32.5 °
	Triaxial shear	506.3	38.4 °
30% lime	Direct shear	775.4	30.4 °
	Triaxial shear	508.5	40.9 °

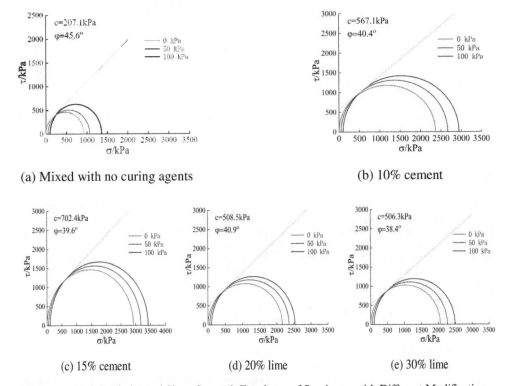

(a) Mixed with no curing agents   (b) 10% cement

(c) 15% cement   (d) 20% lime   (e) 30% lime

Figure 4. Mohr's Circles and Shear Strength Envelopes of Specimens with Different Modification Ratios.

weak. As the friction among particulates of soil, sand, and stone materials is more thorough, the cohesion tested through the large-scale triaxial test is relatively small, and the internal friction angle is slightly larger. Considering the shear properties, the usage of curing agents, and the economic benefits of rammed earth materials with different modification ratios, it is concluded that the shear properties of rammed earth materials mixed with 10% cement are superior.

6 CONCLUSIONS

Based on the results and discussions presented above, the conclusions are obtained as follows:

1) Curing agents can significantly enhance the cohesion of rammed earth materials, but the incorporation of them may slightly reduce the internal friction angles of such materials by 4.7° to 7.2°.
2) Curing agents improve the shear strength of rammed earth materials through cohesion enhancement. Additionally, the normal pressure, to some extent, affects the shear strength and residual shear stress of rammed earth materials.
3) For rammed earth materials with the same modification ratio in the large-scale triaxial test, the cohesion tested is smaller, and the internal friction angle is slightly larger.
4) According to the results of the direct shear and triaxial tests, modification materials enhance the shear strength of rammed earth materials mainly by improving their bonding effect, but the incorporation of curing agents may slightly reduce the internal friction angle.
5) Considering the shear strength parameters, the usage of curing agents, and the economic benefits of rammed earth materials with different modification ratios, it is concluded that the shear properties of rammed earth materials mixed with 10% cement are superior.

ACKNOWLEDGMENTS

This work was supported by Shaanxi Key Industrial Innovation Chain Project in Agricultural Domain (Grant No.: 2020ZDLNY06-03).

REFERENCES

Burroughs S. Recommendations for the Selection, Stabilization, and Compaction of Soil for Rammed Earth Wall Construction. *Journal of Green Building*. 2010, 5(1):101–114.

Burroughs S. Soil Property Criteria for Rammed Earth Stabilization. *Journal of Materials in Civil Engineering*. 2008, 20(3):264–273.

Jiang J.S., Cheng Z.L., Zuo Y.Z. et al. Study on Dilatancy of Coarse-grained Soil through the Large-scale Triaxial test. *Rock and Soil Mechanics*. 2014, 35(11):3129–3138.

Liang N.X. Study on the Principle of Hardening Soil with Lime. *Journal of Xi'an University of Highway*. 1984, 2(4):115–143.

Lu L.L. *Investigation Study of Rammed-earth Technology of Traditional Dwellings*. Xi'an University of Architecture and Technology. 2015.

*Sand for Construction (GB/T14684-*2011). Standards Press of China. 2011.

*Standard for Soil Test Method (GB/T50123-*2019). China Plan Press. 2019.

Sun L.P., Yang Y.C. & Liu S. Advantages and Disadvantages as well as the Scope of Application of the Direct Shear and Triaxial Tests in the Field of Civil Engineering. *Design of Water Resources & Hydroelectric Engineering*. 2013, (4):40–42.

Walker P.J. *New Mexico Adobe and Rammed Earth Building-code*. New Mexico. 1991.

Wang W.J. *Study on Reinforcement Mechanism and Damage Performance of Cemented Soil Stabilized with Nanometer Material*. Zhejiang University. 2004.

Yang H. *Seismic Test of Rammed Earth Wall and Reconstruction of Demonstrate Residence after Panzhihua Earthquake*. Xi'an University of Architecture and Technology. 2010.

Yuan J.Y. *Soil Test and Principle*. Tongji University Press. 2003

Zhang W. *Locally Born and Bred–study on the Rammed Earth-oriented Natural Construction*. China Academy of Art. 2013.

Zhao D.J., Zhang Y., Lu J.L. Research on the Construction of Rammed Earth Buildings. *Advanced Materials Research*. 2011, 243:934–937.

Zhou J.P. & Peng X.Z. Size Effect of Shear Strength of Soil. *Journal of Southwest Jiaotong University*. 2005, 40 (1):77–81.

Zhou T.G., Zhu R.Z., Song L.S., et al. Study on the Modern Pitched-roof Rammed Earth Farmhouse Model through Shaking Table Tests. *Building Structure*. 2016, 46(8):48–52.

Zhou T.G., Zhu R.Z., Zhao S.C., et al. Study on the Modern Single-story, Flat-roof Rammed Earth House Model through Shaking Table Tests. *Journal of Earthquake Engineering and Engineering Vibration*. 2015, 35(6):193–198.

Civil Engineering and Energy-Environment – Gao & Duan (Eds)
© 2023 the Author(s), ISBN 978-1-032-56059-5

Fuzzy matching of historical conditions-based prediction of carbon emission intensity of coal-fired power plants

Fanliang Meng*, Qinpeng Zhang & Huahai Qiu
Hua Dian Zhangqiu Electric Co., LTD, Jinan, China

ABSTRACT: The current emissions and emission intensity calculated by using the carbon emission accounting requirements of the power generation industry rely on monthly elemental carbon detection, so it is impossible to obtain the carbon emissions at the daily level. To better analyze and control the carbon emissions of coal-fired power plants, through the analysis of the condition data of a coal-fired power plant in the past years, the historical conditions are classified according to the key condition indicators, and the carbon emission intensity is calculated in combination with the historical emissions, and a one-to-one correspondence between the distribution of daily conditions and carbon emission intensity is established. Through the analysis of the classification and distribution of the current condition data, the corresponding methods for predicting current and future carbon emission intensity using historical carbon emission intensities are established, the data analysis and validation are carried out and good results are obtained.

1 INTRODUCTION

The carbon emission management of China's power generation industry is mainly based on the methods of verification, contract performance, and trading, of which carbon verification is the main means of carbon asset management in power plants. At present, the carbon emission intensity of power supply and heat supply of coal-fired power plants is mainly calculated through the method of "Guidelines for Accounting and Reporting of Greenhouse Gas Emissions for Enterprises for Power Generation Facilities" (hereinafter referred to as "Accounting Guidelines") using the result data that has occurred, and the current analysis period for the elemental carbon content of the coal entering the boiler is too long so that the carbon emission intensity data can only be calculated once a month, which cannot reflect the real-time situation of carbon emissions and influences.

In view of the above problems, by analyzing the historical data of key indicators of the generator set, defining different working condition types, and constructing a nonlinear programming problem's solution to obtain the power supply carbon emission intensity and heating carbon emission intensity corresponding to each working condition, the current carbon emission intensity is finally realized according to the current working conditions. At the same time, according to the information such as power generation plan and generator set generating power, it can predict the distribution of future working conditions to obtain the prediction result of future carbon emission intensity (li 2020; Wang 2015; Zhu 2018).

---

*Corresponding Author: adef@qq.com

To achieve the projection of carbon intensity, three problems need to be addressed:

1) Clarifying the key indicators of the generator set condition, that is, the basis for the classification of working conditions;
2) The carbon emission intensity accounting directly related indicators data are all about the fuel side, power generation side and heating side, and there are no indicators directly related to the working conditions of the generator set. It is necessary to establish the relationship between carbon emission intensity and working conditions;
3) The known analysis indicators, such as elemental carbon for calculating carbon emission intensity, can only be obtained once a month due to the frequency of coal analysis, while the working conditions indicators of the generator set can be obtained at the frequency of seconds. The corresponding time relationship between the two needs to be solved.

2 WORKING CONDITION INDICATORS

Working condition indicators is the condition indicators that can characterize the impact on carbon emission intensity. According to the Accounting Guidelines, thermal power plants use Formula (1) to calculate carbon emissions.

$$E_b = \sum_{i=1}^{n} \left(FC_i \times C_{ar,i} \times OF_i \times 44/12 \right) \tag{1}$$

where:

E_b—the emissions from fossil fuel combustion, (tCO_2);
FC_i—the consumption of the i-th fossil fuel, (t);
$C_{ar,i}$—the received base element carbon content of the i-th fossil fuel, (tC/t);
OF_i—the carbon oxidation rate of the i-th fossil fuel, (%);
44/12—the ratio of the relative molecular mass of carbon dioxide to carbon;
i—fossil fuel type code.
Thermal power plants can use Formulas (2), (3), (4), and (5) to calculate the carbon emission intensity of the power supply and heat supply.

$$S_{gd} = E_{gd}/W_{gd} \tag{2}$$

$$S_{gr} = E_{gr}/Q_{gr} \tag{3}$$

$$E_{gd} = (1 - a) \times E \tag{4}$$

$$E_{gr} = a \times E \tag{5}$$

where:

S_{gd}—the carbon emission intensity of power supply, that is, the number of carbon emissions produced by the generator set for each MWh of electricity supplied by the generator set, (tCO2/MWh);
E_{gd}—the number of carbon emissions produced by the power supply of the generator set during the statistical period, (tCO2);
W_{gd}—the amount of electricity supplied, (MWh);
S_{gr}—the carbon emission intensity of heat supply, that is, the carbon emission for each GJ of heat supplied by the generator set, (tCO2/GJ);
E_{gr}—the carbon emissions from heat supply during the statistical period, (tCO2);
Q_{gr}—the amount of heat supply, (GJ);
a—the heat-to-power supply ratio, (%);
E—the carbon emissions, (tCO_2).
The standard calculation method of coal (gas) consumption for power supply and heat supply (gas) consumption is based on GB 35574 and DL/T 904. When statistical data is not

available, thermal power plants should first determine the heat-to-power supply ratio, and then calculate the total amount of fuel consumed by the boiler (converted standard coal amount), which is divided into two parts: power supply part and heat supply. Formulas (6) and (7) are used to calculate coal consumption for power supply and heat supply.

$$b_g = (1 - a) \times B_h / W_{gd} \tag{6}$$

$$b_r = a \times B_h / Q_{gr} \tag{7}$$

where:

a—the heat-to-power supply ratio, (%);

b_r—the standard coal consumption of heat supply each generator set, (tce/GJ);

b_g—the standard coal consumption of power supply each generator set, (tce/MWh);

B_h—the total standard coal consumption by the generator set, (tce).

From the Formulas (1)-(7) (Cai *et al.* 2013; Liu *et al*; 2022; Ma *et al.* 2022; Višković & Franki 2022; Yang 2018), it can be found that the carbon emission intensity of power supply and heat supply is similar to the coal consumption of power supply and heat supply, which represent the overall working effect of the generator set. Therefore, the relationship between carbon emission intensity and generator set working indicators can be analyzed according to the indicators that affect the coal consumption of the generator set. Generally, the factors that affect the coal consumption of the generator set the most are the generator set load and heat supply condition. The indicators are the main steam temperature, main steam pressure, vacuum degree, exhaust gas temperature, etc. The working indicators of the generator set are shown in Table 1. By using the main indicators of each load section of the generator set in

Table 1. Analysis indicators of working condition.

Order Number	Type	Indicator
1	Power generation/heat supply	Generator set load
2	Power generation/heat supply	Main steam pressure
3	Power generation/heat supply	Main steam temperature
4	Power generation/heat supply	Ambient temperature
5	Power generation/heat supply	Rehot heat section steam temperature
6	Power generation/heat supply	Condenser vacuum
7	Power generation/heat supply	Condensate water flow
8	Power generation/heat supply	Blower inlet air temperature
9	Power generation/heat supply	The main water supply flow
10	Power generation/heat supply	Total air volume
11	Power generation/heat supply	Transmission fan current
12	Power generation/heat supply	Exhaust gas temperature
13	Power generation/heat supply	Flue gas oxygen content
14	Power generation/heat supply	Fuel oil supply flow
15	Power generation/heat supply	Fuel oil return flow
16	Power generation/heat supply	Coal quantity of coal feeder
17	Power generation/heat supply	Lead fan current
18	Power generation/heat supply	The main motor current of the coal mill
19	Heat addition	Condenser circulating water inlet water temperature
20	Heat addition	Condenser circulating water outlet water temperature
21	Heat addition	Heat supply flow of circulating water
22	Heat addition	Circulating water and return water flow rate
23	Heat addition	Heat supply steam flow
24	Heat addition	Extraction pressure
25	Heat addition	Extraction temperature

the heat/non-heat supply season to divide the working conditions, and calculating the carbon emission intensity corresponding to each working condition, the relationship between the working indicators of the generator set and the carbon emission intensity can be realized.

3 WORKING CONDITION CLASSIFICATION

At present, the methods for classifying working conditions of coal-fired generator sets usually include the load point determination method, equal width method, and K-means clustering algorithm.

Through comparative analysis, the initial cluster center and cluster number calculation method of the partition-based k-means algorithm are suitable for historical working data of coal-fired generator sets, and the determination method of cluster interval number and interval size of the equal-width algorithm is selected. It is applied to determine the combination conditions of external constraint working conditions for generator set operation. Through the analysis of all indicators data in the history of a power plant in the past two years (2018-2019), the clustering of all working conditions of the four generator sets of the power plant (#1 and #2 are all the same 100 MW level coal-fired pulverized coal boiler-condensing steam turbine generator set, #3#4 are 300 MW level coal-fired pulverized coal boiler-condensing steam turbine generator set respectively) has been analyzed, and also special working conditions are also distinguished.

The cluster analysis process and results of each generator set working condition are shown in Figure 1 and Table 2 to Table 6, respectively.

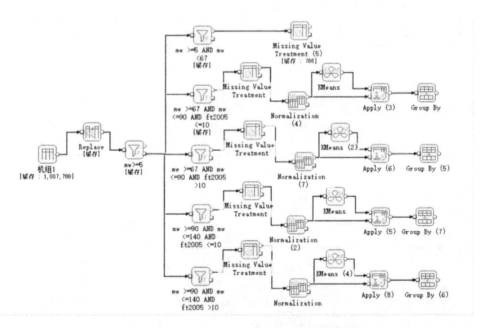

Figure 1. Schematic diagram of the generator set working condition classification and analysis flow.

Table 2. Generator sets classification table.

Set	Classification Number	Effective Classification Value
#1	1612	0~1610
#2	355	0~353
#3	141	0~139
#4	1037	0~1305

Table 3. Classification of 1# generator set.

# 1 Generator Set Load Data Range (MW)	Data Section of Circulating Water Heat Supply and Flow Rate (t / h)	Number of Working Condition Clusters	Working Condition Numbers
5~67	≤10	1	A1
	>10	1	A2
67~90	≤10	6	A3~A8
	>10	4	A9~A12
90~140	≤10	14	A13~A26
	>10	19	A27~A45
	<5	1	A46

Table 4. Classification of 2# generator set.

# 2 Generator Set Load Data Range (MW)	Data Section of Circulating Water Heat Supply and Flow Rate (t / h)	Number of Working Condition Clusters	Working Condition Numbers
5~67	≤10	1	A1
	>10	1	A2
67~90	≤10	6	A3~A8
	>10	4	A9~A12
90~140	≤10	14	A13~A26
	>10	19	A27~A45
	<5	1	A46

Table 5. Classification of 3# generator set.

# 3 Generator Set Load Data Range (MW)		Number of Working Condition Clusters	Working Condition Numbers
5~82		1	A1
82~165		4	A2~A5
165~240	Heating steam flow rate<=10 (t / h)	8	A6~13
	Heating steam flow rate>10 (t / h)	12	A14~25
240~350	Heating steam flow rate<=10(t / h)	10	A26~A35
	Heating steam flow rate>10(t / h)	10	A36~A45
	<5	1	A46

Table 6. Classification of 4# generator set.

# 4 Generator Set Load Data Range (MW)	Number of Working Condition Clusters	Working Condition Number
5~82	1	A1
82~165	4	A2~A5
165~240 (Heating steam flow rate>10(t / h)	10	A6~A15
240~350 Heating steam flow rate<=10(t / h)	15	A16~A30
Heating steam flow rate>10(t / h)	15	A31~A45
<5	1	A46

4 THE RELATIONSHIP BETWEEN THE DISTRIBUTION OF WORKING CONDITIONS AND THE INDEX OF CARBON EMISSION INTENSITY

According to the classification results and the number of working conditions of each generator set every day, the relationship between the distribution of working conditions and the carbon emission intensity of the power supply and heat supply is analyzed. Since the detection of elemental carbon content only has monthly data, by using a concept of the generation coefficient of generator set standard coal consumption proposed by Gao Jianqiang et al. (Gao et al. 2021; 2020; Song et al. 2021; Sun et al. 2018), the linear relationship between the power supply coal consumption and carbon emission intensity of coal-fired generator set is described. The daily emission intensities of the power supply and heat supply are determined according to the ratio of the daily coal consumption for the power supply and heat supply and the average monthly coal consumption for the power supply and heat supply.

The relationship model is constructed from the distribution of daily working conditions and the data of daily power supply carbon emission intensity and heat supply carbon emission intensity, as shown in Formula (8) and Formula (9).

$$(A_1 d_1, A_2 d_2 \ldots A_{46} d_{46}) \rightarrow (P1, P2) \tag{8}$$

$$\sum_{i=0}^{46} d_i = 1440 \tag{9}$$

where:

A_1 to A_{46}—all 46 working conditions of a generator set;

d_1 to d_{46}—the respective numbers of the 46 working conditions of a generator set;

P1—the carbon emission intensity of the power supply of a generator set;

P2—the carbon emission intensity of the heat supply of a generator set.

According to the generator set's minute-level historical data and Formula (8), some daily values of P1 and P2 of the generator set (taking the 1# generator set as an example) are obtained, as shown in Table 7.

By using the data from 2018 to 2019, this step establishes a one-to-one correspondence between the carbon emission intensity and the daily working condition distribution.

The monthly working condition distribution is similar to the analysis method in the daily dimension. The distribution of working conditions corresponding to all historically recorded working indicators in a certain month is analyzed, and the numbers A_1 to A_{46} reflect the actually calculated carbon emission intensity of the power supply and heat supply.

Table 7. Example of Classification of 1# generator set.

Time	Working Mode						Carbon Emission Intensity	
Project	A 1	A 2	A3	...	A45	A46	P1	P2
December 05	0	0	0	...	1066	0	0.5946	0.1205
December 06	0	0	0	...	1123	0	0.6388	0.1209
December 07	0	0	0	...	1089	0	0.6019	0.1246
December 08	0	0	0	...	1071	0	0.5953	0.1147
December 09	0	0	0	...	1064	0	0.5928	0.1119
December 10th	0	0	0	...	1085	0	0.6140	0.1181

5 CARBON EMISSIONS INTENSITY PROJECTIONS

5.1 *Analysis of current daily carbon emission intensity*

Since currently the measurement of elemental carbon in power plants is done monthly, the analysis data can only be obtained in the second month, so the carbon emissions on a certain day of the month cannot be directly calculated. The method corresponding to the fuzzy query of historical working conditions is used to obtain the carbon emission intensity of the daily dimension and then estimate the emission amount so that the power plant can timely know the carbon emission status and adjust the emission reduction strategy. The specific prediction method is shown in Table 8.

Table 8. Daily prediction method.

Predicted Object	Predicting Ideas	Prediction Technique
Daily carbon emission intensity	Each generator set is calculated as the prediction results of the P1 and P2 values according to the P1 and P2 values in the two days of the same period of last year (divided into the heating season and non-heating season).	$\min \sum_{i=0}^{i=46} (x_i - d_i),$ $x_i - d_i \leq a * x_i$ i select the corresponding serial numbers of the 5 with the largest value among the 46 working conditions, a can be taken as 10%, 5%, 3%, and 1%, until the minimum sum value under each proportion is obtained, x_i represents the number of each working conditions under the current working condition distribution, and d_i represents the number of each working conditions under the corresponding time and historical working condition distribution.

5.2 *Future carbon emission intensity projection*

Carbon emission intensity projection for a future period (usually the remainder of the year) can provide power plants with estimates of annual emissions and recommendations. The prediction of future carbon emission intensity is mainly based on various plans for power plants, including power generation plans, heat supply plans, procurement contracts, etc. Because power plant plans are highly variable and are affected by various factors, the monthly projection is more feasible than the daily projection in the time dimension. Here we only predict the carbon emission intensity of the future month. There is a special case.

During the maintenance period of the generator set, the power generation during this period is zero, and it will not be carried out during the heat supply season. It is necessary to pay attention to this when matching the historical power generation.

The distribution of predicted working conditions can be divided into three cases:

1) Heat supply season: The season starts in November and ends in March of the next year;
2) Non-heat supply season: The season starts from April to the end of October;
3) Generator set maintenance period.

The prediction method is shown in Table 9.

Table 9. Prediction methods table.

Type	Month	1#	2#	3#	4#
Heat supply season	1 2 3	According to the working condition type of the same month of last year and the corresponding P1 and P2 values are taken as the prediction results of the working condition type and P1 and P2 values of this month.			
Non-heat supply season	4 5 6 7 8 9 10	Monthly planned power generation calculation: If the planned electricity generation is given this month, the planned generation is used. If the planned electricity generation is not given this month, then it will be calculated according to (Total power generation this year-monthly total power generation this year) / (Last year's total power generation-last year's monthly total power generation corresponding to this year) * Electricity generation this month last year; Firstly the planned power generation of each generator set this month is used to find out the closest power generation month in the Non-heat supply season in history, then the P1 and P2 values corresponding to the closest working condition type are taken as the prediction results of the working condition type and P1 and P2 values of this month.			
Heat supply season	11 12	The P1 and P2 values of the working condition type and the working condition type corresponding to the same month of last year are taken as the prediction results of the working condition type and P1 and P2 values of this month			
Maintenance period	Non-heat supply season month	The working condition type is A_{46}, P1=P2=0, and power generation=0			
Data update	All months	The data of historical months recalculates the working condition category and the corresponding P1 and P2 values according to the indicators of the real condition. The distribution of working condition types and the corresponding P1 and P2 values of the future months are given according to the projection results.			

5.3 *Carbon emission projections results*

According to the prediction methods of 4.1 and 4.2, the four generator sets of the power plant are predicted, and an example of the obtained results is shown in Table 10.

Table 10. Example of daily and monthly projection results.

Current Projection Time	Generator Set Name	Corresponding Time in History	P1	P2
2022-09-01	1#	2021-06-20	0.7006	0.1194
2022-09-01	2#	2021-08-13	0.5946	0.1205
2022-09-01	3#	2021-07-05	0.6388	0.1209
2022-09-01	4#	2021-04-21	0.6019	0.1246
2022-10	1#	2021-08	0.5953	0.1147
2022-10	2#	2021-06	0.5863	0.1175
2022-10	3#	2021-07	0.5911	0.1201
2022-10	4#	2021-06	0.5928	0.1119

5.4 *Prediction algorithm validation*

We use the data from 2018 to 2019 to carry out the projection operation, perform the weighted average processing of the projection value, and compare it with the actual daily/monthly value to verify the accuracy of the projection model. The calculation formula of the accuracy evaluation index is as follows:

1) MAE (Mean Absolute Error)

$$MAE = \frac{1}{n} \sum_{i=1}^{n} |f_i - y_i| \tag{9}$$

2) MSE (Mean Square Error)

$$MSE = \frac{1}{n} \sum_{i=1}^{n} (f_i - y_i)^2 \tag{10}$$

3) MAPE (Mean Absolute Percentage Error)

$$MAPE = \frac{100}{n} \sum_{i=1}^{n} \left| \frac{y_i - f_i}{y_i} \right| \tag{11}$$

The verification results are shown in Table 11.

Table 11. Algorithm verification results table.

Generator Set Name	Metric	MAE	MSE	MAPE
1#	P1	0.019537	0.000918	0.027421
	P2	0.001789	0.000007	0.015645
2#	P1	0.027530	0.001599	0.048874
	P2	0.001550	0.000006	0.012762
3#	P1	0.053722	0.005248	0.060194
	P2	0.004417	0.000068	0.035725
4#	P1	0.035367	0.001910	0.041683
	P2	0.003581	0.000039	0.028646

The results of the actual value P1 and predicted value P1, the actual value P2 and the predicted value P2 of each generator set are shown in Figure 2.

From Table 11 and Figure 2, it can be seen that the carbon emission intensity of the power supply and heat supply predicted by the prediction algorithm used are in good agreement

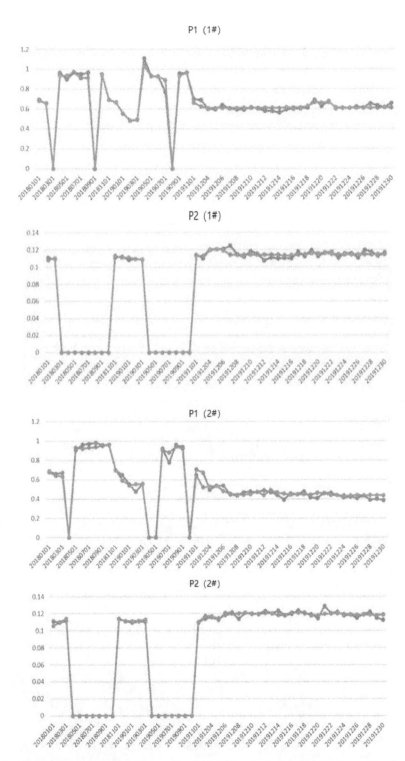

Figure 2. Comparison of the true and predicted P1 / P2 values of each generator set.

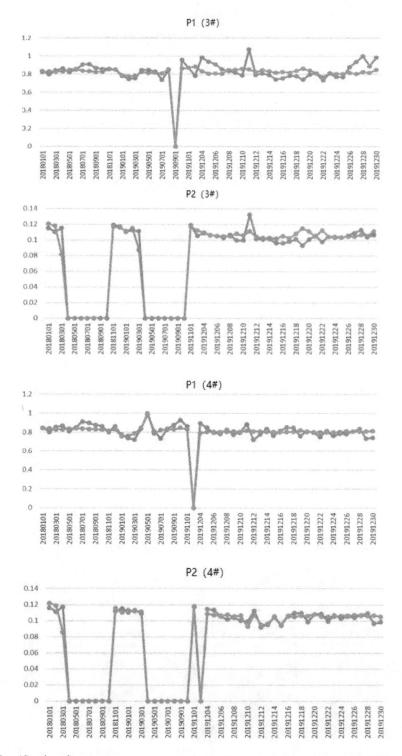

Figure 2. (Continued)

with the real-time calculation of the carbon emission intensity of power supply and heat supply, and the difference is within the allowable error range, the prediction method has good accuracy.

6 CONCLUSION

Through the cluster analysis of key indicators of coal-fired power plant working conditions, the working conditions are classified, and the distribution of working conditions and the corresponding carbon emission intensity are analyzed according to historical data. On this basis, a nonlinear algorithmic model is constructed. The corresponding relationship between carbon emission intensity and various working conditions is obtained by the model, and then the corresponding working conditions are analyzed through current data or future plans. The corresponding carbon emission intensity is predicted by correlation. Finally, the prediction algorithm is verified. The verification results show that the prediction algorithm has relatively high accuracy.

This method can help coal-fired power plants to know the current carbon emission situation in real-time to provide basic support to make carbon management decisions.

REFERENCES

Cai Yu, Li Baowei, Hu Zechun, et al. Calculation of Carbon Emission Index of Coal-Fired Generating Unit and Analysis on Influencing Factors. *Power System Technology*, 2013,37 (5): 11851189.

Gao Jianqiang, Song Tongcopper , Zhang Xue. Analysis and Calculation on Carbon Emission Characteristics of CFB Boiler Units. *Journal of Chinese Society of Power Engineering*, 2021,41 (01): 14–21.

Gao Jianqiang, Song Tongtong, Zhang Qiaobo, Cao Hao. Sensitivity Analysis of Carbon Emission from Coal-fired Power Plant to Variation of Controllable Operating Parameters. *Journal of Chinese Society of Power Engineering*, 2020,40 (07): 517–522 + 555.

Hai-ping Li. Analysis of Carbon Emission of Coal Degree from Coal-fired Power Plants. *Energy Conservation*, 2020,39 (07): 30–33.

Liu Ke, Yang Xingsen, Wang Tai, Dong Xinguang, Zhang Limeng, Zhang Xuhui, Yuan Sen, Xin Gang, Gao Song. Research on Carbon Emission Characteristics of Coal-fired Units based on Real-time Monitoring. *Thermal Power Generation*: 1–7 [2022-09-20].

Ma Xueli, Wang Xiaofei, Sun Xijin, Shi Jing, Chen Jinpeng, Dang Lichen. Influence Factors of Carbon Emission Intensity of Coal-fired Power Units. *Thermal Power Generation*, 2022,51 (01): 190–195.

Song Copper. *Study on Accounting and Influencing Factors of Carbon Emission Intensity of Coal-fired Power Plants*. North China Electric Power University, 2021.

Sun Youyuan, Zheng Zhang, Qin Yaqi, Guo Zhen, Ren Jian. Study on Carbon Emission Characteristics and Suggestions on Carbon Emission Management of Coal-fired Power Plant. *Electric Power*, 2018,51 (03): 144–149 + 169.

Višković A, Franki V. Evaluating and forecasting Direct Carbon Emissions of Electricity Production: A Case Study for South East Europe. *Energy Sources, Part B: Economics, Planning, and Policy*, 2022: 1–21.

Wang Zhixuan. Current Situation of China Low-Carbon Power Development and Its Countermeasures. *China Energy*, 2015,37 (7): 510.

Yang Zheng. The Type of Element Analysis of Coal and its Relationship With Coal Quality. *Chemical Enterprise Management*, 2018 (5): 158–158.

Zhu Dechen. Application of Carbon Ultimate Analysis Into Greenhouse Gas Emissions Accounting for Coal-fired Power Plants. *Power Generation Technology*, 2018,39 (04): 363–366.

Civil Engineering and Energy-Environment – Gao & Duan (Eds)
© 2023 the Author(s), ISBN 978-1-032-56059-5

Influences of fly ash on water and salt transportation of alkali-saline soil

Yue Ma*, Guoli Xie, Qiguang Cao, Hongmei Chen & Peng Lu
Beijing Polytechnic, Beijing, China

ABSTRACT: China's saline-alkali soils cover an area of about 100 million hm^2, of which only about 20% have been improved and utilized. Soil salinization affects the normal growth of plants and causes a significant decrease in soil productivity. The study of soil water and salt transport law is the key to improving saline-alkali soils. Fly ash is a type of powdered solid waste discharged from coal-fired thermal power plants after high-temperature combustion of pulverized coal, and it is one of many types of industrial solid waste with the largest emission in China. Most of the research on improving saline-alkali soils with fly ash focuses on agricultural fertilizers, and there is a lack of research on the effect of fly ash on the vertical distribution of salt in the process of salt washing and salt returning in saline-alkali soils. This paper discusses the water and salt transport after irrigation with different amounts of fly ash by the indoor simulation experiments of water washing desalination and salt return, taking the group without the fly ash as the control. The changes of upward salt transport in saline-alkali soils under three different treatments of adding films to prevent evaporation, natural evaporation indoors, and repeated irrigation once after 30 days of natural evaporation are compared. The results show that the addition of fly ash has a good inhibitory effect on the upward salt transport of saline-alkali soils in the cultivated layer.

1 INTRODUCTION

Saline soil is a general term for saline and alkaline soils and various salinized and alkalized soils. Saline soils are soils containing large amounts of soluble salts that prevent most plants from growing, and their content of salt is generally 0.6% - 1.0% or higher; alkaline soils are soils with a substitutable sodium ion as a percentage of cation substitution (ESP) of more than 20% and a pH of 9 or higher; in fact, saline and alkaline soils are often mixed, so it is customary to call them saline-alkali soils [1]. There are about 100 million hm^2 of various types of saline-alkali land in China, of which 80% have not been exploited and utilized so far.

Fly ash is a type of powdered solid waste discharged from coal-fired thermal power plants after high-temperature combustion of pulverized coal, which is one of many types of industrial solid waste with the largest emission in China [2]. The treatment of fly ash is still mainly conducted in ash storage yards that occupy more land area year by year and cause serious pollution to the surrounding environment with dust flying [3]. Fly ash has the characteristics of looseness and porousness, low bulk weight, high porosity, and large water storage capability, and thus it can reduce soil bulk weight, make soil permeability reasonable, increase pore space for clayey soil, enhance soil water retention capacity, and promote

---

*Corresponding Author: mayuebj@sina.com

112

DOI: 10.1201/9781003433651-15

the growth of beneficial bacteria. In addition, fly ash has low thermal conductivity and thermal insulation properties [4].

The previous experiments on the use of mine fly ash to improve saline land and grow plants in the Huafeng coal mine in the Xinhan mining area of Shandong Province proved that the soil improved by fly ash showed a change in permeability, a decrease in soil bulk, and a significant increase in organic matter, alkaline soluble N, effective P, effective K content and water storage capability [5]. Tests on potted plants in arid and semi-arid areas showed that fly ash, reservoir silt, and sludge applied in appropriate proportions had a significant effect on improving poor soils and promoting plant growth [6]. Tests on organic compound fertilizer for rice and sugarcane developed by composting and fermenting sludge, garbage, and fly ash in 4:5:1 showed that it had a good effect on increasing the yield of both rice and sugarcane and could improve the soil structure and enhance soil fertility [7]. At present, most of the studies on the improvement of saline-alkali soils with fly ash focus on agricultural fertilizers, and there is a lack of research on the effect of fly ash on the vertical distribution of salt during salt washing and salt returning in saline-alkali soils. This paper conducts an indoor simulation study to investigate the effect of fly ash on the physico-chemical properties of saline-alkali soils. In this way, it is of great significance to achieve the effect of "treating deterioration with waste", which can not only alleviate the problem of salinization of land, but also implement the bulk consumption of solid waste of fly ash.

2 MATERIALS AND METHODS

The indoor soil column simulation experiment was adopted. The soil for test came from the topsoil of the Dongmen experimental field of the Chinese Academy of Agricultural Sciences, and the soil was layered into the column. The soil column was a sectional plastic pipe with a height of 80 cm and a diameter of 15 cm (as shown in Figure 1, the bottom section was 20 cm high; the other six sections were 10 cm; and there was a hole on the lowest side of the pipe). The soil in the column was added to 75 cm high, and the top 5 cm was not added with soil. According to the measurement, the indexes of the tested soil were as follows: unit weight 1.26 g/cm^3, porosity 52.41%, specific gravity 2.65 g/cm^3, soil water content 11.40%, saturated water content 41.60%, organic matter content 10.35 g/kg, total N 0.61 g/kg, available P content 5.8 mg/kg, fast available K 81.5 mg/kg, total salt content 81 g/kg and conductivity 2810 μs/cm.

Fly ash was added at 25 cm of the soil in the soil column. The fly ash used in the test was from Anhui Wanbei Mining Bureau, with a particle size of 0.5 to 300 μm (similar to light loam particles), an average specific gravity of 2.14 g/cm^3, an average unit weight of 0.783 g/cm^3, a specific surface area of 1600 to 3500 cm^2/g and a porosity of about 15%.

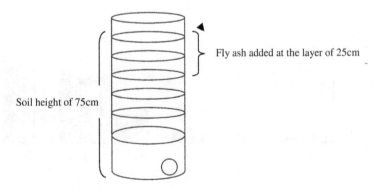

Figure 1. Soil column.

There were four groups in the experiment, and each group had three treatments. The four groups were set according to different fly ash dosages. Tests at home and abroad had proved that the application of 60 to 600 t/hm^2 fly ash soil didn't contaminate food crops. Therefore, the amount of fly ash added to each group of soil columns in this experiment was 0 g, 120 g, 180 g, and 240 g, respectively, which were recorded as FM0, FM120, FM180, and FM240. Three different treatments were designed for each group. The first treatment was to apply films to prevent evaporation after 2.25 L of irrigation. After ten days, each layer of soil was taken to measure its total salt content and the impact of fly ash on the salt distribution under the condition of film covering after the first irrigation was analyzed. The second treatment was to irrigate 2.25 L of water. After 30 days, each layer of soil was taken to measure its total salt content, and the impact of fly ash on the salt distribution after the evaporation of the first irrigation was analyzed. The third treatment was to irrigate 2.25 L water at first and 1 L water again after 30 days of evaporation. After 30 days, the soil of each layer was taken to measure the change of total salt, and the impact of fly ash on the salt distribution after the evaporation of the second irrigation was analyzed. All the soil columns used in the experiment were placed indoors. Based on the actual conditions of agricultural irrigation water in China, the salinity of irrigation water in the test was set at 2 g/L.

Determination items and methods: the semi-micro Kjeldahl method was used to determine the total nitrogen of soil nutrients, the 0.5 mol/LNaHCO$_3$ method was used to determine the fast available P, the 1 mol/LNH$_4$OAc extraction flame photometric method was used to determine the available K, the potassium dichromate volumetric method - external heating method was used to determine the soil organic matter. The total salt was determined by the 1:5 soil water extraction residue drying method. The unit weight was measured by the cutting ring method. The soil porosity was calculated according to the density and unit weight.

3 RESULTS AND ANALYSIS

3.1 *Influences of adding different amounts of fly ash on the transportation of Cl- of alkali-saline soil with a salinity of 0.8%*

3.1.1 *Influence of fly ash on the transportation of Cl- of alkali-saline soil with a salinity of 0.8% under film covering after the first irrigation*

After the first irrigation, covering the upper mouth of the soil column with film can be approximately regarded as water infiltration only in the soil column. After irrigation, most of the Cl- in the soil leached from the upper part to the lower part. The content of Cl- in the surface layer of the soil decreased, and most of the Cl- was leached below 50 cm of the soil. There was little difference in the content of Cl- in the four groups of FM0, FM120, FM180, and FM240 in the surface soil from 0 to 20 cm (Figure 2).

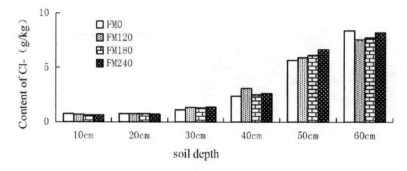

Figure 2. The level of Cl-content in different layers of 0.8% soil column covered by plastic films.

3.1.2 Influence of fly ash on Cl ion distribution of alkali-saline soil with a salinity of 0.8% under 30 days of evaporation after the first irrigation

The 30-day evaporation can be regarded as a process of upward transportation of Cl- with water. The average evaporation of the soil column measured was 0.85 kg. Cl- was transported upward with the evaporation of water. Compared with the film covering condition after the first irrigation, the increase of Cl ion content in the soil of 10 cm at the top surface was 0.71 g/kg in FM0 treatment, 0.57 g/kg in FM120 treatment, 0.54 g/kg in FM180 treatment and 0.53 g/kg in FM240 treatment. The amount of Cl- transported to the upper 10 cm of the soil column was as follows: FM0 > FM120 > FM180 > FM240. Compared with the situation without fly ash, the ratios of inhibiting salt upward transportation of each group with fly ash were as follows: FM120 was 20%, FM180 was 24%, and FM240 was 25%. The addition of fly ash could inhibit the upward transportation of Cl- to a certain extent, and some Cl- was inhibited in the lower layer of the soil column (as shown in Figure 3). The content of Cl- in the soil column with fly ash was higher than that without fly ash in the soil below 40 cm. The inhibition effect on the upward transportation of Cl- was as follows: FM240 > FM180 > FM120. After 30 days of evaporation after the first irrigation, the group with 240 g fly ash had the most obvious inhibitory effect on Cl-.

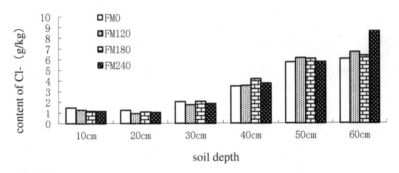

Figure 3. The level of Cl-content in different layers of 0.8% soil column after 30 days of evaporation.

3.1.3 Influence of fly ash on Cl ion distribution of alkali-saline soil with a salinity of 0.8% after evaporation and twice irrigation

After the 30-day evaporation of the first irrigation, the second irrigation was carried out, and then the evaporation lasted for another 30 days, which can be seen as a process of repeating the downward and upward transportation of Cl-. After measurement, the average evaporation of the soil column was 1.7 kg. Cl- was transported upward with the evaporation of water. Compared with evaporation after the first irrigation, the increase of Cl- in the surface 10 cm soil was as follows: FM0 was 2.8 g /kg, FM120 was 2.57 g/kg, FM180 was 2.16 g/kg, and FM240 was 2.12 g/kg, and the amount of Cl- transported to the upper 10 cm soil column was FM0 > FM120 > FM180 > FM240. Compared with the situation without fly ash, the ratios of inhibiting Cl ion upward transportation of each group with fly ash were as follows: FM120 was 0.4%, FM180 was 16%, and FM240 was 18%. Compared with evaporation after the first irrigation, under the treatment of the second irrigation and evaporation, the addition of fly ash had a certain inhibitory effect on the upward transportation of Cl-, and some Cl- was inhibited in the lower layer of the soil column. As shown in Figure 4, the content of Cl- in the soil column with fly ash added was higher than that without fly ash added in most soil below 30 cm. The inhibition effect on the upward transportation of Cl- was as follows: FM240 > FM180 > FM120. After the second irrigation and evaporation, the group with 240 g fly ash also had the most obvious inhibitory effect on Cl-.

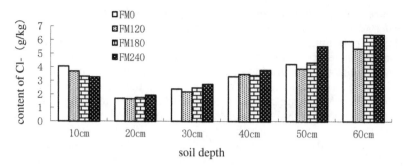

Figure 4. The level of Cl- content in different layers of 0.8% soil column after irrigation twice.

3.2 *Influence of adding different amounts of fly ash on the transportation of total salt of alkali-saline soil with a salinity of 0.8%*

3.2.1 *Influence of fly ash on the distribution of salt of alkali-saline soil with a salinity of 0.8% under film covering after the first irrigation*

After the first irrigation, covering the upper mouth of the soil column with film can be approximately regarded as water infiltration only in the soil column. After irrigation, most of the salt in the soil leaches from the upper part to the lower part, and it can be deemed as a salt-washing process. The content of salt in the surface layer of the soil decreased, and most of the salt was leached below 50 cm of the soil. There was little difference in the content of salt in the four groups of FM0, FM120, FM180, and FM240 in the surface soil from 0 to 20 cm (Figure 5).

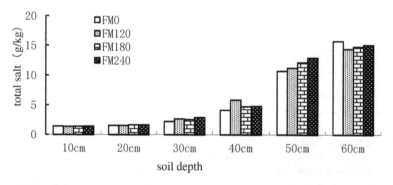

Figure 5. The level of salt content in different layers of 0.8% soil column covered by plastic films.

3.2.2 *Influence of fly ash on the salt distribution of alkali-saline soil with a salinity of 0.8% under 30 days of evaporation after the first irrigation*

The 30-day evaporation can be regarded as a process of salt returning. The average evaporation of the soil column measured was 0.85 kg. The salt was transported upward with the evaporation of water. Compared with the film covering condition after the first irrigation, the increase of salt content in the soil of 10 cm at the top surface was 1.48 g/kg in FM0 treatment, 1.17 g/kg in FM120 treatment, 1.04 g/kg in FM180 treatment and 0.91 g/kg in FM240 treatment. The amount of salt transported to the upper 10 cm of the soil column was as follows: FM0 > FM120 > FM180 > FM240. Compared with the situation without fly ash, the ratios of inhibiting salt upward transportation of each group with fly ash were as follows: FM120 was 21%, FM180 was 30%, and FM240 was 39%. The addition of fly ash could inhibit the upward transportation of salt to a certain extent, and some salt ions were inhibited in the lower layer of the soil column (as shown in Figure 6). The content of salt in the soil column with fly ash was higher than that without fly ash in the soil below 40 cm. The inhibition effect on the upward transportation of salt was as follows: FM240 > FM180 > FM120. After 30 days of evaporation after the first irrigation, the group with 240 g fly ash had the most obvious inhibitory effect on salt.

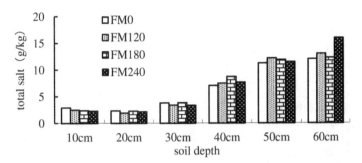

Figure 6. The level of salt content in different layers of 0.8% soil column after 30 days of evaporation.

3.2.3 Influence of fly ash on the salt distribution of alkali-saline soil with a salinity of 0.8% after evaporation and twice irrigation

After the 30-day evaporation of the first irrigation, the second irrigation was carried out, and then the evaporation lasted for another 30 days, which can be seen as a process of repeating the salt washing and the salt returning. After measurement, the average evaporation of the soil column was 1.7 kg. Salt was transported upward with the evaporation of water. Compared with evaporation after the first irrigation, the increase of salt in the surface 10 cm soil was as follows: FM0 was 5.84 g/kg, FM120 was 5.48 g/kg, FM180 was 5.17 g/kg, and FM240 was 4.93 g/kg, and the amount of salt transported to the upper 10 cm soil column was FM0 > FM120 > FM180 > FM240. Compared with the situation without fly ash, the ratios of inhibiting salt upward transportation of each group with fly ash were as follows: FM120 was 6%, FM180 was 11%, and FM240 was 16%. Compared with evaporation after the first irrigation, under the treatment of the second irrigation and evaporation, the addition of fly ash had a certain inhibitory effect on the upward transportation of salt, and some salt ions were inhibited in the lower layer of the soil column. As shown in Figure 7, the content of salt in the soil column with fly ash added was higher than that without fly ash added in most

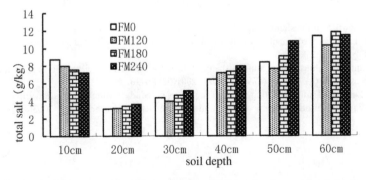

Figure 7. The level of salt content in different layers of 0.8% soil column after irrigation twice.

soil below 30 cm. The inhibition effect on the upward transportation of salt was as follows: FM240 > FM180 > FM120. After the second irrigation and evaporation, the group with 240 g fly ash also had the most obvious inhibitory effect on salt.

4 CONCLUSION AND DISCUSSION

In this paper, we use solid waste fly ash to improve saline-alkali soil. The main conclusions can be summarized as follows: (1) The group without fly ash was taken as the control. Adding fly ash to the soil column tilth layer (0-25 cm) with a salinity of 0.8% can inhibit the upward transportation of Cl⁻ and the total amount of salt in the soil, especially the upward transportation of Cl⁻ and the total amount of salt after the first irrigation and evaporation. (2) The experimental results show that adding fly ash to the tilth layer (0-25 cm) had a certain inhibition effect on the upward transportation of salt, which can inhibit part of the upward transportation of salt in the soil below the tilth layer, thus reducing the salinity of the alkali-saline soil in the tilth layer, making it possible to plant some crops on it. The inhibitory effect is related to the amount of fly ash added. Under the same conditions, the inhibitory effect was FM240 > FM180 > FM120, wherein, the FM240 group (600 t/hm^2) had the best effect.

Using fly ash to improve saline-alkali soil reduces environmental pollution, but fly ash contains a small number of harmful elements, so its dosage should be reasonable. In the application of fly ash, it is still necessary to strengthen the detection of soil quality to avoid secondary pollution. This study is based on the process of two times of salt washing and salt returning, so its conclusion has some limitations and needs further study and verification.

ACKNOWLEDGMENTS

This research was supported by the Research on the Transformation and Development of Capital Vocational Education in the New Era (Grant No. CJGX2022-KY-206) and the Scientific Research Projects of Beijing Polytechnic (Grant No.2022×016-KXZ).

REFERENCES

[1] Lu Hao, Wang Haize. Research Progress of Saline-alkali Soil Treatment and Utilization. *Modern Agriculture*, 2004(08):10–11.

[2] Zhao J., Kang Z.Z., Han Q.Q., et al. Application and Prospect of Fly Ash in Soil Improvement and Remediation. *Jiangsu Agricultural Sciences*, 2017, 45(2): 1–6.

[3] Wang Pengfei. Research Progress of Coal Ash Comprehensive Utilization. *Electric Power Environmental Protection*, 2006, 2(22):42–43.

[4] Li Guibo. Comprehensive Utilization of Fly Ash. Prospect of Fly Ash. *Agricultural Utilization*, 1999 (3):48–52.

[5] Zhou Xuewu, Sun Daisheng, Fang Jianguo, et al. Application of Mine Solid Waste (Fly Ash, Silt and Sludge) in the Improvement of Saline-alkali Land and Planting Experiment. *Resources and Industry*, 2005(03):16–18.

[6] Shen Junfeng, Li Shengrong, Sun Daisheng, et al. Study on the Feasibility of Using Some Solid Wastes to Improve Soil. *Geology and Geochemistry*, 2001(02):86–90.

[7] Wang Dunqiu, Zhang Xuehong, Long Tengrui, et al. Study on the Application Effect of Municipal Solid Waste to Produce Organic Compound Fertilizer. *Water and Sewerage*, 2003(05):34–36.

Civil Engineering and Energy-Environment – Gao & Duan (Eds)
© 2023 the Author(s), ISBN 978-1-032-56059-5

A new natural gas prediction model based on polynomials and its application

Huanying Liu, Changhao Wang, Yulin Liu, Jiaxiang Yang & Bingyuan Hong*
Zhejiang Ocean University, Zhoushan, China

ABSTRACT: The natural gas demand forecast is an important link to analyze the current situation of supply, demand, and intelligent supply guarantee. This paper digs out six important influencing factors of natural gas consumption: GDP, total population, urbanization rate, residential consumption level, the proportion of secondary industrial structure, and urban population. Grey relative correlation degree was used for correlation analysis, and low-weight factors with multicollinearity were eliminated. The Regression Polynomial Model was used to fit the unitary curve between the single factor and the natural gas consumption, and the unitary curve was combined into the multivariate curve based on linear programming to construct a new forecasting model. The model used in this paper has a good fitting effect and fully considers the influence of many factors. Zhejiang province was taken as an example, and the results show that: 1. GDP is the most important factor affecting the demand for natural gas in this area, and residential consumption level is the second most important factor; 2. The annual consumption of natural gas in Zhejiang province in 2031 is expected to increase by 136.9% compared with 2021, and the annual growth volume will increase from 1366×10^6 m^3 in 2022 to 3146×10^6 m^3 in 2031; 3. Zhejiang region should give priority to accelerating GDP growth and raising the level of residents' consumption in order to promote the growth of natural gas consumption.

1 INTRODUCTION

As a green, clean, environmental-friendly, low-carbon, and high-quality energy, natural gas plays a vital role in China's sustainable and healthy development process. The natural gas supply in Zhejiang increased from an initial 20 million cubic meters to 15.063 billion cubic meters in 2021 [1], and it is expected to maintain rapid growth in the future. Accurate prediction of natural gas demand is very important for the adjustment of industrial structure and the development of a green economy in Zhejiang province.

Up to now, scholars have carried out a lot of research on the influencing factors and forecasting models of natural gas demand. In terms of influencing factors, a large amount of literature has proved that the demand for energy is related to many factors, such as economic conditions, industrial structure, population, and so on. Jian chai et al. [2] believe that energy consumption structure, population scale, industrial structure, regional GDP, industrial investment, market share of substitutes and other factors are effective variables affecting natural gas demand, which can help to forecast and analyze the natural gas demand. Hongbing Li et al. [3] used stepwise regression analysis to test the statistical significance of GDP, natural gas consumption level, industrial structure, urbanization rate, and other factors according to the correlation degree and established relevant models to predict the future

---

*Corresponding Author: hongby@zjou.edu.cn

DOI: 10.1201/9781003433651-16

change situation of natural gas demand. In terms of model methods for forecasting natural gas demand, the use of different prediction models can show the change form of the predicted object at multiple levels, such as grey prediction, time series, system dynamics model, BP neural network, and other common methods. However, when these methods are used alone, they are prone to problems such as insufficient accuracy, difficult quantification of influencing factors, slow convergence rate and local optimality. In order to make the prediction result more reasonable, scientific and accurate, more and more methods have been used in the prediction of natural gas demand, and better results are received. Vinayak et al. [4] used machine learning technology to conduct data-driven short-term natural gas demand prediction. Dongkun Luo et al. [5] improved the BP neural network to avoid the shortcomings of local minima, making it more suitable for short-term and medium-term natural gas demand prediction. Hongkun Wu et al. [6] established AR (14) model based on time series to predict the future natural gas demand. In the above mentioned research, the machine learning model needs to be trained with a large amount of data, and there is not enough data to support the prediction in some regions where natural gas utilization starts late. Although the traditional model requires less data, the prediction accuracy of a certain model or a combined model is greatly affected by the region.

The forecast of natural gas demand is constrained by many complex factors. Each factor restricts the other, and its impact is difficult to measure accurately. This paper uses a grey relative correlation degree to determine the influencing factors of natural gas demand from the factors affecting natural gas consumption and fits the data between the basic data and the consumption through the new multiple regression model, and finally realizes the prediction of natural gas demand in Zhejiang province.

2   METHODOLOGY

The core of the natural gas prediction model is the selection of influencing factors. In this paper, six factors that mainly affect the consumption of natural gas, namely GDP, total population, urbanization rate, household consumption level, the proportion of secondary industry structure and urban population, are taken as influencing factors. The correlation analysis between influencing factors and natural gas consumption is carried out by using a grey relative correlation degree to eliminate low-weight explanatory variables. On this basis, a new prediction model is used to fit the data between the basic data and consumption, and linear programming is used to construct the optimization function to reduce the model error. Finally, a new model based on regression polynomials is constructed.

1) Gray relative correlation analysis was conducted among all influencing factors, and multicollinearity factors were eliminated.
2) The remaining influencing factors and natural gas demand are dimensionless.
3) The influencing factors $x_1, x_2, \dots, x_n$; natural gas consumption by polynomial fitting, each fitting model $y_1, y_2, \dots$
4) y_1 and y_2 were combined through the coefficient optimization principle, and a new prediction model $Y=f(x_1, x_2, \dots)$ was constructed.
5) Grey prediction for the remaining explanatory variables was made as the basis for prediction.
6) According to the grey predictive value of explanatory variables and the new multiple regression function $Y=f(x_1, x_2, \dots)$, the predicted value of the explained variable was calculated.

2.1   *Grey relative correlation degree*

Grey relative correlation degree is an important method of grey relational analysis, which is often used to judge the order and size of the correlation between multiple influencing factors and dependent variables and to judge the multiple linearities between influencing factors. Correlation analysis was conducted on natural gas consumption, GDP, household consumption level, urban population, total population, urbanization rate, and the proportion of secondary industry structure. The grey relative correlation degree model was used to analyze, judge, and calculate the influence of a certain factor on natural gas consumption and the grey relative correlation coefficient between two influencing factors.

2.2 *Regression polynomial model*

After removing the low-weight influencing factors with multicollinearity, the remaining influencing factors are important and have no internal collinearity. The regression polynomial model is used to fit the remaining influencing factors, and natural gas consumption, respectively, and the fitting function between the influencing factors and natural gas consumption can be obtained. This unary function can express the data relationship between the influencing factors and natural gas consumption, and the data relationship can be used for prediction calculation. In this paper, polynomial regression is introduced by linear programming to express the functional relationship between the independent variable and the dependent variable, and a polynomial is fitted in the applicable range. The fitting polynomial is set as:

$$y = a_0 + a_1 x + a_2 x^2 + a_3 x^3 + a_4 x^4 + \cdots + a_n x^n \tag{1}$$

After processing, we get:

$$\begin{bmatrix} n & \sum_{j=1}^{n} x_2 & \cdots & \sum_{j=1}^{n} x_i^k \\ \sum_{j=1}^{n} x_i & \sum_{j=1}^{n} x_i^2 & \cdots & \sum_{j=1}^{n} x_i^{k+1} \\ \vdots & \vdots & \ddots & \vdots \\ \sum_{j=1}^{n} x_i^k & \sum_{j=1}^{n} x_i^{k+1} & \cdots & \sum_{j=1}^{n} x_i^{2k} \end{bmatrix} \begin{bmatrix} a_0 \\ a_1 \\ \vdots \\ a_k \end{bmatrix} = \begin{bmatrix} \sum_{j=1}^{n} y_i \\ \sum_{j=1}^{n} x_i y_i \\ \vdots \\ \sum_{j=1}^{n} x_i^k y_i \end{bmatrix} \tag{2}$$

That is, X*A=Y. Linear algebra is used to solve it, and the matrix obtained is the coefficient matrix. Each a value can be calculated to obtain the control variables, and then the fitting function can be obtained.

2.3 *Principle of coefficient optimization based on linear programming*

After using the regression polynomial model for univariate curve fitting, multiple unary fitting functions can be obtained. There are many ways to combine unary functions into multivariate functions. It is particularly important to determine the combined weight to determine the fitting effect. To construct a new prediction model, it is necessary to determine the optimal weight of each univariate fitting function according to the characteristics of the data. There are many methods to determine the weight. In this paper, the linear programming method is used to construct the optimization function to obtain the weight value. This method has the advantages of high scientificity, full use of data characteristics, good combination effect, and so on. The principle of the method is as follows:

2.4 *Grey prediction*

The advantages of the grey prediction model are prominent because it requires only a small number of samples to predict and has high prediction accuracy, so it is often used in the forecasting industry. A new prediction model was constructed in this paper. The multivariate function is used to calculate the predicted value of future natural gas consumption, which can only be calculated when the relevant independent variables are known. The significance test shows that the grey prediction is significantly better than other prediction models in the prediction accuracy of basic data, so the grey prediction model is used in this paper to predict the independent variables. The accuracy test level of GM (1,1) model can refer to Table 1:

Table 1. Reference table of model accuracy test grades.

Prediction accuracy level	Mean relative error (φ_i)	Mean square error ratio C	Small error probability P
Level 1 (good)	<0.01	<0.35	>0.95
Level 2 (qualified)	<0.05	<0.50	>0.80
Level 3 (barely qualified)	<0.1	<0.65	>0.70
Level 4 (unqualified)	>0.1	>0.65	<0.7

3 EXAMPLE ANALYSIS

3.1 *Selection and collection of sample data*

Through the research and comprehensive analysis of the current historical literature on natural gas demand forecasting, five aspects of the high frequency of natural gas demand forecasting, namely GDP, population development level, urbanization rate, residential consumption level and the proportion of the secondary industry structure, are adopted to carry out the direction analysis of influencing factors. Due to the requirement of statistical goodness of fit on the amount of data, it is more appropriate to base on the data of a decade. This paper collects all data from 2012 to 2021. The indicators of natural gas consumption and its influencing factors in Zhejiang province come from the data published by the National Bureau of Statistics, the Statistical Yearbook of Zhejiang province, the Statistical Bulletin of Zhejiang province, and the official website of Zhejiang Development and Reform Commission.

3.2 *Construction of a combined prediction model based on multi-function regression optimization*

According to the calculation in Table 2 by grey relative correlation degree, the grey relative correlation coefficients between natural gas consumption and influencing factors are as follows: GDP, household consumption level, urban population, total population, urbanization rate, and proportion of secondary industry structure. The gray relative correlation between the total population and the urbanization rate is greater than 0.9. Therefore, the five influencing factors can be finally determined by eliminating the smaller influencing factor of the urbanization rate and retaining the larger influencing factor of the total population.

When the five influencing factors determined by the grey relative correlation coefficient were taken as the independent variables and the natural gas consumption as the dependent variable, the polynomial fitting modeling was carried out respectively to determine the control variables, and the polynomial model of every single factor was established. The optimal weight is calculated based on the coefficient optimization principle of linear programming, and a new prediction model is constructed. The secondary precision grade in Table 1 was taken as a constraint, namely $\geq 5\%$ significance level as the standard. After repeated iteration between gas consumption and various single factor regression coefficient of the polynomial model, the optimization calculation of the optimal weights, and eliminating the influence of no significant factors, the final iteration results showed that five

Table 2. Gray relative correlation degree matrix table between variables.

	Natural gas consumption	GDP	The total population	Urbanization rate	Residents' consumption level	The proportion of secondary industry structure	The urban population
GDP	0.776	1					
The total population	0.603	0.649	1				
Urbanization rate	0.602	0.648	0.977	1			
Residents' consumption level	0.717	0.89	0.597	0.605	1		
The proportion of secondary industry structure	0.532	0.534	0.624	0.632	0.533	1	
The urban population	0.641	0.713	0.765	0.783	0.755	0.688	1

factors were included. The fitting error calculation results of the new prediction model, time series model, grey prediction model, and nonlinear prediction model in this paper are shown in Table 3:

Table 3. Table of fitting values and error results of natural gas consumption in Zhejiang province.

Year	New forecasting model	Grey prediction	The time series	Nonlinear prediction model
Error	3.85%	4.86%	22.7%	6.25%

According to the grey relative relational grade, the GDP is the main influencing factor of gas consumption in Zhejiang province, so stimulating the growth of GDP may stimulate the growth of natural gas consumption. The new prediction model in this paper has a good prediction effect on natural gas consumption, and it can be used as a prediction model for natural gas consumption in other regions.

3.3 *Forecast analysis of future natural gas demand in Zhejiang province*

The established new prediction model is used to predict the natural gas demand in Zhejiang province from 2022 to 2031, and the result is shown in Figure 1. With the clean, low-carbon, and high-quality development of energy in China, new requirements are constantly put forward for energy strategy and environmental policy. In order to realize a carbon emission peak within a decade and strive to become carbon neutral after 30 years of green vision, the Zhejiang region will be effectively pushed to realize energy reform, thus promoting the rigid, steady, and sustained growth of natural gas consumption demand [7]. From 2022 to 2031, the demand for natural gas in Zhejiang shows an upward trend, while the growth rate of natural gas demand shows a slow downward trend due to the different urbanization process, economic development, population, and other conditions, but the growth rate still maintains at nearly 10%. The annual growth of natural gas consumption increases from $1366 \times 10^6 \text{ m}^3$ in 2022 to $3146 \times 10^6 \text{ m}^3$ in 2031. The natural gas demand in the Zhejiang region increased by 136.9% compared with 2021, and the net growth of demand is about $241 \times 10^8 \text{ m}^3$.

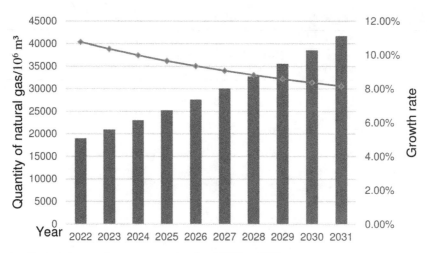

Figure 1. Projected natural gas consumption from 2022 to 2031.

4 CONCLUSION

1) The factors affecting the consumption of natural gas in Zhejiang province are GDP, residential consumption level, urban population, total population, and the proportion of secondary industry structure. GDP is the most important factor affecting the consumption of natural gas in Zhejiang province, and its importance far exceeds other factors. To effectively increase the consumption of natural gas and expand the consumption space of the natural gas market in Zhejiang province, the GDP of Zhejiang province needs to be taken into consideration, and the development and investment of Zhejiang county economy, private enterprises, and enterprise energy transformation must be accelerated.
2) The new prediction model constructed in this paper is superior to the traditional prediction models because it can make full use of the data features to reduce the error effectively, and the fitting result is more accurate.
3) In the next ten years, the natural gas consumption in Zhejiang province will increase from $18,965 \times 10^6$ m^3 in 2022 to $41,685 \times 10^6$ m^3 in 2031. By 2031, the natural gas demand in Zhejiang province will increase by 136.9% compared with 2021. The annual growth of natural gas consumption will increase from 1366×10^6 m^3 in 2022 to 3146×10^6 m^3 in 2031.
4) This indicates that the future natural gas market in Zhejiang province has great potential. On the one hand, we should accelerate the consumption of natural gas; on the other hand, we should develop a network covering transportation to ensure the healthy development of the natural gas industry in Zhejiang province.

ACKNOWLEDGMENT

This work was supported by the National Undergraduate Science and Technology Innovation Project (202110340035) and the Zhoushan Science and Technology Plan Project (2020C21011), both of which are gratefully acknowledged.

REFERENCES

[1] China Natural Gas Development Report (2021). *Publicity and Education Center of the Ministry of Natural Resources [EB/OL]*. [2022-09-25].

[2] Chai Jian, Wang Yaru, Kin Keung-lai Analysis and Forecast of China's Natural Gas Consumption under the "New Normal". *Operations Research and Management Science*, 2019.

[3] Li Hong-bing, Zhang Ji-jun. Analysis of Influencing Factors of Natural Gas Demand and Forecast of Future Demand. *Operations Research and Management Science*, 2021, 30(09): 132–138.

[4] Sharma V., Cali Ü., Sardana B., *et al.* Data-driven Short-term Natural Gas Demand Forecasting with Machine Learning Techniques [J/OL]. *Journal of Petroleum Science and Engineering*, 2021, 206:108979. DOI: 10.1016/j.petrol.2021.108979.

[5] Dongkun Luo, Ping Xu. Natural Gas Demand Forecasting Based on Improved BP Neural Network. *Oil-Gas Field Surface Engineering*, 2008(07): 20–21.

[6] Hongkun Wu, Yuanqi Ji. Natural Gas Forecasting System Based on Time Series Models. *China CIO News*, 2019(11): 82–84+87.

[7] Towards Clean and Low-carbon—A Review of China's Energy Development Achievements_Rolling News_Chinese Government Website [EB/OL]. [2022-09-25]. http://www.gov.cn/xinwen/2021-06/10/content_5616827.htm.

Research on inversion of surface temperature in Hefei City

Kunmeng Zhou
School of Environmental Science and Optoelectronic Technology, University of Science and Technology of China, Hefei, China
Key Laboratory of Optical Calibration and Characterization, Anhui Institute of Optical and Fine Mechanics, Hefei Institutes of Physical Science, Chinese Academy of Sciences, Hefei, China

Xiaobing Zheng*
Key Laboratory of Optical Calibration and Characterization, Anhui Institute of Optical and Fine Mechanics, Hefei Institutes of Physical Science, Chinese Academy of Sciences, Hefei, China

Lan Xiao
Key Laboratory of Optical Calibration and Characterization, Anhui Institute of Optical and Fine Mechanics, Hefei Institutes of Physical Science, Chinese Academy of Sciences, Hefei, China
University of Science and Technology of China, Hefei, China

Qing Kang
Key Laboratory of Optical Calibration and Characterization, Anhui Institute of Optical and Fine Mechanics, Hefei Institutes of Physical Science, Chinese Academy of Sciences, Hefei, China

ABSTRACT: Surface temperature is one of the important indicators for the evaluation of ecological and environmental quality. Remote sensing technology is comprehensive, informative, and widely covered, which provides a powerful means for temperature inversion and ecological environment quality evaluation at this stage. Based on the remote sensing image data of Landsat TM and Landsat OLI, this paper uses the atmospheric correction method model to invert the surface temperature of the whole city of Hefei from 2000 to 2020. Regional differences in temperature and annual variations were analyzed.

1 INTRODUCTION

LST (Land Surface Temperature), the surface temperature, is one of the important parameters of energy absorption and emission among the earth's substances, as well as the circulation between water and gas [1,2]; when the earth's surface energy reaches a state of equilibrium [3,4], the relevant spatiotemporal information can be obtained through the surface temperature. With the development and progress of remote sensing technology, a large number of surface temperature inversion algorithms have appeared, making surface temperature to be one of the more important parameters in the field of scientific research. It has extremely important applications in the evaluation of urban ecological environment quality, urban heat island effect, forest fire monitoring, drought monitoring, and other aspects. In recent years, with the development of remote sensing technology, the relationship between surface temperature and land cover types has been studied [5], applied to the spatiotemporal changes of the thermal environment [6], and combined with precipitation for the

---

*Corresponding Author: xbzheng@aiofm.ac.cn

Figure 1. Hefei.

monitoring of drought and flood disasters [7] and the monitoring of cold disasters [8]; it is one of the important indicators of the evaluation of ecological environment quality [9].

The inversion of the surface temperature is mainly based on the surface radiation observed by the thermal infrared sensor of the satellite sensitive to thermal radiation. Therefore, this paper uses the Landsat TM 5 and Landsat OLI 8 data to invert the surface temperature, thereby qualitatively obtaining the ecological environment changes in Hefei City in recent years. The data were adopted from April 8 of 2000, April 19 of 2004, April 23 of 2011, April 23 of 2017, and April 15 of 2020. The data are all in April, which makes the inversion results more comparable.

2 BACKGROUND AND DATA

2.1 Background of the study area

Hefei City is located in the central part of Anhui Province, at 30°57′~32°32′ north latitude and 116°41′~117°58′ east longitude, with an average altitude of 20 to 40m. Located in the middle latitude, Hefei has a subtropical monsoon climate, four distinctive seasons, a mild climate, and moderate rainfall. The average annual temperature is 15.7°, the average annual precipitation is about 1000 mm, the annual sunshine time is about 2000 h, the average annual frost-free period is 228 days, and the average relative humidity is 77%. The city has four districts, four counties, and one county-level city, with a total area of 11445.1 km^2 and a built-up area of 476.5 km^2. The research area includes the entire scope of Hefei City and the research area is shown in Figure 1.

2.2 Data sources

The data are downloaded from the geospatial data cloud using the Landsat series of satellites; the data are selected for five years from 2000 to 2020, namely: April 8 of 2000, April 19 of 2004, April 23 of 2011, April 23 of 2017, April 15 of 2010; the data for the first three years are Landsat TM data, and the data for the last two years are Landsat OLI data.

The vector map of the administrative boundary of Hefei City is downloaded from the national geographic information resources directory service system, and the remote sensing data obtained by the download is cropped in the area of interest.

2.3 *Data preprocessing*

First of all, the vector diagram obtained by the Arcgis software is fused and saved into SHP format to facilitate subsequent operation. Then the ENVI software is used to use vector graphics to crop the remote sensing data. The cropped data is then radiated, and the DN value is converted into the spectral emissivity of surface features. The atmospheric correction of the remote sensing image after radiation calibration is then performed.

3 SURFACE TEMPERATURE INVERSION THEORY

3.1 *The basic theory of infrared radiation transmission*

According to Planck's law, any object with an absolute temperature greater than 0 K radiates energy outward in the form of electromagnetic waves [9]. The radiation energy of a blackbody in thermal equilibrium at temperature T and wavelength λ can be expressed by Planck's law:

$$B\lambda_{(T)} = \frac{C_1}{\lambda^5[\exp(C_2/T_\lambda - 1]}$$ (1)

where $B_\lambda(T)$ is the spectral radiance ($W \cdot \mu m^{-1} \cdot Sr^{-1} \cdot m^{-2}$) of a blackbody at temperature $T(K)$ and wavelength $\lambda(\mu m)$; C_1 and C_2 are physical constants ($C_1 = 1.191 \times 108$ $W \cdot \mu m^{-4} \cdot Sr^{-1} \cdot m^{-2}$, $C_2 = 1.439 \times 104$ $\mu m \cdot K$).

Since most natural objects are non-black bodies, the surface emissivity ε can be defined as the ratio between the actual thermal radiation of the objects and the blackbody thermal radiation at the same temperature and the same wavelength.

The wavelength λ_{max} means the maximum value of the monochromatic radiation intensity when a blackbody is at a specific temperature $T(K)$, and it can be expressed by Wien's displacement law:

$$T\lambda_{max} = 2897.9\mu m$$ (2)

According to this property, the temperature of the surface is about 250 to 330 K, so the peak wavelength λ_{max} is mainly in the thermal infrared wavelength range (8.8-11.6 μm).

With the conditions of a clear sky and no cloud in local thermal equilibrium, according to the radiation transfer equation, the radiance I_i received by the sensor at the top of the atmosphere can be expressed as:

$$I_i(\theta, \varphi) = R_i(\theta, \varphi)\tau i(\theta, \varphi) + R_{ati\uparrow}(\theta, \varphi) + R_{sli\uparrow}(\theta, \varphi)$$ (3)

where the surface radiance R_i can be expressed as

$$R_i(\theta, \varphi) = \varepsilon_i(\theta, \varphi)B_i(Ts) + [1 - \varepsilon_i(\theta, \varphi)]R_{ati\downarrow} + [1 - \varepsilon_i(\theta, \varphi)]R_{sli\uparrow} + \rho_{bi}(\theta, \varphi, \theta_s, \varphi_s)$$

$$\times E_i\cos(\theta s)\tau_i(\theta_s, \varphi_s)$$ (4)

The atmospheric radiation transmission is shown in Figure 2 [10]:
Path ① represents the near-surface radiance attenuated by the atmosphere, $R_i\tau_i$
Path ② represents the upward thermal radiation of the atmosphere, $R_{ati\uparrow}$
Path ③ represents the upward solar radiation scattered by the atmosphere, $R_{sli\uparrow}$

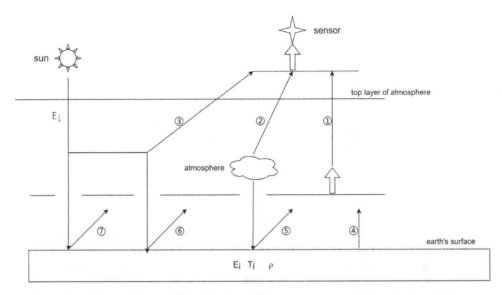

Figure 2. Transform of atmospheric radiation.

Path ④ represents the radiation emitted by the surface itself, $\varepsilon_i B_i(T_s)$

$R_{ati\downarrow}$ represents the descending thermal radiation of the atmosphere, and $R_{sli\downarrow}$ represents the descending solar radiation scattered by the atmosphere;

Path ⑤ represents the descending atmospheric thermal radiation reflected by the surface, $(1 - \varepsilon_i)R_{ati\downarrow}$

Path ⑥ represents the downward atmospheric scattered solar radiation reflected by the surface, $(1-\varepsilon_i)R_{sli\downarrow}$

Path ⑦ represents the direct solar radiation reflected by the surface, namely $\rho_{bi} E_i \cos(\theta_s)\tau_i(\theta_s, \varphi_s)$

ρ_{bi} is the bidirectional reflectance of the surface; E_i is the solar irradiance at the top of the atmosphere; θ_s and φ_s are the solar zenith and azimuth angles, respectively.

Since the solar radiation at the top of the atmosphere is negligible in the daytime and nighttime, when they are in the 8-14 μm spectral region and the nighttime data in the 3-5 μm spectral region, the part about solar radiation in equations (both paths ③⑥ and ⑦) can be ignored without affecting the accuracy.

3.2 *Surface temperature inversion based on atmospheric correction method*

The basic principle of the atmospheric correction method: the thermal infrared radiation value L_λ received by the satellite sensor consists of three parts: the upward radiance of the atmosphere $L\uparrow$, the real radiance of the ground reaching the satellite sensor after passing through the atmosphere, and the energy reflected from the downward radiation of the atmosphere to the ground [11].

$$L_\lambda = [\varepsilon_\lambda B(T_s) + (1 - \varepsilon_\lambda)L\downarrow]\tau_\lambda + L\uparrow \tag{5}$$

where λ is the wavelength, ε is the surface-specific emissivity, T_s is the real surface temperature, $B(T_s)$ is the blackbody radiance, and ε_λ is the atmosphere transmittance at the wavelength λ. The radiance $B(T_s)$ in the infrared band is:

$$B(T_S) = \frac{[L_\lambda - l\uparrow - \tau_\lambda(1-\varepsilon_\lambda)L\downarrow]}{\tau_\lambda \varepsilon_\lambda} \tag{6}$$

T_s can be obtained as a function of Planck's formula:

$$T_S = K_2/\ln[K_1/B(T_s) + 1] \tag{7}$$

For TM data, K_1 = 607.76 W/(m^2 · μm · sr), K_2 = 1260.56 K;
For TIRS Band10 data, K_1=774.89 W/(m^2·μm·sr), K_2=1321.08 K.

Therefore, using the atmospheric correction method to invert the surface temperature, it is necessary to obtain the following data: the radiance value L_λ of the thermal infrared band, the upward radiance of the atmosphere L↑, the downward radiance of the atmosphere L↓, the atmospheric transmittance ε_λ, the surface emissivity ε.

3.3 *Atmospheric transmittance*

Atmospheric transmittance has a very important influence on the conduction of surface thermal radiation in the atmosphere, so it is a basic parameter for getting a surface temperature. Both the atmospheric correction method and the single-window method require extremely accurate atmospheric transmittance to ensure the accuracy of the results [12]. There are many factors affecting the atmospheric transmittance, such as air pressure, temperature, aerosol content, atmospheric moisture content, O_3, CO_2, CO, NH_4, etc., which all have different effects on the thermal radiation conduction so that the conduction of surface thermal radiation in the atmosphere will be reduced.

Studies have shown that the change in atmospheric transmittance mainly depends on the dynamic change of atmospheric moisture content. Other factors have little dynamic change and have no significant effect on the change of atmospheric transmittance. Therefore, moisture content has become the main consideration in the estimation of atmospheric transmittance.

Two situations are considered: summer and winter. Assuming that the air temperature near the ground is 35°C in summer and 18°C in winter [13], the atmospheric transmittance does not decrease linearly with the increase of moisture content, but in the smaller moisture content range; its variation relationship can be considered to be close to linear, for the moisture content in the range of 0.4 ~ 3.0 g/cm^2; atmospheric transmittance estimation equation is shown in Table 1.

Table 1. Atmospheric transmittance estimation equation.

Atmospheric profile	Moisture content	Atmospheric transmittance	Correlation coefficient	Standard error
High temperature	0.4–1.6	τ=0.974290-0.08001 w	0.99611	0.002368
	1.6–3.0	τ=1.031412-0.11536 w	0.99827	0.002539
Low temperature	0.4–1.6	τ=0.982007-0.09611 w	0.99463	0.00334
	1.6–3.0	τ=1.053710-0.14142 w	0.99899	0.002375

3.4 *Surface reflectance*

Before obtaining the surface reflectance, two parameters must be obtained: NDVI and vegetation coverage P_v [14].

The NDVI can be obtained by the following formula:

$$\frac{NIR - R}{NIR + R} \tag{8}$$

where NIR is the near-infrared band, and R is the red band.

The vegetation coverage P_v can be obtained by the following formula:

$$\frac{NDVI - NDVI_{min}}{NDVI_{max} - NDVI_{min}} \tag{9}$$

where $NDVI_{min}$ and $NDVI_{max}$ can be obtained by histogram statistics of the ENVI software, and the value of 3% will be taken as $NDVI_{min}$, and the value of 97% will be taken as $NDVI_{max}$.

3.4.1 *NDVI threshold method*
The NDVI threshold method proposed by Sobrino [15]:

$$\varepsilon = 0.004P_v + 0.986 \tag{10}$$

3.4.2 *Vegetation index method*
Based on the high correlation between NDVI and surface emissivity, a theoretical model was established by Botswana:

$$\varepsilon = 1.0094 + 0.047 * \ln(NDVI) \tag{11}$$

where the empirical formula is summarized based on the Botswana steppe, so the applicable range of NDVI is 0.157-0.727

3.4.3 *Vegetation mixture model*
The main idea of the method is to assume that we know the surface emissivity of vegetation and bare soil. They are 0.985 and 0.96. When NDVI<0.1, it is considered as a pure bare soil pixel; when NDVI>0.72, it is considered as a pure vegetation pixel; when NDVI is between 0.1 and 0.72, it is considered as a mixed pixel, and the following formula is used to calculate:

$$\varepsilon = \varepsilon_v P_v + \varepsilon_s(1 - P_v) + d\varepsilon \tag{12}$$

where $d\varepsilon$ is the emissivity ratio considering the cavity effect of the rough surface, and the calculation formula of the emissivity ratio $d\varepsilon$ is as follows:

$$d\varepsilon = 4 < d\varepsilon > (1 - P_v)P_v \tag{13}$$

where $<d\varepsilon>$ is the geometric shape factor, which can take the mean value of 0.015 according to different geometric distributions.

3.4.4 *NDVI^{TEM}*
Based on previous studies, Sobrino improved the mixed vegetation model. The main idea is: when $NDVI < 0.2$, it is considered that the surface coverage is sparse, and the surface emissivity is 0.98-0.042 ρ, where ρ is the reflectivity of ground objects in the infrared band; when $NDVI > 0.5$, it is considered that the surface cover is in good condition, and the surface emissivity is constant 0.99; when the NDVI is between 0.2 and 0.5, it is considered to be a mixed area, which is calculated by the following formula:

$$\varepsilon = \varepsilon_v P_v + \varepsilon_s(1 - P_v) + d\varepsilon \tag{13}$$

The calculation of vegetation coverage P_v is as follows:

$$P_v = \left[\frac{NDVI - NDVI_{min}}{NDVI_{max} - NDVI_{min}} \right]^2 \tag{14}$$

$$d\varepsilon = (1 - \varepsilon_s)(1 - P_v)F\varepsilon_v \tag{15}$$

where F is a morphological parameter with a value of 0.55.

3.4.5 *Improved NDVITEM algorithm*

Tan Zhihao considered the value of dε under different surface morphology and the problem of urban pixel emissivity estimation. This method considers that: for water body pixels, ε_w = 0.995, it is generally used to estimate; when $P_V \leq 0.5$, dε = 0.003796P_V; when $P_V > 0.5$, dε = 0.003796 (1-P_V). The emissivity estimation formula is as follows:

The emissivity of natural surface cells is calculated as follows:

$$\varepsilon = P_v R_v \varepsilon_v + (1 - P_v) R_s \varepsilon_s + d\varepsilon \tag{16}$$

The urban pixel emissivity is calculated as follows:

$$\varepsilon = P_v R_v \varepsilon_v + (1 - P_v) R_m \varepsilon_m + d\varepsilon \tag{17}$$

where ε_V, ε_S, and ε_m are the emissivity of pure vegetation, pure bare soil and pure building surface pixels, and ε_V = 0.986, ε_S = 0.972, ε_m = 0.970. R_V, R_S, and R_m are the temperature ratios of pure vegetation, pure bare soil and pure building surface pixels, which are related to the vegetation coverage, and the calculation formulas are as follows:

$$R_v = 0.9332 + 0.0585P_v \tag{18}$$

$$R_s = 0.9902 + 0.106P_v \tag{19}$$

$$R_m = 0.9886 + 0.1287P_v \tag{20}$$

4 SURFACE-SPECIFIC EMISSIVITY RESULTS UNDER DIFFERENT CALCULATION MODELS

Since the vegetation index method is not suitable for Hefei City, which contains a large quantity of water, I decided to abandon this method; at the same time, between the NDVI^{TEM} and NDVI^{TEM} improved methods, I chose the improved method, and finally I selected the threshold method, the vegetation mixture method and the NDVI^{TEM} improved method as calculation models for surface emissivity, taking the 2000 data as an example [16]. The calculation results of the three models are shown in Figure 3.

5 LAND SURFACE TEMPERATURE RETRIEVAL BASED ON ATMOSPHERIC CORRECTION METHOD

The results of surface emissivity obtained by the above different models are used to carry out the land surface temperature retrieval in Hefei City. If the calculation results are similar, it means that the methods of calculating land surface temperature retrieval are correct [17]. It can be used for temperature analysis in Hefei and other scientific research related to temperature.

5.1 *Surface temperature inversion based on atmospheric correction method based on different emissivity*

The following results are about the land surface temperature on April 8, 2000, obtained based on the above-mentioned different emissivity models, as shown in Figure 4.

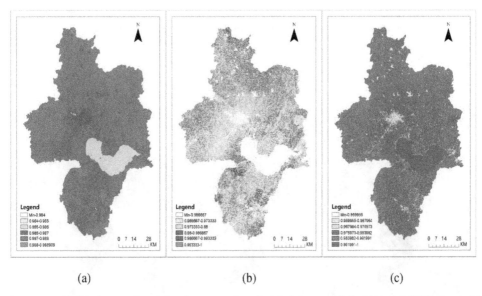

Figure 3. Calculation results of the above-mentioned emissivity models in 2000. (a)NDVI threshold method (b)Vegetation mixed model (c) Tan Zhihao's improved method.

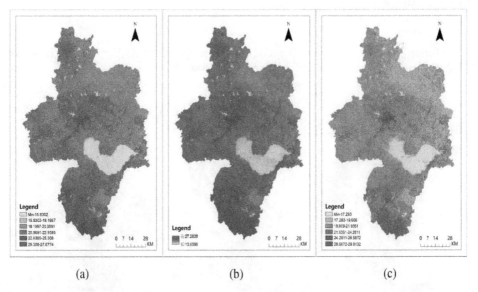

Figure 4. The year 2000. (a) Threshold method for ground temperature inversion (b) Vegetation mixture method for ground temperature inversion (c) NDVI improvement method for ground temperature inversion.

The following results are about the surface temperature on April 19, 2004, obtained based on the above-mentioned different emissivity models, as shown in Figure 5.

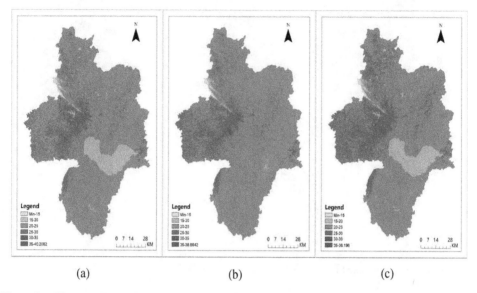

Figure 5. The year 2004. (a) Threshold method for ground temperature inversion (b) Vegetation mixture method for ground temperature inversion (c) NDVI improvement method for ground temperature inversion.

The following results are about the land surface temperature on April 23, 2011, obtained based on the above-mentioned different emissivity models, as shown in Figure 6.

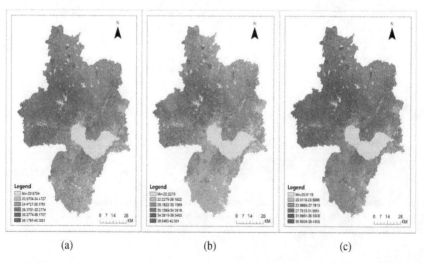

Figure 6. The year 2011. (a) Threshold method for ground temperature inversion (b) Vegetation mixture method for ground temperature inversion (c) NDVI improvement method for ground temperature inversion.

The following results are about the land surface temperature on April 23, 2017, obtained based on the above-mentioned different emissivity models, as shown in Figure 7.

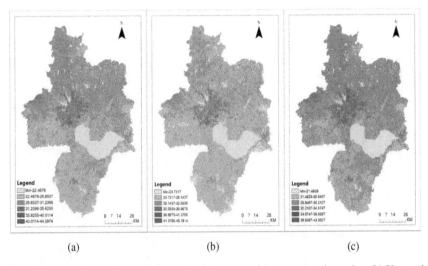

Figure 7. The year 2017. (a) Threshold method for ground temperature inversion (b) Vegetation mixture method for ground temperature inversion (c) NDVI improvement method for ground temperature inversion.

The following results are about the land surface temperature on April 15, 2020, obtained based on the above-mentioned different emissivity models, as shown in Figure 8.

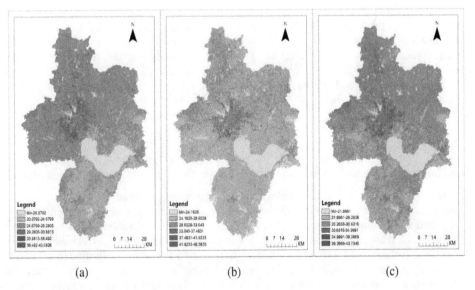

Figure 8. The year 2020. (a) Threshold method for ground temperature inversion (b) Vegetation mixture method for ground temperature inversion (c) NDVI improvement method for ground temperature inversion.

5.2 Geothermal inversion data

In order to remove the error value in the inversion process, including the maximum value and the minimum value, the confidence interval of the inversion result is taken from 3% to 97% according to the results of the histogram statistics. The statistical results are shown in Figure 9:

Methods / Year	2000			2004			2011		
	Maximum	Minimum	Average	Maximum	Minimum	Average	Maximum	Minimum	Average
Threshold method (°C)	23.2	14.45	20.28	25.79	15.52	23.29	31.95	17.86	26.65
Vegetation mixing method (°C)	23.3	16.4	20.41	26.2	19.73	23.57	32.33	19.53	27.07
NDVI improvement method (°C)	22.76	14.94	20.05	26.11	17.74	23.55	32.2	17.33	26.94

	2017			2020					
	Maximum	Minimum	Average	Maximum	Minimum	Average			
Threshold method (°C)	30.25	18.74	26.09	29.7	16.15	25.33			
Vegetation mixing method (°C)	31.06	20.4	26.55	30.7	17.7	25.84			
NDVI improvement method (°C)	30.77	18.18	26.38	30.15	15.64	25.6			

Figure 9. Results.

The emissivity is obtained from the above three models: the threshold method, the vegetation mixing method, and the NDVI improvement method proposed by Tan Zhihao, and then the land surface temperature inversion is performed by the atmospheric correction method. The results are slightly different. The order of inversion results of the five-year data selected from 2011 to 2020 is the same, from high to low: 2011, 2017, 2020, 2004, 2000; from this, we can get the trend of the land surface temperature of Hefei during the past two decades: it rose first and then fell.

6 CONCLUSION

Based on the results and discussions presented above, the conclusions are obtained as below:

1. After consulting the local meteorological data, the temperature of the day obtained from the selected data is compared with the results of the inversion, and the temperature error is found to be about 2°C, and the surface temperature is consistent with the changing trend of the real temperature, indicating that the two have a good correlation.
2. The land surface temperature obtained by different surface emissivity models reflects the trend of temperature in Hefei rising first and then falling; and from the surface temperature distribution map, we can see that the surface temperature in Hefei from the Shushan District, Yaohai District, Luyang District, and other urban areas show a radial decrease to Feidong, Chaohu, and other suburbs.

3. The reasons for the increase in urban temperature in Hefei City are mainly that its urban area belongs to the industrial concentration area, and the impact caused by industrial activities, such as the heat island effect, has a certain positive correlation with the increase in surface temperature. Therefore, the temperature in the urban area is higher than that in the surrounding suburbs.
4. The reason for the temperature rise in Hefei City, the strategic deployment of an "Industrial City" in August 2005, which began to build a series of industrial activities such as a modern industrial base, and the development of industrial production will inevitably bring about an increase in the temperature of the industrial zone; the reason for the downward trend in temperature since 2011 is mainly that Hefei City has intensified its environmental governance in recent years, achieving a record of continuous decline in inhalable particulate matter from 2013 to 2020, the improvement of Chaohu Lake water quality, and the establishment of national forest park, and using technology to create a unified ecological environment monitoring system in the city.

ACKNOWLEDGMENTS

This work was financially supported by the National Key Research and Development Program of China (2018YFB0504601) and the Natural Science Foundation of Anhui Province of China (2008085QA36).

REFERENCES

[1]   Voogt J.A and Oke T.R. (2003). Thermal Remote Sensing of Urban Climates. *Remote Sensing of Environment*, 86(3), pp. 370–384.

[2]   José A. Sobrino and Juan C. Jiménez-Muñoz and Leonardo Paolini (2004). Land Surface Temperature Retrieval from LANDSAT TM 5. *Remote Sensing of Environment*, 90(4), pp. 434–440.

[3]   José A. Sobrino and Juan C. Jiménez-Muñoz. (2014). Minimum Configuration of Thermal Infrared Bands for Land Surface Temperature and Emissivity Estimation in the Context of Potential Future Missions. *Remote Sensing of Environment*, 148 pp. 158–167.

[4]   Andrzej Urbanski Jacek et al. (2016). Application of Landsat 8 imagery to Regional-scale Assessment of Lake Water Quality. *International Journal of Applied Earth Observations and Geoinformation*, 51 pp. 28–36.

[5]   Ding Yuanyuan, Zhao Jianyun, Yang Jing, Zhao QinHao, Wang zushun & Zhao Lijiang (2022). *Spatial and Temporal Variation Characteristics of Surface Temperature and Land Cover in Yongqu River Basin, the Source of the Yellow River Science, Technology and Engineering* (07), 2592–2600

[6]   Guo Yu (2021). *Study on the Spatiotemporal Variation Characteristics and Driving Mechanism of Thermal Environment in Shanghai* (Master's thesis, Shanghai University of Applied Technology)

[7]   Tian Miao & Wang Pengxin (2016). Precipitation Temperature Index and its Application in Drought and Flood Disaster Monitoring Jiangsu Agricultural Journal (04), 810–816

[8]   Wang Chunlin, Tang Lisheng, Chen shuisen, Huang Zhenzhu & He Jian (2006). Research on All-weather Surface Temperature Inversion Method in Cold Disaster Monitoring (eds.) *Proceedings of the "Progress and Application of Satellite Remote Sensing Technology"* Branch of the 2006 Annual Meeting of the Chinese Meteorological Society (pp.175–186)

[9]   Xu Hanqiu (2013). *Remote Sensing Evaluation Index of Regional Ecological Environment Change China Environmental Science* (05), 889–897

[10] Sugita M. and Brutsaert W. (1993). Comparison of Land Surface Temperatures Derived From Satellite Observations with Ground Truth During FIFE. *International Journal of Remote Sensing*, 14(9), pp. 1659–1676.

[11] César Coll et al. (2005). Ground Measurements for the Validation of Land Surface Temperatures Derived From AATSR and MODIS Data. *Remote Sensing of Environment*, 97(3), pp. 288–300.

[12] Tan Zhihao, Zhang Minghua, Arnon karnieli, Pedro berliner (2001). A Single Window Algorithm for Calculating Land Surface Temperature Using Landsat TM6 Data. *Journal of Geography* (04), 456–466

[13] Wang Lixia, Li Jiangang & Zhao Xuan (2015). Research on the Relationship Between Climate Change and Air Pollution in Lanzhou City. *Journal of Tianshui Normal University* (05), 8–11.

[14] Dash P. et al. (2002). Land Surface Temperature and Emissivity Estimation From Passive Sensor Data: Theory and Practice-current Trends. *International Journal of Remote Sensing*, 23(13), pp. 2563–2594.

[15] Liu Fei, Wang Xinsheng, Xu Jing & Gao Shoujie (2012). *Parameter Sensitivity Analysis of Retrieving Surface Specific Emissivity Based on NDVI Threshold Method Remote sensing information* (04), 3–12

[16] Guan Yujie, Liu Shoudong & Cao Chang (2018). Study on the Change of Urban Temperature with Different Urbanization Degrees: Taking Fuzhou and Zhangzhou as examples. *Chinese Journal of Tropical Meteorology* (04), 554–560.

[17] Zhao Huifang & Cao Xiaoyun (2022). Temporal and Spatial Variation of Vegetation Cover in Sanjiangyuan National Park and its Climate Driving Factors. *Plateau Meteorology* (02), 328–337.

Civil Engineering and Energy-Environment – Gao & Duan (Eds)
© 2023 the Author(s), ISBN 978-1-032-56059-5

The engineering geological suitability assessment of Xiaolongtan lignite deposit in Yunnan Province, China

Shuran Yang*
Yunnan Land and Resources Vocational College, Kunming, China
Technical University of Ostrava, Ostrava, Czech Republic

Qianrui Huang
Yunnan Land and Resources Vocational College, Kunming, China

ABSTRACT: Through the investigation of the engineering geological condition, 13 disaster-caused factors of the research field were collected, including the stability of slopes, potential flood, water body, spring, aquifer, aquiclude, gully, lignite spontaneous combustion, building, dump, quarry, road, and step. This paper adopts ArcGIS to carry out the coverage partition of disaster-caused factors in the exposed area, uses an expert-analytic hierarchy processing method to determine the index weight of 13 disaster-caused factors, and finally exhibits the engineering geological comprehensive assessment. According to the evaluation results, the unsuitable area occupies 18.7% of the total research field, the conditionally suitable area occupies 0.9%, the suitable area occupies 32.7%, and the no-disaster area occupies 47.7%.

1 INTRODUCTION

Suitability assessment plays an important role in land reclamation, and other further land uses in the open-pit mine area. Moosavirad et al. [1] divided the comprehensive evaluation condition into four units: open-pit slope, open-pit bottom, dump slope, and the land for roads, and the results showed that the method is more applicable and easier to handle.

The studies of the Xiaolongtan Lignite Deposit mine mainly focused on geological structure [2], formation lithology [3], the liquefaction properties of lignite [4], the mining technique, the influence of vibration load on rock mass [5], and the stability of west slope of Buzhaoba open-pit No.1 [6]. As for engineering geology, the comprehensive research on the suitability assessment of Xiaolongtan Lignite is limited.

The studied locality (GPS: E 103°11′52″, N23°48′45″) is the Xiaolongtan Lignite Deposit in Xiaolongtan Town, which locates beside Gejiu City in Yunnan Province, southwest of China, where two enormous open pits (No1: Buzhaoba open-pit and No2: Xiaolongtan open-pit) could be seen in satellite images, as shown in Figure 1. The Lignite deposit belongs to the largest lignite mine in China. It annually produces 14.9 million tons of lignite, among which 13 million tons are from Buzhaoba open-pit No.1, and 1.9 million are from Xiaolongtan open-pit No.2.

Since Xiaolongtan Lignite Deposit was founded in 1953, due to the continuously expanding production scale and the increasingly growing mining boundary, the accelerated mining of lignite has formed two large open pits through several decades of mining. These

---

*Corresponding Author: yangshuran1988@foxmail.com

DOI: 10.1201/9781003433651-18

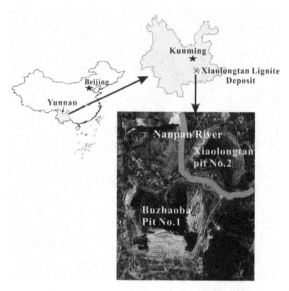

Figure 1. The location and basic information of Buzhaoba open-pit No.1 and Xiaolongtan open-pit No.2.

two pits are surrounded by dense buildings, including residential areas, a thermal power plant and a prison. The mining activity has caused the frequent occurrence of a series of geological hazards, including slope failure, soil failure, water hazard, and fire hazard. This paper takes the geographic information system ArcGIS as the platform to investigate the geological conditions for this research district and applies the comprehensive index model to perform the comprehensive evaluation of the engineering geological suitability for the research district. The evaluation results can provide a reference for the site selection and reclamation of the Xiaolongtan Lignite Deposit.

2 BASIC ENGINEERING GEOLOGICAL CONDITION

The designed exploitation capacity of studied open-pits is 3420×10^6 m^3 in Buzhaoba open-pit No.1 and 930×10^6 m^3 in Xiaolongtan open-pit No.2. Until now, the total extracted capacity of Buzhaoba open-pit No.1 has reached 1180×10^6 m^3, and Xiaolongtan open-pit No.2 has reached 370×10^6 m^3, so the reserved capacity of Buzhaoba open-pit No.1 is 2240×10^6 m^3, and that of Xiaolongtan open-pit No.2 is 560×10^6 m^3. The total lignite capacity of Buzhaoba open-pit No.1 is 730×10^6 m^3, and that of Xiaolongtan open-pit No.2 is 180×10^6 m^3. Until now, the extracted lignite capacity in Buzhaoba open-pit No.1 is 280×10^6 m^3, and that in Xiaolongtan open-pit No.2 is 107×10^6 m^3, so the reserved capacity of lignite of Buzhaoba open-pit No.1 is 450×10^6 m^3, and that of Xiaolongtan open-pit No.2 is 73×10^6 m^3.

Both two open-pits adopt the single bucket excavator as the stripping method and use automobile assisting with belt conveyor as the main transportation technology.

The waste rock excavated from Buzhaoba open-pit No.1 is externally dumped in Longqiao Waste Rock Dump and Xindenger Waste Rock Dump; the former one has located 1 km away from the west border of Buzhaoba open-pit No.1, and the latter one is located 3 km away from the southwest border of the pit. The designed capacity of the Longqiao Dump is 631×10^6 m^3, the designed elevation is from 1225 m to 1525 m, and now the height is reaching 1465 m. Meanwhile, the designed capacity of Xindenger Dump is 176.02×10^6 m^3, the designed elevation is from 1225 m to 1530 m, and now the height is reaching 1410 m. The waste rock excavated from Xiaolongtan open-pit No.2 is externally dumped in Beipingba Waste-rock Dump, which is located 1.4 km away from the north border of Xiaolongtan open-pit No.2. The designed capacity of Beipingba Dump is 53.09×10^6 m^3, the designed elevation is from 1140 m to 1290 m, and now the height is reaching 1275 m. The geological structure of the studied area is diverse and variable, as shown in the geological map in Figure 2; the hydrological condition is also shown in Figure 3.

Based on the geological structure of the studied area, 13 disaster-caused factors were collected from the Xiaolongtan Lignite Deposit, including the stability of slopes, potential flood, water body, spring, aquifer, aquiclude, gully, lignite spontaneous combustion, building, dump, quarry, road, and step.

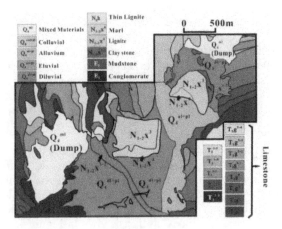

Figure 2. The geological map of the Xiaolongtan Lignite Deposit.

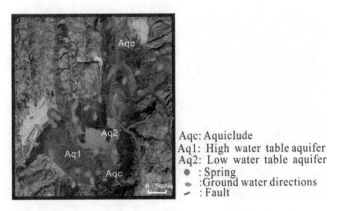

Figure 3. The diagram hydrological map of the Xiaolongtan Lignite Deposit.

3 ENGINEERING GEOLOGICAL SUITABILITY ASSESSMENT

3.1 *Methodology*

This paper applies the comprehensive index model to evaluate the engineering geological suitability for the research district, which mainly involves five comprehensive evaluation factors: slope stability, hydrogeology, fire disaster, and eco-environment. There are 13 evaluation factors in total: unstable slope, potential flood, water body, spring, aquifer, aquiclude, gully, lignite, building, dump, quarry, road, and step.

In terms of the comprehensive analysis of the evaluation index, it is of great importance to determine the hazard weight of the index. Currently, there are a number of weight determination methods such as expert scoring method, investigation & statistical method, analytic hierarchy processing method, and mathematical statistics. Among them, the analytic hierarchy processing method has always been regarded as one reasonable and feasible weight determination method. However, if there is no sufficient recognition of the geological environmental system, overconfidence in the mathematical model of determining weight will cause an unreasonable weight value on the contrary. On the basis of comprehensively understanding the geological background, the expert experience judgment is sometimes more reliable. According to the actual situation of the research district, this paper adopts the expert-analytic hierarchy processing method to determine the index weight, combining the advantages of the qualitative analysis of the expert experience method and the advantages of the quantitative analysis of the analytic hierarchy processing method. The specific methods are as follows.

Step 1: Evaluation factors are determined.
Through analyzing the interrelation, logical belongingness, and importance level among the factors that impact suitability, this thesis merges these factors into different layers so as to develop a multi-layer structure. That is, the engineering geological environmental suitability is taken as the target layer, the four evaluations factors, including slope stability, hydrogeology, fire disaster, and eco-environment, are taken as the intermediate layer, and the 13 evaluation factors are taken as the index layer, as shown in Figure 4.

Step 2: The hazard weight of the evaluation factor is determined.
According to the importance of disasters caused by the evaluation factors to the research district, upon applying the structural model in Figure 4 and expert scoring, this paper

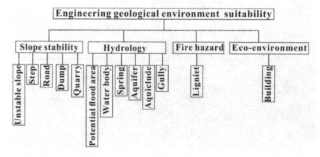

Figure 4. The diagram of the assessment structure model.

constructs the judgment m beloatrix for each layer, and the matrix formula is shown as follows:

$$A = \begin{bmatrix} a_{11} & \cdots & a_{1n} \\ \vdots & \ddots & \vdots \\ a_{n1} & \cdots & a_{nn} \end{bmatrix} \tag{1}$$

According to the judgment matrix, it is able to calculate the weight value for each index. First of all, the maximum characteristic root of judgment matrix λ_{max}, and its corresponding characteristic vector ω are obtained; then the normalization processing is performed on the obtained characteristic vector; the vector after normalization processing is the weight vector, which is the vector composed of the weight of evaluation index.

Then this thesis performs the consistency check on the weight value results. According to the consistency ratio (CR), it is able to evaluate and judge the consistency of the matrix. Usually if CR<0.1, it is able to believe that the judgment matrix A meets the requirements of the consistency check. If CR<0.1, it is required to reconstruct the judgment matrix until CR<0.1. The calculation method for consistency ratio (CR) is shown as follows:

$$CR = CI/RI \tag{2}$$

where CI is the consistency index; RI is the random consistency index, and their values are shown in Table 1.

Table 1. The value of the random consistency index (RI).

n	RI	n	RI
1	0	4	0.89
2	0	5	1.12
3	0.58	6	1.24

The calculation method for CI is shown as follows:

$$CI = (\lambda_{max} \cdot n)/(n \cdot 1) \tag{3}$$

where λ_{max} is the maximum characteristic root for judgment matrix A; n is the order for judgment matrix A.

The weight of the evaluation factor is determined. According to the above-mentioned method, after constructing the judgment matrix, it is able to progressively obtain the weight of the evaluation factor by adopting the analytic hierarchy process, and the final results are shown in Table 2.

Among the four evaluation factors, the hazard weight of slope stability is the highest, which is 0.434662. In it, the evaluation factor unstable slope occupies 0.171792, dump occupies 0.100492, step occupies 0.100492, road occupies 0.024192, and quarry occupies 0.037692. The second highest is hydrology, which is 0.378054. In it, the evaluation factor potential flood area occupies 0.169292, water body occupies 0.082192, spring occupies 0.067292, aquifer occupies 0.020592, aquiclude occupies 0.020592, and gully occupies 0.018092. The fire hazard ranks third, which is 0.169292. In it, there is only one evaluation factor lignite, and it occupies 0.169292. The eco-environment is the lowest one, which only occupies 0.0017992. In it, there is only one evaluation factor building, and it occupies 0.0017992.

Table 2. The weight, area, and area category of each assessment index.

Evaluation elements	Evaluation factors	The disaster happen or not	Hazard weights	Area /km²
Slope stability	Unstable slope	Yes	0.171792	2.69
	Dump	Yes	0.100492	5.02
	Step	Yes	0.100492	2.21
	Road	No	0.024192	1.83
	Quarry	No	0.037692	0.20
Hydrology	Potential flood	No	0.169292	9.05
	Waterbody	No	0.082192	0.73
	Spring	No	0.067292	0.18
	Aquifer	No	0.020592	19.55
	Aquiclude	No	0.020592	14.25
	Gully	No	0.018092	1.65
Fire hazard	Lignite	No	0.169292	3.65
Eco-environment	Building	No	0.017992	2.24

According to the hazard weight, all evaluation factors are ranked as follows: unstable slope>potential flood=lignite>step=dump>waterbody>spring>quarry>road>aqui>aquiclude>gully>building.

Step 3: The area of evaluation factor is determined.
The total area of the research district is 121 km², in which the affected area by slope stability occupies 11.95 km², the affected area by hydrology occupies 45.41 km², the affected area by fire hazard occupies 3.65 km², and the affected area by eco-environment occupies 2.24 km². The statistical results for all impact factor data are shown in Table 5, and the statistical results for occupied areas are shown in Figure 7.

ArcGIS was adopted to complete the comprehensive map of 13 disaster-caused factors in the research area of the Xiaolongtan Lignite Deposit. The detailed scale is shown in Figure 5. The total coverage area for the whole research area is 121 km², of which the slope stability area is 2.69 km², the dump area is 5.02 km², the step area is 2.21 km², the road area is 1.83 km², quarry area is 0.20 km², potential flood area is 9.05 km², water body area is 0.73 km², spring area is 0.18 km², aquifer area is 19.55 km², aquiclude area is 14.25 km²,

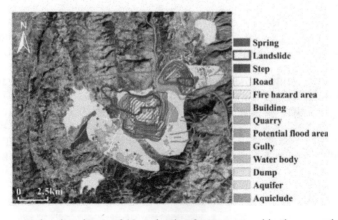

Figure 5. The comprehensive picture of 13 evaluation factors exposed in the research area.

gully area is 1.65 km², lignite area is 3.65 km²· and the building area is 2.24 km². The geological information data is from the high-definition satellite imagery of Google Earth, the geological coordinate system is GCS_Beijing_1954, and the projection coordinate system is Beijing_1954_3_Degree_GK_Zone_36.

3.2 *Results*

As shown in Figures 5 and 6, among 13 evaluation factors, the factors with an occupied area less than 1 km² include spring, quarry, and water body; the factors with an occupied area greater than 1 km² but less than 10 km² include unstable slope, dump, step, road, potential flood, gully, lignite, and building; the factors with an occupied area greater than 10 km² include aquifer and aquiclude.

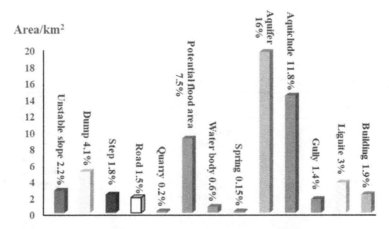

Figure 6. The column diagram of 13 evaluation factors area proportion.

The evaluation factor unstable slope occupies 2.2% of the total research district, dump occupies 4.1%, step occupies 1.8%, road occupies 1.5%, quarry occupies 0.2%, potential flood occupies 7.5%, water body occupies 0.6%, spring occupies 0.15%, aquifer occupies 16%, aquiclude occupies 11.8%, gully occupies 1.4%, lignite occupies 3% and the building occupies 1.9%. The remaining areas are the areas that are not impacted by the above-mentioned 13 evaluation factors, which occupy 47.7%.

As shown in Figure 7, among four evaluation elements, the factor with the largest occupied area is the hydrology, the second largest is the slope stability, and then is the fire hazard, and the smallest is the eco-environment. The evaluation element slope stability occupies 9.9% of the total research district, hydrology occupies 37.5%, fire hazard occupies 3%, and eco-environment occupies 1.9%. The remaining areas are the areas not impacted by the disaster, which occupy 47.7%.

According to the hazard weight of all evaluation factors, the research district can be divided into four areas: unsuitable area, conditionally suitable area, suitable area, and no-disaster area. Among them, the factors with a hazard weight less than 0.03 are attributed as the suitable area; the area occupied by the factors with a hazard weight between 0.03 and 0.1 is the conditionally suitable area; the area occupied by the factors with a hazard weight greater than 0.1 is attributed as the unsuitable area; the area without any hazard is attributed as no-disaster area. The classification results for impacted areas by 13 evaluation factors are shown in Table 3.

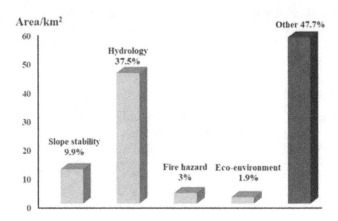

Figure 7. The column diagram of 4 evaluation elements area proportion.

Table 3. The area classification of each evaluation index.

Evaluation elements	Evaluation factors	Area classification
Slope stability	Unstable slope	Unsuitable
	Dump	Unsuitable
	Step	Unsuitable
	Road	Suitable
	Quarry	Conditionally suitable
Hydrology	Potential flood area	Unsuitable
	Waterbody	Conditionally suitable
	Spring	Conditionally suitable
	Aquifer	Suitable
	Aquiclude	Suitable
	Gully	Suitable
Fire hazard	Lignite	Unsuitable
Eco-environment	Building	Suitable
Other	No influence	No-disaster area

ArcGIS was adopted to complete the final comprehensive engineering geological evaluation map of classification areas in the research area of the Xiaolongtan Lignite Deposit. After erasing the overlap area of different classification areas, the detailed scale is shown in Figure 8. Despite the repetitive area, the total coverage area for the research area is 120.950187 km², of which the influenced area is 63.2832 km², the unsuitable area is 22.60606 km², conditionally suitable area is 1.135091 km², the suitable area is 39.24603 km², and no-disaster area is 57.963006 km².

The proportion for the suitability partition in the research district is shown in Figure 9. The unsuitable area occupies 18.7% of the total research district, the conditionally suitable area occupies 0.9% of the total research district, the suitable district occupies 32.7% of the total research district, and the no-disaster district occupies 47.7% of the total research district. The no-disaster district is the largest district, the second largest district is the suitable area, then the unsuitable area, and the conditionally suitable district occupies the smallest district.

Figure 8. The assessment result of 4 different classification areas in the Xiaolongtan Lignite Deposit.

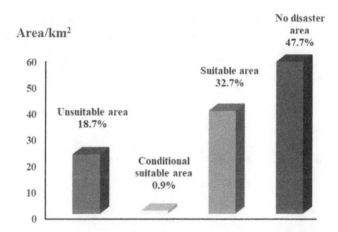

Figure 9. The column diagram of 4 different classification area proportion.

4 CONCLUSION

According to the result of the suitability assessment, the unsuitable area occupies 18.7% (22 km^2), the conditionally suitable area occupies 0.9% (1 km^2), and the suitable area occupies 32.7% (39 km^2), and the no-disaster district occupies 47.7% (59 km^2) of the total research area (121km^2).

This paper raises advice on the further development and rehabilitation of mining areas in the future. The unsuitable area needs to be monitored and treated according to the specific situation. The conditionally suitable area needs to be monitored on a regular basis each year, and the suitable area doesn't need to be monitored and treated. The treatment proposals for both unsuitable and conditionally suitable areas are as follows: the unsuitable area is full of the most serious geological disasters, including landslides, floods, and fires. The disaster-caused factors causing the direct occurrence of these three disasters include unstable slope, step, dump, potential flood area, and lignite.

ACKNOWLEDGMENTS

This work was supported by Yunnan Basic Research Program (2019FD009) and Yunnan Land and Resources Vocational College Research Team Program (2021KJTD04).

REFERENCES

[1] Moosavirad S.M., & Behnia B. (2017) RETRACTED ARTICLE: Suitability Evaluation for Land Reclamation in Mining Areas: Gol-e-Gohar Iron Ore Mine of Sirjan, Kerman, Iran. *International Journal of Mining, Reclamation and Environment*, 31:38–51.

[2] Li S., Deng C., Dong W., Sun L., Liu S., Qin H., & Zhu R. (2015) Magnetostratigraphy of the Xiaolongtan formation bearing Lufengpithecus keiyuanensis in Yunnan, southwestern China: Constraint on the initiation time of the southern segment of the Xianshuihe-Xiaojiang fault. *Tectonophysics*, 655:213–226.

[3] Huang Y., Hao J.Y., Deng G.B., Bai L., & Zhang G.X. (2011) Discovery and Significance of the Seismites in the Neogene Xiaolongtan Formation in Maguan, Yunnan. *Sedimentary Geology and Tethyan Geology*, 31: 79–85.

[4] Shui H.F., Liu J.L., Wang Z.C., & Zhang D.X. (2009) Preliminary Study on Liquefaction Properties of Xiaolongtan Lignite Under Different Atmospheres. *Journal of Fuel Chemistry and Technology*, 37:257–261

[5] Han L., Shu J.S., Hanif N.R., Xi W.J., Li X., Jing H.W., & Ma L. (2015) Influence Law of Multipoint Vibration Load on Slope Stability in Xiaolongtan Open-pit No.2 mine in Yunnan, China. *Journal of Central South University*, 22: 4819–4827.

[6] Peng H.G., Cai Q.X., Shu J.S., & Zhang L. (2007) Application Research of Numerical simulation in the Slope Stability of Bu-zhao-ba Open-pit. *China Mining Magazine*, 6.

Civil Engineering and Energy-Environment – Gao & Duan (Eds)
© 2023 the Author(s), ISBN 978-1-032-56059-5

Toward a sustainable Wyndham: An agriculture-led action for the food security of Wyndham, Australia

Jialing Xie*

Melbourne School of Design, The University of Melbourne, Parkville, Australia

ABSTRACT: As one of the fastest-growing communities in Australia, Wyndham's growth has led to increased urbanization and pressure on the socio-ecological environment. The current situation and associated challenges in the town of Wyndham, Victoria, Australia, suggest that food insecurity is increasingly embedded in the present and future uncertain risks that are disturbed by various socio-environmental variables. Therefore, exploring opportunities for developing sustainable communities in the Wyndham area is necessary. This paper aims to explore resilient community/town development based on agriculture to achieve food security in the town of Wyndham. Through a critical review of the relevant literature and case studies related to entrenching the implementation of the pilot site design approach, this paper presents much-needed and timely action for urban farming practices at their current scale and a critical review of best practices around the world, particularly the opportunities and barriers to scaling up these practices under future risks such as climate change, projected sea-level rise, and increased storm surges.

1 INTRODUCTION

Wyndham District is an urban area in the western part of the city of Melbourne, Victoria, Australia. The district has a population of approximately 217,000, which is proliferating due to the expansion of residential development. It is vulnerable to coastal, river flooding, and urban heat problems. Hundreds of houses are currently at risk of flooding. This risk is likely to increase further as the sea level rises and storm surges increase. Wyndham's population is estimated to exceed 302,650 by 2021. The population of Wyndham is projected to exceed 417,000 by 2030. The City of Wyndham is one of the fastest-growing communities in Australia, which has led to increased urbanization, putting pressure on the community and the environment.

This paper explores the concept of sustainability and its application to the Wyndham settlement's planning, design, and management, focusing on food security. From a socio-ecological system planning perspective, resilience quality defines four key characteristics. Diversity, modularity, feedback, and redundancy are resilience indicators [1]. This concept provides an indicator-based approach to planning and designing urban spaces as an integrated system. For the agricultural system, resilience refers to interventions in response to agricultural lands, such as environmental impacts and changes in agricultural production. Resilience, therefore, encompasses the ability of a system to withstand internal and external pressures and restore necessary functions. In the longer term, it requires the capacity to adapt and renew [2]. The food industry's community and natural resilience actions are becoming

---

*Corresponding Author: jialingx@student.unimelb.edu.au

DOI: 10.1201/9781003433651-19

increasingly essential in contributing to Wyndham's communities' social, economic and environmental resilience.

Based on the challenges identified and critically assessed, the paper discusses how to improve the sustainability of existing Wyndham settlements and relate them to solutions proposed in the relevant literature, case studies, and experimental design sites. It focuses on changes in 'food' in the context of climate change. In summary, there are four objectives to be achieved through several measures to ensure the agricultural system's internal and external security. The objectives are to better prepare the region for future environmental changes, either foreseen or unforeseen, and to respond to planned or unexpected events.

2 PROBLEM ANALYSIS AND CRITIQUE

2.1 *Unbalance between increasing population and declining food production*

For greater land availability and less pressure from the encroachment of other compatible land uses, much of the agricultural production that feeds Greater Melbourne now and in the future occurs on the city's fringe. Based on different activities among urban, peri-urban, and rural areas, Melbourne's food bowl has been split into three distinct regions: inner Melbourne, interface Melbourne, and outer Melbourne.

However, while the Melbourne interface area has relatively lower agricultural land availability than the periphery, its agricultural industries produce a higher gross value of food bowls than their land share would suggest. Wyndham, a highly productive source of food bowls around the city, is on the interface with Melbourne. In Wyndham Municipal Strategic Statement, Werribee South is a significant Victorian asset generating many vegetables that account for a substantial percentage of Victoria's total annual production, including most of the state's cauliflower and broccoli [3]. The Werribee Irrigation District is the closest area of intensive food production to Melbourne, which is essential to the food supply of the rapidly growing city.

It is estimated that Melbourne will be home to 7-8 million people by 2050 and will need 60% more food to feed them [4]. Accordingly, the city of Wyndham is projected to increase by 285,531 people from 2016 to 2041 (an increase of 125.75%), with an average annual change of 3.31% [5].

Hence, the capacity of food bowls to meet local and regional food needs is expected to decline from 41% in 2015 to 18% in 2050. The impact on production in the food bowl is likely to have almost 59% of vegetable production and 64% of fruit production to be lost [6]. As a result, Melbourne's food bowl capacity will not be able to meet the city's vegetable needs and other food types by 2050.

2.2 *Limited food distribution in the region's fringe*

Currently, supermarkets and small fresh food markets are often located in densely populated suburbs. In suburban areas such as prisons and adjacent housing areas, residents can only obtain fresh produce from nearby farmers. However, due to COVID-19, employment opportunities in the area have been significantly reduced, especially in traditional agriculture [7]. The inflexible food distribution system with few employees contributes to restricted access to food for residents in the border area, as shown in Figure 1.

2.3 *Environmental risks including climate changes and salinity*

According to the Werribee Irrigation District, the district faces serious problems related to its dependence on irrigation due to rising groundwater salinity, climate change issues, and a drop in rainfall from 550 mm to 450 mm per year over the past 20 years. The expected rise in

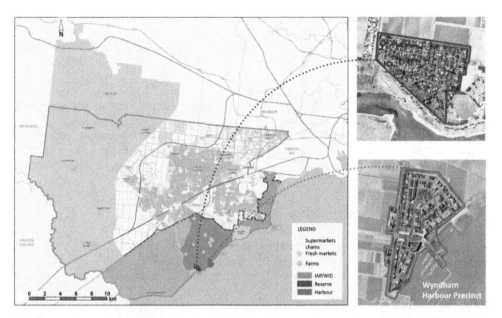

Figure 1. The food distribution system in Wyndham and fringe communities within WID.

groundwater levels indicates that many agricultural lands are constantly threatened by salinization. The Intensive Agricultural Area (IAP), located in the south of Werribee, is an intensive agricultural area of approximately 3000 hectares with an estimated worth of A $200 million per year. However, the diversity of available plant species is low, and salinity tolerance is not very high, so high salinity in groundwater can affect crop production.

3 SUGGESTED SOLUTIONS

3.1 *Literature studies: Agroecology +Traditional Knowledge (TK)*

On this basis, it is appropriate to establish a close relationship and interdependence between agroecology and traditional knowledge (TK) in the agricultural system. Moreover, it is necessary to build their role in adapting to climate variability and change in the region, mitigate the impact of environmental change on agricultural land and restore its arable status.

Agroecology refers to applying ecological concepts and principles to the design and management of sustainable agroecosystems [8]. Agroecological knowledge and practices enable farmers to achieve multi-functionality and optimize, rather than maximize, production. Systems are optimized when farmers reach the highest level of "agroecological integration", that is, the extent to which they apply agroecological principles to managing various resources [9]. When considering productivity, farmers in eco-agriculture do not focus solely on crop production but rather on the degree of agro-biodiversity of the farming system and its ability to provide ecosystem services [10]. Biodiversity conservation is also essential for the sustainability and resilience of the agricultural system. This is because biodiversity can contribute to the ability to absorb shocks and continue to function under changing conditions [11].

Traditional Knowledge (TK) refers to long-standing traditions and practices of adaptive ecosystem management and sustainable use of natural resources. Living in harsh natural environments, indigenous peoples and communities have been forced to cope with extreme weather events and adapt to environmental changes for centuries to survive [12]. The conservation and dissemination of traditional knowledge depend on using different biological resources (wild and domesticated species), and the reintroduction of traditional varieties can restore relevant local knowledge and practices. The identification of different species (crops and livestock) with high agrobiodiversity, based on farmers' local knowledge, is also a response to the risk of uncertainty. In addition, agroecological systems can manage food insecurity, reduce environmental impacts, increase climate resilience and improve agrobiodiversity [10].

Overall, it is suggested to understand what motivates farmers and consumers to make locally based decisions, provide supportive policies and institutions to improve the sustainability and productivity of agricultural systems, and build constructive complementarities between ecological and conventional (TC) agriculture.

3.2 *Case studies - Lai Chi Wo Village, Hongkong*

Agriculture was an essential industry in Hong Kong, especially before the 1980s. The decline of Hong Kong's agricultural sector has been attributed to the glocalization of agricultural products and economic restructuring. Rural revitalization can only be achieved if local and sustainable farming issues are addressed.

Organic farming patterns aim to achieve agricultural productivity and support a viable local economy for local communities while using nature-friendly and renewable farming methods. Mixed farming and agroforestry create more diverse, productive, and ecologically superior land-use systems. Experimental use of agroforestry and biochar increases carbon sequestration in farming systems and improves resilience to climate change. The model involves farmers, environmentalists, agricultural experts, and landowners working together to develop new approaches to sustainable farming models suitable for Hong Kong.

Lack of support for processing is a problem in developing Hong Kong's agriculture. A new model of crop processing has been developed to promote the production of lychee woo crops and, more importantly, to test the socio-economic sustainability of the local crop processing industry. The participation of farmers, producers, sellers, and consumers in developing new agricultural products creates a chain of production, processing, and marketing. Such a partnership can increase production efficiency and promote the social responsibility of producers and consumers.

3.3 *Design solutions - Food-up Hubs*

Food-up streets use the Wyndham Harbour area as a testing ground for integrating agricultural landscape right-of-way, recognizing that the desire and ability to produce food are social. The plan's perspective favors non-transportation modes, such as food production, and allows the use of cars. It addresses food security issues through design work that contributes to the community's decision-making process in future directions.

The Food-up Hubs Plan does not require everyone to be a farmer. Instead, it is to adapt this local food production to the living conditions of modern people. Food production is a local economy and ecosystem that requires a land-use system that balances urban and landscape systems and provides a transferable planning vocabulary that incorporates agricultural potential into urban settlement planning at all scales as shown in Figure 2.

Through the application of multi-purpose planter boxes and frames installed in the unused areas of the site, space is efficiently utilized, and seasonal plants and crops are grown according to the community's needs. Such agricultural landscapes can promote a self-sufficient food cycle and provide alternative food sources for residents.

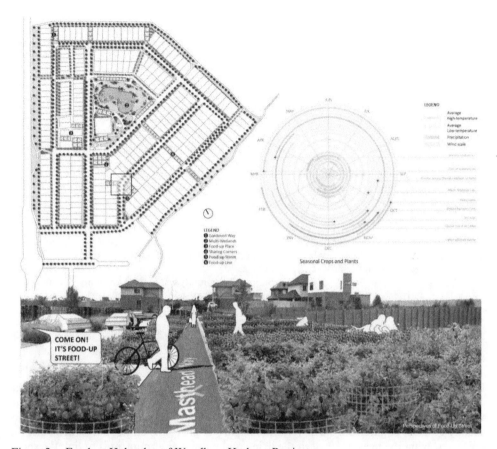

Figure 2. Food-up Hubs plan of Wyndham Harbour Precinct.

3.4 *Action plan*

We generated an action plan based on the above studies and solutions, as shown in Table 1.

Table 1. Action plan.

Outcomes	Objectives	Actions	Timeframe
The inherent security of the resilient agricultural system	To foster sustainability of productive farmlands	Mitigate the impacts of environmental changes on agricultural lands and restore the cultivability of agricultural land.	Years 1, 2, 3
		Make the most efficient use of on-farm resources and advocate for infrastructure and renewable energy.	Years 1, 2, 3
		Improve crop diversity and promote production and viability in local and regional economies.	Years 1, 2, 3
	To develop multi-functionality of	Broaden the range of possible agricultural uses and conditions for diversification in	Years 1, 2

(*continued*)

Table 1. Continued

Outcomes	Objectives	Actions	Timeframe
	agriculture	the area.	
		Recognize the social and local economic benefits of multicultural activities within the Green Wedge and protect the viability of existing sites and businesses.	Year 3
		Combine agricultural land with activities for recreation, tourism development and facilities.	Years 2, 3
The external security of the resilient agricultural system	To achieve high efficiency in food distribution	Develop an efficient food distribution system to buy or sell local products throughout the region.	Year 1
		Encourage community involvement in coordinating food supply and demand.	Year 2
	To create community-based agri-landscapes	Generalize a self-sufficient food cycle within the agri-landscape in the community.	Years 2, 3
		Support community-based food economy and local enterprise in the precinct.	Years 2, 3

4 CONCLUSION

The ability of a social-ecological system to maintain an adaptive response to disruption and other uncertain risks is known as resilience. This degree of resilience implies the ability to self-organize, evolve, adapt and bounce back or recover to another structurally and functionally efficient equilibrium state. With the impact of problems on human settlements, the key to resolving the tension between the living and natural environment is to link the social environment to the ecological system to create a sustainable social-ecological system.

In this study, Wyndham is seen as an example of food security from a social-ecological system's perspective, as it plays a vital role in Melbourne's food bowl. Intensive farming fed the local population and cultivated crops such as broccoli and cauliflower, which provided the Melbourne metropolitan area with many vegetables. The research needs to consider and analyze both social and environmental conditions. Food security can be defined as internal security, ensuring adequate production of high-quality food on a sustainable basis, and external security, ensuring efficient distribution so that people can access sufficient food. Identifying the current situation and related issues in Wyndham Township suggests that food insecurity is increasingly manifested as current and future insecurity risks of insecurity that are perturbed by three socio-ecological variables that are perceived as an imbalance between population growth and declining food production, the limits to food allocation in marginalized communities, and environmental risks associated with declining agricultural production, such as climate change and salinity.

Appropriate literature review and case studies were conducted based on identified problems, and solutions were proposed to consolidate the implementation of the project approach in the experimental area. To ensure the internal and external security of the agricultural system, the project has four objectives and several measures to achieve results.

Objective 1: to promote the sustainability of agricultural production land.

We should mitigate the impact of environmental change on agricultural land and restore its arable status, ensure the most efficient use of farm resources and promote the provision of infrastructure and renewable energy, improve crop diversity and promote the production and viability of local and regional economies.

Objective 2: to develop multifunctional agriculture.

We should expand the range of possible regional agricultural uses and diversify conditions, recognize the social and local economic benefits of multi-purpose agricultural activities in the 'green wedge', protect the viability of existing sites and businesses, and integrate agricultural land with related activities to recreational and tourism development.

Objective 3: to achieve high efficiency in food distribution.

We should build an efficient food distribution system to buy and sell locally produced agricultural products throughout the region, and promote community involvement in coordinating food supply and demand.

Objective 4: to build agricultural landscapes at the community level.

We should summarize the cycle of food self-sufficiency in the local agricultural landscape, and support the local food economy and the operation of local farm shops.

REFERENCES

[1] Biggs R., Schlüter M., Schoon M.L.: *Principles for Building Resilience: Sustaining Ecosystem Services in Social-ecological Systems*. Cambridge University Press, Cambridge, UK. 6–7(2015).

[2] Lundberg J., Moberg F.: Mobile Link Organisms and Ecosystem Functioning: Implications for Ecosystem Resilience and Management. *Ecosystems* 6. 87–98(2003).

[3] Wyndham City.: *Werribee South Green Wedge Policy and Management Plan*. Wyndham City Submission (2017).

[4] Deloitte Access Economics (DAE).: *The Economic Contribution of Melbourne's Foodbowl*. Victorian Eco Innovation Lab (2016).

[5] City of Wyndham.: *Estimated Resident Population (ERP)*. https://profile.id.com.au/wyndham/population-estimate, last accessed 2020/12/05.

[6] Sheridan, J., Larsen, K., Carey, R.: *Melbourne's Foodbowl: Now and at Seven Million*. Victorian Eco Innovation Lab, The University of Melbourne (2015).

[7] City of Wyndham.: Industry sector of employment. https://profile.id.com.au/wyndham/industries, last accessed 2020/12/05.

[8] Altieri M.A.: *Agroecology: The Science of Sustainable Agriculture*. Boulder, CO: Westview Press. 2nd ed. 71–106(1995).

[9] Rosset P.M., Sosa B.M., Roque Jaime A.M.R., Ávila Lozano R.A.: The Campesino-to-Campesino Agroecology Movement of ANAP in Cuba: Social Process Methodology in the Construction of Sustainable Peasant Agriculture and Food Sovereignty. Journal of Peasant Studies 38(1). 161–191 (2011).

[10] Silici L.: *Agroecology: What it is and What it has to Offer*. IIED Issue Paper. IIED, London. 13-15 (2014).

[11] *Oecd.: Sustainable Agricultural Productivity Growth and Bridging the Gap for Small-family Farms*. http://www.oecd.org/agriculture/topics/agricultural-productivity-and-innovation, last accessed 2020/12/05.

[12] Swiderska K., Reid H., Song Y., Li J., Mutta D., Ongugo P. and Pakia M.: *The Role of Traditional Knowledge and Crop Varieties in Adaptation to Climate Change and Food Security in SW China*. Bolivian Andes and coastal Kenya. IIED London. 1–14 (2011).

Civil Engineering and Energy-Environment – Gao & Duan (Eds)
© 2023 the Author(s), ISBN 978-1-032-56059-5

A review of application of deep eutectic solvents as lubricants and lubricant additives

Ting Li*, Zhipeng Zhang, Rui Wang & Junmiao Wu
School of Mechanical Engineering, Shenyang Jianzhu University, Shenyang, Liaoning, China

Qianqian Zou
School of Material Science and Engineering, Shenyang Jianzhu University, Shenyang, Liaoning, China

Yulan Tang*
School of Municipal and Environmental Engineering, Shenyang Jianzhu University, Shenyang, Liaoning, China

ABSTRACT: Deep eutectic solvents (DESs) are emerging mixtures with significantly lower melting points than pure components. These materials are expected to show many adjustable physical and chemical properties. At the same time, the complex hydrogen bond is postulated as the fundamental reason for their depressions in melting points and physicochemical properties. The new application research of deep eutectic solvents has achieved initial results in many fields, such as biology, chemistry, and medicine. Due to its excellent viscosity and other characteristics in mechanical engineering, DES has become a new substitute for lubricants or a better choice of lubricating additives. Due to their sustainable and environmental protection characteristics, natural deep eutectic solvents have become a new favorite of energy conservation and emission reduction research. This review summarizes recent research to clarify the application prospects of DESs, especially natural Deep Eutectic Solvents (NADESs), in lubrication. It covers the latest progress of DESs research in tribology, puts forward prominent scientific problems, and puts forward promising research directions consistent with the basic understanding of these solvents in the field of tribology.

1 INTRODUCTION

Lubricant is an indispensable material on the engineering surface. It can reduce Friction and wear, dissipate heat from the friction surface, keep moving parts separated, transmit force, disperse foreign matters, improve the efficiency of the engine and machine, and make two or more objects move relatively (Harris & Kotzalas 2006). Traditional lubricating oils are pure mineral base oils or mixtures containing additives. In addition to optimizing the lubricating properties of base oil, additives can also optimize the physical properties of lubricating oil, such as corrosivity and oxidation resistance (Donato *et al.* 2021).

Automotive/engine lubricants dominate the liquid lubricants market (Reeves & Menezes 2016). By adjusting and improving the effect of lubricating oil and tribology in various industrial machinery and common engines, a large amount of energy waste and carbon dioxide emissions can be reduced. The new or lubricant additive formula can save energy consumption under various friction conditions (Carpick *et al.* 2016). In recent years, many researchers have attempted to develop a new generation of lubricants and additives to

---

*Corresponding Authors: liting@sjzu.edu.cn and tyl98037@163.com

DOI: 10.1201/9781003433651-20

improve the efficiency of friction lubrication and the durability of engineering surfaces. Using low-viscosity lube base oils and complex solid structure additives has aroused great interest. They can provide sufficient lubrication for various engineering surfaces while minimizing environmental problems.

In the attempt of new lubricants and additives, ionic liquids (ILs) have shown enormous promise due to their special properties. In the recent ten years, many publications, including review articles, have discussed using ILs as lubricants and additives. Mariana *et al.* reviewed this in detail (Donato *et al.* 2021).

DESs were previously considered an ionic liquid. Similarly, they have many common characteristics, including appropriate viscosity, high thermal stability, low vapor pressure, low volatility, and adjustable composition. However, as the processing and synthesis of ILs may cause serious pollution, they cannot be considered environment-friendly compounds. On the contrary, deep eutectic solvents (DESs) are typically inexpensive (Xu *et al.* 2017), biodegradable (Khandelwal *et al.* 2016; Radošević *et al.* 2015), non-toxic (Junaid *et al.* 2016), and raw materials are easier to obtain. For example, common HBA choline is a component of vitamin B. As a nutritional supplement for livestock, it produces millions of tons yearly (Smith *et al.* 2014). Another popular HBD option, urea, is a very common fertilizer (Marsh *et al.* 2005).

There is more and more research on using DES as lubrication, but there is still a lack of summary and discussion of this literature. This review will funnel into the progress made in using DES as a lubricant additive or as a lubricant since it was first proposed as a substitute. It will center on introducing the research achievements related to DESs in recent years and provide new ideas for exploring and researching new lubricants and additives.

2   DESS AS LUBRICANT

2.1   *[Ch]Cl as HBA*

In 2010, Lawes *et al.* proposed for the first time to use choline chloride [Ch]Cl salt and two HBDs (urea and ethylene glycol, EG) at a ratio of 1:2M to lubricate steel/steel contact with DES (Lawes *et al.* 2010). SAE 5W30 was used as the reference lubricant in the experiment.

The result was that the sliding distance before lubricant loss varies with test conditions and surface roughness. In all situations, the combination of [Ch] Cl: urea has relatively good lubrication performance, which was mainly attributed to the high viscosity of this DES.

Abbott *et al.* studied water-soluble DES as a potential lubricant in 2014 (Abbott *et al.* 2014). The author selected [Ch]Cl: urea, [Ch]Cl: oxalic acid, [Ch]Cl: EG, and [Ch]Cl: glycerin, and the viscosity, density, corrosion rate, contact angle, and friction coefficient were compared with those of marine reference oil (Mobil Therm 605).

Their research showed that DES, based on glycol and choline chloride, had correct lubricity, toxicity, and corrosivity and was a feasible choice for basic lubricants. They were miscible with water in various proportions and could inhibit corrosion even when slightly wet, which meant they were particularly suitable for marine lubricants.

Recently, Yuting Li *et al.* published another document about the tribological behavior of choline chloride–urea DES and choline chloride–thiourea DES (Li *et al.* 2022). In that study, they successfully synthesized choline chloride ([Ch]Cl) DES through the hydrogen bond network of urea and thiourea as HBD.

After a detailed characterization experiment (Table 1), they evaluated the tribological properties of [Ch]Cl: urea and [Ch]Cl: thiourea DES at 80 ° C using a reciprocating ball disk friction meter. The thermal stability, kinematic viscosity, and surface tension of the synthesized DESs were also measured as key performance indicators, and the results were summarized in Table 1.

Table 1. Decomposition temperature, melting point, contact angles on 45 steel, and the kinematic viscosities of two DESs [14].

	T$_{oneset}$ (°C)	T$_{endset}$ (°C)	Melting point (°C)	Contact angle on 45steel (°)	Kinematic viscosity (mm2/s)
[Ch]Cl–urea DES	237.7	275.4	69 ± 0.25	83 ± 0.16	23.3 ± 0.64
[Ch]Cl–thiourea DES	223.8	275.6	80 ± 0.04	69 ± 0.13	47.9 ± 0.66

During the whole experiment, the average COF of [Ch] Cl: thiourea DES was much lower than that of [Ch] Cl: urea DES for both 45 steel and Q45 steel friction pairs, that is, the lubricating effect of [Ch] Cl: thiourea DES was relatively better. In the meantime, this study determined that for the friction pair of Q45/45 steel, the sulfur element in the synthetic [Ch] Cl: thiourea DES can react at the friction point to form a special lubricating film mainly composed of FeS, thus further enhancing the lubricating effect.

2.2 Special HBAs

A new exploration study on natural deep eutectic solvents as an alternative lubricant was published by Yuting Li *et al.* in 2022. They successfully synthesized NADESs with betaine (Bet) as HBA and sugar or alcohol as HBD.

Natural deep eutectic solvents(NADESs) are considered "biodegradable" solvents. In addition to the advantages of traditional DES, such as low cost, low toxicity, easy preparation, and adjustable physical and chemical properties, Natural deep eutectic solvents are more environmentally friendly and biocompatible than DES (Feng *et al.* 2019; Lorenzetti *et al.* 2021).

Figure 1a and b showed the coefficient of Friction and the corresponding average coefficient of Friction under oil lubrication of DES and G1830 at 50 N. Besides, Figure 1c shows the wear rate of the steel disc.

The results of this study were exciting because they concluded that the Bet-based NADES prepared had excellent thermal stability, kinematic viscosity, wettability, and non-corrosivity compared with [Ch]Cl: urea DES. The lubrication performance of Bet: Suc NADES was particularly excellent, and the friction and wear rate was far lower than that of the G13830 ester used for comparison experiments. The author attributed the excellent lubricating performance of NADES based on Bet to the formation of lubricated friction film.

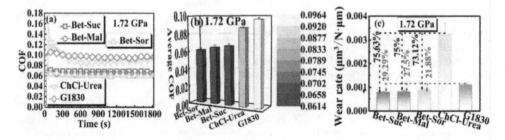

Figure 1. Friction coefficient and wear rate of worn surfaces lubricated by Bet-Sor, Bet-Suc, Bet-Mal, [Ch]Cl-Urea DESs, and G1830 ester under 1.72 GPa (Li *et al.* 2022).

3 ADDITION OF ADDITIVES IN DES

In 2019, Garcia et al. tested DESs lubricating steel/steel contact containing graphene (Garcia et al. 2019). The authors studied low-toxic DES, including [Ch] Cl: urea, [Ch] Cl: EG, and [Ch] Cl: malic acid, and their mixtures with graphene. The PAO6 of commercial synthetic base oil was selected as a reference for comparison. This study's results show that adding graphene to the lubricant can reduce Friction and wear. For this phenomenon, it has been recognized that graphene adsorbed on the steel surface during the friction process and formed a protective lubricating film (Guo & Zhang 2016). In addition, traditional lubricate oil could be replaced by [Ch] Cl: EG deep eutectic solvent, which was biodegradable and harmless to the environment.

Hallet et al. published a study on the friction behavior of [Ch] Cl: EG nano films between mica surfaces in 2020 (Hallett et al. 2020). They studied DES with different water content, including dry, wet (in balance with the environment), and water content of 30% and 50%, respectively. Surprisingly, The results show that dry DES and DES containing 50% water show super lubricity under different loads; Wet DES and DES containing 30% water have relatively low friction coefficients under low load, but the friction coefficient increases to 0.12 under high load. Based on neutron diffraction measurement, the author put forward the following explanation. The author believes that when the water content is low, the hydrogen bond network in DES will be strengthened, making it difficult to slide. On the contrary, adding more than 42% water is because it will destroy the structure of DES, which was conducive to in-plane liquidity and ensures super lubrication.

4 DESs AS LUBRICANT ADDITIVE

In 2021, Ponnekanti et al. published a study on the potential candidate of a new polyol-based deep eutectic solvent as a biological lubricant and tribological performance additive (Nagendramma et al. 2021). They synthesized a new polyol-based eutectic solvent(PDEs), a suitable substitute for industrial lubricants and can be used as an anti-wear and anti-corrosion additive for biodegradable lubricants. When the synthesized TPABr -TMP-based PDES was mixed with cottonseed oil with a concentration of 2 to 5 wt%, it showed admirable tribological properties in terms of antifriction and antifriction agents (Figure 2).

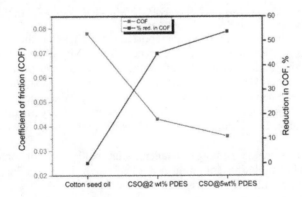

Figure 2. Variation in friction coefficient with PDES dosage (Nagendramma et al. 2021).

They successfully synthesized PDES for the first time and used it as a multi-functional additive for biological lubricants. The results showed that the PDES synthesized from quaternary ammonium salt and polyhydroxy alcohol had good friction reduction and anti-wear properties when mixed with cottonseed oil. Polyol-based eutectic solvents have high viscosity, low corrosion, high thermal stability, and biodegradability. The author suggests that further detailed research can be carried out to determine more applications of PDEs as biodegradable additives in MEMS bio-lubricants.

Another study about DES as a lubricant additive was published by Amzad Khan *et al.* in 2021 (Khan *et al.* 2021). In their research, DES based on aminoguanidine octanoic acid (Ag-C8) was studied as SN 150 lubricating oil additive to improve the tribological properties of steel friction pairs. It could be seen that the COF of DESs was significantly lower than those of SN 150 lubricating oil (Figure 3).

Halogen free, cheap, easy to prepare, proper viscosity, good compatibility with base lubricant, and significant improvement in Friction and wear all indicate that these DESs can be potential candidates for new-generation lubricant formulations.

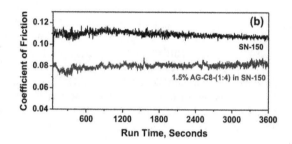

Figure 3. Coefficient of Friction between steel friction pairs in the presence of SN-150 lube oil and its blend with 1.5% AG-C8-(1:4) DES.(Khan *et al.* 2021).

5 CONCLUSION

This paper introduces the latest work on deep eutectic solvents as lubricant additives and pure lubricants.

Generally speaking, DES dissolved in ordinary lubricating oil base oil is superior to traditional lubricating oil. So far, the most studied tribological pair is steel/steel. Many ILS experiments can be used for reference in DES research, such as finding examples of metals or ceramics and some research involving silicon surfaces.

The common conclusion drawn from the reported research is that the adsorption capacity of the components in DES on the friction surface or the ability to form a friction chemical reaction film with the friction surface will affect their lubrication performance. For example, choline chloride thiourea DES has proved an effective lubricant substitute.

At the same time, when designing new DESs for tribological applications, sustainability, simplicity and price production, amplification, and biocompatibility are important factors to be considered. Under this condition, NADESs will undoubtedly become the focus of research.

However, it will be one of the biggest challenges in this field to put some highly efficient DES lubricating oil formulations into industrial use. Perhaps DES, as a lubricant additive, is an effective transitional means to meet this challenge.

ACKNOWLEDGMENTS

The authors appreciate the financial support provided by the Key Research and Development Program of Liaoning Province (No. 2021JH2/10100003).

REFERENCES

Abbott A.P., Ahmed E.I., Harris R.C., Ryder K.S., Evaluating Water Miscible Deep Eutectic Solvents (DESs) and Ionic Liquids as Potential Lubricants, *Green Chem.* 2014, 16, 4156–4161.

Amzad Khan, Raghuvir Singh, Piyush Gupta, Kanika Gupta, Om P. Khatri, Aminoguanidine-based Deep Eutectic Solvents as Environmentally-friendly and High-performance Lubricant Additives, *Journal of Molecular Liquids.* 2021, 339, 116829.

Carpick R.W., Jackson A., Sawyer W.G., Argibay N., Lee P., Pachon A. and Gresham R.M., The Tribology Opportunities Study: Can Tribology Save a Quad?, *Tribol. Lubr. Techno.* 2016, 72, 44.

Feng M., Lu X., Zhang J., Li Y., Shi C., Lu L. and Zhang S. Direct Conversion of Shrimp Shells to O-acylated Chitin with Antibacterial and Anti-tumor Effects by Natural Deep Eutectic Solvents. *Green Chem.* 2019, 21, 8, 7–98.

Garcia I., Guerra S., de Damborenea J. and Conde A., Reduction of the Coefficient of Friction of Steel-steel Tribological Contacts by Novel Graphene-deep Eutectic Solvents (DESs) Lubricants, *Lubricants.* 2019, 7.

Guo Y.B. and Zhang S.W., The Tribological Properties of Multi-layered Graphene as Additives of PAO2 oil in Steel-steel Contacts, *Lubricants* 2016, 4.

Hallett J.E., Hayler H.J. and Perkin S., Nanolubrication in Deep Eutectic Solvents, *Phys. Chem. Chem. Phys.* 2020, 22, 20253–20264.

Harris T.A. and Kotzalas M.N. *Essential Concepts of Bearing Technology*, 5th ed., CRC Press, 2006.

Junaid I., Hayyan M., Mohd Ali, O. Toxicity Profile of Choline Chloride-based Deep Eutectic Solvents for Fungi and Cyprinus Carpio Fish. *Environ. Sci. Pollut. Res.* 2016, 23, 7648–7659.

Khandelwal S., Tailor Y.K. and Kumar, M. Deep Eutectic Solvents (DESs) as Eco-friendly and Sustainable Solvent/Catalyst Systems in Organic Transformations. *J. Mol. Liq.* 2016, 215, 345–386.

Lawes S.D.A., Hainsworth S.V., Blake P., Ryder K.S. and Abbott A.P., Lubrication of Steel/Steel Contacts By Choline Chloride Ionic Liquids, *Tribol. Lett.* 2010, 37, 103–110.

Lorenzetti A.S., Vidal E., Silva M.F. Domini C. and Gomez F.J.V. Native Fluorescent Natural Deep Eutectic Solvents for Green Sensing Applications: Curcuminoids in Curcuma Longa Powder. *ACS Sustainable Chem. Eng.* 2021, 9, 5405–5411.

Mariana T. Donato, Rogério Colaço, Luís C. Branco, Benilde Saramago, A Review on Alternative Lubricants: Ionic Liquids as Additives and Deep Eutectic Solvents, *Journal of Molecular Liquids* 2021, 333, 116004.

Marsh K., Sims G. and Mulvaney R.L. Availability of Urea to Autotrophic Ammonia-oxidizing Bacteria as Related to the Fate of 14 C-and 15 N-labeled Urea Added to Soil. Biol. Fertil. Soils 2005, 42, 137.

Ponnekanti Nagendramma, Praveen Kumar Khatri, Shubham Goyal, Suman Lata Jain, Novel Polyol-based Deep Eutectic Solvent: A Potential Candidate for Bio-lubricant and Additive for Tribological Performance, *Biomass Conversion, and Biorefinery,* 2021.

Radošević K.; Cvjetko Bubalo M. C.; Gaurina Srček V. G.; Grgas D.; Landeka Dragičević T. L.; Radojčić Redovniković I.R. Evaluation of Toxicity and Biodegradability of Choline Chloride Based Deep Eutectic Solvents. *Ecotoxicol. Environ. Saf.* 2015, 112, 46–53.

Reeves C.J. and Menezes P.L., Advancements in Eco-friendly Lubricants for Tribological Applications: Past, Present, and Future, In J. Davim (Ed.), Ecotribology, Springer, 2016, 41–61.

Smith E.L., Abbott A.P. and Ryder K.S. Deep Eutectic Solvents(DESs) and Their Applications. *Chem. Rev.* 2014, 114, 11060–11082.

Somers A., Howlett P., MacFarlane D. and Forsyth M., A Review of Ionic Liquid Lubricants, *Lubricants.* 2013, 1, 3–21.

Xu P., Zheng G.-W., Zong M.-H., Li N. and Lou W.-Y. Recent Progress on Deep Eutectic Solvents in Biocatalysis. *Bioresour. Bioprocess.* 2017, 4, 34.

Yuting Li, Hao Li, Xiaoqiang Fan, Meng Cai, Xiaojun Xu, and Minhao Zhu, Green and Economical Bet-Based Natural Deep Eutectic Solvents: A Novel High-Performance Lubricant. *ACS Sustainable Chem.* 2022, 10, 7253–7264.

Yuting Li, Yuan Li, Hao Li, Xiaoqiang Fan, Han Yan, Meng Cai, Xiaojun Xu, Minhao Zhu, Insights into the Tribological Behavior of Choline Chloride–urea and Choline Chloride–thiourea Deep Eutectic Solvents, *Friction,* 2022.

Civil Engineering and Energy-Environment – Gao & Duan (Eds)
© 2023 the Author(s), ISBN 978-1-032-56059-5

Effect of electric field on combustion characteristics of ethanol-air mixture

Zihao Wang*, Boyun Liu* & Shuai Zhao
School of Power Engineering, Naval Engineering University, Wuhan, China

ABSTRACT: With the rapid development of the world economy, energy shortage, and environmental pollution problems appear increasingly prominent. At the present stage, most of the fire extinguishing facilities and equipment are disposable tools, and halogenated alkane fire extinguishing agents will produce toxic and harmful substances remaining in the fire scene after extinguishing, which is not conducive to the later entry of firefighters. Electric fire extinguishing has become an efficient and low-pollution fire extinguishing measure. In this paper, a simplified model is developed to illustrate the effect of transverse electric fields on diffusion combustion. The results show that the quenching of the electric field on the combustion reaction is mainly reflected in the ion wind generated by the electric field, which causes a sudden increase in the velocity of the local flow field of the flame, eventually leading to the temperature in the flame region to drop below the ignition point and eventually leading to flame extinguishment.

1 INTRODUCTION

At this stage, most of the fire extinguishing facilities and equipment are disposable tools, and halogenated fire extinguishing agents will produce toxic and hazardous residues in the fire scene after extinguishing the fire, which is not conducive to later firefighters to enter. Therefore, reusable, low reloading requirements, less hazardous fire extinguishing materials, and facilities research and development needs are extremely urgent. There are no residual harmful substances in the fire scene after the electric field extinguishing, and the device can be reloaded and used, which has a strong practical value for fire safety (Lu & Li 2018; Shi 2009).

The physical mechanism of the interaction between electric fields and flames has been studied for a long time, and researchers have found that the flame shape, stability, combustion products, and other combustion characteristics of a burning flame under the interaction with electric fields have been elaborated and explained in several ways (Jesse *et al.* 2019; Kuhl *et al.* 2017; Lacoste *et al.* 2016). Chinese scholars such as Yunhua Gan and Yanlai Luo explored the combustion characteristics and chemical reaction mechanism of small flames of ethanol air diffusion under the action of an electric field (Drews *et al.* 2012; Gan *et al.* 2015; Luo 2018). Yan Limin found that certain electric field conditions could strengthen the micro-scale combustion, but the too strong electric field would blow out the small-scale flame (Du *et al.* 2020; Guo *et al.* 2021; Yan 2012).

The main objective of this study is to numerically simulate the effect of the transverse electric field on the flame. In this paper, a numerical model of an ethanol-air diffusion flame was developed using the CFD software FLUENT. The effects of the electric field on flame

---

*Corresponding Authors: 1209800013@qq.com and boyunliu@163.com

shape, combustion rate, and flame temperature were elucidated by comparing the results under different electric field environments.

2 EXPERIMENTAL METHOD AND NUMERICAL SIMULATION

2.1 Experimental setup and method

The experimental setup consists of four main systems, which are the fuel supply system, the combustion system, the applied condition system, and the detection system. As shown in Figure 1, the fuel supply system used in this experiment is a micro-peristaltic pump. The combustion system consists of an atomization nozzle and an igniter. The influencing factor used in this paper is the transverse electric field. The detection system consists of an NPX-GS6500UM high-speed camera, a computer, a wind speed probe, and a S-sex thermocouple.

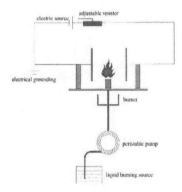

Figure 1. Schematic diagram of the experimental platform.

2.2 External electric field model

In this paper, we study the characteristics of a small ethanol-air diffusion flame under the action of an applied electric field, the supply voltage is connected to the left and right ends of the combustion system, an electric field is formed between the electrode plates, and the electric field distribution can be found by Poisson's equation (Ye & Chen 2012).

$$\frac{\partial^2 \varphi}{\partial x^2} + \frac{\partial^2 \varphi}{\partial y^2} = -\frac{\rho_c}{\varepsilon_0}$$

In this paper, we study that the small flame has very little effect on the electric field of the whole combustion region, i.e., it has very little effect on the whole space charge, then there is a charge density $\rho_c = 0$, and Poisson's equation becomes.

$$\frac{\partial^2 \varphi}{\partial x^2} + \frac{\partial^2 \varphi}{\partial y^2} = 0$$

This equation is also called the Laplace equation for the electric potential. The electric field strength in space can be found from the potential distribution E as follows:

$$\frac{\partial E}{\partial x} + \frac{\partial E}{\partial y} = 0$$

A charged particle in a small flame is subjected to electric field forces in an electric field.

$$F = Een_c = Ee(n_+ - n_-)$$

The calculated electric field forces are added to the source term of the momentum equation through a custom function (UDF) in Fluent software, and the solution is calculated to finally obtain the ethanol-air diffusion small flame combustion under the action of the applied electric field.

2.3 Mesh division and boundary conditions

In this paper, the ICEM meshing software is used to mesh the established physical model. The mesh division and boundary condition settings in the numerical simulation of this paper are shown in Figure 2.

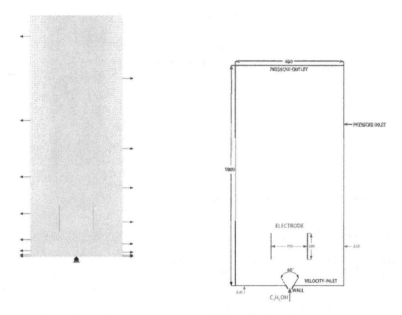

Figure 2. Two-dimensional physical model and grid division of combustion system.

2.4 Solving method

According to the characteristics of the ethanol-air diffusion flame, the numerical simulation of this paper is chosen based on the pressure solver, and the discrete control equations are linearized by an implicit method. The two-dimensional axial plane model is selected, the absolute velocity is selected for the calculation of velocity, and the least squares unit cell is selected for the gradient of the cell center variable. A SIMPLE algorithm is used for pressure and velocity coupling. The finite volume method is used for discretization. The second-order windward discrete format is used for momentum, energy, and each component, and the standard discrete format is used for pressure. The energy residuals are set to 10^{-6} and the residuals of other parameters are set to 0.001. Based on the setup of the solution method described in this section, the results of the combustion process of ethanol-air diffusion flame under the conditions of this study can be calculated.

3 TEST RESULTS AND DISCUSSIONS

3.1 Changes in the flow field around the flame

Figure 3 shows the calculated vector diagram and cloud diagram of the trend of the flow field around the combustion region under the effect of the transverse electric field. From the calculated cloud diagram, it can be seen that at the beginning of the application of the transverse electric field, as shown inFigure 3, the flow field in the combustion region starts to approach the electrode due to a large number of point-carrying ions in the flame, and the flow velocity tends to increase, in which the flow field located above the center of the pole plate, the velocity direction changes from the original vertical upward to vertical downward.

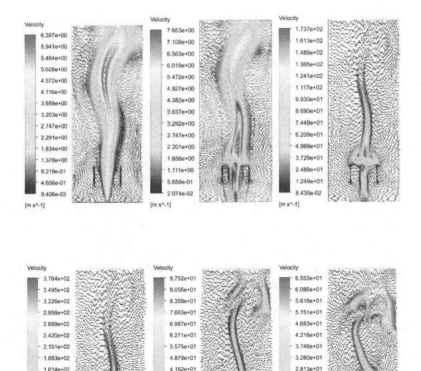

Figure 3. Velocity vector contour of flame flow field under transverse electric field.

As the electric field continues to be applied, Figure 4 shows that, due to the increase in the flow velocity in the region of the pole plate, the pressure in the region between the pole plate is less than the pressure in the region outside the pole plate according to Bernoulli's principle, along with the tendency of charged particles above the pole plate to be attracted by the pole plate, a strong flow with an instantaneous velocity of more than 280 m/s enters between the pole plate, resulting in a sudden drop in the temperature of the combustion region between the pole plate, and the combustion reaction is interrupted because the temperature in the combustion region is lower than the ignition point.

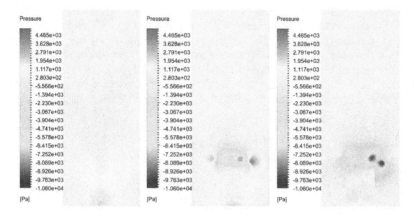

Figure 4. Contour of pressure in combustion area under transverse electric field.

3.2 *Flame quenching experiment under the action of transverse electric field*

Figure 5 shows the image of ethanol-air diffusion combustion flame under the effect of a transverse electric field taken by a high-speed camera. From the flame image without the influence of an electric field to increase the electric field intensity by shortening the spacing between the electrode plates; in the case of constant fuel supply, the charged particles in the flame by the influence of the electric field began to move in a directional manner, and the resulting ion wind accelerated the air injection, so the flame such as flame height decreased, the width increased, and the flame peak appeared bifurcation phenomenon. As the electric field strength continues to increase, the electric field generated by the ion wind speed is increasingly large, and gradually makes the combustion field heat loss speed up, the electric field began to inhibit the combustion situation, and combustion stability becomes worse. Finally, by the influence of the transverse electric field ethanol-air diffusion combustion flame quenching phenomenon process, the image can be seen by the flame by the ion wind, the combustion field temperature is lower than the ignition point, the flame from the bottom to gradually extinguished.

Figure 5. Experimental image of flame quenching under transverse electric field.

3.3 Simulation calculation results and experimental phenomenon analysis

The quenching phenomenon of ethanol-air diffusion flame under the effect of a transverse electric field is due to the flame being weakly ionized plasma, the flame contains a large number of charged particles, and when the external electric field is applied, the charged particles produce directional movement, the directional movement of ions drives the surrounding flow field molecules to move with the formation of ion wind between the pole plate, because Bernoulli phenomenon caused the negative pressure state in the pole plate area, the air above the pole plate is pressed by the atmosphere into the flow field between the pole plate to form a strong airflow, resulting in the top-down temperature of the area between the pole plate to below the ignition point. The air above the pole plate is pressed by the atmospheric pressure into the flow field between the pole plate to form a strong airflow, a large number of high-speed air into the combustion region between the pole plate, making the region from top to bottom temperature drop to below the ignition point, resulting in the flame extinguished from top to bottom. The experimental phenomenon shows that as the electric field strength increases, the flame is extinguished from the bottom up.

4 CONCLUSION

In this paper, a simplified model is developed to elucidate the effect of transverse electric fields on diffusive combustion. The kinetic model is first validated against experimental data collected in the literature. The integrated model including electric field calculation, charged particle distribution, and flow field variation is able to reproduce the experimental observations. The results show that the quenching of the electric field on the production of the combustion reaction is mainly reflected in the sudden growth of the velocity of the local flow field of the flame by the ionic wind generated by the electric field, which eventually leads to the temperature in the flame region dropping below the ignition point and eventually leads to flame extinction.

REFERENCES

Drews A.M., Cademartiri L., Chemama M.L., et al. AC Electric Fields Drive Steady Flows in Flames. *Phys Rev E Stat Nonlin Soft Matter Phys.*, 2012, 86(2):036314.

Du Z., Gao Z., He Z., et al. Effect of AC electric Field on the Combustion Characteristics of Methane/air Premixed Spherical Flame at High Initial Pressure. *Journal of Xi'an Jiaotong University*, 2020, 54(3):9.

Gan, Yunhua, Yan, et al. Effect of Alternating Electric Fields on the Behaviour of Small-scale Laminar Diffusion Flames. *Applied Thermal Engineering: Design, Processes, Equipment, Economics*, 2015.

Guo Zhicheng, Gao Zhongquan, Bao Yantong, et al. Effect of DC Electric Field on Methane/Ammonia/Air Premixed Laminar Flow Flame. *Journal of Xi'an Jiaotong University*, 2021, 55(10):10.

Jesse, Tinajero, Derek, et al. Non-premixed Axisymmetric Flames Driven by Ion Currents. *Combustion & Flame*, 2019.

Kuhl, Johannes, Seeger, et al. On the Effect of Ionic Wind on Structure and Temperature of Laminar Premixed Flames Influenced by Electric Fields. *Combustion & Flame*, 2017.

Lacoste D.A., Xiong Y., Moeck J.P., et al. Transfer Functions of Laminar Premixed Flames Subjected to Forcing by Acoustic Waves, AC Electric Fields, and Non-thermal Plasma Discharges. *Proceedings of the Combustion Institute*, 2016:S1540748916300347.

Lu Jia., Li Jinhe. Research on the Formation of Confined Space Emergency Rescue Fire Extinguishing Device Based on Electric Field Model. *China Safety Production*, 2018(1):2.

Luo Yanlai. Study on the Characteristics of Small Flame and Combustion Chemical Reaction Mechanism of Ethanol Diffusion. *Under the Action of Electric Field South China University of Technology*, 2018.

Qizheng Ye, Dezhi Chen *Electromagnetic Field Tutorial*. Higher Education Press, 2012.

Shi M. Electric Field Cooling and Electric Field Fire Extinguishing. *Big Tech: Science Enigma (A)*, 2009(5):1.

Yan Limin. *Experimental Study and Simulation Analysis of the Interaction Between Electric Field and Small-scale Flame*. Beijing Jiaotong University, 2012.

Civil Engineering and Energy-Environment – Gao & Duan (Eds)
© 2023 the Author(s), ISBN 978-1-032-56059-5

Occurrence characteristics and mining technology of coal seam in Dananhu No.2 Mine

Xiaoqian Yuchi*

College of Energy Science and Engineering, Xi'an University of Science and Technology, Xi'an, China

ABSTRACT: The Dananhu mining area in the Turpan–Hami basin is rich in coal resources. Based on the analysis of coal rock, coal quality characteristics, coal facies, and coal accumulation law in the Dananhu mining area, as well as the study of climate and environment, it is concluded that the coal seam has experienced five evolution stages: dry forest swamp and dry forest peat swamp, wet forest peat swamp, dry forest peat swamp, wet herb swamp, and dry forest peat swamp. Therefore, the occurrence characteristics and mining technology of the coal seam in Dananhu No.2 Mine are revealed.

1 INTRODUCTION

Currently, China is actively promoting the construction of ecological civilization, setting the goal of "double carbon," which indicates the green, clean, and efficient utilization of coal-to-oil and gas and clean coal-fired power generation. It is undoubtedly the main direction of effective utilization of clean coal-fired power generation (Jeffrey 2005). Coal quality and coal facies characteristics determine various treatment and conversion technologies and industrial uses of coal. The Danhu mining area is located in Dananhu sag on the southern edge of the Tuha Basin (Milici et al. 2013). The existing research mainly focuses on paleontology, coal seam correlation, and hydraulic geology, while research on coal quality and coal rock characteristics is relatively rare. Therefore, this paper chooses the No. 25 coal seam in Dananhu Coal Mine as the research object, analyzes its coal quality, coal rock, and coal facies characteristics, and classifies its characteristics to investigate the coal formation law. In addition, there are also coal gathering places that comprehensively utilize regional coal resources, which is also a reliable evaluation standard.

2 RELATED CONCEPTS AND CHARACTERISTICS

2.1 *Geological survey*

Tufu coal belt is located between eastern Xinjiang and Tianshan Mountains and is also the lowest inland mountain area in China. The basin is expanding mainly into fan-shaped land, delta, and sedimentary lake systems. Dananhu Coalfield is located in the southeast margin, distributed in the east-west and northeast directions. The drilling data show that the main strata include: the NeoCretaceous-Paleogene Shanshan Group (K2-E), the Middle Jurassic Xishanyao Formation (J2x), and the Upper Carboniferous Wutongwozi Formation (C2wt). The middle Jurassic Xishanyao Formation is the main coal-bearing stratum in this area, which can be divided into the upper, middle, and lower layers according to lithofacies (Ribeiro 2011). The total thickness of the coal seam is 36.47–143.39 m, including 23 exploitable coal

---

*Corresponding Author: 13484519960@163.com

seams. No. 15 and No. 23 can be mined. No. 25 coal seam is a stable coal seam that can be mined worldwide. The average coal seam thickness of No. 25 coal seam is 9.55 m.

2.2 Coal quality characteristics

According to the industrial analysis results, Raw coal's moisture content (Mad) ranges from 13.19% to 21.17%, with an average of 18.18%. The vertical change is not obvious, the moisture content of the upper coal seam is slightly lower than that of the lower coal seam, and the vertical change is not obvious. Only DNH05, DNH06, and DNH09 have abnormally high values. According to Coal Ash Classification (GB/T15224.1-2010), No. 25 coal seam is ultra-low ash coal and untreated coal. According to Coal Volatile Classification (MT/T849-2000), raw coal's total sulfur (St.d) content is 0.07–1.78%, with an average of 37.49%. According to Coal Sulfur Classification (GB/T 15224.1-2010), No. 25 coal seam belongs to low-sulfur coal. The sulfur content in No. 5 coal is mainly pyrite sulfur, the sulfur content in the lower part of the coal seam increases, and the organic sulfur is abnormally high.

2.3 Content characteristics of macerals

The 25th coal seam in the Dananhu mining area has a high organic content, with an average content of over 93%. Organic inert substances account for the majority, followed by bentonite, pyrite is the least, and clay and silica minerals account for the majority of inorganic components. In addition, the content of the brucellin group is high, with a value of 32.3–91.3%, with an average of 64.7%, while the content of the Wechsler group is low, with a value of 6.1–65.6%, with an average of 30.1%. The proportion of lignin is the lowest, ranging from 0.4% to 5.6%, with an average of 2.4%. Da Nanhu mine area is mainly because Jurassic coal in northwest China is almost all bituminous coal with low metamorphism, followed by clastic inert and oxidized silk (most fragments). However, it is also divided into coarse and fine particles, which may contain fibrinogen.

2.4 Evolution sequence of coal facies

According to the quality characteristics of coal rock and coal, the coal facies of the 25th coal seam in Dananhu No.2 Mine can be classified into five-peat wetlands: dry forest wetland, dry forest wetland, wet peat wetland, wet wetland, and dry forest peat wetland (Watson, Papageorgiou, Jones, Taylor, Symmons, Silman, Macfarlane 2002). Macroscopically, coal is mainly dull coal and semi-dark coal. In the initial stage of coal formation, wetland water is shallow, and oxidation is dominant. The coal types in this stage are mainly dry forest, peat wetland, and peat wetland with relatively dry forest. The macroscopic coal and rock types are white-hot coal and semi-white-hot. According to deep and slow hydrodynamics, the coal-forming vegetation is mainly herbaceous and peat wetland. The coal-forming environment is dominated by blunt coal, the water-covered environment becomes shallow, fluid power is strong, the oxidation environment is dominant, and coal-forming plants are mainly woody. The type of carbon phase in this stage is vanilla wetland type. In the later stage, the macroscopic coal and rock mass are mainly dull, and the coal-forming environment is mainly shallow water, strong hydrodynamic force, and oxidizing environment.

3 ANALYZE

3.1 Working face position and well relationship

A total of 1303 working faces are located in the west wing of the No.1 mining area, within the range of 2075 m to the west of the No. 3 coal belt conveyor alleys, 27 m to the west of the waterproof coal pillar of the old kiln, 545 m to the east of No. 1301 open cut, 18.6 m to the north of No. 1301 working face goaf, and 1305 working faces in the lower section in the south, which has not yet been mined. The underground elevation is +148.0 ~+233.0 m; the mining face strike has a recoverable length of 1937m and an inclined length of 240 m, and the second mining face in the west wing of the No.1 mining area. The mining of No. 1301 mining face in the upper section was completed in August 2013.

3.2 *Coal seam occurrence characteristics and coal seam roof and floor characteristics*

A total of 1303 working faces are fully mechanized upper coal tunnel working faces. Mining of 3# coal seam in Xishan Yang stratum in the middle Jurassic period. The coal seams mentioned above are 1# and 2# that cannot be mined. The coal seam with a low distance is 7.16–63.64 m, with an average of 20.5 m. There are 3–6 layers of gangue, and the thickness of gangue is 0.03–0.25 m. The thickness of the coal seam is 5.52–6.56 m, with an average thickness of 6.2 m. From relatively stable coal seam to stable coal seam. The inclination angle of a coal seam is 8–12, with an average of 10 (Watanabe & Oya 1986).

The roof layer of the third coal seam in this area is very unstable around rocks of type IV and V. Working faces coal seam roof and floor conditions are shown in Table 1.

Table 1. Coal seam roof and floor in working face.

Roof and floor of coal seam	Top and bottom plate name	Rock name	Thickness /m	Rock characteristics
	Loading	Sandstone, mudstone, silty mudstone, Middle coal clamping line. Middle coal clamping line	3.62	Mudstone and siltstone are mostly gray-gray, Obvious bedding, jagged fracture, It is easy to deform when exposed to external force, Local fissures are developed with good caving properties.
	Direct roof	Mudstone, silty mudstone, and siltstone,	2.89	
	Pseudotope	Carbon mudstone	0.01–0.03	The mechanical strength of rock is low, It is easily broken and falls with the mining.
	Direct bottom	Carbonaceous mudstone and mudstone.	1.85	Rock mechanical strength is low, and it will be cemented.
	Old background	Siltstone, locally fine, Medium sandstone with coal seam.	2.70	Hard texture, semi-hard rock, Local fissure development

4 HYDROGEOLOGY OF WORKING FACE

4.1 *Geological structure*

The overall shape of a coal seam in 1303 working face is a wide and gentle syncline structure. The center of the syncline is about 1469.0 m west of the conveying lane of the third coal belt, with an inclination of 131 and an east-wing strike. The dip angle is 105, the dip angle is 195, and the dip angle of the coal seam is 8–12, with an average of 10. Changing the dip angle has a certain influence on mining the working face. The strike slope of the coal seam also changes greatly. Starting from the trial cone hole, the range of 0–710 m is 10–2 downhill, the range of 710–830 m is 0 flat slope, and the range of 830–1950 m is 1–3 slope. 18–2 drilling data shows that the buried depths of coal seams 1, 2, and 3 are 214.4 m, 223.6 m, and 266.8 m, respectively, and those of coal seams 1, 2, and 3 are 260.3 m, respectively. The uppercut 3 is 143 m, 151 m, and 195 m, respectively, the average thickness of coal seam 1 is 1.2 m, and the average thickness of coal seam 2 is 0.4 m, all of which have little influence on the recovery.

4.2 *Hydrogeological classification*

(1) Sandstone aquifer. The water layer in the roof of the coal seam is the gathering place of the weak water layer, which is divided into three sandstone water layers. Water layer in Area V1: Lithofacies are mainly grayish-white-gray medium-coarse sandstone and olivine, most of which are compacted by calcium mud, with a general thickness of 4–20 m and an average thickness of 9.38 m. This layer is generally very thick, locally composed of 4–5 monolayers, and not rich in water. V2 Aquifer: Three coal grids are 40 meters, and the developed cracks are weak water layers. Water layer: No. 3 coal seam with water layer, about 8–11 meters away from No. 3 coal seam, coarse rock, medium sandstone, about 10 meters thick, weak water layer. V3 Aquifer. In this aquifer initially exposed to the coal auxiliary transportation road 2, the water inflow is about 3 m^3/h. According to the existing geophysical data, combined with the hydrogeological data accumulated in the No. 1301 working face mining, the 1303 working face is more affected by the weak zone water layer formation (the upper part of the middle Jurassic Xishanyao formation). In the 743.0–1583.0 m on the west side of the coal belt transportation canal, the surface layer covers the weak water layer at the upper part of the fish layer in Xishan in the middle Jurassic period. In the mining project, the key areas of water treatment in the working face, which are greatly affected by aquifer groups, should be arranged between holes 18-2 and 20-2. Located in the center of the syncline with a wide and gentle action surface, the basin-shaped structure of the whole coal rock also produced water conditions.

(2) Old water and drilling water. Old empty water: the section of 1303 working face is Gaof1301. During the 303 auxiliary transport tunnel excavation, 11×5 groups of drainage holes were constructed continuously, and the MPa was reduced to 0.028 MPa. Drilling water: The holes 18-2 and 20-2 are located at the 1303 working face, 1560 m and 590 m away from the 1303 rubber conveying trough, respectively, and about 15 m away from the south rubber conveying trough. The part of hole 20-2 with a height above 260.6 m is thick mud, and the part below 260.6 m is blocked by cement mortar due to blockage, so the sealing performance is good. No. 18-2 hole is blocked by thick soil sand above 238.3 m high and cement mortar below 238.3 m high, with good air tightness. As 18-2 holes and 20-2 holes are exposed in Tongtang mining, considering the impact of mining, it is necessary to observe the water environment when Tongtang is close to the drilling hole. In order to prevent the borehole from connecting to the aquifer, drilling, and exploration must be carried out to form a water-filled waterway on the working face.

(3) Old kiln water and fire area water. In 303, it opened 880.0 meters to the west, the eastern boundary of the third burning area in the mining area. On the east side of the burning area are five ancient kilns, namely Xinrui, Xiyu, Xingxin, Nanxing, and Dongfang. The underground water hydrostatic reserve with combustion zone is 5004×104 m^3, the highest water level elevation is +414.0 m (hole 15-1), and the elevation of the upper opening of hole 1303 is +236.6 m, the two ends are 177.4 m. The collapse radius is 102.4 m The lower mouth of the kiln opened in 303 is 244.0 m away from the waterline of the ancient kiln, the upper mouth is 30.0 m away from the waterline of the ancient kiln, and the confluence point of the ancient kiln is 150.0 m away from the waterline. The location and elevation of the ancient kiln water are clear. In 303, the coal pillar on the west side

was excavated more than 150.0 m, and its collapse range did not affect the small kiln or burning area. According to the Regulations on Water Control in Coal Mines, small coal mines should be calculated. In which: l-width of coal pillar, m; K: safety factor, generally 2-5, taking 5; M: coal seam thickness or mining height is 6.2 m. P-head pressure, 1.8 MPa according to the actual height difference of 177.4 m, and KP-tensile strength of coal, 0.1 MPa $L = 0.5KM\sqrt{\frac{3P}{Kp}}$

(4) Atmospheric precipitation and other factors affect mining. In addition to the above water injection coefficient, the 1303 working face must also consider the precipitation coefficient. We cannot rule out the possibility of occasionally concentrating precipitation in the form of storms and forming water-filled waterways through cracks in collapsed areas. The normal water inflow of the mining face is 80 m/h (including drilling and production water), and the maximum water inflow is 160 m^3/h. According to the data of the 2013 Annual Gas Appraisal Report of Dananhu No. 2 Mine, the measured gas mainly comes from the process of coal breaking and falling in tunneling and the exposed coal in the transportation lane and return airplane. During the identification, no spontaneous combustion was found. The ventilation, production, and mining, which meet the appraisal conditions, are normal in this process. Through analysis, it is concluded that the main sources of most gas and carbon dioxide in underground mines are gas and carbon dioxide. They are naturally released by coal falling and coal breaking in the production process of coal mining face and tunneling face, and this mine is a gas mine. Therefore, in the production process, it is necessary to do a good job in gas inspection, strengthen the underground ventilation system, and take corresponding prevention measures in case of abnormal gas to avoid gas accidents.

4.3 *Hydrogeological characteristics and calculation model*

Due to coal seam mining, the water-conducting crack area at the upper part of the working face develops into the water layer in the V1 area, and the rocks on the roof and the ground become weak, and the contact with water will bring a great softening effect. The locations of coarse sand and gravel aquifers are shown in Table 2.

Table 2. Location of coarse sand and gravel aquifer.

Aquifer name	Location	Rock character	Distance from coal 3 (m)	Thickness (m)
V1	Aquifer above 1 coal seam	Coarse and medium sandstone	Top plate 90–120	20–40
V2	Aquifer between 1 and 2 coal	Medium-coarse sandstone and glutenite	Top plate 50–60	1–5
V3	Aquifer between 2 and 3 coal	Silt and mudstone	Top plate 20–30	4–7
V4	Aquifer between 3 and 4 coal	Coarse and medium sandstone	Lower part 8–11	7–13

Suppose the model size is 500 × 233 m, the coal pillars are 18 m, 2 m, 6 m, 30 m, 34 m, and 38 m wide, and V1, V2, V3, and V4 are different layers with water layers. Relevant mechanical parameters will be optimized and distributed according to on-site exploration and rock foundation mechanics experiments. To simplify the model, in the Z direction, the average distance between No. 3 coal and the roof is calculated as 155 m, the average buried depth of No. 3 coal is calculated as 220 m, and the surface soil density is calculated as 2500 kg/m^3. When the boundary is subjected to a force of 2.0 MPa, the left and right boundaries of the model are horizontally displaced. Mining plan, Step 1: Excavating the lower lane 1301 and the upper lane 1303; Step 2: Excavate the 1301 working face; Step 3: Excavate the 1303 working face. The initial physical and mechanical parameters of each rock stratum are shown in Table 3.

Table 3. Initial physical and mechanical parameters of each rock stratum.

Rock stratum	E0/GPa	$v0$	$\rho0/kgm^{-3}$	coh0/MPa	fri0/°	tens0/MPa
Siltstone	7.50	0.34	2459	6.35	29.91	1.46
Mudstone	9.80	0.28	2597	4.39	30.41	1.28
Coal	2.26	0.31	1291	2.41	26.44	0.64
Medium sandstone	7.70	0.33	2435	5.78	28.31	1.35
Gritstone	8.30	0.32	2459	6.35	30.28	1.67

5 CONCLUSION

In this paper, the coal seam of Dananhu No. 2 Mine is studied, which belongs to the coal seam with very little ash, high volatile matter, and very low sulfur content. The vertical moisture has no big change, and the ash and volatile matter are lower in the upper part of the coal seam, while the sulfur content is higher in the lower part of the coal seam. It is concluded that the microscopic coal and rock types are mainly inert groups, vermiculite, Shi Ying, and other minerals, followed by silica and pyrite. In order to ensure the safety and stability of the mine, based on the analysis of the geological occurrence characteristics and mining layout of Dananhu No.2 Mine. Combined with the actual demand, the influence degree of mining face on the roadway was investigated, which provided the technical basis for the safe mining of the working face. Finally, the next work was planned and arranged based on the conclusion.

ACKNOWLEDGMENTS

This work was financially supported by the National Natural Science Foundation of China (Grant No. 51874231).

REFERENCES

Jeffrey L.S. (2005). Characterization of the Coal Resources of South Africa. *Journal of the Southern African Institute of Mining and Metallurgy*, 105(2), 95–102.

Milici R.C., Flores R.M., and Stricker G.D. (2013). Coal Resources, Reserves, and Peak Coal Production in the United States. *International Journal of Coal Geology*, 113, 109–115.

Ribeiro M.H. (2011). Naringinases: Occurrence, Characteristics, and Applications. *Applied Microbiology and Biotechnology*, 90(6), 1883–1895.

Watanabe S. and Oya H. (1986). Occurrence Characteristics of Low Latitude Ionosphere Irregularities Observed by Impedance Probe on Board the Hinotori Satellite. *Journal of Geomagnetism and Geoelectricity*, 38(2), 125–149.

Watson K.D., Papageorgiou A.C., Jones G.T., Taylor S., Symmons D.P., Silman A.J., & Macfarlane G.J. (2002). Low Back Pain in Schoolchildren: Occurrence and Characteristics. *Pain*, 97(1–2), 87–92.

Civil Engineering and Energy-Environment – Gao & Duan (Eds)
© 2023 the Author(s), ISBN 978-1-032-56059-5

A framework of carbon emission comprehensive evaluation standard for operation and maintenance of parks

Anshan Zhang, Jian Yang* & Feiliang Wang
School of Naval Architecture, Ocean and Civil Engineering, Hubei, China
Shanghai Key Laboratory for Digital Maintenance of Buildings and Infrastructure, Shanghai Jiao Tong University, Shanghai, China

Xihong Ma, Chaofang Zhuang, Xiwei Qian & Yan Jiang
China Energy Conservation (Huzhou) Science and Technology City Investment Construction Development Co. LTD, Zhejiang, China

ABSTRACT: As one of the critical regions for national environmental protection and carbon emission reduction, parks, which include buildings and green areas, can be used as an example to realize a low-carbon economy. The operation and maintenance (O&M) stage is the period that emits the most carbon in the whole life cycle of a park. However, the application, management, and evaluation system for low-carbon technology are immature. In this study, the state-of-the-art of common low-carbon technology measures in the O&M stage of the park is included in the literature analysis. To improve the accuracy of the existing evaluation methods of low-carbon parks, either qualitatively or quantitatively, a comprehensive evaluation standard framework of emission reduction technologies for the O&M stage in parks is proposed. Finally, the application of the evaluation method is verified by adopting a case study of a university park.

1 INTRODUCTION

To achieve a green, low-carbon, and circular economy and reach the Chinese carbon peaking and carbon neutrality goals under the new development philosophy, all construction industries in China strive to reduce carbon emissions through scientific management. As an important space of social operation, parks can be divided into industrial parks, tourist parks, hospitals, campuses, etc., according to their different functions. It is the basic unit for improving national environmental protection and carbon emission reduction (Fang *et al.* 2017) and an important carrier fulfilling a low-carbon economy (Zong *et al.* 2018).

The whole life cycle of the construction of a low-carbon park includes the design, construction, O&M, and demolition stage. The O&M stage lasts the longest and contributes the most carbon emission in the life cycle (Lu & Wang 2019; Miliutenko *et al.* 2011). Although many costs and efforts for low-carbon park construction have been cast in the pilot studies, the technology application management and evaluation system for low carbon O&M stage is yet very immature. A standard framework for the O&M stage of the low-carbon park is summarized and proposed based on the review of state-of-the-art low-carbon technology to comprehensively evaluate parks' carbon emissions. In addition, a campus is taken as an example to verify the evaluation framework.

---

*Corresponding Author: j.yang.1@sjtu.edu.cn

174

DOI: 10.1201/9781003433651-23

2 METHODOLOGY

To fully obtain the technical methods of carbon emission reduction in the O&M of the park and scientifically evaluate the carbon emission level of the O&M of the park, the carbon emission reduction technologies and evaluation standards of various parks are comprehensively reviewed first.

The Scopus database covers a wider range of scientific literature, including conference papers, research papers, reviews, and patents (Opoku *et al.* 2021; Zhao *et al.* 2019). Meho *et al.* (2008) proposed that Scopus can retrieve recent literature faster than other engines. Therefore, to ensure the pertinence and timeliness of the comprehensive review, Scopus is selected as the search engine in this paper. ("low carbon" OR "carbon emission") AND ("park" OR "building") AND ("operation" OR "maintenance") was used as the search string to search the literature from 2017 to 2022, September 7. The literature type is restricted to *Journal* and *Conference Proceedings*. Finally, 675 publications were retrieved according to the search method(as shown in Table 1). The subject chart of literature distribution (as shown in Figure 1) shows that the literature is distributed in engineering, energy, environmental science, social science, etc., indicating that the relevant issue studied in this paper is a typical cross-disciplinary subject. The 675 publications are analyzed by VOS viewer software. Technical keywords with a frequency of more than 20 times are selected, as shown in Table 2. The keyword network diagram is shown in Figure 2.

Table 1. Search method.

Items	Contents
Search strings	("low carbon" OR "carbon emission") AND ("park" OR "building") AND ("operation" OR "maintenance")
Source type	Journal and Conference Proceeding
Database	Scopus
Period	2017-2022.9.7

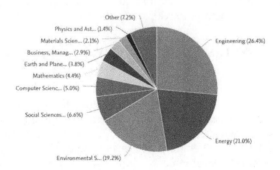

Figure 1. Documents by subject area.

According to the technical keywords obtained by VOS viewers, it can be found that the effective utilization of energy is crucial for the realization of a low-carbon park. In terms of technology, energy recovery and utilization, optimization of energy utilization, energy storage technology and renewable energy sources, including solar energy, wind power, and geothermal energy, etc., life cycle analysis and evaluation, and intelligent and automation technology are common means of low-carbon technologies. And economic means, including carbon trading, investment, cost-effective management, etc., are also important methods to

Table 2. Technical keywords and their count.

Keywords	Count	Keywords	Count
energy utilization	162	electric energy storage	29
optimization	43	cooling systems, energy storage, solar energy	28
renewable energy resource	40	energy use	27
decision making	37	energy management	26
integrated energy systems, intelligent buildings	35	scheduling	25
life cycle analysis	34	sensitivity analysis	24
life cycle assessment (LCA)	33	ventilation, zero energy buildings	22
renewable energies	31	digital storage, greenhouse gas, retrofitting, solar power generation	21
alternative energy, integer programming	30	heat pump systems	20

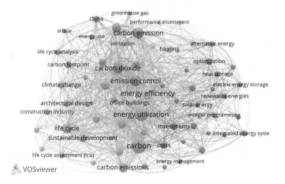

Figure 2. Keywords network of the reviewed papers.

realize the low-carbon O&M of the park. In addition, scientific management of the O&M system is also an important method to realize a low-carbon park.

3 EVALUATE THE SYSTEM ARCHITECTURE.

3.1 *Requirement analysis*

At present, there are two kinds of evaluation schemes for O&M of parks: qualitative assessment and quantitative assessment. In the qualitative assessment, the application and management system of Low-carbon technology is classified and given the scoring weight. For example, the GBT 51356-2019 green campus evaluation standard by the Ministry of Housing and Urban-Rural Development adopts this scheme. The advantages of this scheme are strong systematization of indicators, wide coverage, and easy evaluation. The indicators in the scheme have guiding significance for the low-carbon construction of the park. However, the evaluation cannot quantitatively obtain the carbon emission of the park.

Another evaluation scheme is to evaluate the carbon emissions of the park quantitatively. The value of comprehensive carbon emission is obtained by quantifying the value of carbon emission and carbon sequestration in the park. The advantage of this scheme is that the carbon emission of parks can be quantitatively evaluated. The evaluation results can provide data supporting carbon accounting. However, it has the disadvantages such as a single index, difficulty in comprehensively guiding the low-carbon O&M, and complex operation.

To comprehensively evaluate and guide the carbon emission management and realize the quantitative evaluation of the carbon emission in the O&M stage of parks. A qualitative and quantitative evaluation standard framework is constructed in this paper.

3.2 Construction of qualitative evaluation system for O&M stage of low-carbon parks

3.2.1 Evaluation boundary

Before constructing an evaluation system, the evaluation boundary of a low-carbon park must be clarified. For the evaluation of the O&M stage of the park, the geographical scope of the park's boundary should be taken as the spatial boundary. The time range when the park is put into operation after the completion of construction should be taken as the time boundary. All activities within this temporal and spatial boundary should be included in the scope of carbon emission evaluation for O&M of parks.

3.2.2 Framework of a qualitative evaluation system

Based on the technical keywords about the low-carbon O&M of the park obtained above, a systematical framework of the qualitative evaluation for the low-carbon park operation is proposed. Some items include renewable energy, low-carbon building technology, transportation, energy management, waste treatment and recycling, negative carbon technology, and carbon trading. Each item is divided into several subitems, as shown in Figure 3.

According to the frame diagram in the framework, detailed scoring rules for the O&M stage of low-carbon parks are designed(as shown in Table 3). The higher the score is, the higher the management level of low-carbon park O&M will be.

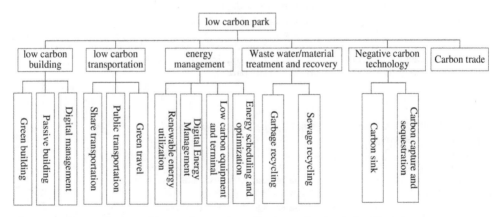

Figure 3. Frame diagram of low-carbon park qualitative evaluation index system.

Table 3. Evaluation and scoring rules for low-carbon campus operation and maintenance.

Items	Subitems	Indicators	Ratio	Score (points)
1 Low-carbon building	1.1 Green building	Score according to the ratio of buildings rated as green buildings to the total number of buildings in the park.	80–100%	4
			50–80%	3
			30–50%	2
			10–30%	1

(continued)

Table 3. Continued

Items	Subitems	Indicators	Ratio	Score (points)
	1.2 Passive House	Score according to the ratio of passive buildings to the total number of buildings in the park.	80–100%	4
			50–80%	3
			30–50%	2
			10–30%	1
	1.3 Digital management	Score according to the ratio of buildings with digital management to the total number of buildings in the park.	80–100%	4
			50–80%	3
			30–50%	2
			10–30%	1
2 Low-carbon transportation	2.1 Share transportation	Setting up shared transportation facilities such as shared bikes and cars.	–	2
	2.2 Public transportation	Setting up public transportation in the park.	–	3
	2.3 Green travel	Advocate green travel, and take green travel education measures.	–	1
3 Energy management	3.1 Utilization of renewable energy	Score according to the ratio of renewable energy generation in the total power consumption of the park.	80–100%	4
			50–80%	3
			30–50%	2
			10–30%	1
	3.2 Digital Energy Management	The park uses a digital energy management system for management.	–	3
	3.3 Low carbon equipment and terminal	Score according to the ratio of low carbon and low energy consumption equipment and terminal in the park.	80–100%	4
			50–80%	3
			30–50%	2
			10–30%	1
4 Wastewater/ material treatment and recovery	4.1 Garbage recycling	Setting up garbage classification.	–	1
		Setting up garbage reuse facilities equipment.	–	3
	4.2 Sewage recycling	Setting sewage treatment and recycling in the park.	–	2
		Setting rainwater collection system in the park.	–	2
5 Negative carbon technology	5.1 Carbon capture and sequestration	Setting up carbon capture and carbon sequestration technology facilities.	–	2
	5.2 Carbon sink	Score according to the ratio of green area.	$\geq 30\%$	3
6 Carbon trade		Score according to the annual use of carbon trading means to purchase clean energy in the annual energy consumption.	80–100%	4
			50–80%	3
			30–50%	2
			10–30%	1

3.3 *Construction of quantitative evaluation system*

The quantitative evaluation system is mainly used to calculate the carbon emission value of the park. The low-carbon construction situation of the park can be directly evaluated through the calculated carbon emission value and the carbon emission per unit area. The carbon emission and reduction projects need to be determined respectively in the quantitative evaluation, as shown in Tables 4 and 5. The sources of carbon emissions and resources consumed by carbon emission projects are shown in Table 4. Moreover, corresponding calculation methods are proposed according to the situation of different carbon emission projects.

Table 4. Carbon emission sources and resources consumed during the O&M stage of the park.

Items	Sources of carbon emissions	Resources consumed
Building	Operational energy consumption	Electric energy, gas, water, etc
Transportation	Vehicle energy consumption	Electric energy, gasoline, diesel, etc
Irrigation	Garden irrigation water	water

Table 5. Negative carbon projects in park operation and maintenance.

Items	Negative carbon absorption
Water recycling	Saving carbon emissions from sewage treatment
Carbon sink	CO_2 is absorbed from the air by green plants

3.3.1 *Carbon accounting for carbon emission items*

(1) Carbon accounting method of energy consumed

The energy consumed in the park mainly comes from building operations and vehicles. The main resources causing carbon emissions include electric energy, gasoline, diesel, and natural gas (Wei *et al.* 2022). The carbon emissions of energy consumed can be calculated by Equation. (1).

$$C_E = \sum_{i=1}^{n} E_i EF_i \qquad (1)$$

Where C_E indicates the carbon emissions generated in the O&M stage of the park (kg); E_i indicates the consumption of energy i(kWh or kg); EF_i indicates the carbon emission factor of the energy i (kg/kWh or kg/kg).

(2) Carbon accounting method of water consumed.

The total carbon emissions of water consumed in the park can be calculated by Eq. (2).

$$C_W = \sum_{i=1}^{n} WWF \qquad (2)$$

where C_W indicates the carbon emissions (kg) generated by water consumed in a park; W indicates the total water consumption (t); WF indicates the carbon emission factor of water (kg/ t).

3.3.2 *Carbon accounting for negative carbon items*

1. Sewage recycling

Sewage recycling mainly refers to establishing sewage treatment devices in the park to realize sewage treatment and form reclaimed water, which can realize the recycling of water resources. Therefore, sewage recycling can reduce the carbon emissions generated by the social treatment of water resources. Based on the sewage treatment volume obtained in a park, the corresponding carbon reduction amount can be calculated by Eq. (3).

$$C_{rw} = \sum_i^n V_w WF \tag{3}$$

where C_{rw} indicates the carbon reduction of sewage treatment (kg), V_w indicates the quality of treated sewage; WF indicates the carbon emission factor of water (kg/t).

2. Carbon sink

Carbon sink mainly refers to the absorption of CO_2 by green vegetation in the park. There are many methods to calculate carbon sink, such as *biomass expansion factor method, remote sensing, model simulation method, planting type-area method*, etc. Among them, the *planting type-area* method proposed by Lin xiande (Zhu *et al.* 2015) has advantages, including simple calculation and easy operation, which is suitable for carbon emission assessment and accounting in large parks. Therefore, this paper adopts this method to conduct carbon emission accounting. Carbon sinks can be calculated by Eq. (4).

$$C_p = \sum_i^n A_i P_i t_i \tag{4}$$

Where C_p indicates the carbon sinks during the O&M stage of the park (kg); A_i indicates the area of the planting method $i(m^2)$; P_i indicates the amount of fixed CO_2 per unit planting surface of the planting method i in one year. t_i indicates the duration of the target period of the planting method i (year).

3.3.3 *Comprehensive carbon emission accounting*

After all carbon emission projects and all carbon reduction projects are calculated. The comprehensive carbon emissions in the O&M stage of the park can be calculated by Eq. (5).

$$C = C_B - C_D \tag{5}$$

where C indicates the comprehensive carbon emissions in the O&M stage of the park; C_B indicates the total carbon emissions generated by each carbon emission project in the park's stage. C_D indicates the total carbon reductions generated by each carbon reduction project in the park stage.

4 CASE STUDY

The college and university campus is a typical type of park with the function of scientific research and teaching. It has the characteristics of a dense population and a long teaching and scientific research equipment operation time. With the continuous development of higher education in China, the number and surface of campuses are expanding. According to statistics, the energy consumption of Chinese colleges and universities accounts for 10% of the total energy consumption of the whole society. The average energy consumption of college students is 4 times the average energy consumption of residents (International energy 2020, Li & Cui 2022). Therefore, constructing low-carbon campuses is very important for the nation to realize the goal of "carbon peaking and carbon neutrality." The environment-friendly campus construction process in China is also gradually changed from the "energy-saving campus" and "green campus" stage to the "low-carbon campus" stage (Zhang & Yan 2021).

To assess the feasibility of the carbon emission evaluation standard framework for the O&M of parks proposed above, a dormitory area of campus is taken as an example. The dormitory area is located northwest of the campus and consists of 12 identical dormitory buildings and a garden. The satellite map of the area is shown in Figure 4.

Figure 4. Overview of the target area scope (Source: AMAP).

4.1 Qualitative evaluation and analysis

Firstly, the qualitative evaluation is applied to the area. In item of low-carbon buildings, the buildings in this area are not green buildings or passive buildings. The energy consumption of the dormitories is counted and paid for via digital management. Therefore, subitem 1.1 and subitem 1.2 will be scored 0 points, and subitem 1.3 will be scored 4 points.

In item of low-carbon transportation, the shared bikes are set up on campus (as shown in Figure 5). The bikes have access to all parts of the dormitory area. So subitem 2.1 will be scored 2 points. The bus site is set up around the area, so subitem 2.2 will be scored 3 points. Green travel education is also conducted on campus, so subitem 2.3 will be scored 1 point.

In energy management, the energy consumed in the dormitory building is mainly electric power and natural gas. There is no renewable energy power generation device set up in the area. A digital energy management system is set up on the campus. The equipment in the dormitory area is also normal, and there are not much carpet equipment and terminals. Therefore, in the item of energy management, subitem 3.2 will be scored 3 points.

In wastewater/material treatment and recovery items, garbage classification is set up (as shown in Figure 6), but there is no special equipment for recycling. No sewage recycling device is set up in this area. Therefore, subitem 4.1 will be scored 1 point.

The park has not set up carbon sequestration technology facilities in negative carbon technology, but the green area has reached 30%. Therefore, subitem 5.2 will be scored 3 points.

Figure 5. Shared bikes on campus.

Figure 6. Garbage sorting on campus.

In carbon trading, electricity in this area is mainly traditional energy, and the purchase of clean energy is limited.

In summary, this area's low carbon O&M management level has scored 17 points.

4.2 *Quantitative evaluation and analysis*

This area's main carbon emission projects come from energy and water consumption in buildings, and the negative carbon projects mainly come from carbon sinks. Accounting for these two items is the focus of the quantitative evaluation of carbon emissions in this area. The following is a quantitative analysis of carbon emissions in this area.

4.2.1 *Calculation of carbon emission*

1. Carbon emission of electricity and water in buildings

The carbon emission calculation is carried out for the carbon emission part. According to the statistics of the digital energy management system set up in the dormitory, energy use is evaluated. The electricity consumption of 40 rooms of dormitories on the campus from June 1st to August 31st is counted to estimate the annual electricity consumption. The average electricity consumption for three months is shown in Table 6.

Table 6. Average electricity consumption of a room in June, July, and August.

Month	June	July	August
Average electricity consumption(kWh)	154.09	184.27	246.75

So the annual electricity consumption of dormitories is about:

$$E = (154.09 + 184.27 + 246.75) \times 4 = 2340.44 \text{kWh}$$

There are about 24 similar rooms in the building. Therefore, the annual electricity consumption of the whole building is about 56170.56 kWh. According to the *GB T 51366-2019, Building Carbon Emission Calculation Standard*, the carbon emission factor of the east China power grid where the park is located is 0.7035t CO_2/MWh. There are 12 similar buildings in the area. According to Equation (1), the annual electricity carbon emissions of the dormitory building are approximate:

$$C_E = 0.7035 \times 56170.56 \times 12 \approx 474191.87 \text{kg}$$

There are about 180 people in a dormitory building. According to the National Bureau of Statistics, the per capita domestic water consumption is about 0.18t per day from 2018 to 2020, and the students' school time is about 300 days. Annual domestic water consumption is:

$$W_1 = 0.18 \times 300 \times 180 = 9720 \text{ t}$$

According to Equation (2), the annual water carbon emissions of the dormitory building are:

$$C_{W1} = 0.168 \times 9720 \times 12 = 19595.52 \text{ kg}$$

(2) Vegetation irrigation water use

According to the reference, the irrigation for shrubs and landscape gardens generally consumes water 1.2-1.8m^3/m^2 per year, and the general lawn irrigation water generally consumes 0.3m³/m² per year. The green space area is estimated based on the satellite map, divided into 1708 cells (as shown in Figure 7). Each cell represents the actual region of about 5 m×5 m. The green space area is determined by counting the grid, and the green space vegetation type is determined according to the graphic color. The dark green area is marked with T indicating tall trees, while the light green area is marked with G indicating low lawn. The schematic diagram of the grid markers of different vegetation types is obtained (as shown in Figure 8). In Figure 8, there are 614 squares marked with T, covering an area of about 15350m². There are 44 squares marked with G, covering an area of about 1100 m².

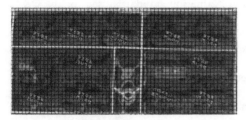

Figure 7. The regional grid division.   Figure 8. Grid marking of different types of vegetation.

The water consumption for tree irrigation in this area is 1.5m³/m² per year, and the water for lawn irrigation is generally 0.3m³/m² per year. The irrigation water consumption is:

$$W_2 = 1.5 \times 15350 + 0.3 \times 1100 = 23355 \text{m}^3$$

According to *GB T 51366-2019*, the carbon emission factor of water is 0.168 kg/m³. Therefore, according to Eq. (4), the carbon emissions due to irrigation water in the area are:

$$C_{W2} = 0.168 \times 23355 = 3923.64 \text{kg}$$

4.2.2 *Negative carbon part calculation*

The main negative carbon source in this region is the fixation effect of green space on CO_2, namely carbon sink. Lin xiande[9] found that low lawn or low grass stem grass can fix about 0.5 kg/m² per year in the subtropical region where the campus is located, and tall broad-leaved trees can fix about 22.5 kg/m² per year. Therefore, according to Eq. (4), the annual carbon sink in the area is:

$$C_p = 15350 \times 22.5 + 1100 \times 0.5 = 345925 kg$$

4.2.3 *Comprehensive carbon emission accounting*

Through the previous accounting, the carbon emission value and the negative carbon project value of the carbon emission project in the region have been obtained. According to Eq. (5), the results of the annual regional comprehensive carbon emission accounting are as follows:

$$C = 474191.87 + 19595.52 + 3923.64 - 345925 = 151786.03 \text{kg}$$

The present area is:

$$S = 1708 \times 5 \times 5 = 42700 \text{m}^2$$

The carbon emissions per unit area are:

$$C_{OM} = C/S = 151786.03 \div 42700 \approx 3.55\text{kg/m}^2$$

The calculation shows that the fixation of CO_2 can be well achieved through a carbon sink, and planting green vegetation is an important part of building a low-carbon park.

5 CONCLUSION

A framework of comprehensive carbon emission evaluation standards used for the O&M stage of the park is proposed to improve the accuracy of the existing evaluation methods of low-carbon parks. Firstly, the common technical methods of low-carbon technologies in the O&M stage of parks are summarized based on the cross-analysis of the existing documents in the Scopus database by VOS viewer. Then, the qualitative and quantitative comprehensive carbon emission evaluation method is set up according to the summary outcomes. Finally, the applicability and feasibility of the proposed framework are verified by a case study. The following conclusions can be drawn from the investigation:

(1) The digital energy management system can provide an efficient supplement for the data collection of carbon emission accounting. The planting type-area method based on a satellite map can adapt well to the accounting of carbon sinks in large-area parks. The calculation result proves that carbon sink can play an important role in the park achieving low carbon.
(2) The qualitative and quantitative comprehensive evaluation framework proposed in this paper can meet the evaluation requirements of the low-carbon O&M of the park. It is proved that the framework has good application potential, which can achieve the unity of operability and science.
(3) In the case study, the differences in the number of rooms, the layout design of different buildings, and carbon emissions from transportation or public area lighting are not considered. The effect of such factors needs to be included in future studies to obtain a more accurate evaluation.

ACKNOWLEDGMENTS

This work was financially supported by the financial support of the Science Research Plan of the Shanghai Municipal Science and Technology Committee (Grant No. 20dz1201301, 21dz1204704), National Natural Science Foundation of China (Grant No. 52078293), the project of Regional Innovation Cooperation in Sichuan Province (2022YFQ0048).

REFERENCES

Fang D., Chen B., Hayat T., et al. (2017) Emergy Evaluation for a Low-carbon Industrial Park. *J. Clean. Prod.*, 163: S392–S400.

Lu K. and Wang H. (2019) Estimating Building's Life Cycle Carbon Emissions Based on Life Cycle Assessment and Building Information Modeling: A Case Study of a Hospital Building in China. *Journal of Geoscience and Environment Protection*, 7: 147–169.

Li P. and Cui P. (2022) Research on the Construction Path of Low-carbon Campus in China. *Resources Economization & Environmental Protection*, 1:146–148.

Miliutenko S., Akerman J. and Björklund A. (2011) Energy Use and Greenhouse Gas Emissions During the Life Cycle Stages of a Road Tunnel - the Swedish Case Norra Länken.Eur. *J. Transp. Infrast.* 12(1): 39–62.

Meho L.I. and Rogers Y. (2008) Citation Counting, Citation Ranking, and-index of Human-Computer Interaction Researchers: A Comparison of Scopus and Web of Science, *J. Am. Soc. Inf. Sci. Technol.* 59 (11): 1711–1726.

Opoku D., Perera S., Osei-Kyei R., et al. (2021) Digital Twin Application in the Construction Industry: A Literature Review. *J. Build. Eng.*, 40: 102726.

The First Comprehensive Energy Service Project of Universities in China, the Benchmark of University Energy use[EB/OL]. *International Energy*.https://www.inen.com/article/html/energy-2299819.shtml, 2020–12-30.

Wei X., Qiu R., Liang Y., et al. (2022) Roadmap to Carbon Emissions Neutral Industrial Parks: *Energy, Economic and Environmental Analysis. Energy*, 238: 121732.

Zong J., Chen L., Sun L., et al. (2018) Discussion on the Construction of Low Carbon Industrial Park. 3S Web of Conferences. *EDP Sciences*, 53: 03007.

Zhao X., Zuo J., Wu G., et al. (2019) A Bibliometric Review of Green Building *Research 2000–2016, Architect. Sci. Rev.* 62 (1): 74–88.

Zhu K., Zhang Q., Wu P., et al. (2015) Progression in Accounting Carbon Sequestration of Urban Green *Space.Shaanxi Forest Science and Technology*, 4: 34–39.

Zhang W. and Yan W. (2021) To Explore the Realization Path of "Carbon Peak" and "Carbon Neutrality" in Universities-Take Beijing University of Science and Technology as an example, *University Logistics Research*, 9: 12–15.

Civil Engineering and Energy-Environment – Gao & Duan (Eds)
© 2023 the Author(s), ISBN 978-1-032-56059-5

Numerical simulation of the swirling effect of an elastic vertical bulkhead in a swirling liquid chamber

Liang Chen* & Wenfeng Wu
School of Ship and Marine Transportation, Zhejiang Ocean University, Zhoushan, China

ABSTRACT: Equating vertical bulkheads in the liquid tank can stop liquid sloshing. Therefore, this paper adopts the VOF method to trace the free liquid surface based on the bi-directional fluid-solid coupling technology and also adopts the dynamic grid to establish the liquid tank model with three-dimensional elastic bulkheads. By analyzing the wave surface elevation of the free liquid surface and the shaking impact pressure of the liquid chamber wall, the effect of elastic bulkheads with different area ratios on the shaking of liquid in the liquid chamber is investigated.

1 INTRODUCTION

Partially loaded liquid tanks are more inclined to produce violent sloshing under certain motion conditions. The motion of a large amount of liquid will produce a highly concentrated slamming pressure at the bulkhead and, in turn, will cause structural damage or even a motion that will affect the stability of the carrying vessel. The liquid cabin will produce a resonance phenomenon when the external excitation frequency is close to or equal to the ship's cabin's inherent frequency. It will cause the liquid in the liquid cabin to climb significantly and have a local impact, so it is important to reduce the liquid sway in the liquid cabin for the safety of marine ship navigation.

The bulkheads are usually set in the ship and marine engineering to stop the liquid sloshing in the cabin. Most of the research on bulkheads at home and abroad is focused on rigid bulkheads, mainly including the location, number, height, form, and solid rate of bulkheads. Based on fluent simulation software, FaltinsenandTimokha (Faltinsen & Timokha 2011) For the rectangular liquid chamber with the addition of vertical baffles, the non-symmetric solution was developed for the asymmetric wobble mode with the addition of vertical baffles, and an approximate solution was derived. Jiao-Yang Tu (Yang 2020) conducted experiments to study the effect of different solid rates of shake-damping bulkheads under different working conditions and carried out fast. Fourier transforms, and wavelet transforms to obtain. Model experiments of Lugni (Lugni *et al.* 2006) were conducted to investigate the overturning mode generated by the impact of the carrier liquid on the liquid chamber wobble structure in a two-dimensional rectangular liquid chamber. The results showed that the rollover mode was related to the air content involved in the liquid. The results showed that the overturning mode is related to the air content involved in the liquid. Souto-Iglesias (Souto-Iglesias *et al.* 2011) elaborated on how to use the shaking platform to perform model tests on liquid cargo tanks and further elaborated on the data processing steps after the experiments. The wave displacement transient amplitude variation characteristics of the tank with a vertical baffle under horizontal excitation. The wave displacement transient amplitude variation characteristics of the tank with a horizontal baffle under transverse shaking excitation were analyzed by Yanmin Guan (Guan *et al.* 2010) using the boundary element method (BEM) to numerically simulate

---
*Corresponding Author: 2482234922@qq.com

186   DOI: 10.1201/9781003433651-24

the liquid sloshing phenomenon in a three-dimensional tank with a baffle. Jin Heng (Jin *et al.* 2017) conducted a numerical simulation study on the resonance shaking problem in the liquid chamber with a local horizontal open plate and compared and analyzed the suppression effect of the local horizontal open plate and local horizontal solid plate, local horizontal open plate and fully covered horizontal open partition on the liquid chamber shaking, and clarified the energy dissipation mechanism of different horizontal plate structures. The energy dissipation mechanism of different horizontal plate structures is clarified.

Therefore, this paper adopts VOF tracking free liquid surface and ANSYS workbench two-way fluid-solid coupling technology to establish a three-dimensional elastic bulkhead swaying model under resonance conditions. It also studies the free liquid surface's pressure distribution and wave surface elevation during the swaying process. It compares the effects of bulkheads with different area ratios on liquid swaying. The study will provide new ideas or methods for the optimal design of the liquid chamber structure.

The results of the study will be presented in the following sections.

2 THEORETICAL ANALYSIS

2.1 *The governing equation*

Based on the general software of ANSYS WORKBENCH and FLUENT, this paper mainly carries out the fluid-structure interaction simulation of the liquid chamber under simple harmonic excitation. The basic control equation is the Navier-Stokes equation, represented by Cartesian tensor form.

The incompressible continuity equation

$$\frac{\partial u_i}{x_i} = 0$$

Incompressible Reynolds equations

$$\frac{\partial \rho u_i}{\partial t} + \frac{\partial \rho \left(u_j - u_j^m \right) u_i}{\partial x_j} = \rho f_i - \frac{\partial p}{\partial x_i} + \rho v \frac{\partial}{\partial xj} \left(\frac{\partial u_i}{\partial x_j} + \frac{\partial u_j}{\partial x_i} \right)$$

ρ is the density of the liquid, u_i is the velocity component in the i direction, u_j is the velocity component in the j direction, and f_i is the volumetric force subjected to the unit volume fluid; This paper deals with both the gas phase and the liquid phase, so the gas-liquid two-phase flow method VOF is used to capture the free liquid level, which is built on top of the Euler mesh, which takes the fluid volume fraction as an interface function

$$F(x, y, z) = \begin{cases} 0 & \text{The calculation domain is filled with gas} \\ 0\text{--}1. & \text{The calculation domin includes both gases and fluids} \\ 1 & \text{The compute domain is filled with fluid} \end{cases}$$

The VOF method completes the nth-phase fluid volume equation for gas-liquid interface tracking by solving a continuous equation for the volume fraction of one phase:

$$\frac{\partial \alpha_q}{\partial t} + \vec{v}_q \bullet \nabla \alpha_q = 0$$

α_q is the volume fraction of the qth phase fluid in the cell; $\vec{v}_q$ is the velocity of the qth phase.

3 NUMERICAL MODEL AND WORKING CONDITION DESIGN

3.1 Modeling

In Figure 1, H is the height of the liquid chamber, h is the depth of the liquid load, L is the length of the liquid chamber and the elastic partition, b is the height of the elastic partition, θ is the rotation angle, H = 0,6 m, L = 1 m, h = 0,24 m, b = 0.01 m in the working condition and apply transverse rocking excitation at the bottom of the liquid chamber, $\theta = 4°*\sin(0.704*2\pi*\text{time})$

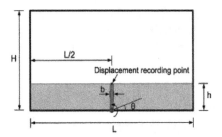

Figure 1. Model.

The test working conditions are shown in Table 1 below.

Table 1. The test working conditions.

	Rotation angle	Liquid loading rate	Bulkhead Young modulus	Area ratio
Case1	4°	40%	50Mpa	25%
Case2	4°	40%	50Mpa	50%
Case3	4°	40%	50Mpa	75%
Case4	4°	40%	50Mpa	100%

4 ANALYSIS OF RESULTS

4.1 Free liquid surface wavefront elevation distribution

During the shaking process, the liquid inside the chamber accumulates energy under external excitation, and the liquid flows through the elastic partition. The elastic partition absorbs the flow field energy, and deformation occurs, producing elastic deformation and the vortex generated in the flow field above the partition, dissipating the energy. It can be seen from Figures 2 and 3 that the wave height of the free liquid surface at X = 0.3 m and X = 0.4 m shows a sinusoidal variation due to the sinusoidal excitation of the external excitation. The wave surface elevation of the free liquid surface at X = 0.3 m and X = 0.4 m shows a sinusoidal variation. With the increasing area ratio of the bulkhead, the wave surface elevation of the free liquid surface decreases continuously. At a 100% area ratio, the wave height reaches the minimum, and the peak is about 0.65 m when the bulkhead has the largest submerged area in the flow field, blocking more liquid and absorbing more flow field energy.

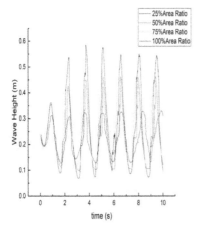
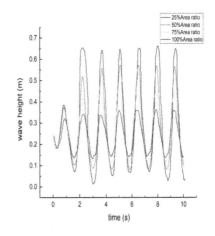

Figure 2. Wave height in x = 0.3 m.   Figure 3. Wave height in x = 0.4 m.

4.2 *Pressure time history curve analysis*

Four pressure monitoring points, p1(0.5.0,0.12), p2(0.5,0,0.18,0.18), p3(0.5,0,0.24), p4 (0.5,0,0.30), were selected at the right bulkhead to monitor the bulkhead pressure in real-time during the shaking process.

Figures 4 to 7 show the pressure distribution of p1-p4, respectively. As can be seen from the figures, the pressure curves all show a cyclic trend, p1–p3 is located at the free liquid surface and the following part, so the pressure value is larger, and p4 is located above the free liquid surface, the pressure is smaller. When the liquid moves from left to right and from right to left, most of the liquid is blocked, the fluid-solid coupling reaction occurs with the bulkhead, and energy exchange occurs. The bulkhead blocks more fluid reduces the kinetic energy in the flow field, and less fluid reaches the liquid bulkhead wall. The pressure impact on the liquid bulkhead wall is reduced, so the pressure amplitude is reduced. It can be seen that the pressure value is greater when located below the free and further away from the free liquid surface when it is mainly realized as the vast majority of the swaying impact pressure and a small amount of hydrostatic pressure, comparing the pressure at different monitoring

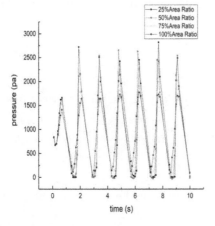
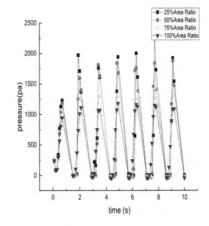

Figure 4. Pressure in p1.   Figure 5. Pressure in p2.

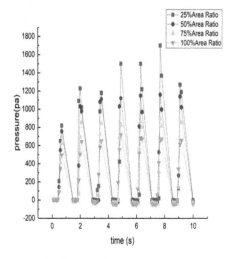 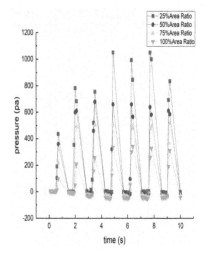

Figure 6. Pressure in p3.   Figure 7. Pressure in p4.

points with the same area ratio. Therefore, the 100% area ratio of the elastic partition can reduce the amplitude of liquid sloshing in the liquid chamber, which is the best effect of sloshing. Using an elastic partition for sloshing, a 100% area ratio is the best choice.

5 CONCLUSION

Using the numerical simulation method at a 40% liquid loading rate, the study of different area ratios of elastic bulkheads for liquid sloshing in the liquid compartment, the following conclusions were drawn.

(1) The area ratio of the elastic bulkhead greatly impacts the shaking impact pressure. When the area ratio is 100%, the shaking effect is good, and the liquid flow field in the chamber can be guaranteed to be smooth.
(2) The area ratio of the elastic bulkhead greatly impacts the wave surface elevation of the free liquid surface. When the area ratio is 100%, the cabin liquid sway amplitude is the smallest.

ACKNOWLEDGMENTS

The Applied Research Project of Public Welfare Technology of Zhejiang Province financially supported this work.

REFERENCES

Faltinsen O.M., Timokha A.N. Natural Sloshing Frequencies and Modes in a Rectangular Tank with a Slat-type Screen. *Journal of Sound and Vibration*, 2011, 330: 1490–1503.

Guan Yanmin, Ye Hengkui, Chen Qingren, Feng Daqui. Numerical Simulation of Liquid Sway in a Three-Dimensional Box with Baffle. *Journal of Huazhong University of Science and Technology (Natural Science Edition)*,2010,38(04):102–104+112. DOI: 10.13245/j.hust.2010.04.011.

Jin Heng, Liu Yong, Li Huajun. Numerical Simulation Study of Liquid Tank Sway with Partial Horizontal Open Plate [C]//. *Proceedings of the 18th China Marine (Shore) Engineering Symposium (Previous).*, 2017:574–578.

Lugni C., Brocchini M. and Faltinsen O.M. Wave Impact Loads The Role of the Flip-through. *Physics of Fluids.* 2006,18(12):1473–1486.

Souto-Iglesias A., Botia-Vera E., Martin A., Perez-Arribas F. A Set of Canonical Problems Sloshing. Part 0: Experimental Setup and Data Processing. *Ocean Engineering*, 2011,38(16):1823–1830.

Tu Jiao Yang. Research on Liquid Sway Resistance Effect in FPSO Liquid Cargo Tanks. *Zhejiang Ocean University*, 2020. DOI: 10.27747/d.cnki.gzjhy.2020.000294.

Civil Engineering and Energy-Environment – Gao & Duan (Eds)
© 2023 the Author(s), ISBN 978-1-032-56059-5

Research on transmission line design under electromagnetic environment impact assessment based on life cycle concept

Yin Du*

Department of Ecological and Environmental Consulting, Guodian Environmental Protection Research Institute Co., LTD., Nanjing, China

ABSTRACT: As an important part of the power grid project, the high-quality construction and safe and stable operation of the transmission line design will determine the safety of the power grid. Therefore, it is particularly important to apply advanced design concepts and ideas. This paper mainly analyzes the application of the life cycle design concept of the substation project.

1 INTRODUCTION

Life cycle design refers to comprehensively considering the requirements and conditions of the project in each stage of the whole life cycle in the design stage and preventing or setting up solutions to the problems that may occur in the subsequent stages in advance. Advanced concepts such as scientific development, sustainable development, and environmental protection are implemented in the project to realize the design concept and method of the project life cycle objective. The whole life cycle design objective of the transmission line design must carry out overall analysis and plan on all costs, resource consumption, environmental cost, and expansion recovery of the whole life cycle of the product so as to satisfy the society, enterprises, and users.

2 RESEARCH STATUS

2.1 *Influence of power frequency electromagnetic field*

With the rapid development of China's electric power industry, the coverage of the power grid is more and more extensive. In recent years, more and more individuals, organizations, and institutions have begun to pay attention to the electromagnetic environment of power transmission and transformation projects. The electromagnetic environment impact of high-voltage and ultra-high voltage power transmission and transformation projects mainly involves power frequency electric field and power frequency magnetic field.

2.1.1 *Some misunderstandings about the electromagnetic environment of power transmission and transformation facilities*

Through research and analysis, the main problem of the electromagnetic environmental impact of power transmission and transformation projects in China is that the public's understanding of power grid environmental protection is biased, which increases the

---

*Corresponding Author: sunnydu@126.com

psychological burden of the public on the negative impact of the power grid and exaggerates the impact of the power grid on environmental pollution. It is precisely because of this psychological factor that more and more people believe that the construction of power transmission and transformation projects and the actual operation of the project will have adverse effects on the public's life and that electromagnetic impact is equivalent to electromagnetic radiation (Ye 2020).

2.1.2 *Influence of power frequency electric field and magnetic field*

The World Health Organization believes that the power frequency electric field and magnetic field that meets international standards are safe for the protection of public health. In contrast, the power frequency electric field and magnetic field standards of domestic power facilities should be stricter than the international standards, Therefore, after the power transmission and transformation project is completed and put into operation, if the test results can meet the standard requirements of the public exposure limit of power frequency electric field strength of 4000 V/m and the public exposure limit of power frequency magnetic induction strength of 100 μt specified in the control limit of the electromagnetic environment (GB8702-2014), it will be safe for the health of the surrounding residents.

2.2 *Power frequency electromagnetic field prediction of transmission line*

2.2.1 *Foreign research*

For the study of the power frequency electromagnetic field around the high-voltage transmission line, foreign countries have been studying it since the 1950s. They have done a lot of calculation and measurement work. At present, foreign scholars mainly use the following methods to calculate the power frequency of electromagnetic fields (Yang 2020):

(1) With the simulated charge method (CSM), Robert G. Olsen calculated the power frequency electric field and magnetic field around the high-voltage transmission line using Maxwell's basic equation and analyzed their respective distribution characteristics.
(2) Boundary element method (BEM): W. Krajewski calculated the two-dimensional power frequency electric field around the 400 kV AC transmission line with BEM and considered the influence of surrounding buildings on the electric field intensity distribution.
(3) Finite element method (FEM). In 1965, a.m. Winslow first applied the FEM to electrical engineering problems, and then P.P. Silvester applied the FEM to solve the steady-state solution of the time-varying field in 1969.

Several mature computing software packages have been developed abroad, such as CDEGS, developed by a Canadian SES company, EFC-400st of Narda company (Yang 2020), and electro and coulomb of ies.

2.2.2 *Domestic research*

The research on the electromagnetic environment of high-voltage power transmission and transformation projects in China started relatively late. However, with the rapid development of power grids, especially the large-scale construction of ultra-high voltage power grids in recent years, more and more scientific researchers have been engaged in the calculation, measurement, and analysis of the electromagnetic environment of high-voltage transmission lines and substations and have made remarkable achievements.

The former high voltage Technology Research Institute of the State Grid Corporation of China carried out a simulation analysis on the electromagnetic environment impact factors such as the ground electric field intensity under the conditions of various tower types and different conductor arrangement modes of the ultra-high voltage AC 500 kV four circuit transmission line on the same tower (Joya & Shirani 2020), and compared them with the actual measured values. At the same time, the design scheme of the electromagnetic field intensity around the conductor was proposed. North China Electric Power University calculated the electromagnetic field intensity of 500 kV and 750 kV transmission lines, respectively, by using CDEGS software. Based on the simulated charge method, Chongqing University has established the power frequency electric field calculation model of an EHV transmission line with a lightning arrester and taken a single-circuit three-phase transmission line as an example.

Reasonable transmission line design can effectively reduce the impact of the electromagnetic environment, such as increasing the height of overhead transmission lines, adopting multi-circuit transmission lines, extending the life cycle of transmission lines as much as possible, reducing the impact of the electromagnetic environment along the lines and reducing the operation cost of transmission lines (Faria & Santos 2020).

3 DESIGN MEASURES FOR THE LIFE CYCLE COST OF THE LINE ENGINEERING

All the following conditions will have an impact on the life cycle cost. Through rational optimization design, its influence can be controlled within the minimum range. The following measures can be considered for transmission line design:

3.1 *Route optimization selection*

During route selection and optimization, we should try to avoid unfavorable geological areas (mining area, goaf area, geological disaster risk area, etc.), bad weather areas, micrometeorological areas, forest areas, and other sections. Various technical measures need to be adopted to ensure the safe operation of the line. Sometimes, even if the cost of line diversion is slightly higher than the cost of crossing and treatment, the diversion scheme shall be selected. However, as for the specific increased investment scheme, the owner, the operator, and the designer shall comprehensively consider the possible reconstruction costs and benefits in the future and determine the final scheme with the maximum investment income.

3.2 *Scientific determination of the meteorological area*

Design meteorological conditions are the basic design parameters of transmission lines. It is directly related to the project cost, safety, and reliability of the power transmission line. Suppose the design maximum wind speed and icing thickness are too small. In that case, the safety and reliability will be reduced, leaving hidden dangers for the safe operation of the line, and tower collapse, line break, or flashover accidents may occur. On the contrary, if the value is too large, the project investment will be increased, and the economy will be poor. In the design process, it is necessary to conduct a comprehensive statistical analysis of the meteorological conditions along the line and adopt different design technical principles for different meteorological areas.

3.3 *Conductor selection and insulation configuration*

The stringing project accounts for a large proportion (about 30%) of the project cost. The selection of a conductor is directly related to the loss of electric energy and has a great impact on the life cycle cost. Combined with the actual situation of the project, the design shall be based on the requirements of the system planning through the consideration of electrical characteristics, environmental characteristics, mechanical performance, power loss, comprehensive technical and economic indicators, etc. A variety of conductors are selected for a comprehensive comparison. In consideration of reducing the cost of operation and maintenance, the insulation standard needs to be properly improved, the appropriate insulation medium and type should be selected, and the project construction cost needs to be increased. The direct and indirect losses that can be avoided far exceed the increased project construction investment in the early stage, with the relationship between the one-time capital investment and long-term production and operation properly handled.

3.4 *Improve the reliability of operation and maintenance*

The transmission lines are distributed in the wilderness, with many points, long lines, and wide areas. The geography, landform, surrounding environment, and meteorology of the lines are complex. If the microtopography and microclimate conditions exceed the design standard, wind bias discharge will be caused. The difficulty of operation shall be considered, and the design conditions shall be increased. For example, the use of iron towers with larger window sizes will increase the construction cost relatively. In comprehensive consideration, the total cost is lower than the cost of operation transformation.

From the perspective of operation and maintenance, if there is a natural disaster exceeding the design conditions, once the accident occurs, it will be difficult and costly to deal with. At the beginning of 2008, the ice disaster in southern China brought huge material and economic losses to the power grids. According to the life cycle cost theory, in the engineering design stage, when it is difficult for the line to avoid unfavorable terrain, it is the most economical measure to properly improve the construction standard, and the direct and indirect losses thus avoided far exceed the project construction investment increased in the early stage. Lightning protection of high voltage transmission lines is the foundation of lightning protection of the whole power system. The transmission line is the weakest link in the system, which is mainly caused by the insulation level, terrain, and grounding of the line. From the perspective of reducing lightning strike losses in the construction stage of the project, it is necessary to improve the design prevention standard, take comprehensive measures, put prevention first, increase the construction project investment, and reduce the life cycle cost. The designer shall strengthen the information communication with the operation unit and refer to the operation experience and lightning protection measures of relevant projects in the same region (Eisner 2020). The lightning protection measures shall be determined after reviewing the data recorded by the operation reclosing and lightning trip rate.

Overhead transmission lines are mostly located in the wilderness, with complex terrain and climate. According to the life cycle cost theory, the line design shall consider the measures against bird damage. According to the bird activity rules mastered by the local operation Department, comprehensive measures are taken according to the frequency of activities, which can effectively reduce the hidden dangers of line operation and thus reduce the operation and maintenance costs.

3.5 *Optimization design of line structure*

In recent years, it has been the consensus of the whole society to protect the ecological balance and forest resources. But in the long run, trees are the biggest hidden danger that directly affects the safe operation of the line. During the construction period of the project, the wasteland has become a forest area. At the initial stage of the line construction, no consideration or inadequate consideration has been given, resulting in the distance to the ground not meeting the requirements. It is bound to make the operation Department transform the line or cut down trees, which is very expensive. This situation is very convenient to solve in the design stage of the construction period. It is enough to avoid or design the ground distance according to the natural growth height of the tree species. The increased cost of the line is limited, but the operation cost and social benefits are very obvious. The loss of tower materials has often occurred before. In recent years, the occurrence of external force damage has become increasingly serious. There are regional sites, and the power tower has gradually become the target of professional thieves. The direct and indirect economic losses are extremely heavy. In order to improve the operation safety of the tower, all connecting bolts (including the cross section) below the tower cross arm shall be provided with anti-unloading bolts. All other bolts shall be provided with anti-loosening fastening nuts (except for the bolts with double caps) to improve the ability of the tower bolts to resist vibration and theft. The increased investment during construction will greatly reduce the repair and transformation costs during operation. Under the premise of reasonable planning of tower type, in-depth research on tower design can further reduce tower material consumption and save project investment. At the same time, selecting a reasonable structural model and reasonable structural construction mode will not only play a key role in the optimization design of the whole structure and the control of the tower weight index but also minimize the damage to the environment.

3.6 *Pay attention to environmental protection and water and soil conservation measures*

According to the life-cycle cost theory, environmental protection and water and soil conservation should be paid attention to at the same time as project construction. The resulting costs will be much smaller than the losses caused by the power outage accident caused by the foundation failure due to the line change and water and soil loss caused by the environmental impact.

4 DESIGN AND APPLICATION OF LIFE CYCLE COST

The main economic and technical principles of a line project, mainly designed by the author, are determined through economic and technical analysis and comparison in strict accordance with the design concept of the whole life cycle of the transmission line and the design requirements of "two types and three new" lines. For example, in the single design of tower grounding, soil electrochemical tests, measured soil resistivity, economic and technical comparisons, and equivalent measures were carried out according to the whole life cycle. Copper-clad steel grounding material with low cycle cost was selected, and the engineering design was optimized.

In order to meet the 30-year maintenance-free grounding grid design of the line, it is particularly important to analyze the corrosiveness of the grounding grid to different grounding materials. According to the analysis of the ground network where the project is located, since the tower is close to the sea and the soil is highly corrosive, if the traditional round steel grounding design is adopted, it cannot meet the requirements of a 30-year safe operation of the line. Therefore, it is necessary to arrange an overhaul or reconstruction in time. According to the life cycle design concept, it is necessary to find new grounding materials, analyze and compare the cost of engineering construction, operation, and maintenance, and determine the optimal grounding design. Therefore, we first analyzed the ground grid, mainly including soil zoning, sampling (Table 1), electrochemical corrosion test (Table 2), etc.

Table 1. Engineering soil collection information.

Number	Specific information	Layer description	Borrow depth (m)
1	#2 Nearby	Alluvial diluvia salty clay	1.2
2	Near phase 6	Diluvia salty clay	1.2
3	#10 Nearby	Plain fill	1.2
4	#13 Nearby	Silt	1.2
5	#16 Nearby	Plain fill or silt	1.2

Table 2. Mineral corrosion life of steel-clad materials with different processes under five soil conditions.

Material type Soil type	1#	2#	3#	4#	5#
Copper layer: 1 mm thick continuous casting copper clad steel	45 years	40 years	38 years	18 years	22 years
Copper layer: 0.254 mm thick, imported electroplated copper clad steel	12 years	10 years	25 years	4 years	6 years
Copper layer: 0.254 mm thick, domestic plated copper clad steel	8 years	9 years	19 years	3 years	5 years
Zinc coating: 0.05 mm thick domestic galvanized steel	5 years	6 years	12 years	2 years	4 years

The corrosion resistance life in the above table shows that in the project area, the corrosion rate of galvanized steel materials is significantly higher than that of copper-clad steel materials. The galvanized layer with a thickness of 0.05 mm can only be used for about half a year, and the corrosion resistance is extremely poor: 1 # The soil in 2 # and 3 # areas is highly corrosive. It is recommended to select 0.8 ^ 1 mm thick copper-clad steel material instead of round steel as the grounding material: 4 # and 5 # soils have very high electrochemical corrosion, and it is recommended to use 1 mm thick copper-clad steel material.

According to the test analysis, in view of the corrosively of the project soil, if the traditional steel grounding material is used in the project, the grounding download of the tower will be operated for about 5-7 years. The reinforcement at the surface will be rusted and broken, and the corrosion degree of the reinforcement at the grounding part will reach 70% after about 8 years of operation. The overall replacement and overhaul must be arranged in time, and the operation and maintenance costs are high: the copper-clad steel has good conductivity and strong corrosion resistance (service life of 50 years), high mechanical strength, low resistivity, and voltage drop. By comparing the life-cycle cost of three grounding materials, such as copper-clad steel (Table 3), according to the life-cycle cost analysis, the new copper-clad steel material is selected for the engineering grounding device.

Table 3. Life cycle cost comparison of three grounding materials.

Comparison items		Hot galvanized steel grounding material	Copper clad steel grounding material	Copper grounding material
Service life (years)		5–8	40	50
Main material cost (10000 yuan)		2.30	10.15	19.75
Installation cost (ten thousand yuan)		3.36	2.12	2.12
Project construction cost (10000 yuan)		5.66	12.27	21.87
Operation in cycle Cost (ten thousand yuan)	Resistance measurement: 2 years/time	3.6	0.5	0.5
	Excavation inspection: 5 years/time	3.84	0.5	0.5
Maintenance cost (overhaul and replacement cost) (10000 yuan)		6.86	Maintenance free	Maintenance-free and easy to be stolen
Cost of power interruption loss caused by fault (10000 yuan)		15.0	High reliability	High reliability
Waste cost (10000 yuan)		Non-recyclable	Recyclable	Recyclable
Total cost in the period (10000 yuan)		34. 94	13.27	22.87
Comprehensive investment comparison (%)		100	38	66

5 CONCLUSION

Life cycle cost is a brand-new economic concept, an advanced and scientific management idea, and the focus of cost control of power enterprises. The life cycle design concept should be applied throughout the whole process of the transmission line design. The relevant theories and methods of transmission line life cycle design should be systematically studied. The impact of various factors and links in the transmission line design process on the entire transmission line service life should be deeply explored. The life cycle management system should be established while taking corresponding countermeasures and measures to reduce the life cycle cost, create economic and social benefits, and ensure the safe and stable operation of power projects.

REFERENCES

Eisner M.J. and Elsharawy H.H. Life Cycle Assessment (LCA) Based Concept Design Method for Potential Zero Emission Residential Building. *IOP Conference Series: Earth and Environmental Science*, 2020, 410 (1):012031 (9pp).

Faria J., Santos I.N. and Rosentino A. et al. *Design of an Underground, Hermetic, Pressurized, Isolated and Automated Medium Voltage Substation*. 2020.

Joya A.J. and Shirani H. *Active Substation Design for Distributed Generation Integration in the Afghanistan's Grid*. 2020.

Yang G. Application of 3D Design Technology in Substation Design. *IOP Conference Series: Materials Science and Engineering*, 2020, 782(3):032086 (4pp).

Yang G. Application of 3D Digital Technology in the Substation Design. *IOP Conference Series: Materials Science and Engineering*, 2020, 782(3):032066 (5pp).

Ye X., Chen H., Sun Q. et al. Life-cycle Reliability Design Optimization of High-power DC Electromagnetic Devices Based on Time-dependent Non-probabilistic Convex Model Process. *Microelectronics Reliability*, 2020, 114:113795.

Civil Engineering and Energy-Environment – Gao & Duan (Eds)
© 2023 the Author(s), ISBN 978-1-032-56059-5

Research on injection, production, and workover engineering technology for fire flooding development pilot test in Menggulin oilfield

Zhonghai Qin* & Xuanqi Yan*
Engineering Technology Research Institute of PetroChina Huabei Oilfield Company, Renqiu, Hebei, P.R. China

Wenjie Wu
PetroChina Engineering & Construction Corp. North China Company, Renqiu, Hebei, P.R. China

Jiancheng Qi
The Second Exploit Factory of PetroChina Huabei Oilfield Company, Bazhou, Hebei, P.R. China

Kejia Wang
Engineering Technology Research Institute of PetroChina Huabei Oilfield Company, Renqiu, Hebei, P.R. China

Zheyong Sun
The Third Exploit Factory of PetroChina Huabei Oilfield Company, Hejian, Hebei, P.R. China

Ying Liu, Lili Wei, Suzhen Guo, Zhi Ma & Zhonghua Shao
Engineering Technology Research Institute of PetroChina Huabei Oilfield Company, Renqiu, Hebei, P.R. China

Fengqun Li
The Third Exploit Factory of PetroChina Huabei Oilfield Company, Hejian, Hebei, P.R. China

ABSTRACT: In order to further improve the recovery efficiency of the Menggulin oilfield, Huabei Oilfield Company has decided to carry out a fire flooding pilot test in the Menggulin oilfield. According to the requirements of fire flooding development, we have researched pilot test engineering technology. Based on the classified research of the old well wellbore and its current situation, we design a reasonable workover project scheme. According to the reservoir scheme parameters, the appropriate gas injection string, ignition technology, and wellhead device are selected to form the gas injection project scheme. Combined with the actual pilot test block and fire drive reservoir parameters, the horizontal and longitudinal comparison and other technical means are used to improve the oil recovery engineering scheme. The research work in this paper is closely combined with the technical problems constantly exposed in the field, which strongly supports the implementation and promotion of the pilot test of fire flooding in Huabei Oilfield.

1 INTRODUCTION

Mengulin Oilfield was formally put into development in 1990. Up to now, the comprehensive water's producing fluid is 91.36%, and the recovery degree of geological reserves is

---

*Corresponding Authors: cyy_qzh@petrochina.com.cn and cyy_yxq@petrochina.com.cn

DOI: 10.1201/9781003433651-26

17.74%. The daily water injection is 1145.5 m^3, the monthly injection-production ratio is 0.65, the cumulative water injection is 2274.3×10^4 m^3, and the cumulative injection-production ratio is 0.95. Now in the "double high" development stage, water flooding to improve recovery is difficult, so it is urgent to change the existing development mode.

The physical properties of crude oil in Menggulin Oilfield are as follows: the surface crude oil density is 0.8996 g/cm^3, the viscosity is 113.63 mPa· s, and the freezing point is 23°C. It is heavy crude oil from the oil properties. In order to improve the recovery rate and change the reservoir development mode, Huabei Oilfield Company decides to use fire flooding to develop the heavy oil reservoir in the Menggulin oilfield and makes the fire flooding development pilot test to realize the economical and effective replacement technology and achieve the goal of sustainable development of the Menggulin oilfield (He *et al.* 2013; Qu 2013).

In order to ensure the smooth implementation of the pilot test, according to the need for fire flooding development, the pilot test engineering technology scheme is formulated, which involves three aspects, including well workover engineering, gas injection engineering, and oil recovery engineering.

2 WORKOVER PROJECT SCHEME

2.1 *The current status of old wells*

The pilot test of fire flooding in Menggulin involves 13 old wells, including 3 injection wells and 10 production wells (7 with screw pumps and 3 with pumping units). The times of drilling and completion vary from 1987 to 1996. The depth of completion drilling is from 865 to 910 m. The table sleeve adopts 273 mm or 339.7 mm completion. The table sleeve steel class is J55, and the downing depth is from 17.16 to 60.38 m. The reservoir casing adopts 139.7 mm casing completion, the depth is from 855.9 to 901 m, and the cement return depth is from 158 to 509 m.

2.2 *The classification of old wells and the design of the workover scheme*

In order to ensure the smooth implementation of the pilot test, the old wells in the test area should be divided into two categories. The first type of well is the single mining conglomerate layer well, which can directly reuse the old production (4 wells). The second type of well is the combined production of sandy conglomerate, which does not meet the pilot test requirements, and the corresponding sealing technology is adopted (7 wells). In view of the above 11 old wells, one observation well, and one conglomerate mining sandstone well with severe shaft deformation, leakage or falling wells (6 wells), milling and repairing sleeves, removing and replacing sleeves, retrieving fallen material, and casing subsidy were adopted (Lei *et al.* 2020).

3 THE GAS INJECTION ENGINEERING SCHEME

3.1 *Perforation process*

At present, the widely used perforation technology in domestic oil fields includes cable transmission perforation and tubing transmission perforation. The advantages and disadvantages of these techniques are shown in Table 1.

The original formation pressure of the perforation layer of the gas injection well in the test area is 7.77 MPa, the pressure coefficient is 0.972, and the well depth is less than 1000 m. Combined with Table 1, we recommend the wireline transmission perforation technology.

Table 1. Comparison of advantages and disadvantages of the two perforating methods.

Name	Advantages	Disadvantages	Scope of application
Cable transfer perforation	1. Adapt to all kinds of gunless perforators, fast construction 2. Negative pressure cleaning, penetration is deeper.	1. It is necessary to use dynamic sealing during construction. If a blow-out occurs, it is not easy to control. 2. Wax or oil in the well is thick, and it is difficult to go down the well	It is mainly used in low-pressure oil reservoirs
Tubing delivered perforation	1. High negative pressure value perforation can be achieved to protect the production capacity of oil and gas reservoirs 2. It can simultaneously shoot the formation of a long well section or multilayer section in one run 3. It can perforate high-pressure oil and gas Wells.	1. Need to drive down the string 2. Long operation cycle 3. High operation cost	Exploration wells, evaluation wells (interpreted as a gas reservoir by logging)Wells containing hydrogen sulfide, wells with unknown formation fluid performance and pressure, and wells with high deviation (>35°)

3.1.1 *Pipe diameter selection*

The gas injection well adopts tubing injection. According to the prediction result of the fire flooding development index of oil reservoir engineering in Table 2, the highest injection volume is expected to be 4×10^4 Nm^3/d, and the maximum injection pressure is expected to be 10.8 MPa.

Table 2. Prediction results of fire flooding development index.

		Plan 1		Plan 2		Plan 3		Plan 4
Time (year)	Gas injector (mouth)	Daily gas injection in a single well (x 10^4 Nm^3/d)	Gas injector (mouth)	Daily gas injection in a single well (x 10^4 Nm^3/d)	Gas injector (mouth)	Daily gas injection in a single well (x 10^4 Nm^3/d)	Gas injector (mouth)	Daily gas injection in a single well (x 10^4 Nm^3/d)
1	3	1.80	5	1.80	6	1.80	10	1.80
2	3	4.00	5	4.00	6	4.00	10	4.00
3	3	4.00	5	4.00	6	4.00	10	4.00
4	3	4.00	5	4.00	6	4.00	10	4.00
5	3	4.00	5	4.00	6	4.00	10	4.00
6	3	4.00	5	4.00	6	4.00	10	4.00
7	3	3.75	5	3.75	6	3.75	10	3.75
8	3	3.50	5	3.50	6	3.50	10	3.50
9	3	3.00	5	3.00	6	3.00	10	3.00
10	3	2.75	5	2.75	6	2.75	10	2.75

The washout yield is calculated using Beggs's formula as follows (Shen et al. 2005):

$$q_{SC} = 40538.17 \times D^2(P_{wh}/(Z \cdot T \cdot \gamma_g))^{0.5} \qquad (1)$$

Type: q_{sc} – critical erosion gas production, 10^4 m³/ d;
D – tubing inner diameter, m;
P_{wh} – wellhead pressure, MPa;
γ_g – relative density of natural gas; Z – natural gas deviation coefficient; T – air flow temperature, K.

Equation 1 is used to calculate the critical erosion flow rate of D73 mm tubing at 10.8 MPa, which is between 31.25×10^4 m³ /d to 34.80 m/d³ (see Table 3 for details). According to this, the maximum injection is 4×10^4 Nm³/d will not cause erosion to the tubing. Therefore, D73 mm tubing is recommended for gas injection wells in the Menggulin conglomerate reservoir.

Table 3. Calculation results of critical erosion flow rate in the Menggulin conglomerate reservoir.

Wellhead pressure (MPa)	Wellhead temperature (K)	Tubing outer diameter/ inner diameter(mm)	Gas density(kg/m³)	Maximum washout flow rate of a gas well($\times 10^4$m³/ d)
1.00	373	73/62	7.572	9.43
2.00	373	73/62	15.143	13.34
3.00	373	73/62	22.715	16.34
5.00	373	73/62	38.944	21.40
10.00	373	73/62	83.098	31.25
12.00	373	73/62	128.799	34.80

3.1.2 Tube selection

The reservoir engineering scheme predicts that the highest injection pressure in the injection well is 10.8 MPa, and CO_2 is injected into the air when the content is less than 0.033%. The calculated P_{CO2} obtained is 0.0036 MPa, so CO_2 corrosion is negligible. But O_2 content is about 21%, and the ignition temperature is 450°C. For the rest of the well section, because the gas injection tubing is above the igniter and is cooled by the injected air, the temperature should be lower than 150°C; there may be oxygen-rich corrosion as the gas injection tubing is in high-pressure air for a long time (Gao 2006).

Combined with the oxygen corrosion rate curve in Figure 1 under different humidity, it can be seen that the corrosion rate of pipe is related to the relative humidity of compressed air. When the relative humidity is less than 50%, the iron and steel materials will not be

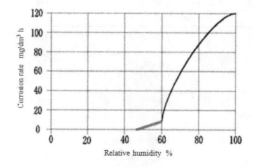

Figure 1. Relation curve between relative humidity and oxygen corrosion rate.

corroded by oxygen absorption. In the Menggulin conglomerate reservoir, the injected air needs to be compressed to 40°C below the dew point, corresponding to a relative humidity of about 15%, for oxygen-rich corrosion does not occur.

Therefore, it can be concluded that oxygen-rich corrosion of materials in a dry environment basically does not occur under high temperature and pressure conditions. Thus, it is determined that only high-temperature oxidation of pipes needs to be considered for gas injection pipe string. Since the 90 H pipe can withstand temperatures up to 500°C and resist oxidation, it is recommended to use 90 H tubing for 100 m in the lower part of the gas injection well and steel N80 tubing for other well sections.

3.2 *Ignition process*

Table 4. Comparison of application conditions and technical characteristics of different ignition processes.

| Applicable conditions and characteristics | Ignition way | Steam preheating | | Electric ignition |
		Spontaneous ignition	Chemical combustion ignition	
The well conditions	Well depth, M	< 1200	< 1500	–
	The net is better than	Small	Small	Big
	Permeability difference	Small	Smaller	Big
	Crude oil viscosity, mPa· s	< 10000	< 20000	< 100000
The characteristics of process	Ground supporting equipment	Steam generator	Steam generator	-
	Downhole string requirements	No	No	More
	Workload and operation difficulty	Less	Less	More
	Controllability	Low	Low	High
Economic suitability	The cost	Low	Lower	High
	Ignition timing	Slow	Faster	Fast
	Heating utilization, %	> 90	> 90	< 80
	Lighting effect	The oil layer vomiting is serious, and the number of ignited layers is less	Vomiting is weak, and the number of igniting layers is large	

At present, there are three main fire drive ignition methods, which are spontaneous ignition, chemical combustion ignition, and electric ignition process. According to the comparison of applicable reservoir conditions, process characteristics, economic applicability, and other aspects (Table 4), the electric ignition process is recommended for the conglomerate reservoir in Menggulin.

3.3 Wellhead device

The reservoir engineering scheme predicts that the maximum gas injection pressure is 10.8 MPa, and QG/HBYT 156-2017 and SY 6561 "Safe Technical Regulations for Gas Injection in Oil and gas fields" (see Table 5) are implemented. It is recommended to install AA grade gas production trees at the wellhead.

Table 5. Wellhead device selection.

Material level	Relative corrosiveness	CO_2 The partial pressure		H_2S partial pressure	Minimum material requirements	
		PSI	MPa	PSI	Body, cap, end and Connect the export	Controlled pressure parts, stem, and Mandrel type hanger
AA Generally run	No corrosion	< 7	< 0.05	< 0.05	Carbon steel or low alloy steel	Carbon steel or low alloy steel
BB Generally run	CO_2 Mild corrosion	7 to 30	0.05-0.21	< 0.05	Carbon steel or low alloy steel	Stainless steel
CC Generally run	High CO_2 Don't include H_2S Moderate to high corrosion	> 30	> 0.21	< 0.05	Stainless steel	Stainless steel
DD Acid run	Low H_2S acid corrosion	< 7	< 0.05	> 0.05	Carbon steel or low alloy steel	Carbon steel or low alloy steel
EE Acid run	H_2S Brittle and containing CO_2 Mild corrosion	7 to 30	0.05-0.21	> 0.05	Carbon steel or low alloy steel	Stainless steel
FF Acid run	H_2S Brittle, high CO_2 Moderate to high corrosion	> 30	> 0.21	> 0.05	Stainless steel	Stainless steel
HH Acid run	H_2S Brittle, high CO_2 Highly corrosive	> 30	> 0.21	> 0.05	Corrosion resistant alloy CRA	Corrosion resistant alloy CRA

According to the Well Control Implementation Rules for oil and natural gas downhole Operations, the rated working pressure of the wellhead device must be more than 130% of the predicted maximum wellhead injection pressure. The rated working pressure of the wellhead device is 14.04 MPa, calculated from the maximum injection pressure of the gas injection well (10.8 MPa). According to GB/T 22513-2013 "Oil and gas industry drilling and production equipment wellhead device and tree" standard (Table 6), the recommended maximum rated working pressure of the wellhead is 21 MPa.

Table 6. Maximum rated working pressure of wellhead equipment.

Pressure (MPa)	Pressure (PSI)
14	2000
21	3000
35	5000

4 OIL PRODUCING ENGINEERING SCHEME

4.1 *Wellhead device*

In order to save the investment cost and refer to the actual production situation of brother oil fields, the old production oil wellhead can meet the actual production requirements. It is suggested that the new wells adopt a DD grade tree (Table 5), which is unified with the old well tree (Zhang *et al.* 2015).

4.2 *Lifting process*

At present, there are 93 pumping wells and 91 progressing cavity pump wells in the production wells of the Menggulin reservoir. The progressing cavity pump wells are mainly direct drive progressing cavity pumps which is more energy-saving. In the main process of fire flooding development, the selection of lifting mode focuses on the influence of crude oil physical properties, temperature, and gas changes (Zhao 2018).

According to the reservoir engineering scheme, the viscosity of formation crude oil is 102.1 mPa·s, the temperature of the produced fluid is 50-60°C, the amount of the produced fluid is 30-40 m^3/d, and the lower pump depth is less than 800 m. Table 7 is a comprehensive comparison of the parameters of the two lifting methods. We combine the current use of equipment in the Menggulin oilfield (Table 8) and comprehensive consideration of various

Table 7. Parameter comparison table of conventional progressing cavity pump and tubular pump.

Pump type	Drive way	Common displacement range (M^3)	Pressure of work (MPa)	Recommended operating temperature (°C)	Down to the depth (m)	devi (°)	note
Screw pump	Direct drive/ non-direct drive	2–86.	< 15	< 100	< 1500	< 30	The full Angle change rate of the
Tubular pump	Pumping unit	5–250.	< 30	< 150	< 3000	< 30	application section of the
Pump type	Recommended viscosity (mPa· s)	Sand bearing effect	Gas impact	Invest in maintenance fees	Commonly used motor (kw)	Conventional pump efficiency (%)	progressing cavity pump should not
Screw pump	< 5000	Small	Small	Low	15 and 22	40–90	be greater than 1/10 m
Tubular pump	< 800	Big	Big	High	15, 22, 37	30–80	
Pump type	Pump diameter (mm)	The largest diameter (mm)	Matching tubing specifications (mm)	Matching casing specifications (mm)	Sucker rod specifications (mm)		
Screw pump	73, 90, 102	90, 107,	62, 89,	> 114	19, 22, 25		
Tubular pump	60, 73, 89, 102	73, 89, 102, 114	62, 73, 89	> 95	16, 19, 22, 25		

Table 8. Basic overview of conglomerate pumping Wells in Menggulin.

Main pumping unit models	Maximum load (kN)	Stroke (m)	Rush time (min⁻¹)
Model 6, Model 8	64	3–4.2 -	3 to 7
Median moving fluid level (M)	Pump depth (m)	Reservoir limit (m)	Pump diameter (mm)
500	530-850.	730-1080.	38, 44, 56

factors. We suggest that the old well oil production equipment is obsolete, and the new wells should choose a pumping unit plus tubular pump lifting process.

5 CONCLUSION

1) The workover project of the pilot test of fire flooding in the Menggulin Oilfield should be combined with the specific well conditions and wellbore conditions of a single well, and the targeted workover process plan should be designed.
2) According to the reservoir scheme parameters, D73 mm flat tubing is recommended for the gas injection string in the gas injection engineering scheme. Electric ignition is recommended for the ignition process. AA grade gas production tree is recommended for the wellhead.
3) DD tree is recommended for the wellhead in the oil production engineering scheme. The lifting process is recommended for old wells, and the rod pump lifting process is for new wells.

ACKNOWLEDGMENTS

This work was financially supported by the Huabei Oilfield Company Research Project: Monitoring process and scheme optimization of fire flooding improvement in Menggulin Oilfield (2022-HB-D16).

REFERENCES

Gao Hongbin. Prediction of Carbon Dioxide Corrosion on Oil Gathering Pipelines. *Journal of Oil and Gas Technology*, 2006(04):410–413.

He Jiangchuan, Wang Yuanji, Liao Guangzhi, et al. *Existing and Emerging Strategic Technologies for Oilfield Development. Beijing: Petroleum Industry Press*, 2013: 48.

Lei Qun, Li Yiliang, Li Tao, et al. Technical Status and Development Direction of Workover Operation of PetroChina. *Petroleum Exploration and Development*, 2020, 47(1): 155–162.

Qu Yaguang. Development Influence on Heterogeneity by In-situcombustion in Heavy Oil Field. *Petroleum Geology and Recovery Efficiency*, 2013, 20(6): 65–68.

Shen Dehuang, Zhang Yitang, Zhang Xia, et al. Study on Cyclic Carbon Dioxide Injection After Steam Soak in Heavy Oil Reservoir. *Acta Petrolei Sinica*, 2005, 26(01):83–86.

Zhang Xialin, Guan Wenlong, Diao Changjun, et al. Evaluation of Recovery Effect in Hongqian-1 Well Block by In-Situ Combustion Process in Xinjiang Oilfield. *Xinjiang Petroleum Geology*, 2015, 36 (4):465–469.

Zhao Liming. *Research on Injection-Production Relation of In-Situ Combustion Based on Oxidation Kinetics. Xi'an Shiyou University*, 2018.

Civil Engineering and Energy-Environment – Gao & Duan (Eds)
© 2023 the Author(s), ISBN 978-1-032-56059-5

Study on the sensitivity of mechanical properties of two-dimensional random granules matter to hole depth

Zhongshan Lu, Yuan Liu* & Yun Lei

School of Civil Engineering, Lanzhou Jiaotong University, Lanzhou, China

ABSTRACT: In view of the sensitivity of mechanical properties to the hole depth, a random granules matter model with circular pores was established based on the discrete element method, combined with complex network theory. The structure of the force chain network was analyzed, and the distribution and evolution of the second invariant of the contact force chain and stress bias in the random particle system were revealed in a compartmental manner from the perspective of mesoscopic mechanics. The results showed that the maximum contact force of the system decreases in the ExpAssoc function with the increase of the hole depth. The mean degree value and mean cluster coefficient of the force chain network are sensitive to the burial depth when the hole depth is shallow but not sensitive to the hole depth when the hole depth is large. The average shortest path is always sensitive to the hole depth. The deformation of the system mainly occurs in the main pressure area, the deformation of the main pressure area and the secondary pressure area are the most sensitive to the hole depth, and the J_{2max} decreases exponentially with the increase of the hole depth. It can be seen that both the permeability and deformation of the system are more sensitive to the burial depth, and the sensitivity of different areas is very different.

Chinese library classification number: 0347.7 **Document code: A**

1 INSTRUCTIONS

The particle system is formed by a large number of discrete particles through contact. Due to the unique deformation and motion characteristics of the particles, the system is highly flexible and dynamic (Shinbrot & Muzzio 2001). As the experimental technique gradually becomes mature with the development of age, scholars have achieved some outcomes on the law of internal force transfer of granular matters in various aspects. Silva and Rajchenbach (2000) conducted photoelastic tests and showed that the strain area inside the two-dimensional particle system under the action of point load has a parabolic boundary, which verified that the force transmission mode of granule discrete medium is different from that of traditional continuum medium. Kruyt (2016) redefined the strong force chain and the weak force chain by comparing the contact force between particles with the average force in the corresponding contact direction and found that the pressure and the shear stress were mostly borne by the weak force chain. Sun and Wang (2008) conducted quantitative analysis on the distribution of force chains in the granular stacks by changing the influencing factors, such as the geometrical properties of the granular medium and the types of external loads, and constructed a force chain network, finding that the force chains were extremely sensitive to the changes of the local force. Yi *et al.* (2008) and Yi *et al.* (2009) calculated the size and

---

*Corresponding Author: liuy@mail.lzjtu.cn

DOI: 10.1201/9781003433651-27

shape of the force chain in the two-dimensional particle system under isotropic and pure shear boundary conditions. They studied the geometric morphology of the force chain network using a complex network method, obtaining insights into the patterns. Fang *et al.* (2020) learned that the arching effect significantly impacted mechanical properties, such as granule deformation and load transfer. Vijayan *et al.* (2022) identified the critical fail points by studying the force chain network in the defected hexagonally packed granule system and discussed the impact of the defects on the formation of the critical fail points. They found that the hexagonally packed structure had two stable force chain network structures. With the continuous development of the application of complex networks, many domestic and oversea scholars studied the surface topological properties of the granular medium using the complex network theory (Liu *et al.* 2020; Ostojic *et al.* 2006; Walker & Tordesillas 2010) and obtained the topological properties of some force chain networks.

Analysis of the existing measurement results showed that the sensitivity of mechanical properties to the hole burial depths varied significantly. In the test simulation, tunnel excavation or shallow pipeline burial inevitably encounters troubles, such as test difficulty, high cost, long duration, and difficulty in changing the burial depths. Due to the development of computers, numerical modeling has become a very convenient choice. Zheng *et al.* (2010) analyzed the location of the tunnel rupture plane and the stability of surrounding rocks under different burial depths using simulation tests and the finite element numerical method. Cao *et al.* (2014) established a model with the discrete element software and discussed and analyzed the energy changes caused by deep rock excavation. Jiang *et al.* (2018) established a discrete element model of the two-dimensional rocks, which transformed the question of different burial depths into the question of different initial stresses. They analyzed the mechanical properties of the surrounding rocks after tunnel excavation under different initial stresses. Most of the existing outcomes only studied the mechanical properties of a small number of regular granule stack systems (Yang *et al.* 2014; Wang *et al.* 2018), and few studied the partitioning of the random particle systems. In engineering, only the stress and strain around the surrounding rocks were studied, and the sensitivity of the internal mechanical properties of the structure was rarely analyzed. In the actual situation, the internal contact force of the system has a great impact on the structure, and different areas play different roles. Therefore, this paper used the discrete element software PFC^{2D} to establish a two-dimensional random particle system with circular holes and simulated and analyzed the mesoscopic mechanical properties of the entire system under the action of local pressure. This paper also extracted the force chains and analyzed the full force chain and the strong (3 times) force chain network. It further studied the characteristics of the force chain network and the sensitivity of the second invariant of stress bias to the burial depths. The study hopes to provide a useful research method for studying the sensitivity of the mechanical properties of the particle system to the depth of the hole.

2 DISCRETE ELEMENT MODEL

To obtain a uniform random distribution and reduce the internal contact force between particles at the beginning of generation, this paper adopted the improved layered compression method (IMCM) (Zhao *et al.* 2020) to generate a two-dimensional random particle model with a height (H) of 0.75 m and a width (L) of 1.5 m through PFC^{2D}. As shown in Figure 1, a loaded plate with a width of a = 0.32 m was set at the midpoint of the upper surface of the model, and the loaded plate was loaded by moving the wall downward, with a descending velocity of v = 0.01 m·s^{-1} and a drop for 0.3s. A circular hole with a radius of r = 0.075 m was set directly below the loaded plate, and the distance between the center of the hole and the upper surface was h. The entire model comprised more than 20,000 two-dimensional disk particles with a particle size of 3-5 mm and a particle porosity of 0.15. For details of other mesoscopic parameters of the model, please see Table 1.

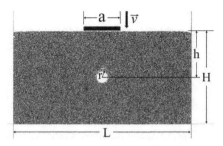

Figure 1. Schematic diagram of a two-dimensional particle model.

Table 1. Mesoscopic parameters of the PFC^{2D} particle model.

Model parameters	Value
Normal stiffness k_n/N·m^{-1}	5×10^7
Shear stiffness k_n/N·m^{-1}	2×10^7
Friction coefficient μ_1	0.8
Gravitational acceleration g/m·s^{-2}	9.8
Damping coefficient ζ/N·(m·s^{-1})	0.3

3 ANALYSIS OF INFLUENCING PATTERN OF HOLE DEPTH

On the condition that other parameters were kept unchanged, the burial depths were changed, obtaining the contact force distribution and the invariants of stress bias of the system at different burial depths $\Delta = h/r$.

3.1 *Analysis of contact force*

Figure 2 shows the force chain diagram of the system under different burial depths. Taking the loaded plate and the hole as the boundary, the system was divided into five areas according to the distribution characteristics of the force chains. Force chains concentrated in the area between the bottom of the loaded plate and the top of the hole, which was the main pressure area. Force chains around the hole were distributed in an arched shape on the top area, which was the secondary pressure area. Force chains under the hole gradually converged, which was the low-pressure area. Force chains spread to both directions in the area on both sides of the loaded plate and above the hole, which was the upper diffusion area. The area under the hole was the lower diffusion area. For the convenience of description later, these five areas were numbered I-V in the figures. It can be seen from the figures that the force chains in Area I were mainly strong force chains and were always the densest, and the distribution of the strong force chains larger than three times the average force became more uniform with the increase of the hole depth Δ. The force chains formed an arch structure at the hole. The top of the hole in Area I is born with most of the force, and the upper side wall of the hole in Area II is also born part of the force due to the mutual extrusion of particles. However, due to the faulted particles, the lower sidewall of the hole in Area II and the bottom of the hole in Area III were hardly affected by force. That is, an arched non-force chain area similar to a vortex structure was formed around the hole in Areas II and III. The strong force chains were distributed sparsely in Areas II and III and became sparser as the burial depth increased. The force chains in the diffusion area were mainly weak force chains. Weak force chains in Area V tended to aggregate near the boundary wall with the increase of

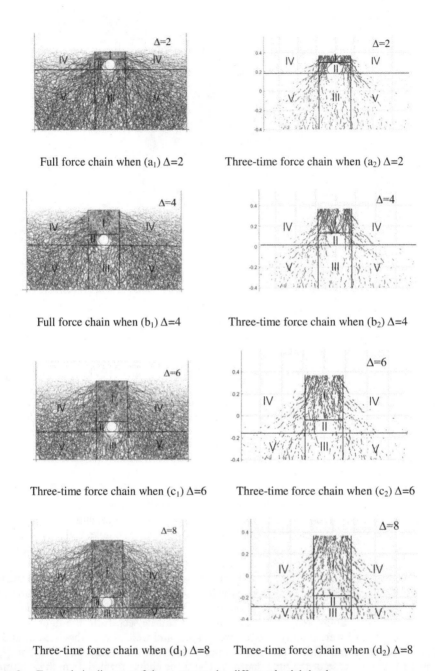

Figure 2. Force chain diagram of the system under different burial depths.

the maximum value of the system contact force occurring at the upper boundary of the hole when $\Delta \leq 4$ and at the lower side of the loaded plate after $\Delta > 4$.

The maximum contact force F_{max} of the system and the changes of the average contact force $<F>$ of the five areas with burial depths Δ were further analyzed. Figure 3 shows that the maximum contact force F_{max} of the system decreased with the increase of the hole burial depths Δ. When Δ increased from 1 to 9, the value of F_{max} decreased from 98477 N to 54301 N, with an amplitude decrease of 44.86%. The decreasing trend approximately conformed to

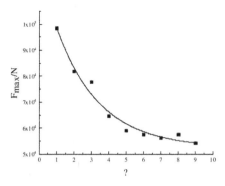

Figure 3. Maximum contact force varies with burial depth.

the ExpAssoc function Fmax=121741.15-69243.6(1-e$^{-\Delta/2.47}$). Further analysis indicated that possibly because the granular medium was different from the general elastic medium, and its force transmission method was related to the arch structure formed by the circular holes, the maximum contact force was more sensitive to the burial depth. Figure 4 shows the variation of the average value of the contact force <F> with the burial depths in different areas. It can be seen that with the increase of Δ, the average value of the lateral contact force <Fx> in Area I increased first and then decreased and was the largest when Δ = 5. The amplitude of variation was between 36.12-65.18 N. The average value of the longitudinal contact force <Fy> increased approximately linearly and increased from 72.21 N to 159.27 N, with an amplitude of an increase of 54.66%. The average value of lateral and longitudinal contact forces in Area II decreased when Δ<6 and slightly increased after Δ>6. Both <Fx> and <Fy> in Area III decreased linearly, and their fitting formulas were <Fx>=-8.65Δ+74.07 and <Fy>=-16.9Δ+149.67, respectively. In Area IV, both <Fx> and <Fy> increased as quadratic functions. The fitting formula <Fx> =4.45Δ^2-24.08Δ+50.3 obtained the minimum value when Δ was 2.71, and <Fy> =2.58Δ^2-9.9Δ+11.97 achieved the minimum value when Δ was 1.96. The change of <Fx> in Area V was similar to that of Area II, and the change of

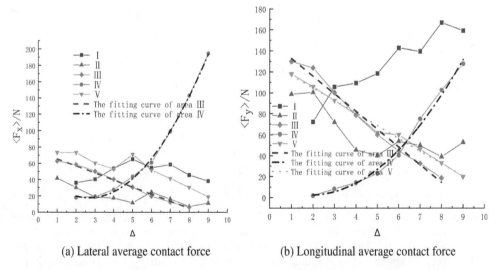

(a) Lateral average contact force   (b) Longitudinal average contact force

Figure 4. Diagram of changes of average contact forces in different areas with the burial depths.

<Fy> was similar to that of Area III. The fitting formula was <Fy> =-12.02Δ+128.68, and the decay rate was slightly lower than that of Area III.

3.2 *Analysis of characteristics of force chains*

Data were extracted, and pajek was applied to calculate the geometric characteristics of the full force chain network. The changing curve of the network characteristic parameters is obtained (mean degree value, mean cluster coefficient, and the mean shortest path) with the hole burial depths Δ, as shown in Figure 5. It can be seen that when Δ increased from 2 to 3, the mean degree value, mean cluster coefficient, and the mean shortest path of the force chain network all dropped sharply, in which the mean degree value decreased by 0.56%, and the mean cluster coefficient decreased by 49.97%, and the mean shortest path decreased by 4.74%. As Δ further increased to 4, the mean degree value further decreased, but the magnitude of the decrease was smaller, which was 0.11%. The mean cluster coefficient increased slightly, with an amplitude of increase of 8.26%. The average shortest path increased sharply by 6.84%, reaching a maximum value of 30.24. When Δ increased again, the mean degree value slightly increased linearly with Δ first and then slightly decreased when Δ= 9. The mean cluster coefficient remained unchanged at around 0.011. The mean shortest path first decreased, then increased at a relatively large margin, reaching a minimum value greater than Δ = 3 when Δ = 7. This means that when the holes were located near the top, the contact between particles was the largest, and the degree of aggregation was the highest. The force transmission efficiency was relatively low. After the holes shifted downward, the contact between particles decreased, the network became sparser, and the degree of aggregation was only half of the original value. The contact between particles and the degree of aggregation did not change significantly with the increase in burial depths. However, the permeability of the network fluctuated back and forth. When the holes were at a distance of three times the diameter from the top or the bottom, the permeability was better, and the force transferred easily in it. When the holes were located near the middle and the bottom, the network permeability was poor, and force could not transfer easily.

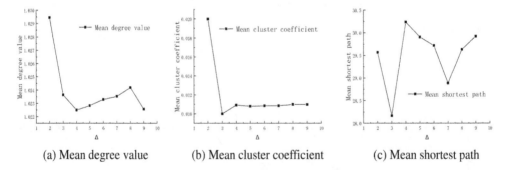

(a) Mean degree value (b) Mean cluster coefficient (c) Mean shortest path

Figure 5. Variation of characteristic parameters of force chain network with burial depth.

3.3 *Analysis of second invariant of stress bias*

The deviatoric stress tensor characterizes the deformation of the object without volume change, and the deviatoric stress tensor invariant does not change with the change of the coordinate system. It is applied by researchers to characterize the essence of the shape change of the object. When the shear modulus was determined, the relationship between the shape-changing energy density of the object and the second invariant J_2 of the stress bias differed by a 1/2 G constant. The second invariant of stress bias has the physical meaning of distortion energy (Xu 2014), which can be used to understand the elastic-plastic state of an object. Therefore, studying the second invariant of stress bias can help us understand the physical meaning of system deformation from a deep level. The distribution diagram of the second invariant J_2 of the system stress bias under different burial depths is shown in Figure 6. It can be seen from the figure that around the hole, the distribution of larger J_2 formed an obvious arch bridge structure, and the value on the main pressure area (Area I) located in the bridge deck was larger. When $\Delta \leq 6$, the arch axis coincided with the upper curved surface of the hole. When $\Delta > 6$, the arch axis was located at the upper curved surface, which gradually became higher than the hole, and the radius of the arch ring was significantly reduced. With the increase of Δ, the maximum value of the upper boundary of the hole gradually disappeared, the value of the lower diffusion area (Area V) gradually increased, and the maximum value of J_2 always occurred near the loaded plate. This shows that the deformation of the system mainly occurred in the main pressure area, and the low-pressure area (Area III) under the hole had almost no deformation and aggregated toward the middle of the bottom boundary.

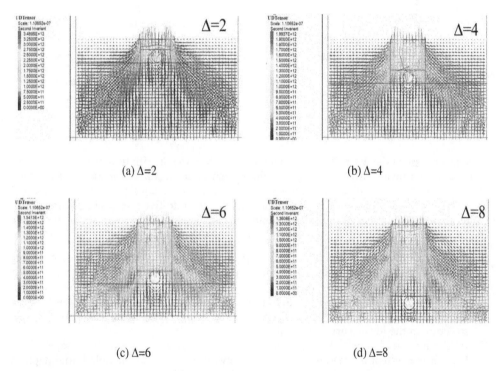

(a) $\Delta=2$ (b) $\Delta=4$

(c) $\Delta=6$ (d) $\Delta=8$

Figure 6. Second invariant distribution of stress bias under different burial depths.

A rectangular coordinate system was established with the center of the system as the origin, and the value and position of the maximum value J_{2max} of the second invariant of stress bias in the five areas were analyzed as a function of the burial depth Δ, as shown in Figure 7. The J_{2max} values of Area I and Area II decreased significantly with the increase of Δ, and the decreasing trend changed exponentially. The fitting equations were as follows: $J_{2max, I} = 0.67\Delta^{-0.79}$, and $J_{2max, II} = 0.29\Delta^{-0.89}$. The value of Area I was always significantly higher than the other four areas, up to 2-8 times that of the other areas. When $\Delta<5$, the value of Area II was larger than the values of the other three areas, and it gradually became smaller than the other areas after Δ>5 and was the smallest when Δ = 9. The J_{2max} values of the other three areas changed slightly. J_{2max} always appeared near the loaded plate in Area I and gradually moved down in Areas II, III, and IV with increasing hole burial depths. When the hole was close to the bottom of the model, the position of J_{2max} in Area V shifted to the middle of the bottom.

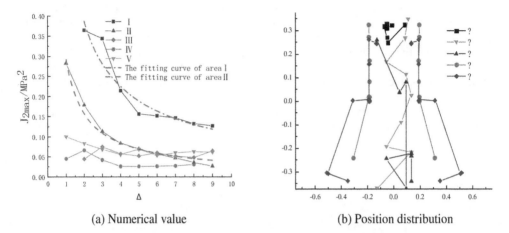

(a) Numerical value (b) Position distribution

Figure 7. The second invariant change of the maximum stress bias under different burial depths.

4 CONCLUSIONS

This paper applied the discrete element method to obtain the distribution of the force chains and the second invariant of stress bias in the system under local compression for the two-dimensional random particle system with circular holes. According to the distribution characteristics of the force chains, the system was divided into five areas. The sensitivities of the mechanical properties of the entire system and each area to the hole depths were studied. Below are the conclusions:

(1) Taking the upper boundary of the hole as the apex of the arch, the distribution of strong chains formed an arch structure. The main pressure area (Area I) located on the upper part of the apex was dominated by the strong force chains and was always the densest. The low-pressure area under the arch (Area III) and the diffusion area (Area V) were dominated by weak force chains. There was a non-force chain area similar to a vortex structure in the lower part of the hole.

(2) As the hole gradually shifted near the bottom boundary, the maximum contact force F_{max} in the system decreased, and the law between its size and the hole burial depth Δ approximately conformed to the ExpAssoc function. The average longitudinal contact force <Fy> of the main pressure area (Area I) and the upper diffusion area (Area IV) located in the upper part of the hole increased. The average longitudinal contact force <F_y> of the low-pressure area (Area III) and the lower diffusion region (Area V) in the lower part decreased linearly. The decay rate of Area III was higher than that of Area V. The average value of the lateral contact force <Fx> in Area IV was the most sensitive to the burial depth and was in a quadratic function relationship with the burial depth Δ. The F_{max} of the entire system occurred at the upper boundary of the hole when Δ ≤ 4 and at the lower side of the loaded plate after Δ>4.

(3) The mean degree value and the mean cluster coefficient of the force chain network were sensitive to the burial depth when the burial depth was small ($\Delta < 4$), and both decreased abruptly with the increase of Δ. When the burial depth further increased ($\Delta \geq 4$), both were less sensitive to the burial depth and hardly changed with the change of Δ. The mean shortest path was always sensitive to the burial depth, but its variation did not have an obvious pattern, and it reached the minimum value when $\Delta = 3$.

(4) Similar to the strong force chain, the larger value of the second invariant J2 of stress bias of the system formed an arch structure around the hole, the main pressure area (Area I) was the main deformation area of the system, and the low-pressure area under the hole (Area III) barely deformed. The deformation of the lower diffusion area (Area V) was sensitive to the hole burial depth. With the increase of Δ, the range of the large deformation area gradually expanded and shifted to the middle of the bottom boundary. The J_{2max} values of the two pressure areas (Areas I and II) were the most sensitive to the burial depth, and both decreased exponentially with the increase of Δ.

REFERENCES

Akhil Vijayan, Arnab Banerjee and Raghuram Karthik Desu. Role of Packing Defects in Force Networks of Hexagonally Packed Structures Using Discrete Element method. *Granular Matter*, 2022, 24(1): 1–17.

Cao W.Z., Li X.B., Zhou Z.L. et al. Energy Dissipation of High-Stress Hard Rock Excavation Disturbance. *Journal of Central South University: Science and Technology*, 2014, 45(08): 2759–2767.

Fang Y., Guo L. and Hou M. Arching Effect Analysis of Granular Media Based on Force Chain Visualization. *Powder Technology*, 2020, 363.

Jiang M.J., Zhang P., Chen T. et al. Discrete Element Simulation of Tunnel Excavation at Different Depths. *Chinese Journal of Underground Space and Engineering*, 2018(S2): 9.

Kruyt N.P. On Weak and Strong Contact Force Networks in Granular Materials. *International Journal of Solids & Structures*, 2016, 92–93: 135–140.

Liu X.F., You S.H. & Xie C.K. Study on the Instability of Clay Granular Slope Piles Based on the Complex Network. *Applied Mathematics and Mechanics*, 2020, 41(9): 12.

Ostojic S., Somfai E. and Nienhuis B. Scale Invariance and Universality of Force Networks in a Static Granular Matter. *Nature*, 2006, Vol 439: 828–829.

Sliva M.D. and Rajchenbach J. Stress Transmission Through a Model System of Cohesionless Elastic Grains. *Nature*, 2000, 406: 708–710.

Sun Q.C. and Wang G.Q. Force Distribution in the Static Granular Matter in two Dimensions. *Acta Physica Sinica*, 2008, 57(8): 4667–4674.

Troy Shinbrot and Fernando J. Muzzio Noise to Order. *Nature*, 2001, 410; 251.

Walker D.M. and Tordesillas A. Topological Evolution in Dense Granular Materials: A Complex Networks Perspective. *International Journal of Solids and Structures*, 2010, 47(5): 624–639.

Wang D., Du W. and Wu Y. Tuning Mechanical Response in Granular Layers by Adding Rigid Particle Channel. *Powder Technology*, 2018, 329: 85–94.

Xu L. Distribution and Application of Floor Deviatoric Stress Tensor Invariants Under Close-Distance Multiple Pillars. *Beijing: China University of Mining and Technology (Beijing)*, 2014.

Yang Y., Wang D. and Qi Q. Quasi-static Response of Two-Dimensional Composite Granular Layers to a Localized Force. *Powder Technology*, 2014, 261(261): 272–278.

Yi C.H., Mu Q.S. and Miao T.D. Discrete Element Method Simulation on the Force Chains in the Two-Dimensional Granular System Under Gravity. *Acta Physica Sinica*, 2009, 58(11): 7750–7755.

Yi C.H., Mu Q.S. and Miao T.D. The DEM Simulation for a Two-Dimension Granular System with Point Defects. *Acta Physica Sinica*, 2008, 57(6): 3636–3640.

Zhao L.L., Ning G.X. and Li S., et al. Discrete Element Analysis of Earth Pressure and Influencing Factors for the High-Filled Cut-and-Cover Tunnel with Different High-Span Ratios Based on EPS Load Reduction. *Journal of Engineering Geology*, 2020, 28(6): 9.

Zheng Y.R., Xu H. & Wang C. et al. Failure Mechanism of Tunnel and Dividing Line Standard Between Shallow and Deep Bury. *Journal of Zhejiang University: Engineering Science*, 2010(10): 7.

Civil Engineering and Energy-Environment – Gao & Duan (Eds)
© 2023 the Author(s), ISBN 978-1-032-56059-5

Multifractal characteristics of Cu element grade distribution in Jiama porphyry copper deposit

Hui Liu & Li Wan*
School of Guangzhou University, Guangzhou, China

ABSTRACT: The mineralization intensity of the mining area is determined by the grade distribution of ore bodies. Taking the Cu grade sequence of No.28 and No.32 exploration lines of the Jiama porphyry copper deposit in Tibet as an example, MFDFA was used to discuss the spatial distribution characteristics of the Cu grade sequence under different mineralization levels. The results show that the Cu grade of all drifts has multifractal characteristics. There are differences in the multiple parameters of element distribution under different mineralization levels, which the size relationship of Δh and $\Delta\alpha$ are intensely mineralized drifts > moderately and barely mineralized drifts. The $\Delta\alpha_L$ of intensely mineralized drifts is significantly greater than $\Delta\alpha_R$, indicating that the high-grade Cu element in intensely mineralized drifts are relatively concentrated. It is instructive to study the multifractal characteristics of Cu grade distribution in different mineralization levels, which is conducive to providing an evaluation for further regional metallogenic potential.

1 INTRODUCTION

A multifractal is a set defined on a fractal, which is composed of singular measures of multiple scalar indexes. It can effectively characterize the complexity and heterogeneity of a fractal. The Multifractal Detrended Fluctuation Analysis (MFDFA) method proposed by Kantelhardt (2002) analyzes the scaling behavior of time series at different levels through different order fluctuation functions, which is often used to analyze nonstationary time series. MFDFA is widely used in nonlinear complex signal processing and analysis in various disciplines such as physics, biology, and the environment. This method can effectively reveal the multifractal scaling behavior of sequences (Li *et al.* 2021; Wang *et al.* 2016; Zhang *et al.* 2021; Zhang *et al.* 2019).

Jiama is the most important copper polymetallic resource base in the Gangdise metallogenic belt. The Cu element in this region has a high grade and good continuity. As a typical representative of porphyry deposits in the Gangdise metallogenic belt, it has superior prospecting prospects (Zhang *et al.* 2019). In recent years, scholars have carried out a series of studies on the Jiama copper deposit. The results show that the distribution of the Cu grade has obvious spatial distribution characteristics of porphyry-skarn type deposit genesis (Deng *et al.* 2017; Zheng *et al.* 2010).

In this paper, MFDFA is introduced into the field of geochemical element analysis, taking Jiama porphyry copper polymetallic deposit in Tibet as an example. The MFDFA method is used to study the distribution characteristics of the Cu grade, which can provide an effective function for revealing the geological and metallogenic process.

---

*Corresponding Author: wanli@gzhu.edu.cn

216   DOI: 10.1201/9781003433651-28

2 MATERIALS AND METHODS

2.1 *Study area and data source*

The Jiama mining area is located in the eastern part of the Gangdise metallogenic belt in the Tethys tectonic domain of Tibet. The tectonic position is located in the central and southern parts of the Gangdise-Nyainqing Tanggula plate in Tibet. Jiama copper polymetallic deposit is a super large porphyry-skarn deposit discovered and developed in Tibet in recent years, of which the amount of copper is about 7.52 million tons. In this paper, the Cu grade of 10 drifts on exploration lines No.28 and No.32, located in the middle of the mining area, is selected as the research object.

Based on the proportion of the Cu grade greater than 0.3 wt%, the drifts are classified into three levels: (I) intensely mineralized drifts, in which the proportion is larger than 50%, and the mineralization is very thick; (II) moderately mineralized drifts, in which the proportion is between 20% and 50%, and the mineralization is relatively intermittent and thin; (III) barely mineralized drifts, in which the proportion is lower than 20%, and the orebodies are barely developed (Wan *et al.* 2015). Therefore, S2801, S2802, and S2803 belong to the intensely mineralized drifts; S3201, S3202, S3203, and S3204 belong to the moderately mineralized drifts; S3205, S3206, and S3207 belong to the barely mineralized drifts.

2.2 *MFDFA method*

MFDFA can effectively eliminate interference trends and estimate multifractal spectrum. For a sequence $\{x_i\}(i = 1, 2, \ldots, n)$ with a given length of n, the steps are as follows:

Step 1: Calculate the cumulative deviation sequence $y(i)$ of sequence $\{x_i\}$ by

$$y(i) = \sum_{k=1}^{i}[x_k - \bar{x}], i = 1, 2, \ldots, n \tag{1}$$

Where $\bar{x}$ is the mean value of $\{x_i\}$; x_k is the kth data value of the original sequence.
Step 2: The cumulative deviation sequence $y(i)$ is equally divided from the first data into $N_s = \text{int}(n/s)$ non-overlapping segments with equal length s. Since the length n is often not an integer multiple of s, in order not to discard the remainder of the tail, the segmentation process is repeated from the last data of the sequence to obtain $2N_s$ segments.
Step 3: Let $y_v(i)$ be the m-order fitting polynomial at each segment v, and m is a positive integer. When $v = 1, 2, \cdots N_s$, the variance $F^2(s, v)$ is determined by

$$F^2(s, v) = \frac{1}{s}\sum_{j=1}^{s}\left\{y[(v-1)s + i] - f_v(i)\right\}^2 \tag{2}$$

When $v = N_{s+1}, N_{s+2}, \ldots, 2N_s$, the variance $F^2(s, v)$ is determined by

$$F^2(s, v) = \frac{1}{s}\sum_{i=1}^{s}\left\{y[N - (v - N_s)s + i] - f_v(i)\right\}^2 \tag{3}$$

Step 4: The qth order fluctuation function of all $2N_s$ segments is calculated by

$$\begin{cases} F_q(s) = \left\{\dfrac{1}{2N_s}\sum_{v=1}^{2N_s}[F^2(s, v)]^{q/2}\right\}^{1/q}, q \neq 0 \\ F_q(s) = \exp\left\{\dfrac{1}{4N_s}\sum_{v=1}^{2N_s}[\ln F^2(s, v)]\right\}, q = 0 \end{cases} \tag{4}$$

Step 5: The scaling exponent of the fluctuation function based on the relationship between $F_q(s)$ and s is determined by

$$F_q(s) \propto s^{h(q)} \tag{5}$$

By analyzing log-log plots of $F_q(s)$ versus s for each value of q, the slope of the fitting line is considered as the generalized Hurst exponent $h(q)$. When the sequence is stationary, $h(q)$ has constant values. Generally, the fluctuation function value $F_q(s)$ is an increasing function of s, and the generalized Hurst exponent $h(q)$ is a monotone decreasing function varying with q.

Step 6: The relationship between the multifractal scaling exponent $\tau(q)$ and the generalized Hurst exponent $h(q)$ can be calculated by

$$\tau(q) = qh(q) - 1 \tag{6}$$

Then the multifractal singularity index α and singularity spectrum $f(\alpha)$ can be determined from $\tau(q)$ via the Legendre transform:

$$\alpha = d\tau(q)/dq \tag{7}$$

$$f(\alpha) = q\alpha(q) - \tau(q) \tag{8}$$

The singularity index α is used to describe the singularity degree of different intervals in the observation sequence, and the singularity spectrum $f(\alpha)$ is used to describe the fractal dimension of the singularity index in different intervals. When $f(\alpha)$ is independent of α, the sequence shows monofractal characteristics; When the shape of $f(\alpha)$ is unimodal convex distribution, the sequence shows multifractal characteristics.

The multifractal intensity can be expressed by the range of $h(q)$:

$$\Delta h = h(q_{\min}) - h(q_{\max}) \tag{9}$$

The maximum width of the multifractal spectrum is defined as

$$\Delta \alpha = \alpha_{\max} - \alpha_{\min} \tag{10}$$

The left and right branch width of the multifractal spectrum can be defined as

$$\Delta \alpha_L = \alpha(0) - \alpha_{\min}, \ \Delta \alpha_R = \alpha_{\max} - \alpha(0) \tag{11}$$

Where $\alpha(0)$ is the special value of the singularity index at $q = 0$.

3 ANALYSIS

3.1 *Statistical description and normality test*

The study calculates the mean value, standard deviation, skewness, and kurtosis of the Cu grade sequence of 10 drifts. It carries out the Jarque-Bera (J-B) normality test at a 5% confidence level to analyze the normality of element content distribution.

The results show that the skewness of the Cu grade sequence in all drifts is greater than 0, showing the characteristics of right-skewed. The kurtosis values are higher than 3, except for S2802 and S3202. Compared with the normal distribution, their distribution is characterized by sharp peaks and fat tails. The J-B normal test statistics are significant at the 5% confidence level except for S2802, which shows that the Cu grade sequence does not obey the normal distribution. This nonlinear complex feature is not suitable to be described by the traditional statistical method of normal distribution hypothesis, so the nonlinear fractal method is used.

3.2 Results and discussion

According to the nonlinear fluctuation trend of the Cu grade sequence, the polynomial fitting order $m = 3$ is taken. The length s ranged from 10 to $L/5$, with an increment of 2 for sequences length with less than 300 and an increment of 3 for sequences length with more than 300. The qth-order ranged from -5 to 5 with an increment of 0.05. The corresponding multiple parameters are calculated by Formulas (1) - (11). The calculation results are shown in Table 1.

Table 1. Multifractal parameters calculated by MFDFA method.

Borehole	Rank	Δh	$\Delta \alpha$	$\Delta \alpha_L$	$\Delta \alpha_R$
S2801	III	1.2899	1.7217	0.3302	1.3915
S2802	III	1.8000	2.4281	0.2170	2.2112
S2803	III	1.0456	1.6815	0.6577	1.0238
S3201	II	1.0066	1.3144	0.5343	0.7801
S3202	II	0.4311	0.6193	0.2744	0.3448
S3203	II	0.6547	0.9140	0.3991	0.5149
S3204	II	0.4455	0.7763	0.3343	0.4420
S3205	I	0.3939	0.6793	0.4166	0.2627
S3206	I	0.5714	0.7840	0.4340	0.3500
S3207	I	0.8645	1.2272	0.8845	0.3427

Δh and $\Delta \alpha$ are unequal to 0, indicating that the Cu grade sequence in all drifts are multifractal, and their intensity is different; all drifts have $\Delta \alpha_L \neq \Delta \alpha_R$, which represents an asymmetric multifractal spectrum; Δh and $\Delta \alpha$ of intensely mineralized drifts are greater than those of moderately and barely mineralized drifts; $\Delta \alpha_L$ of intensely mineralized and moderately mineralized drifts is smaller than $\Delta \alpha_R$, while $\Delta \alpha_L$ of barely mineralized drifts is larger than $\Delta \alpha_R$, and the difference between $\Delta \alpha_L$ and $\Delta \alpha_R$ of moderately mineralized drifts is relatively small, indicating high-grade local aggregation of drifts with high mineralization intensity.

Figures 1(a) – (b) show the curve of the generalized Hurst index $h(q)$ changing with q of 3 drifts in the No.28 exploration line and 7 drifts in the No.32 exploration line, respectively,

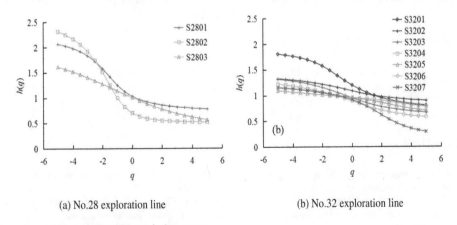

(a) No.28 exploration line　　　　　　　　(b) No.32 exploration line

Figure 1. The generalized Hurst index curve.

which demonstrate that $h(q)$ of the Cu grade sequence in all drifts decreases with the increase of q, showing a nonlinear change. Figures 2(a) – (b) show the curve of the multifractal spectrum of 3 drifts in the No.28 exploration line and 7 drifts in the No.32 exploration line, respectively. The multifractal spectrum curves of all drifts are convex and asymmetrical, indicating that they are multifractal.

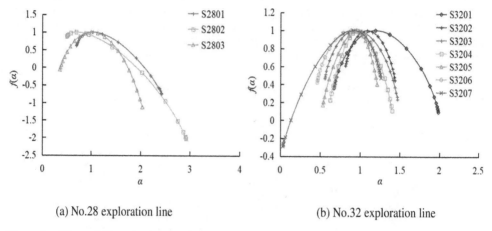

(a) No.28 exploration line (b) No.32 exploration line

Figure 2. The multifractal spectrum curve.

4 CONCLUSION

MFDFA was used to study the multifractal characteristics of the Cu grade sequence in multiple drifts of different mineralization levels of the Jiama porphyry copper deposit in Tibet.

The results show that all Cu grade sequences have multifractal characteristics. Δh and $\Delta \alpha$ of the Cu grade sequence with intensely mineralized drifts are larger than those of moderately and barely mineralized drifts, and the sequence fluctuation is more significant. $\Delta \alpha_L < \Delta \alpha_R$ in intensely and moderately mineralized drifts indicate that high grade is relatively uniform and favorable for the mineralization and contains metallogenic potential. The study discusses the differences in multifractal parameters of different mineralization levels, which is helpful in revealing the formation mechanism of the metallogenic system. It is of great scientific significance for understanding the complex structure of the super-large porphyry copper system.

ACKNOWLEDGMENTS

This research was financially supported by the National Science Foundation of China (41872246); Innovation Research for the Postgraduate of Guangzhou University (2020GDJC-M33).

REFERENCES

Deng Jun, Wang Qingfei, Li Gongjian (2017) Tectonic Evolution, Superimposed Orogeny, and Composite Metallogenic System in China. *Gondwana Research*, 50:216–266.
Jan W. Kantelhardt, Stephan A. Zschiegner, Eva Koscielny-Bunde et al. (2002) Multifractal Detrended Fluctuation Analysis of Nonstationary Time Series. *Physica A: Statistical Mechanics and its Applications*, 316(1):87–114.

Li Yanhui, Wu Bo, Zhang Jiao et al. (2021) Impact of COVID-19 on the Evolution of PM2.5/PM10 in Zhangjiajie Forest Park. https://kns.cnki.net/kcms/detail/51.1699.N.20210604.1634.006.html.

Wan Li, Deng Xiaocheng, Wang Qingfei et al. (2012) Method of MF-DFA and Distribution Characteristics of Metallogenic Elements: Example from the Dayin'gezhuang Gold Deposit, *China. Journal of China University of Mining*, 41(01):133–138.

Wan L., Zhu Y.Q., Deng X.C. (2015) Multifractal Characteristics of Gold Grades Series in the Dayingezhuang Deposit, Jiaodong Gold Province, *China. Earth Sci Inform*, 8(4): 843–851.

Wang Xianxun, Mei Yadong, Li Weinan et al. (2016) *Influence of Sub-Daily Variation on Multi-Fractal Detrended Fluctuation Analysis of Wind Speed Time Series. PloS one*, 11(1).

Zhang Shi, She Lihuang, Wang Yafan et al. (2019) Retinal Image Segmentation Based on Multifractal Detrended Fluctuation Analysis. *Journal of Northeastern University (Natural Science)*, 40(02):158–163.

Zhang Shuqing, Zhang Yun, Liu Haitao et al. (2021) Application in Power Quality Analysis Based on Multifractal Detrended Fluctuation Analysis and Improved Decision Tree. *Acta Metrologica Sinica*, 42 (04):424–431.

Zhang Zebin, Tang Juxing, Tang Pan et al. (2019) The Origin of the Mafic Microgranular Enclaves from Jiama porphyry Cu Polymetallic Deposit, Tibet: Implications for Magma Mixing/Mingling and Mineralization. *Acta Petrologica Sinica*, 35(03):934–952.

Zheng Wenbao, Chen Yuchuan, Song Xin et al. (2010) Element Distribution of Jiama copper -polymetallic Deposit in Tibet and its Geological Significance. *Mineral Deposits*, 29(05):775–784.

Civil Engineering and Energy-Environment – Gao & Duan (Eds)
© 2023 the Author(s), ISBN 978-1-032-56059-5

Analysis of the experience and enlightenment of EU carbon market construction

Wenjing Ruan*
State Grid Energy Research Institute, Beijing, China

ABSTRACT: The European Union ETS (EU ETS) was launched in 2005. It is the world's largest carbon market, has the longest running time, and has the most mature system design. Its construction experience has important reference significance for constructing China's carbon market. The fourth phase of the EU carbon market will be launched in 2021. This article combs the EU carbon market dynamics, summarizes the EU carbon market construction experience, and puts forward the experience and enlightenment for China's carbon market construction and development.

1 INTRODUCTION

The EU Carbon Market (EU ETS), launched in 2005, includes 27 EU member states and three countries: Iceland, Norway and Liechtenstein, covering electricity, energy-intensive industries and aviation, accounting for about 40% of the EU's total carbon emissions. In 2020, the trading volume of the EU carbon market reached 8.1 billion tons, accounting for about 78% of the total global carbon trade; 201.4 billion euros, accounting for about 88% of the total global carbon trade.The operation of the EU carbon market follows the principle of "Cap-and-Trade", that is, the EU sets quotas, each country sets caps for their own emissions, determines the industries and enterprises included in the emission trading system, and allocates a certain number of emission permits to them. If the actual emission of the enterprise is less than the quota, the remaining quota can be sold, otherwise it needs to be purchased in the trading market.

2 THE FOUR STAGES OF THE EU ETS DEVELOPMENT

The long-term emission reduction target was determined at the early stage of the establishment of the carbon market and gradually promoted according to four stages:

The first phase (2005-2007) is a pilot phase, covering power and energy-intensive industries. The total quota is determined using a "bottom-up" approach, and more than 95% of the quota is allocated free of charge. This stage mainly includes companies with power plants and internal combustion engines exceeding 20 MW, as well as industrial companies such as oil refining, coke ovens, steel, cement, lime, ceramics, glass, and paper (Wang *et al.* 2019). The total quota is formulated by EU member states and submitted to the European Commission for review and implementation after review and approval. In 2005, the total annual quota approved by the European Commission was 2.122 billion tons, and more than 95% of the quota was allocated free of charge. The allocation method was based on the

---

*Corresponding Author: ruanwenjing@sgeri.sgcc.com.cn

222   DOI: 10.1201/9781003433651-29

"grandfather rule." (Chevallier 2009). At this stage, the free quotas obtained by power generation companies cover most of the emissions.

The second stage (2008-2012) is a transitional stage, the coverage is gradually expanded to the aviation industry, and the proportion of free quota is reduced to 90%. At the end of 2007, the total annual quota approved by the European Commission was 2.082 billion tons. The quota allocation method is consistent with the first stage, and the free allocation proportion shall be at least 90%.

The third stage (2013-2020) is the stabilization stage. The national distribution plan is canceled. The total quota is determined by the European Union and tightened at an average annual rate of 1.74%. The free distribution is gradually replaced by auctions (Demailly & Quirion 2006). The power industry fully implements auctions to obtain quotas. Include petrochemical, aluminum, ammonia, and other industries, and eliminate small facilities with annual emissions of less than 25,000 tons. Starting from the third phase, a unified total emission control at the EU level has been adopted to replace the national distribution plan. Since 2013, the carbon emission ceiling has been reduced by 1.74% yearly to ensure that greenhouse gas emissions in 2020 will be reduced by 21% compared to 2005. The quota allocation method has undergone significant changes. The overall auction ratio is not less than 50%, and 100% of the quotas for the power industry are obtained through auctions. The allocation of free quotas has been changed from the "Grandfather's Rule" to a unified EU benchmark law.

The fourth stage (2021-2030) is a mature stage to accelerate the pace of emission reduction, establish a variety of support mechanisms, promote low-carbon investment, adjust the distribution mode, and prevent carbon leakage more targeted. In 2021, the fourth ten-year phase will be launched, increasing the annual reduction of quotas from 1.74% in the third phase to 2.2%. At the same time, the benchmark value of free distribution in the manufacturing industry was revised, and the modernization fund and innovation fund were established. On July 14, 2021, the European Commission issued a "fit for 55%" package proposal to align its carbon market with the EU's climate ambition and plan for 2030. These measures include increasing the planned emission reduction in 2030 (from 43% to 61%), phasing out the free quota of the aviation industry, and integrating the maritime industry into the European carbon market.

Table 1. Four stages of the development of the EU carbon market.

Stage	Stage I (2005-2007)	Stage II (2008-2012)	Stage III (2013-2020)	Stage IV (After 2021)
Carbon emission reductiontarget	Getting carbon trading experience does not require Kyoto Protocol emission reduction commitments.	By 2012, 8% on a 1990 basis.	By 2020, 21% reduction from 2005.	By 2030, 43% reduction from 2005 (likely adjusted to 61% in the future).
Coveragetrade	Enterprises with power plants and internal combustion engines of over 20 MW, as well as oil refining, coke ovens, steel, cement, lime, ceramics, glass, and paper-making.	On the basis of the first stage, increase the aviation industry.	On the basis of the second stage, to increase the petro-chemical, aluminum, ammonia and other industries.	On the basis of the third stage, gradually included in the shipping industry.

(continued)

Table 1. Continued

Stage	Stage I (2005-2007)	Stage II (2008-2012)	Stage III (2013-2020)	Stage IV (After 2021)
Totalsetting	The Member States propose aggregate control targets from the bottom up.		The European Commission sets an aggregate target and distribution program.	
Quotadistribution	Free at "grandfather law" and 5% at auction.	The "grandfather law" gives free 90% and auction 10%.	The auction proportion shall not be less than 50%, and the free part is mainly according to the 100% auction of the "benchmark line law" power industry.	57% of the quota auction was allocated, and the aviation industry free quota was phased out.

In 2021, the EU carbon market entered the fourth stage. Under the 55% emission reduction target, European carbon prices have repeatedly hit new highs. In May, the European carbon price broke the historical high of 56.43 euros/ton, an increase of more than 50% compared to January, and doubled the price of 20-25 euros/ton at the beginning of the EU carbon market in 2005.

3 EU CARBON MARKET CONSTRUCTION EXPERIENCE

Firstly, the free quotas have been declining year by year, and the allocation of quotas has gradually transitioned from free allocation to auction. Since 2013, 100% of the quotas for the power industry have been obtained through auctions, and 15% of the quotas for energy-intensive industries have been obtained through auctions. The auction proportion has increased in stages, and 100% of the quotas will be auctioned by 2020. The auction revenue is mainly used to support projects such as emission reduction, energy efficiency, and renewable energy and to provide energy subsidies for low-income groups (Beck & Kruse-Andersen 2020).

Secondly, the market mechanism has been improved year by year, and a market stability reserve mechanism has been established to reduce excess quotas. After the 2008 financial crisis, the EU's carbon market had excess quotas and low prices. To solve the problem of supply and demand imbalance in the carbon market, the EU adopted measures such as postponing the auction of quotas and establishing a market stability reserve mechanism (MSR). In 2019, about 397 million quotas were put into the MSR to reduce supply and stabilize the price of quotas at 25 euros/ton in 2019.

Thirdly, the power industry can effectively transmit carbon prices downstream through electricity prices. The EU electricity market is relatively mature (Tan *et al.* 2020). In the electricity market environment, carbon prices are included in the cost of power generation and ultimately transmitted to end users through electricity prices. According to estimates, in the European electricity market, power generation companies included in the EU carbon market can transfer 60%-100% of their carbon emission costs to electricity prices.

4 CONCLUSION

Generally speaking, the construction and operation of the EU carbon market are based on a relatively mature electricity market. The growth of electricity consumption in the EU economy and society has slowed down, the structure of power generation energy is diversified, the proportion of coal power is low, and the retirement period is generally approaching. The actual conditions are quite different. China's economy has entered a stage of high-quality development, and electricity demand is still growing. The energy structure is still dominated by coal. With the rapid development of new energy, coal power is also required to continue to play a role in ensuring a reliable power supply and providing system flexibility. Therefore, the construction of China's carbon market should be designed in accordance with national conditions, and the EU's experience should not be copied. The relationship between the construction of the carbon market and the guarantee of coal power supply should be properly handled.

(1) The national carbon market gradually introduces an auction mechanism and expands the proportion of paid distribution. Learning from the EU carbon market experience, we can choose specific industries, such as steel and electricity, to try a certain percentage of auction models after the national carbon market runs smoothly. After the market matures, we can gradually reduce the free allocation ratio and increase the auction ratio simultaneously. The auction revenue can support investment in energy efficiency, renewable energy, smart grid, and other projects through carbon emission funds and other methods.

(2) A carbon price and electricity price transmission mechanism could be established. Unlike the European Union, China's planned electricity and market electricity coexist, and the transmission of carbon prices to electricity prices faces a "dual track" problem, which is difficult to transmit in the short term. Coal prices have been high recently, and carbon prices should be reasonable. It is necessary to ensure the basic survival of thermal power companies. With the improvement of the electricity market pricing mechanism, the carbon price can be transmitted to the market-based electricity price. The transmission mechanism must be designed for the planned electricity generation and consumption to effectively share the cost of emission reduction.

ACKNOWLEDGMENTS

This work is supported by the science and technology project of State Grid Co., Ltd (Compliance risk assessment of the company's international business under the new international situation; No.: SGHEXT00FZJS2100017)

REFERENCES

Beck U. and Kruse-Andersen P.K. (2020) Endogenizing the Cap in a Cap-and-Trade System: Assessing the Agreement on EU ETS Phase. *Environmental and Resource Economics*, pp.77(4):781–811.

Chevallier J. (2009) Carbon Futures and Macroeconomic Risk Factors: A View from the EU ETS. *Energy Economics*, pp.31(4):614–625.

Demailly D. and Quirion P. (2006) CO2 Abatement, Competitiveness and Leakage in the European Cement Industry Under the EU ETS: Grandfathering Versus Output-based Allocation. *Climate Policy*, pp. 6 (1):93–113.

Tan D., Gao S. and Komal B. (2020) Impact of Carbon Emission Trading System Participation and Level of Internal Control on Quality of Carbon Emission Disclosures: Insights from Chinese State-Owned Electricity Companies. *Sustainability*, pp. 11–12.

Wang Q., Guan H., Wan R. et al. (2019) The Impact of European Union Emission Trading Scheme on the Development of China's Carbon Market. *Foreign Economic Relations & Trade*, pp. 101–112.

Civil Engineering and Energy-Environment – Gao & Duan (Eds)
© 2023 the Author(s), ISBN 978-1-032-56059-5

Statistics and restoration governance research on gradation geological environment problems in limestone mines

Lei Cheng & Dongdong Li*
School of Engineering Management and Real Estate, Henan University of Economics and Law, Zhengzhou, China

ABSTRACT: Eleven limestone mines were investigated in the study area. Gradient-type geological environment problems were mapped existing in the research area. The gradient geological environment problems were summarized by combined survey data of gradient geological environment problems in 11 limestone mines were counted, and relevant geological environment data of the research area were systematically analyzed. These gradually changing geological environment problems could be summarized as land resource destruction, terrain with landscape destruction, and ecological resource destruction. The causes of the gradual geological environment problem were objectively analyzed from the geological environment system input, and output perspective by the limestone mine current situation were combed in the study area. It was found that gradual geological environment problem was a gradual process, and the impact on the environment was distributed in the whole disaster effect. Its genetic mechanism was an external manifestation of the mine's primary geological environment system structural change process under the external mining activities action. The geomorphological remodeling, soil reconstruction, vegetation system with biodiversity, biodiversity conservation, and monitoring were effective paths for restoration of gradually variable geological environment problems combined with the current status of regional land use. The stacked topsoil had a natural seed bank role as an important part of soil reconstruction.

1 INTRODUCTION

Limestone was a building material mineral resource. Limestone mines played an important role in infrastructure construction (Prakash 1995; Li 2013; Wu 2009). At the same time, the large-scale and high-intensity development and utilization of limestone mines also produced a series of geological and environmental problems. These geological and environmental problems include sudden geological disasters such as collapses, landslides, and mudslides as well as gradient mine geological environment problems such as soil erosion and land occupation caused by the destruction of topographic landforms and vegetation (Tang 2013). The gradual evolution of geological environmental problems attracted people's attention compared to sudden geological disasters. The analysis and restoration plan for the graded geological environment problems causes still need to be further improved, although the existing mine geological environment protection policies and measures also involve the gradation geological environment of limestone mines (Bai 2014; Diamantis 2021; Ewa Desmond 2022; Gupta 2022; Hemeda 2019; Nyamsari 2022; Tang 2013). This paper took a limestone mine in a certain area as an example to carry out research based on the above. The paper has

---

*Corresponding Author: ldd@huel.edu.cn

226

DOI: 10.1201/9781003433651-30

identified the gradual geological environment problems through the combined on-site investigation with mapping, analyzed causes of the gradual geological environment problems by relevant data such as land use status, and proposed corresponding restoration and governance plan based on research area geological environment data systematic analysis.

2 OVERVIEW OF THE MINING AREA AND BACKGROUND OF THE GEOLOGICAL ENVIRONMENT

2.1 Overview of the mining area

The limestone mine was close to villages and towns in the study area. The surrounding traffic lines were in all directions. The infrastructure was perfect. The Pinglian Line runs through the east and west. The open flat line, semi-flat line, and Dongping line run through the north and south. There were 1 middle school, 1 primary school, 2 kindergartens, 2 natural scenic areas, 1 new Jiangnan ecological park, and some villages and towns were scattered around the mining area, which was a residential area. There were 11 limestone mines in the research area. These mining rights were independent and densely distributed. Since December 2007, the limestone mines in the mining area have begun mining operations. Some mines have been closed at present. Some mines had been mined close to recoverable reserves. These mining of limestone mines are shown in Table 1 (Lei 2015; Zhao 2014). Number 1 represents the Yiyang Jinshan calcium products limestone mine, 2 represents Yiyang Luoyuan Calcium Oxide Plant, 3 represents the Yiyang Hengyuan limestone mine, 4 represents the Yiyang Bee Sugar Ridge limestone mine, 5 represents Yiyang Xiangyuan Calcium Oxide Plant, 6 represents Yiyang Dongsheng active calcium oxide plant, 7 represents Yiyang Li Zhiqiang limestone mine, 8 represents Yiyang poplar limestone mine, 9 represents Yiyang Hongyuan Calcium Oxide Plant, 10 represents Yiyang Hengda limestone mine, 11 represents Yiyang Liliu limestone mine. A total land area of 9.4877 km^2 was affected and mainly watered land due to limestone mine production before field investigation and mapping. These were destroyed land, both including dry land and include facility agricultural land, both include pit pond water surface and include other grasslands, as well as formed towns, mining land, villages, and bare land.

Table 1. Limestone mine present situation.

Mine Number	Footprint/ km^2	Recoverable Reserves/m^3	Mining Scale (t/y)	Mining History Years	Mine Service Total Years
1	0.1455	38,000	14,000	7.0	7.3
2	0.1465	35,850	14,000	6.5	6.9
3	0.0800	30,950	13,500	6.1	6.1
4	0.1569	45,100	15,000	8.0	8.0
5	0.0855	31,090	13,500	6.0	6.2
6	0.1465	42,000	14,000	8.0	8.0
7	0.0755	28,800	12,000	6.0	6.4
8	0.1475	41,500	14,000	7.5	7.9
9	0.0855	31,000	11,000	7.0	7.5
10	0.1459	35,200	12,500	7.0	7.5
11	0.0955	31,200	11,000	7.5	7.6

2.2 Geological background

The study area was located in the North China stratum Shichi-Qishan Community. The geotectonic division belongs to the monopoly area of Bears Ears Mountain. It was controlled by the Sanmenxia fault with obvious tectonic properties. The geological structure form was mainly monoclinic. The overall trend of the stratum was north-west-south-east, and the tendency was 205° with an inclination angle was 21°. It was formed as a small undulating low mountain by controlling the Longbo-Huashan anticline and the Sanmenxia fault. The terrain was high in the north, west, south, and east. The height of the valley landmark was 280 m~310 m. The elevation of the mountain peak was 650 m~690 m, and the terrain height difference was 380 m~410 m. The exposed strata mainly include Cambrian, Devonian and Carboniferous strata. The Cambrian to Devonian subsystem was carbonate rock with anisotropy and rheology. The characteristics of softening in contact with water were more obvious. The middle Devonian to the carboniferous system was dominated by non-soluble clastic rocks with complete rock masses, dense rocks, hard, high vertical compressive strength, and high mechanical strength. In addition, a small number of Mesozoic intrusive veins had emerged in the periphery of the mining area. It was a fourth-series distribution in gullies and mountain basins. The lithology was residual slope accumulation and thin layers of reddish-brown clay sandy soil.

The mining area was dominated by carbonate fissure karst water. There was a local distribution of fracture karst water and loose soil pore water sandwiched between carbonate rock and clastic rock. The karst in this area was moderately developed with a linear levitorosity of 5% to 15%. There were steep rock cliffs formed by quartzite-like sandstone, which had collapsed phenomena. The surface slope was large. The topographic cutting was relatively strong, and the trenches were developed. The crisscrossing was staggered, which was conducive to the runoff and discharge of atmospheric precipitation. The main factors of water filling in mining were atmospheric precipitation. Rainfall was the leading factor in geological disasters' frequent occurrence caused by landslides, rockfalls, and debris flow due to mining which had great destructive power on traffic routes.

The mining area belongs to the warm temperate continental monsoon climate with four distinct seasons. It was westerly and northwesterly winds mainly and larger winds in spring and winter with a maximum wind speed of 20 m/s. Statistics from 1972 to 2021 hydro-meteorological data show the average annual temperature was 14.8°C; the annual precipitation range was 288.6 mm~1,022.6 mm; the average annual precipitation was 694.9 mm, and mostly concentrated in 7~9 months, accounting for about 60% of the annual precipitation; the maximum frozen soil depth was 16 cm; the annual frost period was October to April of the following year and the annual frost period was 145 days. The soil types were cinnamon soil and brown loam, mainly with high organic matter content, good water retention, and fertilizer retention ability. The vegetation type was mainly deciduous broad-leaved forest and shrubland. Locust and Sour Jujube were distributed at the bottom of the valley mostly. Thorn strips were mostly mixed with weeds on the slopes. Crops were wheat, corn, and sweet potatoes mainly. Common wildlife included rabbits, sparrows, and insects.

3 SORTING OUT THE GRADIENT GEOLOGICAL ENVIRONMENT PROBLEMS IN THE MINING AREA

The mine production excavated and damaged cultivated land, another grassland, and urban and village land based on occupying industrial and mining land and bare land in the mining area according to the geological environment data of the research area. It destroyed the closed water storage environment of the pit pond water surface and destroyed the original vegetation in the mining area on a large scale. This destroyed the terrain and landscape and the destruction of the surface water system. The land occupied by Mine No. 1 in the research area was all arable land and village land, which affects the development of the New Jiangnan Ecological Park. Mines No. 2, 3, 10, and 11 were close to the school, which not only occupies the cultivated land but also causes environmental pollution to the school. Mine No. 4 occupies cultivated land while directly excavating and damaging the land of the established town. Mine No. 6, 7, 8, and 9 were close to the waters causing damage to the surface water system. The waste slag in the mining area was easy to form a slag slope flow which was washed by the rainwater to the low-lying area, destroying the surrounding watering land, dry land, and agricultural facilities under the erosion of rainwater. The open-pit limestone mine strips the rock mass by blasting operation, and the blasting crumbs were irregularly distributed. The rock body of the mining surface was exposed, which seriously damaged the natural landscape of the original green mountains and green waters. The Huoshanzi Scenic Area and Banpo Mountain Scenic Area have become important growth points for the local economy. However, the devastating limestone mines along the Banping Line and Pinglian Line seriously affected the overall landscape of the study area by causing strong visual pollution and greatly restricting the development of the tourism industry in the study area. The rock walls of the mine bracing face were exposed, the rock was broken, the soil was lost, and the terrain and landform landscape were gradually destroyed.

The residents of the surrounding residential areas, school teachers and students, and scenic staff responded strongly during the on-site investigation. Mine production drove the local economic development at the same time, was quietly worsening the local livable living environment. Destroying land resources and mining palm surfaces exposed broken rocks, waste rock piles, and abandoned mining buildings not only destroyed the surrounding beautiful natural landscape but also destroyed the vegetation and other ecological factors, resulting in the gradual deterioration of geological disasters, serious soil erosion, and affecting the sustainability of local natural ecological functions. The total land resources destroyed by limestone mining in the mining area were 1.1431 km^2, according to statistics. These gradual geological environment problems included land resource destruction, terrain, landform landscape damage, and ecological resource destruction. The gradual geological environment problems in the mining area could be summarized using the inductive summary method. The graded geological environment problems were summarized in Table 2 in the study area.

Table 2. Limestone mine area geological environment problem summary table in the study area.

	Gradation Geological Environment Problem		
Mine Number	Destruction of Land Resources (km^2)	Destruction of Topographic Landscapes	Destruction of Ecological Resources
1	0.1455	The rock wall of the mine brace was exposed, the rock was broken, the soil erosion was lost, and the terrain and landform were gradually destroyed; the mining area was close to the scenic spot, and the exposed surface of the mine brace on the rock wall causes visual pollution.	It destroys the local natural ecological function and its sustainability, resulting in a gradual geological environmental disaster.
2	0.1109		
3	0.0534		
4	0.1569		
5	0.0002		
6	0.1457		
7	0.0755		
8	0.1378		
9	0.0855		
10	0.1362		
11	0.0955		

4 CAUSE ANALYSIS OF GRADIENT GEOLOGICAL ENVIRONMENTAL PROBLEMS

The use of blasting, mechanical, and other operational technologies to mine limestone mines was the direct cause of the gradual geological environment problems in the open air.

These result in the fragmentation of the rock structure, the opening of the structural surface, the formation of exposed slopes of manual excavation, the destruction of slope vegetation, and the destruction of the plant growth vegetative layer. The excavation of open-pit stops with slag and construction plants to occupy land was an original form that led to the destruction of arable land resources and topography. Land resource protection was not considered in the process of mining, so the abandoned ore and slag were disorderly piled to form waste rock piles. Soil erosion and gradual geological disasters induced by rainfall were the indirect reasons for the gradual destruction of surrounding arable land, roads, and construction land. The phenomenon of private mining and illegal mining in the research area was one of the important reasons for the continuous expansion of the gradual geological environmental problem scale.

The primary limestone mine was a relatively stable geological environment system formed by geological historical sedimentation and tectonic action. The main components of the system were limestone rock layers, rock mass structures, slopes, small undulating low mountain landforms, ecological environment, water and soil environment, and human activities. The study area belongs to the warm temperate continental monsoon climate according to the above-mentioned discussion of geological and environmental conditions. The regional vegetation was dominated by other grasses and shrubs. The vegetation cover on the slopes was better, the surface precipitation was abundant, the survival of vegetation depended on the infiltration of rainfall, and the plants grew in the crevices of the rock in the thin layer of residual soil before mining. The thorns and vegetation on the mining operation surface were completely damaged due to the open-pit mining of ore. The rock mass excavation was exposed and formed barren slopes and many unstable steep slopes. The local rock and soil body washed surface soil under the action of rainwater and causing soil erosion. There was a tendency further to damage the ecological environment of the mining area. At the same time, the greening of plants was difficult to survive due to the bare rock face lacking the soil for plant growth which greatly increased the damage to ecological resources. Limestone mining activity injected the original geological environment system, which transformed the original geological environment system structure and formed a new geological body.

The process of mining was an environmental input effect that the native mine geological environment system then responds to this steady stream of input, and the gradual geological environment problem has been formed at this time which includes land resource destruction, terrain, landform landscape damage, and ecological resource destruction. The mining area's geological environment system has been in the extreme degradation stage of the ecosystem at this time and then superimposed on other external environment inputs, such as rainfall, earthquakes, etc. The gradual geological environment problems were more prominent, and the possibility of restoring the geological environment system was getting smaller and smaller in the mining area.

The gradual geological environment problem itself was a gradual process, and it can be considered that its impact on the environment was distributed in the entire disaster effect by the perspective of the limestone mining impact on the geological environment from the mining to the closed pit. There was not only the transformation of the original geological environment system but also the formation of new geological bodies due to this transformation in the process of interaction between mining activities and the geological environment of the mine. It can be considered that the gradual geological environment problems' causative mechanism was a structural change process external manifestation that primary geological environment system under the action of external mining activities of the mine.

5 RESTORE GOVERNANCE SCENARIOS

Its restoration needs to be adapted to local conditions, not only in line with the original land use planning but also to ensure the sustainable use of the restored mining area. It should have the best comprehensive benefits and adhere to the principles of economic feasibility and reasonable technology according to the classification and cause analysis of the gradual geological environment problems in the research area. The land damaged by mining was mainly watered along other grasslands and pit pond water surfaces. So, the gradient geological environment problems in the mining area had good ecological restoration potential. For example, abandoned industrial building land could be restored to cultivated land, open pits could be restored as pit pond water surface land, and increased water area and exposed rock walls could be restored to other grasslands by climbing plants (Mustafa 2019; Nyamsari 2019; Steinmann 2020). There were three major restoration goals in this restoration plan. One was geomorphological reconstruction. The other was soil reconstruction, non-polluting, and fertile land resources remediation. The third was the systematic beautification of the environment with biodiversity. The fourth was the protection and monitoring of biodiversity.

5.1 *Terrain and landform landscape restoration*

Intercepted drainage ditches were set up in the upper part of the open stope. The stope was cleaned of dangerous rock, the slope surface was trimmed, and vegetation-type concrete was used on the limestone slope. A truncated drainage ditch was set up in the upper part of the slag discharge field, and a stone retaining dam was set up at the lower level to cover the slag discharge field with soil and grass.

5.2 *Land resource restoration*

The land remediation in the mining area required soil reconstruction according to the causes analysis of the gradient geological environment problems and the land damage. The construction machinery as the mainstay artificially supplemented topsoil cover and land leveling mode with the plot as the unit for leveling according to the size of the plot. In the topsoil and drainage direction, a certain slope ratio should be set so that the rainwater is smooth to reduce topsoil erosion. The thickness of the covering layer was 30 cm~50 cm with mechanical compaction and artificial loosening combined.

5.3 *Ecological resource restoration*

It was necessary to achieve the principle of anti-pollution, harm prevention, beautiful appearance, and certain economic benefits. Both areas select excellent varieties with simple planting methods, low cost, fast early growth, soil improvement, soil erosion prevention effect, adaptability, and strong resistance to vegetation restoration, under the premise of giving full play to the comprehensive functions of forest and grass protection and ornamentation according to the mining area itself characteristics and climatic conditions. Suitable trees and plants selected in the study area are pine, locust, and cypress. Suitable herbaceous plants are ryegrass, knotweed grass, artemisia grass, alfalfa, and sheep grass. Suitable shrub plants are small-leaved privet, boxwood ball, purple locust, thorn, and sour date. The methods used for vegetation restoration include planting technology, direct seeding technology, and transplanting technology.

5.4 *Biodiversity monitoring*

The monitoring methods of the research area were mainly based on investigation and inspection monitoring and supplemented by positioning observation. Investigation and inspection monitoring is conducted 4 to 5 times a year to strictly prevent the arbitrary expansion of disturbing the surface area. Sampling is used to investigate the survival rate, depression closure, coverage rate, and growth of forests with grass measures. In the key areas of the project area, set up dynamic monitoring points. The method of combining fixed-point regular observation and investigation was adopted with setting up observation points for waste rock fields and stopping serious land damage.

6 CONCLUSIONS AND RECOMMENDATIONS

(1) Limestone mining was an environmental input. Gradient geological environmental problems accompanied by this steady stream of input have been formed. The impact of gradient geological environmental problems on the environment further deterioration and is distributed in the entire disaster effect when superimposed rainfall, earthquakes, and other factors.
(2) The gradual geological environment problems both included land resource destruction and topographic landscape damage and included ecological resource destruction in the 11 limestone mines in the study area. Landform remodeling, soil reconstruction, vegetation systems with biodiversity, biodiversity conservation, and monitoring were effective ways to repair the problems of the gradual geological environment.
(3) More scholars judged the restoration effect of gradual geological environment problems from were qualitative perspective for limestone mines. The restoration of the gradual geological environment problems, whether they meet ecologically sustainable development requirements or not, still needs to be further tracked and studied. Therefore, it was necessary to strengthen the study of disaster carrier vulnerability.

ACKNOWLEDGMENTS

This study was supported by the horizontal project from Henan University of Economics and Law in Henan province, China (Technical consultation on the construction of information models with simulation and dual prevention platforms development for safety risks belonging to Coal Mine and Limestone Rock, Project NO. 20220236; technical consultation on mine geological environment protection with restoration and land reclamation, Project NO. 20220089). The authors would like to thank the teacher Dongdong Li, who works in a geotechnical laboratory at the School of Engineering Management and Real Estate at Henan University of Economics and Law.

REFERENCES

Bai Z.K. Industrial *and Mining Land Reclamation and Ecological Safety*. Beijing: The Ministry of Land and Resources Regulation Key Laboratory of Land, 2014. (In Chinese)

Ewa Desmond E., Egbe Enang A., Ukpata Joseph O., Etika Anderson. Sustainable Subgrade Improvement Using Limestone Dust and Sugarcane Bagasse Ash. *Sustainable Technology and Entrepreneurship*, 2022, 2 (1).

Gupta Sandeep, Mohapatra B.N., Bansal Megha. Development of Portland Limestone Cement (PLC) in India Using Different Compositions of Cement and Marginal Grade Limestone: A Sustainable Approach. *Journal of the Indian Chemical Society*, 2022, 99 (11).

Konstantinos Diamantis, Davood Fereidooni, Reza Khajevand, *et al*. Effect of Textural Characteristics on Engineering Properties of Some Sedimentary Rocks. *Journal of Central South University*, 2021, DOI: 10.1007/s11771-021-4654-5, (926 - 938).

Lei L., Lu F.L., Wang H *et al*. *1:50000 Geological Disasters Detailed Investigation Report Yiyang County of Henan province*. Zhengzhou: Geotechnical Engineering LTD of Henan Province, 2015: 25–28. (in Chinese)

Mustafa, Gurhan, Yalcin, *et al*. Geomedical, Ecological Risk, and Statistical Assessment of Hazardous Elements Inshore Sediments of the Iskenderun Gulf, Eastern Mediterranean, Turkey. *Environmental Earth Sciences*, 2019, 78 (15): 1–28.

Nyamsari D.G. *Natural Radioactive Risk Assessment in Topsoil and Possible Health Effect in Minim and Martap villages*, Cameroon: Using Radioactive Risk Index and Statistical Analysis. Kerntechnik, 2019, 84 (2): 115–122.

Nyamsari D.G., Yalçin M.G., Wolfson I. The Alteration, Chemical Processes, and Parent Rocks of Haléo-Danielle Plateau Bauxite, Adamawa–Cameroon. *Lithology and Mineral Resources*, 2020, 55 (3): 231–243.

Nyamsari Daniel Ganyi, Yalcin Mustafa Gurhan. Possible Geo-environmental Influences and the Chemical Content Variation Model of Limestone Within Zn Ultrabasic Complex. *Arabian Journal of Geosciences*, 2022, 15 (6).

Prakash K.J., Suresh N., GoPinathan M.C. *et al.* Suitability of Rhizobia-inoculated Wild Legumes Argyrolobium Flaeeidum, Astragalus Graveolens, Indigofera Ganglia, and LesPedeza Stenoearpa in Providing a Vegetational Cover in an Unreclaimed Limestone Quarry. *Plant and soil*, 1995, 177 (2): 139–149.

Sayed Hemeda, Abdulrahman Fahmy, Alghreeb Sonbol. Geo-Environmental and Structural Problems of the First Successful True Pyramid, (Snefru Northern Pyramid) in Dahshur, Egypt. *Geotechnical and Geological Engineering*, 2019, 37(4).

Steinmann J.W. Assessing the Application of Trace Metals as Paleo Proxies and a Chemostratigraphic Tool in Carbonate Systems: A Case Study from the "Mississippian Limestone" of the Midcontinent, United States. *Marine and Petroleum Geology*, 2020, 112(C): 104061–104061.

Tang Z H., Chai B., Luo C. *et al.* Design Ideas of Controlling Geo-Environmental Problems of Mines: a Case Study of a Limestone Mine in Fengshan County, Guangxi. *Hydrogeology and Engineering Geology*, 2013, 40(2): 123–128. (In Chinese)

Tang Z.H. Research on Geo-Environmental Risk Analysis and Management of Limestone Mine: a Case Study of the Limestone Mine in Fengshan County. Wuhan: China University of Geosciences (Wuhan) College of Engineering, 2013.

Wu L.L. Study on Ecological Restoration of Abandoned Mine of Yima city. Zhengzhou: Henan Agricultural University, 2009. (in Chinese)

Yuan Yao Li, Chen Xin Zhou, Zhi Wei Li. Risk Analysis of a Closed Limestone Mine Geological Environment in Fengshan County, *Southwest China. Advanced Materials Research*, 2013, 2695(807–809).

Zhao Hui Tang, Bo Chai, Yuan Yao Li *et al.* Design Engineering to Control Geological Environmental Problems of Limestone Mine in Fengshan, Guangxi. *Advanced Materials Research*, 2013, 2695(807–809).

Zhao Q.S., Lei L., Lu F.L *et al. Hengda Limestone Mine Land Reclamation Project Report Baiyang Town Yiyang county*. Zhengzhou: Geotechnical Engineering LTD of Henan Province, 2014: 12–18. (in Chinese)

Civil Engineering and Energy-Environment – Gao & Duan (Eds)
© 2023 the Author(s), ISBN 978-1-032-56059-5

Study on pretreatment and leaching technology of a fine disseminated refractory gold mine

Zhongbo Lu*, Guangsheng Li, Xingfu Zhu, Mingming Cai & Yanbo Chen
Shandong Gold Mining Technology Co., Ltd. Smelting Laboratory Branch, Laizhou, Shandong

ABSTRACT: This study focuses on the gold leaching process of a fine disseminated refractory gold ore in Gansu, mainly including the whole slime cyanidation leaching process, flotation concentrate pretreatment cyanidation leaching+flotation tailings cyanidation leaching process, among which the pretreatment process includes ultra-fine grinding, ultra-fine grinding peroxidation, oxidation roasting, and pyrometallurgical lead smelting for gold capture. The results showed that the leaching rate of all slime cyanide gold was only 61.67%, and the gold recovery after pretreatment was higher than that of all slime cyanide gold. The highest gold recovery rate of the flotation concentrate pyrometallurgy lead gold capture +flotation tailings cyanidation leaching process is 76.38%. Through comparison of pretreatment processes, it is considered that the flotation concentrate pyrometallurgy lead gold capture+flotation tailings cyanidation leaching process is the most appropriate gold recovery process for this ore.

1 INTRODUCTION

The gold in the finely disseminated gold ore is wrapped by sulfide minerals or gangue, so even fine grinding cannot fully expose the gold, resulting in a low gold leaching rate in the leaching process (Guo *et al.* 2015; Zheng 2009). At present, the main pretreatment methods are generally divided into the fire method and wet method, and the roasting pretreatment is the fire method; the wet process can be divided into the pressure oxidation method, biological oxidation method, and other chemical oxidation methods (Gold 2000; Hydrometallurgy 2003; Han 2013). The oxidation roasting method is characterized by strong adaptability, high energy consumption, and high waste gas treatment cost; the pressurized oxidation process has a high gold recovery rate and no flue gas pollution problem, but it has high requirements on equipment design and material quality and has dangerous factors; biological oxidation process is characterized by the friendly environment, simple process, small investment, low cost, and its disadvantages are demanding production conditions; the chemical oxidation method is characterized by no air pollution, no high-pressure problem and no harsh requirements of biological method (Dai et al. 2010; Nie & Liu 2006; Shen & Tang 2014; Song 2009).

2 ORE PROPERTIES

The gold grade of ore is 3.0 g/t, and the main gold ore is natural gold, with a small amount of electrum and tellurium gold, as well as pyrite, pyrrhotite, magnetite, hematite, and other

---

*Corresponding Author: 379195351@qq.com

metal minerals; gangues mainly include quartz, feldspar, kaolinite, mica, calcite, chlorite, dolomite, andalusite, etc. Gold mainly exists in ores as exposed and semi-exposed gold, accounting for 60.81%, followed by sulfide-wrapped gold, accounting for 24.67%, and quartz-wrapped gold, accounting for 14.52%. Under the electron microscope, it is found that the particle size of gold is all 5 μ Below m, partially less than 0.5 μ m. See Table 1 for the chemical composition analysis results of raw ore and Table 2 for metallographic analysis results. It can be seen from Table 1 that the available elements in the ore are gold.

Table 1.   Results of chemical multi-element analysis of raw ore/%.

Au(g/t)	Ag(g/t)	Cu	Pb	Zn	Fe	S	As	Hg
3.00	2.20	trace	trace	trace	1.54	1.16	0.019	0.05
SiO2	Al2O3	MgO	CaO	Na2O	K2O	C		
94.18	1.53	0.06	1.32	0.12	0.40	0.33		

Table 2.   Results of phase analysis of raw ore gold.

Phase name	Content, g/t	Distribution rate,%
Exposed and semi-exposed gold	1.82	60.81
Sulfide wrapped gold	0.74	24.67
Gold wrapped in quartz	0.44	14.52
Total	3.00	100.00

3   TEST EQUIPMENT AND TEST METHODS

3.1   *Test equipment*

The test equipment mainly includes RK/XJT air charging multi-function leaching mixer, DMQX-1 grinder, and pre-oxidation device, single cell flotation machine, muffle furnace, and conical ball mill (RK/ZQMφtwo hundred and forty×90), etc.

3.2   *Test method*

The ore properties show that the gold particles are too fine and difficult to dissociate. Given the associated relationship between gold and sulfide, sulfide can be effectively recovered by flotation, and gold will be enriched accordingly. The pretreatment of flotation concentrate can not only improve the gold recovery rate but also reduce the pretreatment cost; the cyanide leaching of flotation tailings can further recover the dissociated gold particles that cannot be recovered by flotation. To sum up, the flotation concentrate pretreatment cyanidation leaching+tailings cyanidation leaching process and the all slime cyanidation leaching process are used for index comparison to determine a better gold recovery process. See Figure 1 for process flow and Table 3 for flotation indexes.

After flotation, the gold grade of the flotation concentrate is 14.60 g/t, the yield is 10.35%, and the recovery is 50.36%; the gold grade of flotation tailings is 1.66 g/t, and the yield is 89.65%.

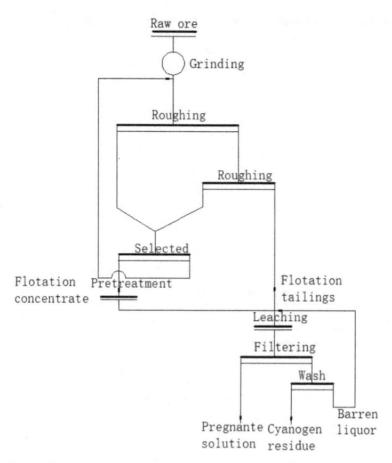

Figure 1. Process flow.

Table 3. Flotation indexes.

Product name	Au grade (g/t)	Productivity (%)	Rate of recovery(%)
Flotation concentrate	14.60	10.35	50.36
Flotation tailings	1.66	89.65	49.64
Total	3.00	100	100

4 TEST RESULTS AND DISCUSSION

4.1 *All slime cyanidation leaching process*

For the fine disseminated refractory gold ore, the all-slime cyanidation leaching process is usually used to recover gold. Therefore, the best leaching conditions are determined through grinding fineness and leaching time conditions tests.

Under the conditions of grinding fineness - 200 mesh content of 95%, pH value of 11.5, sodium cyanide concentration of 5/10000, liquid-solid ratio of 3:2, and leaching time of 24 hours, the gold leaching rate is 61.67%. See Figure 2 for grinding fineness test results and Figure 3 for leaching time test results.

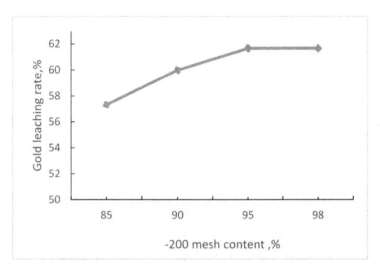

Figure 2. Grinding fineness test results.

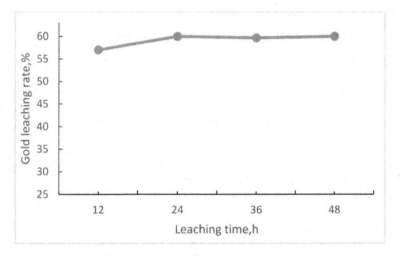

Figure 3. Leaching time test results.

4.2 *Ultrafine Grindin-cyanidation leaching process*

4.2.1 *Ultrafine grinding of flotation concentrate (pre-oxidation of ultra-fine grinding) - cyanide leaching process*

Ultrafine grinding can dissociate the gold minerals with fine embedded particle size and pre-oxidize them in acidic or alkaline environments, which can further expose the gold covered by sulfide, thus improving the gold leaching rate (Tian *et al*. 2016; Wang 2007).

Ultrafine grinding of flotation concentrate (grinding fineness: P (90)=9 μ m) And super-fine grinding pre-oxidation (grinding fineness P (90)=9 μ m. Oxidation for 8 h under pH=11), cyanide leaching is carried out under the conditions of pH 11.5, sodium cyanide concentration of 5/10000, the liquid-solid ratio of 3:2, and leaching time of 24 h. See Table 4 for the test results.

Table 4. Ultrafine grinding cyanidation leaching results.

Name	Gold grade of flotation concentrate(g/t)	Leaching residue Au grade (g/t)	Au leaching rate (%)
Ultrafine grinding-cyanidation leaching	14.6	5	65.74
Superfine grinding pre-oxidation-cyanidation leaching	14.6	2.97	79.65

4.2.2 Cyanide leaching of flotation tailings

As the gold grade in the flotation tailings is relatively high (1.66 g/t), a cyanide leaching test was carried out. The specific test conditions are pH value 11.5, sodium cyanide concentration 5/10000, liquid-solid ratio 3:2, leaching time 24 hours, and gold leaching recovery 57.25%. Combined with flotation and cyanide leaching of flotation tailings, the total recovery rate of gold from ultrafine grinding cyanide leaching+cyanide leaching of flotation tailings is 63.04%, and the total recovery rate of gold from ultrafine grinding pre oxidation cyanide leaching+cyanide leaching of flotation tailings is 68.53%.

4.3 Oxidation roasting-cyanidation leaching process

As a pretreatment process before cyanidation and gold extraction of refractory gold ores, oxidation roasting has a wide range of adaptability to ores. To further improve the gold recovery, the gold inclusion is opened by roasting to improve further the gold leaching rate (Pei 2000; Xiong 2011). If the raw ore is roasted, the roasting cost and waste gas treatment cost are high, so only the flotation concentrate is subject to oxidation roasting pretreatment for 2 h, and the roasted ore after oxidation roasting is subject to cyanidation leaching at pH 11.5, sodium cyanide concentration of 5/10000, the liquid-solid ratio of 3:2, and leaching time of 24 hours. See Table 5 for the test results.

It can be seen from the test data that the highest gold leaching rate is 85.33% at the roasting temperature of 500°C. The total gold recovery rate is 71.39% after comprehensive flotation and cyanide leaching of flotation tailings.

Table 5. Results of oxidation roasting-cyanide leaching test.

Baking temperature (°C)	Gold grade of flotation concentrate (g/t)	Leaching residue Au grade (g/t)	Au leaching rate (%)
500	14.60	2.14	85.33
600	14.60	2.88	80.67
700	14.60	5.69	61.00

4.4 Pyro metallurgical lead gold capture process

The flotation concentrate is treated by the lead gold capture process. See Table 6 for the test results.

From the test results, it can be seen that the lead gold capture rate of flotation concentrate is 95.23%, and the total gold recovery rate is 76.38% through comprehensive flotation and cyanide leaching of flotation tailings.

Table 6. Test results of lead capture process for flotation concentrate.

Flotation concentrate				Residue			
Weight (g)	Grade (g/t)	Metal quantity (mg)	Amount of gold in the lead gold alloy(mg)	Weight (g)	Grade (g/t)	Metal quantity (mg)	Lead gold capture rate(%)
100	14.6	1.46	1.4	580	0.12	0.10	95.23

4.5 *Pretreatment process comparison*

See Table 7 for the recovery rate of all slime cyanidation leaching of ores, ultrafine grinding cyanidation leaching of flotation concentrates+cyanidation leaching of flotation tailings, pre-oxidation cyanidation leaching of ultrafine grinding of flotation concentrates+cyanidation leaching of flotation tailings, pyrometallurgical lead capture of flotation concentrates +cyanidation leaching of flotation tailings.

It can be seen from Table 7 that the gold recovery rate of cyanide leaching after ore pretreatment is higher than that of all slime cyanide leaching. Although the gold recovery rate of the oxidation roasting cyanidation leaching process is high due to the high roasting cost, high energy consumption, and unfriendly environment, it does not conform to the current national environmental policy of "double carbon and double high." The ultrafine grinding cyanidation leaching process has a higher treatment cost, less significant improvement in gold recovery and less obvious benefit. Pyrometallurgy is the most suitable process for recovering gold from the ore because of its highest gold recovery rate, mature process, and no impact on the environment.

Table 7. Statistical table of gold recovery rate of different pretreatment processes.

Process name	Gold recovery rate, %
All slime cyanidation leaching	61.67
Ultrafine grinding of flotation-concentrate cyanide leaching+cyanide leaching of flotation tailings	63.04
Pre oxidation cyanidation leaching of the superfine mill in flotation-concentration+cyanidation leaching of flotation tailings	68.53
Flotation concentrate roasting-cyanidation leaching+flotation tailings cyanidation leaching	71.39
Pyro metallurgical lead capture of flotation concentrate+cyanide leaching of flotation tailings	76.38

5 CONCLUSION

1. Gold mainly exists in ores as exposed and semi-exposed gold, accounting for 60.81%, followed by sulfide-wrapped gold, accounting for 24.67%, and quartz-wrapped gold, accounting for 14.52%. Under the electron microscope, it is found that the particle size of gold is all 5 μ below m, partially less than 0.5 μ m. It belongs to micro fine refractory ore;
2. The gold recovery rate of all slime cyanide leaching of raw ore is 61.67%, and the gold leaching rate is low;

3. The gold recovery rate of cyanide leaching after ore pretreatment is higher than that of all slime cyanide leaching. Although the gold recovery rate of the oxidation roasting cyanidation leaching process is high, due to the high roasting cost, high energy consumption, and unfriendly environment, it does not conform to the current national environmental policy of "double carbon and double high." The ultrafine grinding cyanidation leaching process has a higher treatment cost, less significant improvement in gold recovery and less obvious benefit. The highest gold recovery rate of pyrometallurgy is 76.38%, the process is mature and has no impact on the environment, so it is considered to be the most appropriate process for recovering gold from the ore.

REFERENCES

Current Situation and Development of Pretreatment Process for Refractory Gold Ores. *Hydrometallurgy*, 2003, (1): (1–8)

Dai Shujuan, Hu Zhigang, Meng Yuqun, et al. Gold Flotation, and Cyanide Leaching Test in a Gold Ore. *Metal Mines*, 2010 (8): 75–78

Guo Yuwu, Zhang Yu, Zhu Enling, et al. Study on Beneficiation Process of a Finely Disseminated Gold Mine in Inner Mongolia. *Nonferrous Metals (Beneficiation)*, 2015 (4): 46–50

Han Yuexin. Research on Key Technologies of Ore Pretreatment and Preconcentration. *China Conference*, 2013, 6: 220–244

Nie Guanghua, Liu Chunlong Experimental Study on the Beneficiation of Fine Gold Ores. *China Mining*, 2006, 15 (11): 76–78

Pei Hongming. Mineral Processing Test of Micro-fine Disseminated Primary Gold Ore. *Gold Science and Technology*, 2000, 8 (4): 45–47

Review of Pretreatment Technology for Refractory Gold Ores. *GOLD*, 2000,1: (38–45)

Shen Shubao, Tang Minggang. Research Progress in the Flotation of Arsenic-bearing Refractory Gold Ores. *Gold Science and Technology*, 2014, 22 (2): 63–66

Song Xin. China's Refractory Gold Resources and Their Development and Utilization Technology. *Gold*, 2009, 30 (7): 46–49

Tian Runqing, Liu Yunhua, Tian Minmin, et al. Study on Beneficiation Test of a Finely Disseminated Gold Mine in Shaanxi. *Gold Science and Technology*, 2016 (12): 102–106

Wang Xuejuan. Research on New Beneficiation Technology for Carlin-type Gold Deposits in Guizhou. *Kunming: Kunming University of Science and Technology*, 2007. (10)

Wei Bangfeng. Study on the Combined Process of Beneficiation and Metallurgy of a Fine Disseminated Refractory Gold Ore in Xinjiang. *Xinjiang Nonferrous Metals*, 2014 (s1): 111–112

Xiong Wenliang. Experimental Study on Gold Extraction from a Refractory Gold Mine [J]. *Comprehensive Utilization of Mineral Resources*, 2011 (2): 19–21

Zheng Hua. Overview of Pretreatment Technology for Refractory Gold Ores. *Gold*, 2009, 1: (36–41)

Civil Engineering and Energy-Environment – Gao & Duan (Eds)
© 2023 the Author(s), ISBN 978-1-032-56059-5

Research progress of electrochemical oxidation and its coupling technology to remove algal in water

Di Jia, Li Lin* & Yueqi Cao
Basin Water Environmental Research Department, Changjiang River Scientific Research Institute, Wuhan, P.R. China
Key Lab of Basin Water Resource and Eco-Environmental Science in Hubei Province, Wuhan, P.R. China

Sheng Zhang
State Key Laboratory of Pollution Control and Resources Reuse, School of Environment, Nanjing University, Nanjing, P.R. China

Xiong Pan, Lei Dong & Yuting Zhang
Basin Water Environmental Research Department, Changjiang River Scientific Research Institute, Wuhan, P.R. China
Key Lab of Basin Water Resource and Eco-Environmental Science in Hubei Province, Wuhan, P.R. China

ABSTRACT: An algal bloom is an important ecological disaster that seriously threatens the water quality safety of drinking water sources electrochemical oxidation technology has the advantages of clean energy, a friendly environment, and effective avoidance of secondary pollution, and the algae removal effect is remarkable. This paper introduces the mechanism of algae removal by electrochemical oxidation, discusses the main factors affecting its algae removal effect, outlines the evaluation of its effect on algal toxin removal, and integrates its research status with the coupled algae removal by aeration, cavitation, adsorption, ultrafiltration. Previous studies have shown that coupling technology can make up for the shortage of electrochemical oxidation and improve the efficiency of algae removal. Finally, the shortcomings of the current electrochemical oxidation and its coupling technology for algae removal are summarized, and the research prospects prospect.

1 INTRODUCTION

In recent decades, many large lakes and reservoirs have experienced frequent water blooms, which produce undesirable tastes and odors and form algal toxins (Zheng *et al.* 2012), causing global water quality degradation, seriously affecting water safety and security, and endangering human and animal health (Zhang *et al.* 2014). At the same time, algae in the water can adversely affect drinking water treatment processes, so the removal of algae is of great concern (Dittmann & Wiegand 2006). Especially in tropical and subtropical regions, sufficient nitrogen and phosphorus nutrients create favorable conditions for algal overgrowth due to the discharge of domestic, agricultural, and industrial wastewater (Gao *et al.* 2010) and lead to a series of ecological and environmental problems. In addition, the

---

*Corresponding Author: linli1229@hotmail.com

DOI: 10.1201/9781003433651-32

presence of algae in water treatment plants interferes with physical or chemical water purification processes (Henderson *et al.* 2008).

Currently applied algae removal techniques are mainly physical, chemical, and biological algae removal techniques, which have their advantages and disadvantages. Physical algae removal is usually time-consuming and costly and cannot be used on a large scale, and it may also cause some damage to the device (Liang *et al.* 2005); chemical algae removal is the most developed and commonly used method to add chemical reagents to the water body, this method is simple, fast, and efficient, but it is easy to harm the ecological environment and cause secondary pollution. While biological algae removal technology is through aquatic plants and animals, bacteria, etc., to remove algae (Narasinga *et al.* 2018). It usually needs long periods and has a slow effect, thus cannot be the emergency treatment of sudden water bloom.

Electrochemistry covers electrochemical oxidation (EO), electrocoagulation (EC), electro floatation (EF), electrodeposition (ED), electro-adsorption (EST), electrodialysis (ED), etc. (Zhong *et al.* 2019), among which electrochemical oxidation is based on the destruction of algal cells by generating a large amount of strong oxidizing substances, supplemented by a variety of physical and chemical effects. The research and improvement of electrochemical oxidation and its coupling technology make the new technology not only can take the length of the chemical method to eliminate algae efficiently, but also can remove the pollutants in the water body in a targeted way, and at the same time, it has the advantages of clean energy, friendly environment and can avoid secondary pollution effectively (Zhang *et al.* 2019), which has a broad application prospect. This paper introduces the mechanism of electrochemical oxidation for algae removal, reviews the current research status around the main factors affecting electrochemical oxidation for algae removal, and finally lists the current coupling technologies for electrochemical algae removal and outlooks its development trend.

2 MECHANISMS OF ALGAE REMOVAL FROM WATER BODIES BY ELECTROCHEMICAL OXIDATION

2.1 *Direct role*

Direct action means that the electrode directly adsorbs, penetrates, and oxidizes the algal cell itself, destroying its membrane structure, intracellular enzyme (COA) activity (Okochi *et al.* 1999), etc., so that it suffers the loss of activity.

2.1.1 *Electric field breakdown*
Electric current acts on both sides of the cell membrane, and the membrane structure electric field forces the cell membrane to produce membrane potential, forming a penetrating membrane potential difference. With the increase in electric field strength, the membrane potential gradually increases, exceeding the threshold value after the current penetrates the cell wall and cell membrane, causing damage to the membrane structure, and increased permeability, leading to thinning of the cell membrane until rupture, the flow out of cell contents, strong oxidation produced in the water column enters, oxidization of algal cell organelles and enzyme, metabolic, the entering of water molecules and the rupture and death of algal cells (Zimmermann 1986).

2.1.2 *Electrode adsorption*
Some of the algal cells are attached to the cathode, presumably because polysaccharides, proteins, peptides, an amino sugar, and other organic acids in algal cells dissolve and adhere to the surface of intact and ruptured cells, which can be protonated in an acidic environment, causing the surface of algal cells to take on a positive charge and move toward the cathode (Huang *et al.* 2009). Also, some of the electrodes used in the study contain a large number of

oxygen-containing functional groups, such as carboxyl, carbonyl, and hydroxyl groups on the surface, which have a good adsorption effect on organic substances (Xiang *et al.* 2020). Algal cells are adsorbed on the cathode surface due to the electrostatic interaction, van der Waals forces, and hydrogen bonding between the cell contents and the electrode surface (Cloirec *et al.* 1997), and when the cell surface material is oxidized, the algal cells fall off and are electrochemically oxidized or subjected to electrical levitation and rupture and die. Matsunaga *et al.* (1994) concluded that direct electron exchange between algal cells and the electrode is possible, and determined that intracellular coenzymes, as well as Respiration-related enzymes, were destroyed by oxidation, and algal cells were inhibited and died.

2.2 *Indirect role*

The indirect effect of electrolysis refers to a series of strong oxidants generated on the electrode during electrolysis, mainly including reactive oxides (ROS), such as hydrogen peroxide (H_2O_2), hydroxyl radicals (•OH), ozone (O_3), and superoxide radicals (•O_2^-); reactive chlorine generated from the electrolysis of chlorides in the water, such as hypochlorous acid (HClO), chlorate ions (ClO$^-$), and chlorine gas (Cl_2); peroxides generated in the electrochemical reaction of a certain acid salt in water, etc. (Martínez-Huitle & Brillas 2008). On the other hand, it refers to the removal of phosphorus and nitrogen nutrients that cause algal blooms (Wang *et al.* 2022).

2.2.1 *Generation of active substances*

Water electrolysis produces hydroxyl radicals (the most powerful oxides known in water (Urs von Gunten 2003), hydrogen peroxide, ozone, and other reactive oxidation groups. When Cl$^-$ is present in water, the •OH formed oxidizes with chloride to form chlorine, which is then converted into reactive chlorine with continuous oxidation, such as hypochlorite and hypochlorite ions (the exact amount depends on the pH) (Polcaro *et al.* 2009). The reaction involved in the electrolysis of the solution is as follows equations (1)-(10) (Nath *et al.* 2011):

$$H_2O \; - \; 2e^- \rightarrow 2 \; \bullet OH + 2 \; H^+ \tag{1}$$

$$H_2O + 2 \; e^- \; \rightarrow H_2 + 2 \; OH^- \tag{2}$$

$$O_2 + 2 \; H_2O \; + \; 4 \; e^- \rightarrow 4 \; OH^- \tag{3}$$

$$OH^- - e^- \; \rightarrow \; \bullet OH \tag{4}$$

$$H_2O - 4e^- \rightarrow O_2 \; + \; 4H^+ \tag{5}$$

$$O_2 + 2H^+ \; + \; 2e^- \rightarrow H_2O_2 \tag{6}$$

$$Cl^- \; - \; e^- \; \rightarrow \; \bullet Cl \tag{7}$$

$$2Cl^- \; - \; 2 \; e^- \; \rightarrow \; Cl_2 \tag{8}$$

$$Cl_2 \; + \; 2 \; OH^- \rightarrow ClO^- + H_2O \; + \; Cl^- \tag{9}$$

$$Cl_2 \; + \; H_2O \rightarrow HClO \; + \; H^+ + Cl^- \tag{10}$$

One of the main reactive substances produced by the electrolysis process is H_2O_2, when its concentration is in the order of μmol/L, that will play a certain effect on the growth of Microcystis aeruginosa cells, but cannot cause the death of undamaged algal cells. While at the same concentration, the algal cells in the discharge bed reactor that have been damaged by the electric field will increase the degree of damage and tend to die more easily (Wang *et al.* 2008). And when the concentration of H_2O_2 increases, the lipid peroxidation of algal cells will be enhanced subsequently, and the stronger their antioxidant defense system, the more damaged algal cells will be (Qiu & Wang 2017). Meanwhile, H_2O_2 has a selective inhibitory effect on cyanobacteria (Hans *et al.* 2012). Also, the oxidants produced by electrochemical oxidation can effectively disrupt the structure of algal toxins (MCs) and achieve degradation (Brooke *et al.* 2006; Rodríguez *et al.* 2007).

3 COUPLING OF ELECTROCHEMICAL OXIDATION WITH OTHER ALGAE SUPPRESSION TECHNIQUES

3.1 *Electrochemical oxidation - aeration*

During electrolysis, if oxygen (O_2) or air is introduced into the solution, O_2 can be reduced to H_2O_2 at the cathode, but usually, H_2O_2 is easily decomposed into O_2 and H_2O by further catalytic reduction. So, if we want to produce as much H_2O_2 as possible during electrolysis, we need to select suitable electrode materials on the one hand and continuously introduce oxygen or air into the solution on the other hand (Tan & Li 2006). Zuo (2012) demonstrated that the oxidation effect was better when oxygen was passed into the electrolyte, and the bicarbonate formed in the water by the carbon dioxide in the air might trap the free radicals and affect the oxidation effect when the air was passed in. Naoyuki Kishimoto *et al.* (2007) conducted coupling experiments using nitrogen, oxygen, ozone, and ruthenium oxide (RuO_2) coated titanium anode and stainless steel (SUS304) cathode, respectively. They found that electrolysis and nitrogen-filled electrolysis were characterized by a rapid decrease in biochemical oxygen demand (BOD_5) and total nitrogen and a slow decrease in chemical oxygen demand (COD) in 5 days; O_2 electrolysis and O_3 electrolysis induced the conversion of persistent organic substances into biodegradable substances. The following reactions occur when ozone is introduced (Kishimoto *et al.*2008; Zhou *et al.* 2010):

$$O_3 + e^- \rightarrow \bullet O_3 - \tag{11}$$

$$\bullet O_3^- + H_2O \rightarrow \bullet OH + O_2 + OH^- \tag{12}$$

$$O_2 + 2H^+ + 2e^- \rightarrow H_2O_2 \tag{13}$$

$$2O_3 + H_2O_2 \rightarrow 3O_2 + 2\bullet OH \tag{14}$$

Aeration to promote the production of active substances can play a better effect on algae inhibition, while aeration has a significant effect on the inhibition of anode passivation, and has a better enhancement effect on electrolytic phosphorus removal (Ouyang *et al.* 2009), phosphorus removal not only solves the problem of eutrophication of water bodies but also inhibits the outbreak of water blooms.

3.2 *Electrochemical oxidation - cavitation*

Hydrocavitation has been used in microbial cell destruction, water disinfection and sterilization, and wastewater treatment (Gogate 2011). Li *et al.* (2014) found that a brief treatment of Microcystis aeruginosa by hydrocavitation for 10 min reduced the cell density of the sample by 88% after three days of incubation, mainly because the bubbles inside the algal cells were destroyed, and the photosynthetic system was damaged, and the free radicals generated by the cavitation process could sustain damage. The cavitation effect of ultrasound can revive the electrode, and the moment the cavitation bubble breaks, the water molecules entering the cavitation bubble will be evenly split into highly reactive free radicals (hydrogen atoms ($\bullet H$), hydroxyl radicals) under extremely high temperatures and pressure conditions (Ciawi *et al.* 2006; Kanthale *et al.* 2007), strengthening the mass transfer process of reactants from the main body of liquid phase to the electrode surface, eliminating the concentration polarization, etc.

The use of coupled electrochemical oxidation-ultrasonic cavitation technique for the treatment of organic wastewater increased the removal rate by 10% to 20% compared to electrochemical methods alone (Guo *et al.* 2003). Pradhan and Gogate (2010) combined ultrasonic cavitation and Fenton (Fenton) processes and showed that this process could generate higher concentrations of hydroxyl radicals, thus improving the degradation efficiency of pollutants. By comparing cavitation alone, electrolysis alone, and coupled algae removal techniques, Ye *et al.* (2007) found that the algae removal efficiencies of the three were 27.7%, 27.8%, and 97.5%, respectively, with the remaining conditions unchanged, indicating that cavitation and electrolysis have synergistic effects and can significantly improve the algae control effect. It was also found that this system has a more significant effect on algae growing in a single-cell state, which is more easily destroyed by the impact of external forces (Figure 1).

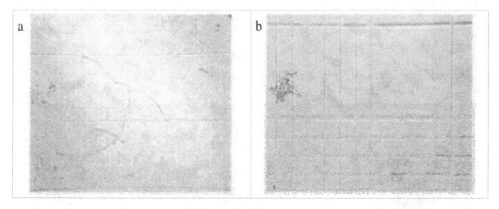

Figure 1. Comparison of the microstructure of Anabaena flosaquae before (a) and after (b) treatment (Wang 2005).

3.3 *Electrochemical oxidation-adsorption*

Xue *et al.* (2020) used tourmaline and titanium to form a three-dimensional electrode system, tourmaline adsorbed algae cells, degraded them under the action of electrolysis, and then adsorbed them again to cycle this process, forming a dynamic balance of "adsorption-degradation-adsorption," which can effectively improve the algae removal rate and reduce the treatment time. Wang (2005) studied the electrode-adsorption synergistic process, choosing to add a mixture of tourmaline and activated carbon, and found that it had a better treatment effect on the water containing algae, and the algae elimination rate could reach 96.6% when the current density was 30 mA/cm^2 and the treatment time was 15 min. Meanwhile, the technology also had a better removal effect on ammonia nitrogen and total phosphorus, which could better solve the eutrophication problem of water bodies.

3.4 *Electrochemical oxidation-ultrafiltration*

Water quality purification, desalination, and other scenarios have certain use for microfiltration (MF) or ultrafiltration (UF) membranes, and algal cells can rapidly clog microfiltration (MF) or ultrafiltration (UF) membranes and reduce permeability. Liu *et al.* (2020) combined electrochemical pretreatment with ultrafiltration for algae removal technology.

They found that when the oxidation time exceeded 30 min, regardless of the membrane material, BDD anodic oxidation could effectively reduce membrane contamination. The electrochemical pre-oxidation process greatly reduced the contamination resistance of the filter layer and reduced the interaction between contaminants and the membrane, which was beneficial to reduce membrane contamination and improve algae removal efficiency.

3.5 *Photocatalytic oxidation*

Liang et al. (2012) combined the photocatalytic performance of TiO_2 and the electrocatalytic activity of RuO_2 in the electrode coating to compare the efficiency of photocatalysis, electrocatalysis, and photocatalysis for the removal of algal toxins, and the results showed that the photocatalysis was better than the first two, indicating that the coupling of the two produced some synergistic effects. Using TiO_2/Ti plates as photoanodes, the degradation rate of MC-LR could reach more than 95% after 2 h of reaction. Chen et al. (2022) concluded that in the boron-doped diamond (BDD) anode system, the generation of •OH is the main degradation mechanism of electrochemical oxidation, and UV light can decompose the H_2O_2 formed on the surface of BDD anodes, hinder the conversion of O_2 and promote the free •OH (free) production so that the surface-bound •OH (surface) is converted to free •OH (free) (Figure 2). Wang et al. (2020) inactivated ballast water microalgae with a combined electrochemical oxidation-ultraviolet (UV) system and found that the inactivation effect was better than that of the electrocatalytic system alone, and the two showed some synergy, overcoming the disadvantages of the high-energy consumption of the electrocatalytic system alone, reducing the requirements of the UV radiation system alone regarding the depth of UV lamp radiation and the degree of attenuation, and having better algae removal effects.

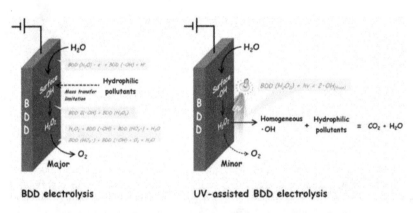

Figure 2. Action mechanism of the BDD-UV system (Wang et al. 2020).

4 CONCLUSION AND OUTLOOK

The research on electrochemical oxidation and its coupling technology is mostly focused on the degradation of organic matter in wastewater, but relatively little research has been done on the inhibition of water bloom, and the research on the coupling technology for algae removal is relatively limited. From the above water bloom prevention and control measures and the current research status of electrochemical oxidation and its coupling technology, this technology is mainly through the production of active substances to destroy algae cells and has a good algae inhibition effect. And the electrode material and coupling method can be selected and combined according to different water quality conditions in a more targeted manner. However, the following shortcomings exist in domestic and international research:

1) Research is mostly focused on small-scale experiments, lacking large-scale applications. 2) Research on the development and exploration of coupling technology is weak. 3) The ecological effects of the technology are not evaluated. Therefore, the future development trend of water bloom control is to integrate different control measures based on different water body conditions, try large-scale field experiments, and explore a friendly method of algae removal for aquatic plants and animals.

ACKNOWLEDGMENTS

This work was financially supported by Wuhan City Applied Basic Frontier (No. 2020020601012285)

REFERENCES

Brooke S., Newcombe G., Nicholson B. et al. (2006). Decrease in Toxicity of Microcystins LA and LR in Drinking Water by Ozonation. *Toxicon.* 48 (8), 1054–1059.

Chen P., Mu Y., Chen Y. et al. (2022). Shifts of Surface-bound •OH to Homogeneous •OH in BDD Electrochemical System via UV Irradiation for Enhanced Degradation of Hydrophilic Aromatic Compounds. *Chemosphere.* 291, 132817.

Ciawi E., Rae J., Ashokkumar M. et al. (2006). Determination of Temperatures Within Acoustically Generated Bubbles in Aqueous Solutions at Different Ultrasound Frequencies. *The Journal of Physical Chemistry B.* 110 (27), 13656–13660.

Cloirec P.L, Brasquet C., Subrenat E. (1997). Adsorption onto Fibrous Activated Carbon: Applications to Water Treatment. *Energy & Fuels.* 11 (2), 331–336.

Dittmann E., Wiegand C. (2006). Cyanobacterial Toxins - occurrence, Biosynthesis, and Impact on Human Affairs, Mol. *Nutr. Food Res.* 50, 7–17.

Gao S., Du M., Tian J., Yang J., Yang J., MaF., Nan J. (2010), Effects of Chloride Ions on the Electro-coagulation-Flotation Process with Aluminum Electrodes for Algae Removal, *J. Hazard. Mater.* 182, 827–834.

Gogate P.R (2011). Hydrodynamic Cavitation for Food and Water Processing. *Food Bioprocess Technol.* 4, 996–1011.

Guo Z.B, Zheng Z., Yuan S.J et al. (2003). Application of Ultrasound in Wastewater Treatment in Combination with Other Technologies. *Industrial Water Treatment.* 23 (7), 8–12. (Chinese)

Henderson R., Parsons S.A. and Jefferson B. (2008), The Impact of Algal Properties and Peroxidation on Solid-liquid Separation of Algae, *Water Res.* 42, 1827–1845.

Huang J., Graham N., Templeton M.R et al. (2009). A Comparison of the Role of Two Blue-green Algae in THM and HAA Formation. *Water Research.* 43 (12), 3009–18.

Kanthale P.M., Ashokkumar M. and Grieser F. (2007). Estimation of Cavitation Bubble Temperatures in anIonic Liquid. *The Journal of Physical Chemistry C.* 111 (50), 18461–18463.

Kishimoto N., Morita Y., Tsuno H et al. (2007). Characteristics of Electrolysis, Ozonation, and their Combination Process on the Treatment of Municipal Wastewater. *Water Environment Research.* 79 (9), 1033–1042.

Kishimoto N., Nakagawa T., Asano M et al. (2008). Ozonation Combined with Electrolysis of 1,4-Dioxane using a two-compartment Electrolytic Flow Cell with a Solid electrolyte. *Water Res.* 42 (1/2), 379–385.

Li P., Song Y. and Yu S (2014). Removal of Microcystis Aeruginosa using Hydrodynamic Cavitation: Performance and Mechanisms. *Water Research.* 62 (Oct.1), 241–248.

Liang W.Y., Qu J.H., Chen L.B., Liu H.J. and Lei P.J. (2005), Inactivation of Microcystis Aeruginosa by Continuous Electrochemical Cycling Process in the Tube using Ti/RuO$_2$ Electrodes, *Environ. Sci. Technol.* 39, 4633–4639.

Liang Z.X., Liang W.Y., Wang L., Yu J. and Xu J (2012). Photocatalytic Oxidation for the Degradation of Algal Toxin MCLR. *Journal of Environmental Engineering.* 6 (11), 3817–3821. (Chinese)

Liu B., Zhu T.T, Liu W.K, Zhou R., Zhou S.Q., Wu R.X, Deng L., Wang J. and Van der Bruggen Bart. (2020). Ultrafiltration Pre-oxidation by Boron-doped Diamond Anode for Algae-laden Water Treatment: Membrane Fouling Mitigation, Interface Characteristics, and Cake Layer Organic Release. *Water Research.* 187, 116435.

Martínez-Huitle C.A. and Brillas E. (2008). Electrochemical Alternatives for Drinking Water Disinfection. *Electrochemical Water Purification*. 47, 1998–2005.

Matsunaga T., Nakasono S. and Kitajima Y. et al. (1994). Electrochemical Disinfection of Bacteria in Drinking Water Using Activated Carbon Fibers. *Biotechnology and Bioengineering*. 43 (5), 429–433.

Hans C.P.Matthijs, Petra M. Visser, Bart Reeze, Jeroen Meeuse, Pieter C. Slot, Geert Wijn, Renée Talens, Jef Huisman (2012). Selective Suppression of Harmful Cyanobacteria in an Entire Lake With Hydrogen Peroxide. *Water Research*. 46 (5), 1460–1472.

Narasinga R.H.R., Russell Y., Michael W. et al. (2018). The Role of Algal Organic Matter in the Separation of Algae and Cyanobacteria using the Novel "Posi" - Dissolved air flotation process. *Water Research*, 130, 20–30.

Nath H., Wang X., Torrens R. et al. (2011). A Novel Perforated Electrode flow through Cell Design for Chlorine Generation. *Journal of Applied Electrochemistry*. 41 (4), 389–395.

Okochi M., Nakamura N. and Matsunaga T. (1999). The Electrochemical Killing of Microorganisms Using the Oxidized Form of Ferrocene Monocarboxylic Acid. *Electrochimica Acta*. 44, 3795–3799.

Ouyang C., Shang X., Wang X.Z. and Kong H.N. (2009). Study on the Effect of Aeration on Electrolytic Phosphorus Removal. *Environmental Pollution and Prevention*. 31 (07), 30–33. (Chinese)

Polcaro A.M., Vacca A., Mascia M., Palmas S., Rodiguez Ruiz J. (2009). Electrochemical Treatment of Water with BDD Anodes: Kinetics of the Reactions Involving Chlorides, *J. Appl. Electrochem*. 39, 2083–2092.

Pradhan A.A. and Gogate P.R. (2010). Degradation of P-nitrophenol using Acoustic Cavitation and Fenton Chemistry. *Journal of Hazardous Materials*. 173 (1–3), 517–522.

Qiu C.E. and Wang W.D. (2017). Effect of Hydrogen Peroxide on the Growth and Physiological Characteristics of Microcystisaeruginosa (Copper Green). *Journal of Hubei Normal University (Natural Science Edition)*. 37 (04), 1–5. (Chinese)

Rodríguez E., Majado M.E., Meriluoto J. et al. (2007). Oxidation of Microcystins by Permanganate: Reaction Kinetics and Implications for Water Treatment. *Water Research*. 41 (1), 102–110.

Tan L., Li B.G. (2006). Research Progress of Electrochemical Sterilization Technology in water Treatment. *Industrial Water Treatment*. 02, 1–5. (Chinese)

Urs von Gunten (2003). Ozonation of Drinking Water: Part I. Oxidation Kinetics and Product Formation. *Water Research*. 37 (7), 1443–1467.

Wang C.H., Li G.F., Wu Y. and Wang Y. (2008). Mechanism of Inactivation of Microcystis Aeruginosa by a Packed Bed Discharge Reactor. *High Voltage Technology*. 05, 1051–1056. (Chinese)

Wang H.F. (2005). *Study of Micro Electrolysis-Adsorption Synergy for Algae Control*. Dalian University of Technology. (Chinese)

Wang M., Ao Z., Gong Z. et al. (2022). Deactivation of Cyanobacteria Blooms and Simultaneous Recovery of Phosphorus through Electrolysis Method. *Environmental Science and Pollution Research*. 1–10.

Wang X.W., Li Z.P., You H., Su T., Xu Z., Chen Q.W. (2020). Ti/SnO2 -RuO$_2$ Anodic Electrocatalytic-UV Inactivation of Microalgae in Ballast Water. *Water Treatment Technology*. 46 (03), 105–109. (Chinese)

Xiang P., Jiang Y.Z., Jiang W.C., Zhang Z., Xue Y.H., Li M.Y. and Mo J.Y. (2020). Efficacy and Mechanism of Algae Removal by Iron-carrying ACF/Ni Cathode Electrochemical System. *China Environmental Science*. 40 (11), 5010–5019. (Chinese)

Xue X.D., An Z.H., Pan C.Y., Li L., Shao T.B. and Su H.C. (2020). Experimental Study of Micro Electrolysis-Adsorption for Seawater Algae Removal. *Salt Science and Chemical Engineering*. 49 (08), 25–28. (Chinese)

Ye D.N., Jia J.P., Xu Y.F. and Xu D.H. (2007). Preliminary Study on the Performance of Hydrocavitation Coupled Electrolytic Algae Suppression Process. *Journal of Environmental Engineering*. 12, 67–71.

Zhang H., Yang L.F., Yu Z.L., Huang Q. (2014), Inactivation of Microcystis Aeruginosa by DC Glow Discharge Plasma: Impacts on Cell Integrity, Pigment Contents and Microcystins Degradation, *J. Hazard. Mater*. 268, 33–42.

Zhang R., Zhao X., Li Q.W., Wei J.F. and Li Z.H. (2019). Research and Application Progress of Electrochemical Water Treatment *Technology*. Water Treatment Technology. *45 (04), 11–16. (Chinese)*

Zheng B., Zheng Z., Zhang J., Luo X., Liu Q., Wang J., Zhao Y. (2012). The Removal of Microcystis Aeruginosa in Water by Gamma-ray Irradiation, Sep. *Purif. Technol*. 85, 165–170.

Zhong Z.Y., Huan H.Q., Miao L., Dong K.M. (2019). A Review of Electrochemical Oxidation for the Treatment of Organic Wastewater. *Contemporary Chemical Research*. 13, 42–44. (Chinese)

Zimmermann U. (1986). Electrical Breakdown, Electropermeabliziatoin, and Electrofusion. *Rev. Physiol. Biochem. Pharmacol*. 105, 175–256.

Zuo Y.B. (2012). Photoelectric Synergistic Catalytic Oxidation for the Treatment of Algal Toxins in Drinking Water-LR. *Water Treatment Technology*. 38(07), 49–52. (Chinese)

Civil Engineering and Energy-Environment – Gao & Duan (Eds)
© 2023 the Author(s), ISBN 978-1-032-56059-5

Advances in treatment technology of microcystins in water

Jun Wang
*China Coal Science & Industry Chongqing Design & Research Institute (Group) Co., Ltd,
Chongqing, China*

Dong Liang* & Fei Wang
Chongqing University of Science and Technology, Chongqing, China

Jing Li
Laojunmiao oil production plant of Yumen Oilfield Branch, CNPC, Yumen, China

Tao Wang
*Fracturing and Acidizing Research Institute of Engineering Technology Research Academy in Tuha
Oilfield, CNPC, Qingdao, China*

Zhongwu Zhang
Engineering Technology Research Academy of Tuha Oilfield, CNPC, Turpan, China

ABSTRACT: In recent years, with the deepening of global warming and water eutrophication, cyanobacterial blooms occur frequently. Among all the identified cyanobacterial variants to date, microcystins are the most harmful cyanobacterial toxins produced and released, posing a serious threat to the ecological environment and public health. Therefore, microcystins become one of the global water pollution control problems to be solved. Biodegradation, chemical oxidation, and activated carbon adsorption are still the main strategies for repairing microcystin pollution in water. This paper reviews the current research status of microcystin treatment, aiming to provide a valuable reference for researchers to identify potential cooperation opportunities in related fields and seek research directions for future interdisciplinary and multi-view methods of microcystins.

1 INTRODUCTION

With the rapid development of industry and urbanization, a large amount of organic matter containing nitrogen and phosphorus nutrients is discharged into the water body, and the eutrophication of freshwater resources is becoming more and more serious. Because eutrophication is constantly occurring around the world, it has seriously threatened the aquatic ecosystem. Among them, the blue-green algae bloom caused by eutrophication has become a major challenge for water resources. When blue-green algae bloom in large quantities, it will emit odors that affect the normal physiological and biochemical functions of the water body, making the physical and chemical indicators of the water body exceed their standard value. Microcystins (MCs) are synthesized by large multi-enzyme complexes composed of different modules, including polyketide synthase and nonribosomal peptide synthase, as well as several custom enzymes. It is a class of bioactive monocyclic heptapeptide toxins. MCs are soluble in water and have stable chemical properties, heating and boiling can not destroy

---

*Corresponding Author: 1632413306@163.com

DOI: 10.1201/9781003433651-33

249

their structure, so the degradation process is extremely slow. How to effectively remove MCs in water has become a research hotspot.

At present, the commonly used treatment methods for MCs are divided into the physical method, chemical method, and biological method. In the traditional physical treatment process, coagulation sedimentation, ultrasound, adsorption, and membrane filtration are mainly used to remove MCs, but membrane filtration and ultrasound are expensive and not suitable for large-scale promotion. The removal effect of coagulation sedimentation on MCs is not obvious. The adsorption method is limited by its large surface area of nanomaterials and easy to causes secondary pollution. The chemical treatment methods of MCs mainly include oxidation methods to remove MCs, advanced oxidation technology, etc. The oxidation removal method often uses strong oxidants such as ozone, chlorinating agent, potassium permanganate, and ferrate to eliminate microcystins in micro-polluted water. However, the original balance of these strong oxidants will be destroyed, and the algae cells will be broken in advance. Compared with the traditional oxidation method, advanced oxidation technology has a weaker selectivity. Most of the pollutants are directly oxidized to CO_2 and inorganic acid. There are the Fenton method, the photocatalytic method, and so on. The Fenton method is widely used to remove MCs because of its environmental friendliness and high removal efficiency. The biological method of removing MCs uses a biofilter, ecological floating bed, and membrane biofilm reactor. The removal effect of MCs is no secondary pollution and no toxicity, but it is easy to be limited by the scope of application, long reaction cycle, and difficulty in resisting extreme environments.

In the past few decades, a series of studies have been conducted on the most common microcystins. According to the current research results, the physical methods, membrane treatment, and ultrasound for removing MCs are costly. Biological methods can be effectively removed but take too long, and chemical agents and Fenton reactions are limited in terms of secondary pollution and pH. Any single process is difficult to remove them, requiring a combination of methods to effectively remove algae toxins in the water. This paper summarizes and analyzes the physical, chemical, and biological methods commonly used to remove MCs. Finally, the current problem of removing MCs is summarized, which is helpful for future research directions and removal technologies.

2 MATERIALS AND METHODS

2.1 *Physical methods removal method of MCs*

2.1.1 *Removal of MCs by adsorption*
The adsorption method is commonly used in various types of water treatment. Due to the considerable types of adsorbents, most of the adsorbents for microcystins currently use activated carbon, which is divided into powdered activated carbon (PAC) and granular activated carbon (GAC). Activated carbon is a relatively low-cost adsorbent and has been widely used to remove a variety of pollutants. Due to its porous structure, a high surface area per unit mass is generated and has a high adsorption capacity. In addition, activated carbon may be used for large-scale MCs removal in waterworks. In addition, it has also been widely used in the adsorption of metal ions in previous studies; the adsorption capacity of metal ions is determined by the number of hydrophilic groups on the surface of activated carbon. Dai (Dai 2018) et al. prepared PAC-Fe (III) for rapid and effective removal of MCs in water. Compared with unmodified PAC, PAC-Fe (III) showed excellent MCs-LR removal ability and efficiency. In the range of pH 4.3 $\sim$ 9.6, the removal efficiency of MCs-LR by PAC-Fe (III)increased with the decrease in pH. PAC-Fe(III)can be reused three times by methanol elution, while the removal rate of MCs-LR is still more than 70%.

2.1.2 *MCs removal by membrane technology*

Membrane treatment technology is feasible for MCs treatment, which is divided into microfiltration, ultrafiltration, and nanofiltration according to different pore sizes. The membrane acts as a physical barrier, allowing water to pass through while retaining suspended solids or even dissolved substances, depending on the type of membrane. In this specific application, microfiltration (MF) and ultrafiltration (UF) will be sufficient to remove cyanobacterial cells, but cannot remove cyanobacterial toxins because these membranes have large pore sizes and high molecular weight cut-offs. Therefore, membrane technology is the best among the many methods for removing MCs in the current physical method, but the fatal determination of this method is that the membrane needs to be changed frequently during the removal process, the utilization rate is not high, and the recovery cost is high. Water treatment cannot be used effectively. It can only be used in a small range or with other methods.

2.2 *Chemical methods removal of MCs*

2.2.1 *Conventional oxidant removal method*

As a high molecular weight organic matter, MCs usually choose an oxidant with a strong oxidizing function to oxidize MCs in water and inactivate it naturally. According to the cyclic structure of MCs, the use of strong oxidants to oxidize toxic groups directly targeted to inactivate or lose toxicity of MCs, to achieve the purpose of removal. Commonly used strong oxidants are ozone, chlorinating agents, potassium permanganate, etc. Studies have shown that chlorine and potassium permanganate can destroy algae cells, make algae cells rupture in advance, and release MCs. Therefore, when these strong oxidants are used, the original balance will be destroyed, and the concentration of MCs in water will increase sharply in a short time, causing more serious pollution.

2.2.2 *Photocatalytic oxidation removal method*

Photocatalytic degradation of pollutants is an economical and green technology. The development of photocatalysts is very important for the efficient removal of pollutants. Under certain light intensity, the double bond structure of MCs containing the Adda group will undergo partial variation, making it lose toxicity. Moreover, the photocatalyst heterojunction can inhibit the recombination of photoelectrons and holes, broaden the absorption range of visible light, and thus improve the removal efficiency of MCs. Cheng (Cheng 2018) et al. used 5×104 cells/mL Microcystis aeruginosa irradiated by 20 mg/L TiO_2 and 5 mg/L H_2O_2 under visible light for 5 h. The lethality rate of Microcystis aeruginosa was 96%, and the microcystins were completely released to the extracellular, from 3.2 μg /L to 1 μg/L. Guo (Guo 2018) et al. found that when 200×104 cells/mL Microcystis aeruginosa was treated with 150 mg/L TiO_2 for 10 h, the content of chlorophyll decreased to 15%, while the degradation rate of MCs-LR reached 89.9%.

2.3 *Biological methods removal of MCs*

2.3.1 *Microbial degradation of MCs*

The natural degradation process of MCs in water is realized by some specific bacteria, and microorganisms can produce one or more special enzymes through themselves. These enzymes can act on MCs, accelerate the hydrolysis of MCs with toxic ring structures and a double bond, and play a catalytic role. Another important reason is that the use of microbial degradation of MCs will not cause more serious water pollution and at a low cost. MCs and bacteria have high biodegradation efficiency, eco-friendly and non-toxic, but some bacteria need strict environmental conditions to cultivate and also need to be properly domesticated in natural water.

Table 1. Comparison of microcystins removal methods.

type	name	principle	limitation
Physics method	activated carbon adsorption	Removal of Algae Toxins from Water by Activated Carbon	Activated carbon consumption is large; adsorption efficiency was affected by many factors; adsorbed algal toxins may be released again
	membrane filtration	A certain pressure was applied to the algal toxin solution to permeate the solvent through the membrane to obtain the toxin concentrate and to realize the separation of algal toxin and water.	High cost, high technical requirements, unstable system operation; isolated microcystin can cause secondary pollution
Chemistry method	basic oxidation method	Oxidation of Adda Groups in Microcystin to Reduce or Deplete Toxicity	It is easy to produce toxic intermediates and has a great influence on water quality
	advanced oxidation method	Destroying the conjugated double bond of the microcystin Adda group to reduce or detoxicate the toxin	High technical requirements, high cost, high energy consumption, and unstable system operation
Organism method	microbial action	Degradation of microcystin in water by breaking conjugated double bonds on the Adda group	It takes a long time, and microorganisms have strict requirements on the environmental conditions for survival.
	vegetation effect	Secretion of inhibitory algae species; destruction of Adda conjugated double bond	Long time consuming

2.3.2 *Degradation of MCs by plant biological floating bed*

Plant biological(Song 2009) floating bed is a simple and low-cost way to remove algae toxins, but the removal rate is limited by plant growth rate and plant growth rate per unit area. Some special bacteria in natural water reduce the toxicity of microcystins by the unsaturated double bond of the Adda group. Three hydrolases (MlrA, MlrB, MlrC) are present in the bacterium, which degrade it into tetrapeptides and hydrolyze the compounds into smaller amino acids, thereby eliminating MCs toxicity.

3 CONCLUSION

Microcystins are the most widespread cyanobacterial toxins in nature. Traditional physical, chemical, and biological methods can remove algae cells in water, but the removal rate of MCs is not high and even cause algae cells to rupture and cause toxin leakage, causing water pollution, as shown in Table 1. In addition, although a small amount of MCs can cause great harm to the environment, the low content of microcystins in natural water is often lower than the detection limit of high-performance liquid chromatography, so it is necessary to purify and enrich them. Biodegradation, chemical oxidation, and adsorption are the main treatment technologies for microcystins. The combined application of bacterial degradation, photocatalytic degradation, and various degradation technologies to maximize the advantages of a single method will be a promising research direction. The future research directions of algal toxins mainly focus on the following:

1. The method of long-term governance and reduction of MCs investment is huge. Therefore, the focus of research should be on the development of efficient, non-toxic, and environmentally friendly natural biomaterials to remove MCs.

2. The release mechanism and influencing factors of intracellular microcystins in the traditional water treatment process were studied to avoid secondary pollution.
3. The efficient and practical water purification technology was studied. The traditional process was properly optimized. The activated carbon adsorption treatment, biological environment treatment, and conventional and advanced oxidation technology were combined. Based on the actual situation, in the process of removing MCs, the cost and secondary pollution were ensured to be as low as possible, which was helpful to improve their application in the actual elimination of MCs.

REFERENCES

Cheng C.W., Huo X.C., Lin T.F. (2018) Exposure of Microcystis Aeruginosa to Hydrogen Peroxide and Titanium Dioxide Under Visible Light Conditions: Modeling the Impact of Hydrogen Peroxide and Hydroxyl Radical on Cell Rupture and Microcystin Degradation. *Water Res.*, 141, 217–226.

Dai G.F., Nan Q.G, Song L.R., Fang S.W., Peng N.Y. (2018). Fast Adsorption of Microcystin-LR by Fe(III)-Modified Powdered Activated Carbon. *J. Oceanol. Limnol.*, 36(4): 1103–1111.

Guo Y.F., Wu S.S., Hu X.D., Li Y. (2018) Study on the Inhibition of Cyanobacteria and Degradation of Microcystin (MCS-LR) by Carbon and Nitrogen co-doped TiO_2 with a Visible Light Response. *Environmental Engineering*, 36 (6): 35–41.

Song H.L., Li X.N., Lu X.W., Y I. (2009) Investigation of Microcystin Removal from Eutrophic Surface Water by the Aquatic Vegetable Bed. *Ecol. Eng.*, 35(11): 1589–1598.

Civil Engineering and Energy-Environment – Gao & Duan (Eds)
© 2023 the Author(s), ISBN 978-1-032-56059-5

Remote sensing ecological environment assessment based on GF image data – Taking Anshan open-pit mine as an example

Zhiwen Hu, Shunbao Liao* & Yuna Qi
Institute of Disaster Prevention, Langfang, Hebei, China

ABSTRACT: In the process of obtaining natural resources, human beings will inevitably destroy the environment and affect the ecological balance of nature. The establishment of an ecological environment evaluation method for the mining area and surrounding area can protect the environment of the mining area and realize the construction of ecological civilization. In this paper, taking Anshan open-air iron mine as the study area, the spatial resolution image of 2 m was obtained by fusion of GF-1 satellite remote sensing image data and DEM image data of the study area, and the remote sensing index information of the mining area, such as NDVI, vegetation coverage, water body information, and soil index, was extracted to normalize the ecological impact index. Finally, the comprehensive index method is used to construct the mining area ecological environment comprehensive evaluation model to analyze and evaluate the mining area ecological environment.

1 INTRODUCTION

Mineral resources are the necessary resources for social development and also the most basic resources. In the early years of mining, environmental protection was neglected in the pursuit of quantity. The ecological environment damage in mining areas is easy to cause vegetation destruction and soil erosion, which will lead to natural disasters, harm human life and human society, and bring about property losses. The ecology of the mining area is a problem that governments all over the world attach great importance to, and the research on it has been continuous.

In most of the mining areas established at an earlier time, the mining area in the process of development has a certain scale, but in the operation of the mining area, the method is relatively single, coupled with insufficient awareness of the mining area's ecological environment protection, it is easy to cause ecological damage in mining areas, and there is no a unified system of protection. As a result, the ecological environment of the mining area is still worrying (Wang & Zhang 2015). With the development of science and technology, the number of remote sensing satellites and their functions is becoming more and more powerful. Through satellites, we can obtain image data with high time resolution and high spatial resolution. These data provide a basis for us to study the ecological environment of mining areas. Yin Jianping made use of multi-temporal Landsat images and principal component analysis to establish remote sensing ecological index model to evaluate the ecological environment of Pingshuo mining area (Yin 2021);Hu Kehong et al. made use of Landsat images and GDM to make attribution of spatial differentiation of ecological environmental quality (Hu & Zhang 2021);Zhu Quan selected the samples with the separation degree over 1.8 to classify the images by the maximum likelihood method, and applied the results to the

---

*Corresponding Author: liaoshunbao@cidp.edu.cn

254

DOI: 10.1201/9781003433651-34

ecological assessment of land reclamation in mining areas (Zhu 2021); Xiu Liancun et al. used airborne hyperspectral imager to establish a comprehensive survey technique, method and operation process, and expounded the demonstration of the application of ecological and environmental geological survey in some areas of the Yangtze River Economic Belt (Xiu et al. 2021); Song Qifan et al. extracted the mining information by using WorldView-2 image and using the method of inter-spectral relationship, normalized differential water index and supervised classification (Song et al. 2011); in these studies, due to the data collection accuracy, evaluation methods, funds and other reasons, the mining area environmental information can not be accurately obtained and applied to the ecological environment assessment.

GF-1 satellite data has the advantages of high spatial resolution, high temporal resolution, and multi-spectrum. The GF-1 satellite contains two sensors: the PMS sensor with a spatial resolution of 2 m/8 m and a temporal resolution of 4 Days; the WFV sensor has a 16 m spatial resolution and 4Days temporal resolution. In this paper, the Anshan open-pit iron mine is taken as the study area. Based on GF-1 satellite data, multiple indexes are calculated to establish a comprehensive evaluation model to evaluate the ecological environment of the mining area.

2 STUDY AREA AND DATA

2.1 *Overview of the study area*

Anshan Iron Mine, located in the southeast of Dagushan Town at the foot of Qianshan Mountain, Anshan City, is the earliest iron mine in Anshan. The main product is magnetite, with an annual output of 7 million tons. At present, the mine is the deepest open-pit iron mine in Asia, with a closed elevation of 78 meters. The mine is 1,620 meters long and 1,200 meters wide along the direction, covering an area of 10.6 square kilometers. The Dagushan mine has proven reserves of 340 million tons and an ore body 700 meters deep, which can only be reached to minus 450 meters by open pit mining.

2.2 *Data*

The data used in this study include GF-1 data, ASTGTM-DEM (30 m), mining location and geographical data, etc. Since this paper is to establish a comprehensive evaluation model to evaluate the ecological environment of the mining area, there is no special requirement at the time of data shooting. The GF-1PMS data dated April 30, 2014, is selected. The GF-1 satellite was launched into orbit on April 26, 2013, and is China's first launch satellite for the National Science and Technology Major Project of the High-Resolution Earth Observation System. The data of administrative divisions are from the National Geographic Information Resource Catalogue Service System.

3 THE RESEARCH METHODS

3.1 *Ecological evaluation process of the mining area*

In the study of the mining area ecological environment, we determine the evaluation factors of the ecological environment quality, extract the image feature information through remote sensing and GIS, and then adopt the method of ecological factor normalization unified outline quantity of each factor by establishing a comprehensive index method, so as to determine the weight of each influence factor index, establish the evaluation model of eco-logical environment system, and realize the comprehensive evaluation of the ecological environment of the mining area. The technical flow chart is shown in Figure 1.

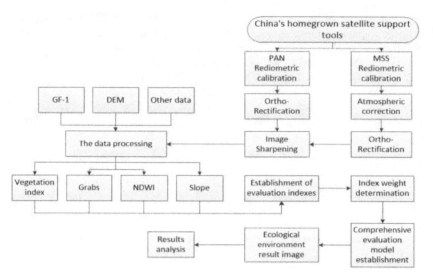

Figure 1. Flow chart of ecological evaluation in a mining area.

3.2 *Remote sensing data processing*

Firstly, the GF-1PMS data were opened and preprocessed by the Chinese satellite support tool, and then the high-resolution image of the study area was obtained by image fusion and image cropping.

Radiometric calibration, atmospheric correction, and orthographic correction were carried out for GF-1 multispectral data. Radiometric calibration and atmospheric correction were performed on GF-1 panchromatic data. Radiation calibration is to convert the recorded original DN value into radiation brightness value to eliminate the error of the sensor itself. For radiometric calibration of multispectral data, the dimension of the data should be consistent with that of the subsequent atmospheric correction processing. Atmospheric correction means that the total radiation brightness measured by the sensor is not a reflection of the real reflectivity of the surface, which includes the radiation quantity error caused by atmospheric absorption, especially the scattering effect. Atmospheric correction eliminates the radiation errors caused by atmospheric influence and inverting the real surface reflectance of ground objects. The aerosol model was selected as a city. Through the historical weather query, the weather was clear and cloudless on the day of the data shooting, so the visibility was set as 40 km. Orthophonic correction is the process of using digital elevation model (DEM) data to simultaneously correct the tilt and projection error of the image, resampling the image into an orthophonic image, and using the RPC orthophonic correction process tool to correct the image. Panchromatic data preprocessing is similar.

The NNDiffuse Pan Sharpening method is used to fuse data in image fusion. This method can preserve the color, texture, and spectral information of the image well. A 2 M spatial resolution multispectral remote sensing image was obtained by processing. Through visual judgment, the image is cut manually to get the appropriate image of the study area. After the GF-1 data is processed by the above method, the high-resolution multispectral image of the study area is obtained (Figure 2).

Figure 2. Data preprocessing result map.

3.3 *Mining environmental information extraction*

3.3.1 *Vegetation coverage*

Firstly, the difference between the reflection value of the near-infrared band and the reflection value of the red band is used to obtain the normalized vegetation index (NDVI), and then the vegetation coverage is calculated. The calculation formula is as follows:

$$FC = \frac{NDVI - NDVI_{min}}{NDVI_{max} - NDVI_{min}} \quad (1)$$

FC is the vegetation coverage, NDVI is the normalized vegetation index, $NDVI_{max}$ and $NDVI_{min}$ is the maximum and minimum of NDVI. The NDVI value of the image was calculated by the NDVI tool, and the NDVI range was calculated by using computational statistics. According to the statistical results, the cumulative percentage was selected as the standard, and the corresponding pixel values at 5% and 95% were read, respectively, to determine the maximum and minimum values of effective NDVI, and finally, the vegetation coverage map of the study area was obtained (Figure 3a).

3.3.2 *GRABS*

The soil brightness index and greenness index can be used to judge bare soil and vegetation. Therefore, the bare soil vegetation index is formed through the linear combination of the greenness and soil brightness indexes. Its calculation formula is as follows:

$$GRABS = VI - 0.9178BI + 5.58959 \quad (2)$$

GRABS is the bare soil vegetation index, VI is the greenness index of ear cap transformation, and BI is the soil brightness index of ear cap transformation. Calculated soil index diagram (Figure 3b).

3.3.3 *NDWI*

Normalized water index is the difference between the green band and the near-infrared band of remote sensing images. Its calculation formula is as follows:

$$NDWI = \frac{G - NIR}{G + NIR} \quad (3)$$

NDWI is the normalized water body index, G is the green light band and Nir is the near-infrared band. Figure 3c of the water index in the research area was obtained by calculation.

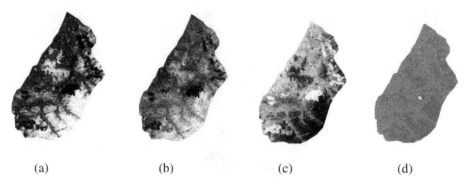

(a)   (b)   (c)   (d)

Figure 3. Image classification results.

3.3.4 *Slope*

The slope is the degree of steepness of surface units. The ratio of vertical height and horizontal distance of slope surface is usually called slope. Its calculation formula is:

$$\text{Slope} = \text{Elevation difference/horizontal distance} \qquad (4)$$

Calculated Slope Map of the Research Area (Figure 3d)

3.4 *Ecological environment assessment*

3.4.1 *Ecological factor normalization*

Vegetation coverage and soil index can be used as ecological factors for research, but each factor has a corresponding dimension. Therefore, data need to be normalized for comprehensive evaluation. The density segmentation method is used to divide it into ten levels (1-10). The coding value is proportional to the pixel value. The larger the coding value is, the larger the pixel value is. Using the above methods, vegetation coverage, slope, and soil index were divided into 10 grades. After the grading is completed, the data pixel value is normalized to 1-10.

3.4.2 *Establishment of index system*

Ecological environment evaluation is to quantitatively determine the ecological environment of the study area. In this paper, the comprehensive index method is adopted. Under an evaluation index system, the weighted average of all kinds of indexes is carried out, and the comprehensive value is calculated to evaluate the ecology of the mining area. The evaluation model is:

$$E = W_1 * S_{v1} + W_2 * S_{v2} \qquad (5)$$

E is the comprehensive index value, W is the weight value, and S_v is the number of various indicators. According to Meng Xiangliang et al. 'Technical Specification for Assessment of Ecological Environmental Condition', the weight ratio of vegetation coverage, soil index, and the slope is 8:2 (Meng et al. 2020). The resulting graph is divided into four categories to construct the environmental evaluation table.

Table 1. Grading table of ecological environment evaluation.

Rating	Comprehensive evaluation index	instructions
Excellent	9~10	The environment is not damaged and the structure is reasonable.
Good	6~9	The environment is not damaged and the structure is reasonable.
Medium	3~6	Environmental damage, unreasonable structure.
Poor	1~3	Serious environmental damage, unreasonable structure.

4 EVALUATION RESULTS AND ANALYSIS

The vegetation coverage and soil index extracted from the GF-1PMS image were used as the normalized ecological image index, and the comprehensive index method was used to calculate ENVI to obtain the ecological environment index results of the study area. Finally, the percentage of each gray value was calculated by Compute Statistics. Finally, the ecological environment evaluation results of the Anshan open-pit mining area are obtained. The ecological environment of the Anshan open-pit mining area can be divided into four levels: excellent, good, medium, and poor (Figure 4).

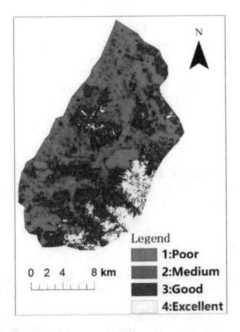

Figure 4. Results of ecological environment evaluation.

According to the area statistics of the four ecological environment zones in the mining area, the proportion of each zone in the mining area is obtained (Table 2). The evaluation results show that the percentage of the evaluation index within the range of 1-3 is 41.83%, which is the highest value, indicating that most of the natural ecology in this region is in a poor state, The main area is the mining area and the surrounding areas of the mining area, and the rest is concentrated in the urban construction land. The evaluation index of 9-10 is 20.41%, which indicates that the natural ecology of this region is in an excellent state. Most of these regions are concentrated in mountainous areas with good vegetation growth. The area with an evaluation index from 3 to 9 accounted for 37.76%.

Table 2. Statistical results.

DN	1	2	3	4
Percentage(%)	41.83	19.25	18.51	20.41

5 CONCLUSION

This study takes Anshan open-pit iron mine as the research area and uses GF-1 (PMS) satellite data, DEM data, and other data to process and analyze to get the ecological environment information of the mining area, such as vegetation coverage, water body information, soil index. On this basis, the ecological factors were normalized, and the comprehensive index method was used to analyze and evaluate the ecological environment of the mining area. The research results show that about 60% of the ecological environment in the Anshan open-pit iron mining area is in good condition, and about 40% of the regional comprehensive index is in poor condition. The Gaofen-1 satellite data used in this study has the advantages of high resolution and multi-spectrum compared with other remote sensing data, and the evaluation method is simpler and easier to implement than the traditional method. For the next step, it is necessary to increase the extraction of the index information of the mining area to make a more comprehensive evaluation of the mining environment. The most important thing is to establish a general evaluation system for the mining areas with different landforms and landforms, which is helpful to the environmental monitoring of mining areas and the prevention of disasters in advance.

ACKNOWLEDGMENTS

This study was supported by the Fundamental Research Funds for the Central Universities, "Pre-research on Geographic Information Big Data Platform for Earthquake Preparedness and Disaster Reduction" (grant number ZY20180101).

REFERENCES

Hu K.H. and Zhang Z. Spatial and Temporal Characteristics and Influencing Factors of Ecological Quality in Liuba County, Qinling Mountains, Shaanxi Province. *Journal of Ecology and Rural Environment*, 2021,37 (06): 751–760.

Meng Xiangliang, Liu Wei, Kong Mei, Wang Qi, Shi Tongguang. Evaluation of Ecological Management Utility in the Revised Technical Code for Assessment of Ecological Environment Condition - A Case Study of Shandong Province. *Environmental Monitoring and Early Warning*,2020,12(02): 56–62.

Song Qifan, Wang Shaojun, Zhang Zhi *et al.* Study on Extraction Method of Water Body Information in Tungsten Mine Area based on World View II image – Taking Dayu County of Jiangxi Province as an example. *Remote Sensing of Land and Resources*, 2011, (02): 33–37.

Wang Yao and Zhang Ying. Construction and Research on Evaluation Index System of Ecological Environmental Protection in Mining Area. *Heilongjiang Science and Technology Information*,2015,{4}(30): 70.

Xiu Liancun, Zheng Zhizhong, Yang Bin, Yin Liang, Gao Yang, Jiang Yuehua, Huang Yan, Zhou Quan-Ping, Shi Jian-Long, Dong Jinxin, Chen Chunxia, Liang Sen, Yu Zhengkui. The Effect of Hyperspectral Imaging Technology on Environmental Protection in the Yangtze River Economic Belt. *China's geological*, 2021,{3}: 1–27.

Yin Jianping. Ecological Environment Assessment of Pingshuo Open-pit Mine Based on Remote Sensing Ecological Index. *Open Pit Mining Technology*,2021,36(01): 45–47.

Zhu Quan. Ecological Assessment of Mining Land Reclamation Based on Remote Sensing *Technology. Geological and Mineral Surveying and Mapping*,2021,37(02): 7–11.

*Ecological data modeling and
environmental technology*

Civil Engineering and Energy-Environment – Gao & Duan (Eds)
© 2023 the Author(s), ISBN 978-1-032-56059-5

Maintenance and treatment technology of oil-based drilling fluid polluted by high-pressure brine

Shuanggui Li & Sheng Fan*
Key Laboratory of Enhanced Oil Recovery for Fractured-Vuggy Carbonate Reservoirs & Sinopec Northwest Oilfield Company, China

Guohe Xu
Oil & Gas Field Productivity Construction Department of PetroChina Tarim Oilfield Co., China

Zhong He, Shafei Shu & Cheng Zhai
Key Laboratory of Enhanced Oil Recovery for Fractured-Vuggy Carbonate Reservoirs & Sinopec Northwest Oilfield Company, China

ABSTRACT: The drilling fluids are easily polluted by the saline water in high-pressure salt-gypsum formations, and the properties of the drilling fluid are damaged. Therefore, accidents easily occur. In this research, the mechanism of drilling fluid contamination by the brine was analyzed, and the capacity limits of the contamination were found. The results showed that a small amount of high-pressure brine entered the oil-based drilling fluid, and the drilling fluid could remain stable, while large amounts of brine invasion led to oil-based drilling fluid demulsification. When the saltwater content of the contaminated drilling fluid accounted for less than 10% of the drilling fluid volume, the drilling fluid performance remained good. When the amount of salt water that polluted the drilling fluid accounted for more than 10%, but less than 30% of the drilling fluid volume, the performance of the drilling fluid became worse and needed maintenance and treatment. When the amount of salt water that polluted the drilling fluid accounted for more than 30% of the drilling fluid volume, the performance of the drilling fluid became very poor, and the maintenance and treatment costs were high, which explained why the drilling fluid was often discarded. This proposal has been successfully used in 8 wells in the Kuqa foreland area, and the drilling cycle was saved by 35 days on average.

1 INTRODUCTION

A large number of salt-gypsum layers are developed in the Tarim basin, in which high-pressure brine is developed between the layers. During drilling, a lot of brine and salts will invade or dissolve into the drilling fluid, which will damage the coalescence stability of the drilling fluid itself, deteriorate the performance of the drilling fluid, and even cause accidents such as sticking, seriously affecting the drilling safety (Wang & Zhou 2019; Yin *et al.* 2012). Table 1 shows the number of drilling accidents and complications caused by high-pressure saline contamination of drilling fluid.

High-density oil-based drilling fluid (2.45-2.59 g/cm^3) is generally used to handle the pollution of high-pressure brine on drilling fluid. When drilling with high-pressure brine, the invasion performance of drilling fluid will worsen, leading to drilling trouble and accidents

---

*Corresponding Author: liji3211979@163.com

DOI: 10.1201/9781003433651-35

263

Table 1. The number of drilling accidents and complications caused by high-pressure brine contaminated drilling fluid.

Well	Lost circulation	Brine overflow	Blocking	Other accident
D5	13	1	42	
D206	6	3	17	
D303	6	2		
D304	22	1	28	2
D306	17	4	11	
K7	11	1	40	1
K2-1-4	1	2	11	
K2-1-11	4	1	30	
K2-1-18	17	4	15	
B101	14	1	44	12

such as blocking and sticking. However, the mechanism of brine invasion and the capacity limit of high-density oil-based drilling fluid affected by brine are unknown, making drilling operations ineffectively guided. As a result, laboratory experiments and mechanism analysis are used to improve understanding of the mechanism and capacity limit of brine contamination of drilling fluids, as well as to guide field operations.

2 THE MECHANISM OF BRINE CONTAMINATION

To study the mechanism of oil-based drilling fluid contaminated by brine, a certain amount of brine is added to the oil-based drilling fluid, and its performance is measured. The influence mechanism of the oil-based drilling fluid on the drilling fluid is analyzed based on the change in performance. The experimental oil-based drilling fluid system was taken from a K2-1-4 well with a depth of 5546 m. The drilling fluid is: diesel oil + 2%–3% main emulsion + 2%–3% secondary emulsion + 0.5%–0.8% organic soil + 0.3%–0.6% suspending agent + 1.5%–2.5% duratone + barite.

2.1 *The rheology and filtration properties of drilling fluid polluted by brine*

The oil-based drilling fluid was added to 5%, 10%, 15%, 20%, 30%, 40%, 50%, and 60% brine to test its rheology and filtration properties. The experimental results are shown in Figures 1 and 2.

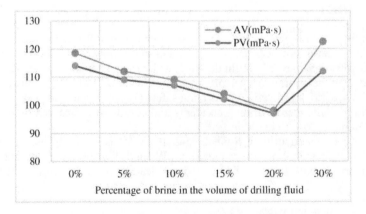

Figure 1. The rheology properties of oil-based drilling fluid polluted by brine.

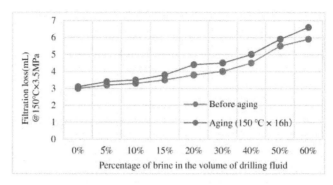

Figure 2. The filtration properties of oil-based drilling fluid polluted by brine.

The results show that when the percentage of brine is less than 20%, the viscosity decreases as the percentage of brine increases; when the percentage of brine is more than 30%, the viscosity increases; and when the percentage of brine is more than 50%, the viscosity of the system exceeds the measurement range of the instrument. The high temperature and high pressure (HTHP) filtration loss increase slightly.

2.2 *The settlement stability of drilling fluid polluted by brine*

The settlement stability of high-density oil-based drilling fluid polluted by brine is bound to be affected since it is one of the important parameters to be concerned with after the drilling fluid is contaminated by brine (Li *et al.* 2013; Shen *et al.* 2016). To determine the settlement stability of drilling fluid by brine contamination, a steel cylinder with an inner diameter of 6.35 cm, an outer diameter of 7.6 cm, an inner height of 35 cm, a total height of 39 cm, and a volume of 1107 ml was used.

The drilling fluid contaminated by brine settled in the steel cylinder, and the steel cylinder was heated by the incubator to keep the drilling fluid at a certain temperature. After a certain amount of settling time, the upper and lower parts of the drilling fluid in the steel cylinder were taken, respectively, to measure their densities. Then the density difference between the upper and lower parts was used to measure the settling stability of the drilling fluid at different temperatures.

The settling stability of oil-based drilling fluid at temperatures of 120°C and 150°C after adding 5%, 10%, 15%, 20%, 30%, 40%, 50%, and 60% brine was tested. The density difference test results are shown in Figure 3, and the amount of precipitated emulsified salt water is shown in Figure 4.

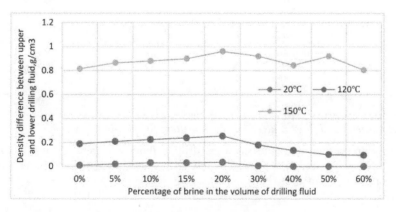

Figure 3. The settlement stability of drilling fluid polluted by brine.

265

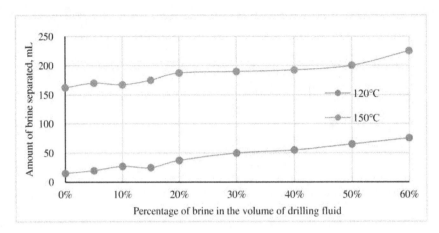

Figure 4. Amount of precipitated emulsified salt water.

The results in Figure 3 show that the density difference of the drilling fluid first increases and then decreases with the increase in brine percentage; with the increase in temperature, the settling stability of the drilling fluid decreases; the performance of the oil-based drilling fluid has become worse at 150 °C, and more emulsified salt water is precipitated from the upper part.

2.3 The demulsification voltage of drilling fluid polluted by brine

Demulsification voltage (DV) is one of the indices of oil-based drilling fluid stability (Wang et al. 2016). The demulsification voltage of drilling fluid contaminated by 0%, 5%, 10%, 15%, 20%, 30%, 40%, 50%, and 60% brine was tested. The test results are shown in Table 2.

Table 2 shows that the demulsification voltage gradually decreases after brine addition and slightly increases after aging (150 °C, 16 h).

Table 2. The demulsification voltage (DV) of drilling fluid polluted by brine.

Percentage of brine (%)	DV of drilling fluid after aging (v)	DV of drilling fluid by aged (v)
0	468	445
5	396	410
10	340	380
15	225	305
20	180	295
30	175	263
40	150	225
50	121	180
60	90	143

2.4 The lubricity and filter cake of oil-based drilling fluid polluted by brine

The lubrication coefficient was tested when drilling fluid was polluted by 5%, 10%, 15%, 20%, 30%, 40%, 50%, and 60% brine. The test results are shown in Table 3. The mud cake measured in the experiment is shown in Figure 5.

Table 3. The lubricity and filter cake of oil-based drilling fluid polluted by brine.

Percentage of brine (%)	0	5	10	15	20	30	40	50	60
Lubrication coefficient	0.075	0.081	0.094	0.125	0.138	0.152	0.170	0.191	0.217

Figure 5. Mud cake of drilling fluid polluted by brine.

The data in Table 3 and Figure 4 show that with the increase in brine, the filter cake tends to thicken, resulting in an increase in the lubrication coefficient, but the overall lubricity is still good.

2.5 *Mechanism of oil-based drilling fluid polluted by brine*

By analyzing the settling stability, filtration, demulsification voltage, filter cake, and lubricity of oil-based drilling fluid polluted by brine, the mechanism of brine pollution in oil-based drilling fluid is as follows:

1) When a small amount (10%) of brine is added to oil-based drilling fluid, the viscosity gradually decreases, and the filtration loss gradually increases, but the system still has a high demulsification voltage, indicating that the immersed brine and the excessive emulsifier in the system form a relatively stable emulsion.
2) When a large amount (> 10%) of brine is invaded by the oil-based drilling fluid system, the stability of the emulsion is broken, and the viscosity and filtration loss continue to increase. This indicates that the brine enters the oil-based drilling fluid in the form of the water phase, reduces the oil-water ratio of the system, increases the viscosity of the oil-based drilling fluid, and worsens the mud cake.
3) When the volume of brine in the system increases to more than 30%, the viscosity and filtration loss increase a lot, yet it remains watery in an oil emulsion. When the actual experiment reaches 60%, the system is demulsified. Therefore, the possibility of wetting reversal is relatively small in actual drilling.

3 THE MAINTENANCE AND PROCESS OF DRILLING FLUID POLLUTED BY BRINE

According to the mechanism of drilling fluid contaminated by brine and the actual requirements of drilling in the Tarim basin, the key performance indices of oil-based drilling fluid were selected. The capacity of drilling fluid contaminated by brine was determined to guide field drilling fluid performance and maintenance based on performance changes and maintenance after brine contamination.

3.1 Index of drilling fluid contaminated by brine

Combined with the influence and mechanism of brine on the rheological properties, filtration rate, temperature resistance, and sedimentation stability of the drilling fluid system, the first dividing point is whether the drilling fluid can contain the pollution of brine, and the second dividing point is whether the drilling fluid is suitable for maintenance. Therefore, the amount of brine that pollutes the drilling fluid is below the first dividing point, and the drilling fluid does not need maintenance and treatment to maintain stable performance. When the amount of salt water that pollutes the drilling fluid is between the first and second dividing points, the drilling fluid needs maintenance to maintain stable performance. When the salt water volume of the contaminated drilling fluid is greater than the second boundary point, the maintenance and treatment costs of the drilling fluid are high, and the drilling fluid can only be discarded. See Table 4 for the data for the first and second dividing points.

Table 4. The dividing points of oil-based drilling fluid polluted by brine.

Index	DV (V)	Gel strength of 10 min. (Pa)	AV (mPaS)
The first dividing points	>400	≤ 8	100–120
The second dividing points	400–260	8–15	<100

3.2 The dividing points of oil-based drilling fluid

The key performance characteristics of drilling fluid are divided. If the index value of the drilling fluid is still within the design range after contamination, it is recorded as a safe area. If there is not much beyond the design range, the index value can be adjusted to the design range as the warning area after simple treatment. Those that may lead to drilling accidents and complications are classified as dangerous areas, from which the salt water invasion amount and drilling fluid safety chart are drawn, as shown in Table 5.

According to the analysis, the first dividing point in oil-based drilling fluid is 10%, and the second dividing point is 30%.

Table 5. Safety chart of drilling fluid by brine Invasion.

Percentage of brine (%)	Density g/cm^3	Gel strength of 10 minute(Pa)	Apparent viscosity (mPaS)	Demulsification voltage (V)	Lubricity Coefficient	Safety level
0%	2.51	5.5	118.5	445	0.075	Safe
5%	2.45	6	112	410	0.081	
10%	2.40	7	109	380	0.094	
15%	2.34	5.5	104	305	0.125	Warning
20%	2.29	6	98	295	0.138	
30%	2.21	6	122	263	0.152	
40%	2.13	9	133	225	0.170	Dangerous

3.3 Maintenance and treatment of drilling fluid polluted by brine

In the experiment, 30% brine was added to the oil-based drilling fluid, and then diesel oil, the main emulsifier, the auxiliary emulsifier, organic soil, a suspending agent, and a fluid loss reducer were added to the oil-based drilling fluid polluted by brine. Their properties were later experimented with, as shown in Table 6.

Table 6.   Maintenance of oil-based system polluted by 30% brine.

No.	Density (g/cm^3)	AV (mPa·s)	PV (mPa·s)	YP (Pa)	Gel10″/ 10′ (Pa/Pa)	API.FL (mL/mm)	HTHP@ 150°C (mL/mm)
1#	1.813	20	18	2	1/1.5	0/1	5.9/1.5
2#	1.610	15	13	2	1.5/2.5	0/1	4.4/1.5
3#	2.499	55	53	2	2.5/3	0/1	4.5/1.5
4#	2.510	113	102	11	6/7	0/1	3.2/2

Maintain the drilling fluid formula as follows:

1. Oil-based drilling fluid 30% brine + 50% diesel oil + 1.5% main emulsifier + 1.5% auxiliary emulsifier + 0.3% organic bentonite + 0.2% suspending agent + 1% fluid loss reducer;
2. Oil-based drilling fluid 30% brine+ 100% diesel oil + 1.5% main emulsifier + 1.5% auxiliary emulsifier + 0.4% organic bentonite + 0.3% suspending agent + 1% fluid loss reducer;
3. Oil-based drilling fluid+ 30% brine+ 100% diesel oil+ 1.5% main emulsifier+ 1.5% auxiliary emulsifier+ 0.4% organic bentonite+ 0.3% suspending agent+1% fluid loss reducer + barite;
4. Oil-based drilling fluid+ 30% brine+ 100% diesel oil+ 1.5% main emulsifier+ 1.5% auxiliary emulsifier+ 0.4% organic bentonite+ 0.3% suspending agent+ 1.5% fluid loss reducer+ barite;

For the oil-based drilling fluid system, 30% salt water, 100% diesel oil, 1.5% main emulsion, 1.5% secondary emulsion, 0.4% organic bentonite, 0.4% suspending agent, and 1.5% fluid loss additive were used for maintenance. The performance of the system was restored to the same level as that of the field drilling fluid.

4   CONCLUSIONS

1) When a small amount of brine (10%) is immersed in the oil-based drilling fluid, the viscosity slowly decreases while the demulsification voltage remains high, indicating that the immersed salt water forms a relatively stable emulsion.
2) When a large amount of salt water invades the oil-based drilling fluid system, the salt water enters the drilling fluid in the form of water, reducing the oil-water ratio;
3) When the amount of brine that pollutes the drilling fluid is below 10%, the drilling fluid does not need maintenance or treatment to maintain stable performance. When the amount of brine that pollutes the drilling fluid is between 10% and 30%, the drilling fluid needs maintenance to maintain stable performance. When the salt water volume of the contaminated drilling fluid is greater than 30%, the maintenance and treatment cost of the drilling fluid is high, and the drilling fluid can only be discarded;
4) For the drilling fluid with 30% salt water, diesel oil, an emulsifier, organic soil, a suspending agent, and other treatment agents should be added in proportion and increased to the original density so that the performance of the drilling fluid can be restored to the same level as that of the original field drilling fluid.

ACKNOWLEDGMENTS

This work was supported by the National Major Science and Technology Projects of China titled *Key engineering technologies of ultra-deep oil and gas wells in marine carbonates* (No. 2017ZX05005-005) and the Sinopec Science and Technology Key Project titled *Study on quality and speed-up drilling and completion technology in No. 5 fault zone of Shunbei 1 Block* (No. P2002). The authors would like to thank them.

REFERENCES

Jiaxin Wang, Yan Zhou. Study on Remaining Oil Distribution of Fractured Vuggy Carbonate Reservoir in Tahe 4 Area [J]. *Petrochemical Technology*, 2019, 26(09):302–303 + 356.

Li Junwei, Zhao Jingfang, Yang Hongbo, *et al.* Drilling Technology of Salt Gypsum Layer in MISSAN Oilfield [J]. *Journal of Changjiang University (science edition)*. 2013, 10(16): 92–94.

Shen Wenqi, Yin Da, Wang Zhilong, *et al.* Study on Settling Stability of Organic Salt Drilling Fluid[J]. *Chemistry and Bioengineering*. 2016, 33(5):56–58.

Wang Jan, Peng Fanfan, Xu Tongtai, *et al.* Research Progress on Test and Prediction Methods of Drilling Fluid Settling Stability[J]. *Drilling Fluids and Completion Fluids*. 2016, 29(5):79–63.

Yin Da, Ye Yan, Li Lei, *et al.* High-pressure Salt-water Treatment Technology of Keshen 7 well in Tarim Piedmont Structure[J]. *Drilling Fluids and Completion Fluids*. 2012, 29(5):6–8.

Zhang Yan, Xiong Hanqiao, Ding Fenget, *et al.* Effect of Oil-water Ratio on the Rheology of Oil-based Drilling Fluid[J]. *Science Technology and Engineering*. 2016, 16(12):238–242.

Civil Engineering and Energy-Environment – Gao & Duan (Eds)
© 2023 the Author(s), ISBN 978-1-032-56059-5

Research on physical quantity pricing method for new mechanical excavation of transmission line foundation under rock geology

Xuemei Zhu*, Fangshun Xiao, Cong Zeng, Ye Ke & Ying Wang
Economic and Technological Research Institute of State Grid Fujian Electric Power Co., Ltd, Jin'an, Fuzhou, Fujian Province, China

ABSTRACT: Based on the new electric construction rig currently used in the construction of power transmission lines, the typical construction scheme and process flow of foundation excavation under different weathering degrees of rock geology are analyzed. On-site human, material, and machine consumption were counted through the physical quantity method. The reference standard for the new electric construction rig in the power industry under rock geology has been studied and completed, which is 36%–82% more costly than the mechanical hole-digging cost in loose sandstone geology.

1 INTRODUCTION

With the sustained and rapid development of the economy and society, and as against the backdrop of constructing a strong smart grid, investors set higher standards for the quality and duration of project construction and stated unequivocally that we must vigorously promote mechanized construction innovation and change the traditional construction mode of "people-oriented, machinery supplemented." The construction mode of a transmission line with the lowest degree of mechanization is basic engineering. Manual excavation is often required due to inconvenient traffic, equipment being unable to reach the tower location, and other factors. In the case of rock geology, it is necessary to combine blasting with manual construction. These methods are slow or dangerous. The electric construction rig developed at this stage reduces the risk of manual excavation or blasting during the construction of foundation works, improves the construction efficiency, improves the construction accuracy, reduces the pollution to the human body, and greatly reduces the safety cost.

Based on the design features of transmission line foundations and Zhang's (2013) proposal, special equipment should be created for mechanically drilling transmission tower foundations. Zhang came to the conclusion that the new mechanical excavation is more advantageous from an economic and social standpoint than manual excavation. After analyzing the rock strength, pile length, and hole-forming strategy, Zhou et al. (2019) came to the conclusion that while some structural improvement is needed, the hole-forming efficiency of rotary drilling rigs in hard rock strata essentially satisfies the construction requirements. Through detailed analysis and research on the drilling combination process of a rotary drilling rig in sandy gravel strata and by using specific engineering construction cases as references, He (2021) further improved the application of a rotary drilling rig in this kind of stratum. Cheng (2017) investigated the effectiveness of rock breaking and the ideal TBM disc hob penetration. Han (2019) modified the cone drill's structure to make it suitable for rotary

---

*Corresponding Author: 8862585@qq.com

DOI: 10.1201/9781003433651-36

drilling into rock and achieved positive results in production tests. Shi and Li (2019) claimed that the supplementary quota sub-items were divided using the main construction procedures and that the supplementary quota for the pipe curtain freezing method was prepared using field measurement and the original quota data extracted. Based on a comparison and analysis of the norm method and the real object quantity method, Liao and Yang (2004) discovered that they both have advantages and disadvantages when it comes to the work cost management of hydraulic and hydroelectric projects. According to Liu et al. (2021), the investigation into the mechanism of conical pick rock cutting aids in structural design optimization and boosts drilling effectiveness for rotary drilling rigs.

Although research on mechanized construction under rock geology has been carried out, the corresponding construction costs vary greatly due to different mechanical equipment models, construction conditions, geological conditions, and other factors, and no corresponding pricing standard has been formed in the power industry. There is no unified cost standard for the application of electric construction rigs in power transmission line projects. In this paper, through the collection and investigation of the physical quantity data of manpower, material resources, machinery, and so on in the construction of electric power construction drilling rigs under rock geological conditions with different weathering degrees, the construction cost pricing standard suitable for the mechanized hole digging of transmission line foundations in the power industry is studied to improve the accuracy of project construction investment management.

2 STUDY ON CONSTRUCTION SCHEME OF MECHANICAL HOLE DIGGING

According to the mechanical construction principle of an electric power construction rig combined with the transmission line project, Figure 1 depicts the precise operation process, which includes construction preparation, hole leading, drilling promotion, hole forming inspection, hole cleaning, site cleaning, tool and instrument transfer, etc.

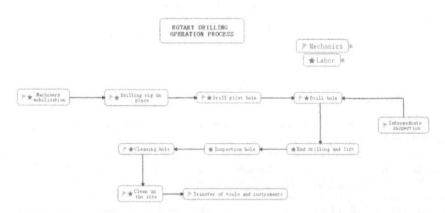

Figure 1. Flow chart of rotary drilling operation.

In electric power construction, different drilling tools and construction methods are used for rotary drilling. For example, large and small bits are used to achieve large-diameter construction, and different types of bits are used to achieve drilling progress under different geological conditions. Table 1 lists the relevant performance parameters for different models.

Table 1. Performance parameters of electric construction rotary drilling rig.

Performance	Unit	KR50D	KR100D	KR110D	KR150D
Maximum diameter of one-time hole forming in rock stratum below 30 MPa	m	1.2	1.5	2	2
One time hole forming efficiency of rock stratum below 30 MPa	m/h	2	3	3.5	4
Maximum diameter of one-time hole forming in 30-60 MPa rock stratum	m	0.6	0.8	1	1.2
One time hole forming efficiency of rock formation below 30-60 MPa	m/h	1	2	3	3

3 RESEARCH ON VALUATION BY PHYSICAL QUANTITY METHOD

The physical quantity method is used to thoroughly record the amounts of labor, materials, and machinery used during implementation in accordance with reasonable construction technology, a labor plan, and under typical construction conditions. The recorded data cannot, however, be used directly due to the level of professionalism and expertise of the recording staff. Therefore, to evaluate reasonable consumption, scientific data processing techniques are needed. This paper investigates the consumption of people, materials, and equipment based on the standard construction scheme when the saturated uniaxial compressive strength of rock is less than 30 MPa.

3.1 Composition of quota base price

Quota base price = ordinary working days × daily unit price of ordinary labor

An 8-hour workday is a standard workday. A technical working day costs 112 yuan per working day, while an ordinary working day costs 70 yuan per working day. Under the assumption that the rotary excavation construction can be completed normally, the consumption of man days is based on the number of man days needed to finish each linear meter of drilling depth. The time used as the quota time is advised to be the time that must be consumed, excluding the lost time, because the working time includes both the time that must be consumed and the time that was lost. For more information, see Figure 2.

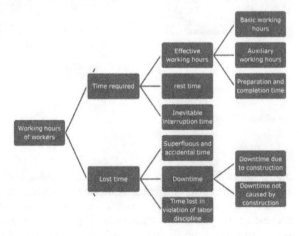

Figure 2. Composition of manual working hours.

Pricing material cost = Σ (construction consumption + loss) × unit price of materials

Pricing materials refer to the construction process or tool materials consumed to complete the construction of qualified products in the unit of measurement specified in the quota, such as picks, steel casings, skids, and other construction processes or tools that need to be amortized. The quantities include reasonable construction consumption, construction loss, on-site transportation loss, and stacking loss on the construction site.

Machinery cost = Σ (machine shift quantity × unit price of machine shift)

The machine shift adopts an 8-hour working system. The time that must be consumed, excluding lost time, is proposed as the quota time (see Figure 3 for details). The unit price of each shift consists of depreciation cost, overhaul cost, maintenance cost, installation and disassembly cost, off-site transportation cost, labor cost, fuel and power cost, and other costs.

Figure 3. Time composition of machine shift.

3.2 *Quota base price results*

After the collection and summary of human, material, and machine consumption, through the statistical analysis of construction efficiency, equipment investment, material loss, and other data, the statistical method is used to process some abnormal data, the fixed consumption data applicable to the power construction drilling rig excavation rock geology in the power transmission line project are finally formed, as shown in Table 2.

Table 2. Quota result table.

Number		BG-01	BG-02	BG-03	BG-04
		Hole depth within 12 m hole diameter			
Project		Within 1.4 m	Within 1.6 m	Within 1.8 m	Within 2 m
Unit		m	m	m	m
Base price		1171.47	1338.47	1365.58	1488.88
Labor cost (yuan)		303.87	308.01	312.84	317.21
Material cost (yuan)		7.21	8.20	9.25	10.31
Machinery cost (yuan)		860.39	1022.26	1043.49	1161.36
Name	Unit	Quantity			
Ordinary worker	Work day	2.36	2.39	2.74	2.99
Technician	Work day	1.52	1.54	1.77	1.92
Steel casing machined parts	Kg	1.36	1.55	1.98	2.36
Pick	Unit	0.15	0.18	0.22	0.28
Crawler hydraulic excavator capacity 1 m^3	One-shift	0.27	0.27	0.31	0.34
Special power transmission truck	One-shift	0.25	0.25	0.28	0.31
Crawler rotary drilling rig hole diameter 800	One-shift	/	/	/	/
Crawler rotary drilling rig hole diameter 1000	One-shift	/	/	/	/
Crawler rotary drilling rig hole diameter 1200	One-shift	/	/	/	/
Crawler rotary drilling rig hole diameter 1500	One-shift	0.22	/	/	/
Crawler rotary drilling rig hole diameter 1800	One-shift	/	0.23	0.26	/
Crawler rotary drilling rig hole diameter 2000	One-shift	/	/	/	0.29

4 COMPARISON WITH OTHER SIMILAR QUOTAS

The cost standard of each hole diameter is 36%-82% higher based on the quota cost standard of electric drilling rigs under rock geology studied above, compared to the cost unit price of rotary drilling rigs under loose gravel geology in the budget quota of *Electric Power Construction Projects Volume IV: Overhead Transmission Line Project* (2018 Edition).

5 CONCLUSION

The construction environment of a transmission line project is complex, and differences in rock-saturated uniaxial compressive strengths have a significant impact on mechanical excavation efficiency. Based on the typical construction scheme and the set saturated uni-axial compressive strength of rock and according to the cost standard formed by the collation of measured data, the quota base price is 1241.4 yuan per meter to 2430.77 yuan per meter, which is 36%–82% higher than the *Quota Base Price Under Loose Sandstone Geology*

(2018 Edition). The economy is reasonable and can be used as a reference for subsequent construction projects.

REFERENCES

Cheng Y.L. Numerical Simulation on Optimal Penetration of TBM Disc Cutter's Rock Fragmentation. *Journal of Central South University (Science and Technology)*, 2017, (4):936–943.

Han Y.Q. Discussion on the Structure and Application of Rotary Drill, *Exploration Engineering (geotechnical drilling and Excavation Engineering)*, 2019, 46(6):64–69.

He W.Y. Research and Application of Drilling Combination Technology of Rotary Drilling Rig in Sandy Conglomerate Stratum, *Architectural Design (Building Technology Development)*, 2021, 48(3):32–33.

Liu Hongliang, Yin Li, Fu Ling, Wang Xiaoteng Numerical Study and a New Analytical Calculation for the Rock Cutting with the Conical Pick, *Chinese Journal of Construction Machinery*, 2021, (2):38–43.

Liao S.T and Yang L.N. Analysis on Quota Method and Physical Quantity Method, *Jiangxi Water Conservancy Science and Technology*, 2004, (2):122–124.

Shi L.M. and Li J., Research and Application of Supplementary Quota for Tunnel with Pipe Curtain Freezing Method, *Construction Economy*, 2020, (7):74–78.

Zhang W. Application of Rotary Drilling Rig in Manual Digging Pile Foundation Construction, *Research on Urban Construction Theory*, 2013, 31(12):360.

Zhou W.N. and Miao Y.L. Drilling Technology of Rotary Drilling Rig for Bored Piles in Hard Rock Strata, *World Bridge*, 2019, 4:5.

Civil Engineering and Energy-Environment – Gao & Duan (Eds)
© 2023 the Author(s), ISBN 978-1-032-56059-5

Construction of carbon budget management mechanism for power generation enterprises amid national strategy of carbon peaking and carbon neutrality

Hongji Li*
Huadian Electric Power Research Institute Co., Ltd., Hangzhou, Zhejiang Province, China

Lin Hu*
Hangzhou Tonglu Ecological Environment Inspection Station, Hangzhou, Zhejiang Province, China

ABSTRACT: As the national strategy of "carbon peaking and carbon neutrality" is enacted amid the launch of the national carbon market, it is urgent for power generation enterprises not only to pay attention to structural and technological emission reduction but also to consider achieving economic emission reduction through the management and institutional innovation. This paper aims to study and construct the carbon budget mechanism for power generation enterprises in terms of preparation, organization, implementation, and evaluation and puts forward relevant suggestions to help them reduce emissions and eventually achieve the above-mentioned goals.

1 INTRODUCTION

1.1 *The national strategy of "carbon peaking and carbon neutrality" and promotion of the national carbon market*

After the national strategy of "carbon peaking and carbon neutrality" had been put forward, the construction of the corresponding "1 + N" policy system was accelerated, and the top-level documents were also promulgated one after another, making it possible for the strategy to be decomposed from top to bottom. It will be a broad and profound change for the economy and society if carbon peaking and carbon neutrality can be realized. As one of the high-emission industries, the power generation industry needs to make greater efforts in terms of economic structure, technology, and management mechanism.

The national carbon online trading market was officially launched on July 16, 2021. The first batch of power generation enterprises included in it is those with an annual greenhouse gas emission of about 4.5 billion tons of carbon dioxide. It is foreseeable that constructing a carbon trading budget and correspondent strategy will become critical for power generation companies with the gradual tightening of quotas, increasing proportion of auctions year by year, as well as the availability of financial instruments such as CCER, green power, and options.

1.2 *Current status of the carbon budget research*

At present, the carbon budget is divided into four levels, namely the global carbon budget, national carbon budget, corporate carbon budget, and personal carbon budget (Zhang

---

*Corresponding Authors: 124013976@qq.com

DOI: 10.1201/9781003433651-37

2021). The carbon budget was initially used to calculate the cumulative carbon emissions in a certain region for a certain period and later was applied to deal with climate change, becoming a policy tool for the government to control emissions of the greenhouse gas.

It is important to notice that enterprises must not only fulfill the social responsibility of energy conservation and emission reduction, respond to the national carbon target assessment and compliance requirements, but also pay close attention to the increased cost caused by emission reduction and carbon offset and the economic benefits brought by carbon trading. Tu Jianming pointed out there is a vertical arrangement of "national carbon budget - regional carbon budget - corporate carbon budget" in the carbon budget system (Liu 2022; Tu 2016). The corporate carbon budget is a micro-mechanism for the realization of national and regional carbon budget goals. A budget is used to guide enterprises to carry out carbon reduction activities in a planned way. Generally, domestic scholars tend to take industries related to power generation, cement, steel, etc. as cases to establish a corporate carbon budget system, and propose to embed it into the enterprises' comprehensive budget management system to achieve emission reduction while considering management accounting (Jiang 2021; Wu 2019; Zhou *et al.* 2016).

1.3 *Significance of carbon budget management for power generation enterprises*

A carbon budget can help realize the planning, restraint, and evaluation of carbon emissions for power generation enterprises. Specifically, carbon quotas, carbon emissions, carbon emission reductions, etc. are all key indicators. The preparation of the carbon budget in the early stage facilitates the planning of the carbon emission activities and incorporation of them into the overall strategy. Besides, the mid-term budget for carbon emissions and reductions can have a substantial impact on the business activities of power generation enterprises, as well as quantitative guidance and binding force on the implementation of annual emission reduction. Lastly, its comparison and implementation results can help enterprises get a clear understanding of carbon management efficiency, evaluate the economic feasibility of carbon emission reduction plans, and manage the remaining carbon budget.

2 CONSTRUCTION OF A CARBON BUDGET MANAGEMENT MECHANISM FOR POWER GENERATION ENTERPRISES

2.1 *Basic idea*

The following four carbon budget balance formulas are involved in the establishment of the enterprise carbon budget management mechanism (Tu *et al.* 2014):

(A)   Carbon quota estimation - carbon emission demand = expected carbon emission surplus and deficit
(B)   Carbon emission demand - carbon emission reduction budget = carbon emission budget
(C)   Carbon emission demand - carbon emission reduction budget - carbon quota = carbon trading budget
(D)   Total benefits of carbon emission activities - total cost of carbon emission activities = total net benefits

Table 1 shows the carbon emission demand calculation. The statistical data is relatively complete and standardized since power generation enterprises have been included in the verification of the carbon market. The annual emissions are estimated through annual

Table 1. Carbon emission demand accounting table of power generation enterprises.

Resource Type		Net resource consumption before emission reduction (a)	Activity level data (b)	Emission factor (c)	Estimated demand for carbon emissions (d)
Fossil fuel	Coal				
	Natural gas				
	Fuel oil				
	Diesel fuel				
	...				
Subtotal					
Net purchase of electricity	Electricity				
Subtotal					
Total					

Note: (1) Fossil fuel: The resource type is added according to the actual situation (same as in the table below). (a) is the net resource consumption, t 10^4 Nm^3; activity level (b) = resource consumption (a) × low calorific value, GJ; emission factor (c) = carbon content per unit calorific value × carbon oxidation rate × 44/12, t CO_2/GJ;

budgets for the consumption of different resources such as fossil fuels and purchased electricity according to the Guidelines for Corporate Greenhouse Gas Accounting and Reporting for Power Generation Facilities (2022 Revision).

(2) Net purchased electricity: activity level (b) = net resource consumption (a), (unit MWh); emission factor (c) adopts 0.5810 t CO_2/MWh;

(3) Estimated carbon emission demand (4) = activity level (2) × emission factor (3), (unit t CO_2).

China's national carbon market is an intensity-based carbon emission trading market. The future free allowance ratio is set, and the annual carbon emission allowance is calculated according to the latest *National Carbon Emission Trading Allowance Total Setting and Distribution Implementation Plan (Power Generation Industry)*.

Enterprises can make a judgment on the surplus and shortage of carbon emission rights in advance according to balance A, and estimate whether it is advisable to purchase or sell carbon emission rights in the national carbon market. At the same time, the planning department can make decisions for carbon emission reduction and formulate an annual emission reduction plan based on the historical annual corporate emission laws and carbon-neutral strategic goal of the peaking carbon.

2.2 Compilation system

The carbon budget preparation is divided into a carbon emission sub-budget, a carbon emission reduction sub-budget, and a carbon trading sub-budget based on the identified carbon emission activities and the above-mentioned carbon budget balance formulas.

2.2.1 Carbon emissions sub-budget

The annual carbon emission budget is different between the carbon emission demand and carbon emission reduction, as shown in Table 2.

Table 2. Carbon emission sub-budget for power generation enterprises.

Process	Resource Type	Carbon emission demand	Carbon emission reduction	Carbon budget	Explanation
Fossil fuel	Coal Natural gas Fuel oil ...				
Subtotal					
Net purchase of electricity	Thermal power Hydropower Wind power				
Subtotal					
Total					

2.2.2 *Carbon reduction sub-budget*

The carbon reduction sub-budget indexes the cost-benefit of emission reduction provides a reference for the decision-making of emission reduction projects, and establishes a basis for the subsequent evaluation of the implementation of emission reduction measures. The planned various emission reduction schemes are divided into two categories according to the capital and operation calculating carbon emission reduction and cost-benefits, as shown in Table 3.

Table 3. Carbon emission reduction sub-budget for power generation enterprises.

Process	Resource Type	Operational expenditure and emission reduction	Capital expenditure and emission reduction	Operational reduction cost	Annual capital reduction cost	Explanation
Fossil fuel	Coal Natural gas Fuel oil ...					
Subtotal						
Net purchase of electricity	Thermal power Hydropower Wind power					
Subtotal						
Total						

Among them, capital emission reduction involves the research and use of low-carbon technologies, the purchase of energy-saving equipment, new energy development, etc. The long cycle of construction and revenue needs the cost to be apportioned on an annual basis, such as an investment in energy efficiency improvement technologies and CCUS technologies. Operational emission reduction involves activities related to procurement, production, sales, etc., such as enterprises replacing high-emission coals with low-emission coals and purchasing renewable electricity. The main formulas are:

Operational emission reduction cost = the cost of using low-carbon resources - the cost of using original resources

Annual capital emission reduction cost = adopted low carbon technology/emission reduction cost of equipment/lifetime of technology or equipment

2.2.3 *Carbon trading sub-budget*

The final emission budget is calculated by the difference between the carbon emission demand and emission reduction amount, which is then compared with the quota to obtain the surplus or shortage of carbon emission rights. The surplus carbon emission rights are sold or held according to the market conditions, generating carbon trading income. The shortage of carbon emission rights is purchased in the trading market to complete the compliance, resulting in carbon trading expenses. The carbon trading price can be determined according to the carbon price trend offered by professional institutions or the average market price of the previous year as a reference. The product of the carbon trading price and transaction volume is the carbon emission trading budget, as shown in Table 4.

Table 4. Carbon trading sub-budget for power generation enterprises.

Carbon allowance (10,000 tons)	Carbon budget (10,000 tons)	Carbon trading volume (10,000 tons)	Carbon trading income/expense (10,000 yuan)	Carbon trading fees (transaction fees, etc.) (10,000 yuan)	Net carbon trading profit/ loss (10,000 yuan)

2.2.4 *Total budget*

It can be judged from Table 5 whether it is economical for power generation enterprises to carry out all carbon emission activities according to the balance formula D.

Table 5. Total carbon budget for power generation enterprises.

Total income (ten thousand yuan)			Total cost (ten thousand yuan)			Total net income/ expense (ten thousand yuan)
Net income from carbon trading	Total government subsidies for emission reduction	Other income	Carbon trading expense	Total carbon reduction cost	Other expenses	

Note: (1) Total income = net income from carbon trading + total government subsidies for emission reduction + other incomes (including patent fees for the use of emission reduction technology and equipment and sales of carbon assets such as CCER, etc.);
(2) Total cost = carbon trading expense + total carbon emission reduction cost + other expenses (including special expenditures other than carbon trading and carbon emission reduction);
(3) Total net income = total income - total cost.

2.3 *Organizational system*

The carbon budget management and organization mechanism should be established according to the actual situation of an enterprise. This paper simplifies it and builds an enterprise that consists of a leadership organization, decision-making organization, management organization, and executive organization. Specifically, the leadership organization refers to the board of directors, which is mainly responsible for deliberating and approving the annual carbon budget plan submitted by the carbon budget management committee, in addition to the plan of enterprise operation and investment. The decision-making organization, led by the board of directors, refers to the carbon budget management committee, which mainly reviews and approves carbon budget management issued through meetings or documents to ensure the authority and standardization for the implementation of carbon budget management. The management organization comprises the working group for carbon emission management and financial management, which is responsible for the daily work of carbon budget management, decomposition organization and coordination, preparation, execution, and analysis of carbon budgets. The executive organization refers to the functional departments and subsidiaries of an enterprise, which are responsible for the specific planning and implementation of the carbon budget, such as issues related to profits, costs, and expenses.

2.4 *Executive system*

As an important link to check the rationality of budget management, the executive process is the embodiment of the responsibility of each organization, directly affecting the effectiveness of carbon budgets, and controlling emissions and funds.

2.5 *Evaluation system*

Based on the execution control for carbon budget and PDCA theory, power generation enterprises should first determine the evaluation dimensions for the preparation of the carbon budget, control of the carbon budget process, and performance of carbon emission reduction, including qualitative and quantitative indicators. Then they should evaluate by comparing the actual data of carbon accounting and carbon emission reduction operations with the carbon budget data, as shown in Table 6.

Table 6. The evaluation index system of carbon emission budget for power generation enterprises.

Dimensions	Qualitative evaluation index	Evaluation standard	Score
Carbon budgeting	Carbon budget submission time and format		
	Carbon budget adjustment		
	Carbon budget adjustment frequency		
Carbon budget execution and control	Carbon budget execution ledger registration		
	Carbon budget target breakdown		
	Annual carbon budget excess emissions		
	Annual carbon budget excess expense		
	Carbon budget implementation analysis report		

Dimensions	**Quantitative evaluation index**	**Budget value**	**Actual value**	**Score**
Carbon reduction performance	Carbon emission reduction cost input per unit			
	Emission reduction cost per unit and emission reduction amount			
	Full-caliber carbon emission intensity			
	Thermal power carbon emission intensity			
	Carbon emission reduction intensity			
	Carbon emission			
	Carbon emission reduction			
	Carbon trading volume			

3 CONCLUSIONS OF CARBON BUDGET MANAGEMENT MECHANISM FOR POWER GENERATION ENTERPRISES

It is found through the above research that the carbon budget, a management mechanism for emission reduction, enables enterprises to scientifically estimate the economic benefits of carbon emission reduction and emission performance, thus obtaining the trend of expected development. Power generation enterprises should prepare carbon budgets according to the framework of national and regional carbon budgets and design a carbon pre-management system suiting their needs. The specific suggestions are as follows:

3.1 Thorough execution of the carbon emission reduction and carbon trading assessment to ensure economic optimality

The carbon budget system of an enterprise is based on its demand for actual emissions and thus should be established through emission reduction measures and carbon trading planning, which makes early economic evaluation very important. Although the carbon market price is relatively low and many emission reduction technologies have a high cost, the decreasing number of carbon allowances year by year will force enterprises to carry out activities for emission reduction if we consider time and control factors. Enterprises, therefore, should consider how to deploy and respond to the technologies for carbon emission reduction and carbon trading in the medium and long term through the cost-benefit assessment.

3.2 Improvement of the carbon budget mechanism and national strategy for "carbon peaking and carbon neutrality"

The realization of corporate carbon neutrality depends on carbon emission reduction and carbon offset. In the initial stage of carbon budget management, carbon emission reduction is mainly considered. Carbon offset can be considered when the implementation and evaluation are gradually matured, so every link of the management mechanism for the carbon budget will be improved. Under the guidance of the national strategy of "carbon peaking and carbon neutrality", enterprises will be able to use the management mechanism for carbon budget to achieve the linkage between the upper level (treatment and control for the target management) and lower level (decomposition for the emission target) and formulate standard to test its performance according to the carbon budget process.

3.3 Improvement of the organizational structure for carbon management and cultivation of corresponding talents to carry out the national strategy of "carbon peaking and carbon neutrality"

The management mechanism for the carbon budget involves coordination and communication among different departments. Power generation enterprises, therefore, should establish and improve their organizational structure for carbon management to achieve overall planning, guidance, and management. In addition, carbon trading-related policies are being promulgated one after another. Power generation companies should cultivate more talents who match the national strategy of "carbon peaking and carbon neutrality" to track policy directions on time and be familiar with cross-domain knowledge such as carbon emissions accounting, trading, and emission reduction technologies. It is critical to take the lead in formulating an administration guide relevant to the carbon budget.

3.4 Establishment of an information platform for carbon budget and improvement of carbon information disclosure

With the accumulation of time and projects, the corporate carbon budget will generate more and more carbon information and data related to energy and capital. It is recommended that enterprises add a carbon budget section to the original ERP system or carbon information system and design budget-related functions to fundamentally analyze, evaluate, supervise and control the implementation of carbon budgets, so that the errors of carbon budget can be reduced and arbitrary revisions are effectively avoided, eventually improving the scientific nature of the carbon budget assessment. Meanwhile, enterprises should link the management mechanism for carbon budget with the system of carbon information disclosure, so that carbon information will be regularly disclosed, thus providing value orientation for the overall green and high-quality development.

REFERENCES

Jiang Yan and Tang Qingliang: Carbon Information Disclosure from the Perspective of Accounting: Research Review and Prospects. *Financial Management Research*, 2021 (12): 1–16.

Liu Fengwei: Analysis and Interpretation of "Carbon Budget". *New Finance*, 2022, (01): 18–20.

Tu Jianming, Deng Ling, and Shen Yongping: Management Design and Institutional Arrangement of Enterprise Carbon Budget - Taking Power Generation Enterprises as an Example. *Accounting Research.* 2016 (3): 147–160.

Tu Jianming, Li Xiaoyu, and Guo Zhangcui: Enterprise Carbon Budget Concept Embedded in a Comprehensive Budget System amid Low-carbon Economy. *China Industrial Economics*, 2014 (3): 147–160.

Wu Shaoyan: *Design and Application of Carbon Budget System for Iron and Steel Enterprises.* Chongqing University of Technology, 2019.

Zhang Zizhu: Function and Structure Analysis of Carbon Budget System. *Transportation World*, 2021, (14): 37–38.

Zhou Zhifang, Li Cheng, and Zeng Huixiang: Construction of Enterprise Carbon Budget System Based on Product Life Cycle. *Jiangxi Social Sciences*, 2016, 36 (11): 65–72.

Civil Engineering and Energy-Environment – Gao & Duan (Eds)
© 2023 the Author(s), ISBN 978-1-032-56059-5

Summary of energy management technology for electromechanical system of more electric aircraft

Pengyu Wang, Wei Li* & Wei Liang
AVIC First Aircraft Institute, Xi'an, Shaanxi, China

ABSTRACT: Multi-electrific is one important direction of aircraft development. The basis of energy management is the multi-electrification of aircraft. This paper studies energy management technology. First, it summarizes the development of related technologies of foreign more electric aircraft, takes their characteristics with three typical more electric aircraft as examples, studies the energy management methods of more electric aircraft, and divides them into two categories. Then, it analyzes the energy management methods of power grids, hybrid vehicles, ships, and other industrial fields and summarizes their enlightenment to the energy management technology of more electric aircraft.

1 INTRODUCTION

Since the 1940s, the secondary energy in an aircraft has been mixed with hydraulic energy, pneumatic energy, and electric energy (Sun 2015). The air pressure energy mainly comes from the engine, which is used for anti-icing and environmental control. The hydraulic pressure can drive most of the flight actuation surfaces through the hydraulic pipeline. The electric energy provides energy for the electric load. In the 1970s, the aviation industry proposed the concept of electric aircraft. However, due to the limitations of the technical level at that time, the concept of more electric aircraft was put forward; that is, the original hydraulic, pneumatic, and mechanical drive systems are replaced by electrical systems.

In fact, the more electric aircraft realize the unified planning, management, and centralized control of the aircraft electromechanical system by converting it into electrical energy. Therefore, the top-level architecture design of electromechanical system energy management (Lin 2011) and electromechanical system energy management are important directions for the development of more electric aircraft.

This paper first summarizes the development of related technologies abroad and analyzes the existing three typical more electric aircraft, studies the energy management methods of more electric aircraft and classifies them into two categories, then analyzes these two types of methods, and finally analyzes the energy management methods in other industrial fields such as microgrid, hybrid vehicle, high-speed rail, and analyzes their enlightenment to the energy management technology of more electric aircraft.

2 DEVELOPMENT HISTORY OF FOREIGN MORE ELECTRIC AIRCRAFT-RELATED TECHNOLOGIES

Foreign countries have made major technological breakthroughs in the field of multi-electricity technology by carrying out a series of special research programs, such as ITAPS,

---

*Corresponding Author: liw036@avic.com

DOI: 10.1201/9781003433651-38

MEA, and INVENT in the United States, and TIMES, POA, and MOET in the European Union.

The more electric aircraft (MEA) plan (Zhou 2019) is to explore the technology of electromechanical components with electric energy as the main operating energy and carry out tests on C-180, F-16, and other aircraft.

In 2002, the EU implemented the energy-optimized aircraft (POA) program, which aimed to optimize onboard energy management. In view of the success of the POA plan, the MOET project was implemented in 2006 to apply multi-electricity technology to the next generation of aircraft, realize the integration of aircraft electromechanical systems, and complete digital verification.

In 2007, the US air force launched the invest (aircraft energy integration technology) program (Daniel 2012) to make aircraft energy more efficient, reduce aircraft thermal constraints, expand aircraft power growth space, and increase aircraft combat capability. Invest plans to develop a new type of 'energy-optimized aircraft.' From the point of view of design, energy-optimized aircraft, as a kind of aircraft with comprehensive performance optimization, can ensure the highest energy utilization rate while meeting the minimum system complexity. From the perspective of use, energy-optimized aircraft means minimizing the use of fuel while ensuring the completion of tasks.

In addition, in the investment plan, the concept of load-on-demand management is proposed for the previous cut-off load management mode, that is, dynamic management of various equipment under different flight states, so as to improve the energy utilization rate.

Abroad, through a series of plans, from the verification of electrification of various subsystems to the research of electromechanical system integration, and then to the energy management of more electric aircraft, the more electric aircraft is gradually realized.

Take three typical types of more electric aircraft as examples:

1) A380 aircraft is designed completely according to the electric power system of more electric aircraft, and its prominent feature is the multi-electric hydraulic system and actuation system, which makes the design of the aircraft more concise, reduces ground support equipment, and greatly improves the performance of the aircraft.
2) Boeing 787 aircraft uses a large amount of electric energy in various electromechanical systems, the most important of which is the elimination of engine air bleed and the use of two compressor air bleed electric environmental control systems, which can control the state of the motor according to the power demand, thereby reducing fuel consumption and reducing the structural weight of the aircraft.
3) In the F-35 fighter, centralized power distribution has greatly improved reliability, reduced weight, and improved the performance of the aircraft.

These three types of more electric aircraft choose one or more of the multi-electric schemes such as electric environmental control, electric anti-ice, electric brake, and electro-hydraulic actuation to replace the traditional aircraft architecture and adopt different degrees of integrated thermal / energy management technology, which shows that the energy management technology of more electric aircraft can indeed improve the aircraft performance.

3 OVERVIEW OF ENERGY MANAGEMENT TECHNOLOGY

The rapid development of various avionics equipment provides a hardware basis for the energy management of more electric aircraft: in terms of load, some electrical equipment is no longer in the on-and-off states, but in the low-voltage state or other states. In terms of load management, the integrated switching solid-state power controller SSPC is an intelligent solid-state appliance integrating protection, switching and remote control appliances, which can more conveniently manage the load. In terms of hydraulic pressure, EHA, EMA, and other equipment replace some hydraulic equipment. In terms of energy storage, the emergence of electric energy storage devices such as lithium batteries and supercapacitors makes it possible to improve the power peak (Wang 2019).

From the perspective of hardware, this paper divides the energy management modes of electromechanical systems into two types: the energy management mode based on aircraft secondary energy and the energy management mode based on hybrid energy storage.

3.1 *Energy management mode based on aircraft secondary energy*

Before energy management, it is necessary to analyze and select the operation scenarios of the electromechanical system, analyze the needs and importance of each function, and then divide the functions hierarchically. On this basis, the functions of the electromechanical system are reasonably allocated.

Direct energy management is to manage the load directly or indirectly by controlling various controllers, power distribution centers, and other measures, so as to achieve the predetermined purpose.

During energy management, the measures taken for load and corresponding purposes are mainly divided into the following three points,

1) Give priority to providing energy to functions with high priority, so as to improve security;
2) The load at some time is started in time-sharing and used in peak shifting to reduce the peak power at that time;
3) Turn off the load under specific conditions, or switch the load to a half-load state, so as to reduce energy consumption and peak power as a whole.

When compiling the algorithm, the objective function can select the lowest fuel consumption and peak power in the whole cycle according to the purpose of energy management. The control variable can select the load state, bus bar switching, etc., according to the control mode. The constraint function is mainly determined by the flight state of the aircraft.

On the Boeing 787, a typical more electric aircraft, advanced electric environmental control, electric braking, and electric anti-ice systems are used. The main engine and APU are started by high-power frequency conversion. 16 high-power equipment such as engine starting, APU starting, electro-hydraulic pump, electric drive fan, air circulation fan, etc., are directly connected by the engine, and eight general-purpose motor drivers (CMSC) of the same model can realize time-sharing starting of high-power equipment in different flight stages. This energy management method can greatly improve the efficiency of energy utilization.

3.2 *Energy management based on hybrid energy storage*

The use of a large number of electrical equipment in more electric aircraft may cause sudden power changes in the aircraft energy network (mainly referring to the power grid), thus affecting the stability of the energy network. At present, there are two solutions. One is to increase the rated capacity of the engine. The other is to provide peak power by means of energy storage equipment.

Electric energy storage devices include lithium electronic batteries, supercapacitors, and flywheel engines. These storage devices have their advantages and disadvantages:

1) Lithium battery has obvious advantages in power density, energy density, and cycle times;
2) Supercapacitors have the advantages of fast charging and discharging speed, high power density, small temperature influence, long service life, etc. (Zhang 2017);

3) Flywheel energy storage realizes bidirectional conversion of electrical energy and mechanical energy through physical methods. In the case of large current charging and discharging, it has little impact on its service life, simple structure, and almost no maintenance, and its service life is equivalent to that of supercapacitors. However, the self-discharge rate of flywheel energy storage is high. Long-term energy storage will cause large energy loss, and the unit energy storage cost is high.

Therefore, to absorb sudden changes in the power grid through energy storage equipment, it is necessary to comprehensively consider working conditions, load characteristics, and other factors, and find out a reasonable configuration and control mode.

In foreign countries, Jeff *et al.* (2012) analyzed the application of batteries in energy-optimized aircraft, analyzed factors such as safety and cost, and proposed the method of battery chemical selection to improve the performance of aircraft batteries. Tim *et al.* (2010) proposed energy management by using batteries to absorb energy and resistors to consume energy for the electrical system of more electric aircraft, and analyzed the advantages and disadvantages of these two methods. Thomas *et al.* (2012) designed a new type of super-capacitor energy storage unit, which can effectively absorb the peak energy in the power grid. The development of these energy storage devices makes it possible to reduce peak power. The related energy management methods come into being.

At present, the research on the energy management method of hybrid energy storage equipment in China is still at the theoretical research and experimental stage. Pei Wang (2019) applied supercapacitors to the load network and then selected the typical electrically driven load EMA for analysis and modeling. Experiments show that the load energy management technology based on supercapacitors can improve the response speed of load startup and effectively reduce the impact of load braking and disturbance on the power grid. Yu Wang (2019) proposed a hybrid energy storage control strategy based on adaptive energy management for three energy sources: generator, battery, and supercapacitor, which proved that battery and supercapacitor could effectively eliminate power peak. Tianxiang Huang (2017) used the fuzzy control method to realize energy management and carried out simulations under different working conditions, which proved that this method could improve electrical efficiency.

4 OVERVIEW OF ENERGY MANAGEMENT TECHNOLOGY IN OTHER INDUSTRIAL FIELDS

The research on energy management in the power grid, high-speed railway, hybrid vehicles, and other fields is relatively mature. Next, the content and characteristics of energy management technology in these fields and the feasibility of applying it to aircraft will be analyzed.

4.1 *Hybrid electric vehicle*

A hybrid vehicle refers to a vehicle equipped with two or more power sources (Wang 2017), one of which is a traditional engine, and the other is provided by the motor battery pack, fuel cell, flywheel, etc. This hybrid power system is similar to the hybrid energy storage system of the aircraft electromechanical system.

As the core of the hybrid electric vehicle, the quality of the energy management system directly affects its reliability, controllability, economy, and emission performance.

Figure 1 shows the classification of energy management strategies for hybrid vehicles. The early energy management methods used rule-based control methods, and later the energy management algorithm based on dynamic programming was widely used (Zhao 2016). In recent years, a variety of advanced intelligent energy management methods have been

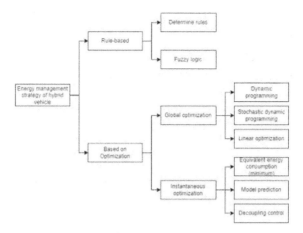

Figure 1. Classification of energy management strategies for hybrid vehicles.

emerging. Representative methods include real-time equivalent fuel consumption control algorithm, model predictive control algorithm, optimal control algorithm, fuzzy control algorithm, neural network control algorithm, genetic algorithm, and other intelligent control methods.

The rule-based energy management strategy mainly determines the working state of the system by setting the operating rules of the automobile power system in advance. This control strategy has little computation or difficulty, but the control parameters are set according to experience, which makes optimal control difficult.

The optimization-based control strategy seeks the global or local optimal solution of the system by solving the minimum of the objective function under the constraint conditions. Among them, the most typical global optimization is dynamic programming (DP), and the representative of instantaneous optimization is the equivalent energy consumption minimization strategy (ECMs).

DP strategy can achieve ideal optimization results, but the algorithm itself has the problem of dimension disaster; that is, the optimization results will be biased if the state dimension is too large. In addition, dynamic planning must know the vehicle running state and driving conditions, and it is difficult to achieve real-time control.

ECMS strategy is a kind of instantaneous optimization algorithm, which obtains the best control after predicting the road conditions, and has high timeliness. However, if the working condition prediction deviation is large, the optimal control cannot be achieved, and even the opposite effect will be obtained.

Global optimization can achieve good optimization results but poor timeliness, and instantaneous optimization has strong timeliness but poor optimization results. Artificial intelligence and other algorithms are introduced to identify and predict operating conditions to make up for poor timeliness, and the introduction of these algorithms will also bring other problems. Therefore, energy management focuses include balancing timeliness, optimization effect, and other factors, synthesizing different methods, and learning from each other is the focus of.

4.2 *Microgrid*

The microgrid is a comprehensive integration technology, including renewable energy and other distributed power sources. Its energy management system is a set of energy management software with functions such as power generation optimization dispatching, load management, microgrid synchronous automatic monitoring, and implementation (Wu

2014). The research on the energy management of microgrids at home and abroad shows that the current energy management of microgrids mainly includes the power generation side and the demand side. Generation side management includes distributed generation management, energy storage system management, and distribution network side management, while demand side management is mainly classified as load management.

Microgrid is similar to the power grid of solar-powered aircraft. Renewable energy, such as photovoltaic power generation they contain depends on the natural environment and has randomness and volatility. The microgrid energy management system forecasts the output of renewable energy units, arranges the charging and discharging of energy storage, manages controllable loads, and maintains the stability of the system. Figure 2 is the functional diagram of the microgrid energy management system, mainly including four functional modules: a human-machine communication module, a data analysis module, a prediction module, and a decision optimization module.

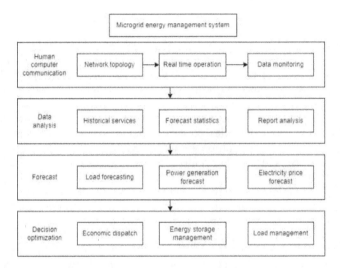

Figure 2. Functional diagram of microgrid energy management system.

The prediction module is used to predict the load, renewable energy, market electricity price, etc. It can be divided into short-term (1D to 1 week) and ultra-short-term (minutes or hours). The ultra-short-term prediction generally requires online prediction and real-time rolling. The decision optimization module is a module for distributed generation scheduling, power purchase plan from the power grid, energy storage and output distribution, load arrangement, and decision optimization according to the predicted value of load and renewable energy, user power demand, scheduling rules, market electricity price, and other information.

The energy management of microgrids optimizes decision-making through data collection, data analysis, and then information prediction.

4.3 *High-speed railway*

Energy management in the field of high-speed railway mainly occurs when the train runs to the uphill section and the downhill section (Luo 2021). When the train enters the downhill section, the locomotive performs regenerative braking, and the energy storage system works in the energy storage state and absorbs the regenerative braking energy. When the train enters the uphill section, the train enters the traction condition, and the energy storage system switches to the energy feeding state to release the stored energy.

The energy management scheme of the energy storage system of the high-speed railway needs to solve the problem of energy coordination and interaction of the energy storage system under traction and regenerative braking conditions and improve the "peak shaving and valley filling" ability of the energy storage system as much as possible. Therefore, it is necessary to formulate corresponding energy management schemes according to different working conditions to meet the power demand. At present, the rule-based energy management strategy is mainly used in the high-speed railway field.

4.4 *Enlightenment on energy management of more electric aircraft*

1) In terms of architecture, the energy management technology of the electromechanical system of the more electric aircraft can learn from the energy management system architecture of the microgrid and choose centralized control or distributed control in combination with various factors of the aircraft itself.
2) In the direction of hybrid energy storage, the hybrid energy storage system of the aircraft is similar to the hybrid power system of the automobile and the hybrid energy storage system of the high-speed railway and can be managed by referring to the energy management algorithm based on optimization or instantaneous optimization in these fields.
3) In the direction of the algorithm, the algorithms based on global optimization and instantaneous optimization can achieve energy management, but the timeliness of global optimization is poor, and the optimization of instantaneous optimization is relatively poor. Therefore, the global optimization algorithm can be implemented in more electric aircraft under normal conditions, and the instantaneous optimization algorithm can be implemented under special conditions.
4) In the field of microgrids, the energy management of renewable energy with volatility and randomness is similar to that of solar aircraft, so the research on microgrids can be used for reference in the energy management of solar aircraft.

5 CONCLUSION

This paper summarizes the reasons for the development of more electric aircraft, then analyzes the progress of relevant foreign projects, classifies and the methods of energy management in electromechanical systems. It studies the energy management in hybrid vehicles, microgrids, and high-speed rail fields, respectively, because energy management has been developed in other industrial fields, and analyzes the enlightenment of technologies in these fields to the energy management of more electric aircraft. It provides a reference for the research of energy management of more electric aircraft.

REFERENCES

Daniel Schlabe, Jens Liensing. Energy Management of Aircraft Electrical Systems-State of the Art and Further Directions [C]. *50th AIAA Aerospace Sciences Meeting including the New Horizons Forum and Aerospace Exposition.* Nashville, 2012.

Eric A. Walters, Steve Iden. Invent Modeling, Simulation, Analysis and Optimization [C]. *48th AIAA Aerospace Sciences Meeting including the New Horizons Forum and Aerospace Exposition. Orlando,* 2010.

Huang Tianxiang, Xue Longxian, Wang Xinggang, Chen Junxiang. Research on Energy Synthesis Method of More Electric Aircraft Based on Fuzzy Control Rules [C] *Proceedings of the 2017 Academic Annual*

Meeting and the First More Electric Aircraft Forum of the Professional Committee of Avionics Engineering of the Chinese Academy of Aeronautics. 2017

Jeff Knowles. Electrical Energy Storage for Energy Optimized Aircraft [C]. *SAE Power Systems Conference. Seattle,* 2012.

Lin Ming, Cai Zengjie, Zhu Wufeng. Development Trend and Enlightenment of Aircraft Electromechanical System [C]. *Mechanical Research and Application.* 2011

Luo Jiaming, Wei Xiaoguang, Gao Shibin, Huang Tao, Li Duo. Overview and Prospect of Capacity Configuration and Energy Management Technology of High-speed Railway Energy Storage System [J]. *Proceedings of CSEE,* 2021

Sang Bo, Zhang Tao, Liu Yajie, Chen Yandong, Liu Lingshun, Wang Rui. Review of Research on Multi Microgrid Energy Management System [J]. *Proceedings of CSEE.* 2020,40 (10): 3077–3093

Steven C. Griggs, Steven Mark Iden, Peter T. Lamm. Energy Optimized Aircraft: What is it and how do we make one? [C]. *SAE Power Systems Conference. Seattle,* 2012.

Sun Youshi, Yu Xiao, Huang Tieshan. A Review of the Research Progress of the Energy-optimized Aircraft of the US Air Force [C]. *Proceedings of 2017 (third) China Aerospace Science and Technology Conference (Volume II).* 2017:38–41

Sun Youshi. From More Electric Aircraft to Energy Optimized Aircraft—an Analysis of the Development Plan of the US Air Force in the Field of Avionics [C]. *Proceedings of the second China Aerospace Science and Technology Conference,* 2015: 503–506

Thomsa Wu, Tony Camarano, Jon Zumberge, Mitch Wolff. Design of Electrical Accumulator Unit (EAU) Using Ultracapacitor [C]. *50th AIAA Aerospace Sciences Meeting including the New Horizons Forum and Aerospace Exposition.* Nashville, 2012.

Tim O'Connell, Greg Russell, Kevin McCarthy, Eric Lucus, Jon Zumberge, and Mitch Wolff. Energy Management of an Aircraft Electrical System [C]. *46th AIAA Aerospace Sciences Meeting including the New Horizons Forum and Aerospace Exposition.* Nashville, 2010.

Wang Li, Dai Zehua, Yang Shanshui, Mao Ling, Yan Yangon. A Review of the research on Intelligent Design of the Electric Power System of Electrified Aircraft [J]. *Acta aviaciae Sinica,* 2019,40 (2): 1–15

Wang Pei. Research on Energy Management Technology of Electric Drive Load of More Electric Aircraft [D]. *Nanjing University of Aeronautics and Astronautics,* 2019

Wang Qinpu, You Sixiong, Li Liang, Yang Chao . Summary of Research on the Energy Management Strategy of Plug-in Hybrid Vehicles [J]. *Journal of Mechanical Engineering,* 2017, 53 (16): 1–19

Wang Yu, Xu Fang. Adaptive Energy Management Method Based on Hybrid Energy Storage System of More Electric Aircraft [C] *Proceedings of the More Electric Aircraft Forum of the 2019 Academic Annual Meeting of the Aviation Electrical Engineering Professional Committee of the Chinese Academy of Aeronautics,* 2019

Wu Xiong, Wang Xiuli, Liu Shimin, Zhu Zhenpeng, Liu Chunyang, Duan Jie, and Hou FeiF. Overview of Microgrid Energy Management System [J]. *Power Automation Equipment,* 2014, 34 (10): 7–14

Xun Qian, Qin Haihong, Yan Yangon. Research Status and Development of the Electrical System of More Electric Aircraft [C]. *National Academic Conference on Aviation Electromechanical, Human Body and Environmental Engineering.* 2013

Yang Zhou, Lei Tao, Lin Zicun. Overview of Energy Management Strategies for New Energy Hybrid UAVs [C]. *Proceedings of the 2017 Academic Annual Meeting and the First More Electric Aircraft Forum of the Professional Committee of Avionics Engineering of the Chinese Academy of Aeronautics.* 2017

Zhang Lei, Hu Xiaosong, Wang Zhenpo. Overview of Supercapacitor Management Technology and its Application in Electric Vehicles [J]. *Journal of Mechanical Engineering,* 2017, 53 (16): 32–69

Zhao Xiuchun, Guo Ge. Summary of Research on Energy Management Strategy of Hybrid Electric Vehicles [J]. *Journal of Automation,* 2016, 42 (3): 321–334

Zhou Hao, Shen Zhengbin, Wang Xinggang, Xu Jie, Chen Junxiang Research on Power System Development of More Electric Aircraft. *Proceedings of the 2019 Academic Annual Meeting of the Aviation Electrical Engineering Professional Committee of the Chinese Academy of Aeronautics and the First More Electric Aircraft Forum.* 2019

Civil Engineering and Energy-Environment – Gao & Duan (Eds)
© 2023 the Author(s), ISBN 978-1-032-56059-5

Layout and construction organization design of river ecological control project — A case study of urban section of Shichuan River

Kairen Yang & Jiwei Zhu*
State Key Laboratory of Eco-hydraulics in Northwest Arid Region, Xi'an University of Technology, Xi'an, China
Research Center of Eco-hydraulics and Sustainable Development, New Style Think Tank of Shaanxi Universities, Xi'an, China
Department of Engineering Management, School of Civil Engineering and Architecture, Xi'an University of Technology, Xi'an, China

Liang Li
Research Center of Eco-hydraulics and Sustainable Development, New Style Think Tank of Shaanxi Universities, Xi'an, China

Wangyu Luo
Xi 'an Water Conservancy Planning Survey and Design Institute, Beilin District, Xi 'an City, China

Chao Zou
State Key Laboratory of Eco-hydraulics in Northwest Arid Region, Xi'an University of Technology, Xi'an, China
Research Center of Eco-hydraulics and Sustainable Development, New Style Think Tank of Shaanxi Universities, Xi'an, China
Department of Engineering Management, School of Civil Engineering and Architecture, Xi'an University of Technology, Xi'an, China

ABSTRACT: As a key project of the Shaanxi Provincial Department of Ecology and Environment, the comprehensive treatment of the Sichuan River plays an important role in improving the flood control standard of the urban section of the Sichuan River in Yanliang District, safeguarding people's life and property in Yanliang City, promoting the social and economic development of the region, and forming an ecological and low-carbon urban environment. It is a project that integrates flood control, social, environmental, and economic value-added benefits for the people. This paper focuses on the flood calculation of the agency for embankment design, slope protection design, and barrage layout, and from the embankment earthwork calculation, slope protection engineering and barrage construction, and other aspects of construction organization design and analysis, to provide a practical case for urban river ecological management project. It enriches the related engineering experience of urban river ecological control project layout and construction organization design. It also shows the necessity and importance of urban river ecological management projects in the current background, which has reference significance for the subsequent similar projects.

1 INTRODUCTION

At present, urban eco-systemization has become a major trend in urban development in China, and urban rivers play a pivotal role in urban ecological construction (Ning *et al.*

---

*Corresponding Author: xautzhu@163.com

DOI: 10.1201/9781003433651-39

2012). Therefore, the ecological governance of urban rivers is of great significance for realizing the harmonious operation and sustainable development of the urban system and satisfying people's yearning for a better ecological environment (Luan 2010).

In 2019, the ecological protection and high-quality development of the Yellow River basin were elevated to a national strategy. Carrying out ecological management of urban rivers is a concrete practice to implement General Secretary Xi Jinping's thoughts on ecological civilization and the spirit of the Yellow River Symposium speech (He *et al.* 2006). The layout and construction organization design of the river regulation project is regarded as the ecological regulation project of urban rivers. As an important part of the project, its rationality and compilation quality play a key role in project implementation.

Supported and driven by relevant national policies, urban river ecological governance and its construction organization and design have attracted the attention of many scholars, and a lot of scientific research has been carried out.

Zhang Yicai *et al.* believe that a comprehensive river regulation project is necessary for the face of the grim situation that the ecological environment of rivers in modern cities is deteriorating due to development and construction (Zhang 2012). Zhou Xuemin *et al.* pointed out the importance of construction organization design in hydropower projects through detailed analysis of the implementation scheme of downstream river control engineering of shaping first-class hydropower stations (Yang *et al.* 2012).

To sum up, many scholars are currently paying more and more attention to river management engineering and have proposed various management measures for its existing problems. However, there are still some problems that need to be further solved: the incomplete investigation of the conditions in the project area at the pre-construction stage of the urban river management project leads to unsafe construction and affects the progress of the project; the unreasonable engineering layout; and the lack of detailed construction organization design. This study focuses on the ecological management project of the urban section of the Sichuan River in terms of both engineering layout and construction organization design, which further indicates the importance of river management engineering layout and construction organization and provides experience for subsequent similar projects.

2 PROJECT OVERVIEW

2.1 *Watershed overview*

Sichuan River is the first-grade tributary of the Weihe River on the left bank. It starts at the bottom of the east of Fenghuang Mountain, Yintai District, Tongchuan City, with an elevation high in the northwest and low in the southeast, and flows through Tongchuan, Yaozhou, Fuping and Yanliang counties (districts) from north to south, starting from Jiaokou Town, Lintong District and entering the Weihe River. The total length of the river channel is 137 km, the drainage area is 4,478 km^2, and the channel gradient is 4.6‰.

The scope of the project is the Sichuan River Yanliang Fuping junction (entrance to the tableland) to the Yan Guan road bridge downstream 300 m, a total length of 7.7 km. River desilting and beach clearing are done to repair the flooding section of the river to ensure smooth flood discharge. Embankments are constructed to minimize the scouring damage to the banks of the Sichuan River by flooding occurring in the agency and to maximize the utilization of the beaches on both banks. Ecological slope protection projects are carried out to improve the regional ecological environment. Barrage dams are constructed to impound river water and create urban water surfaces to improve the well-being of the surrounding people's lives (Zhou & Gao 2022).

2.2 Hydrological and geological conditions

2.2.1 Hydrometeorology

The project area belongs to the mid-latitude warm, subhumid continental monsoon climate zone, with obvious seasonal characteristics. The average annual temperature is 13.0°C, and the extreme maximum and minimum temperatures are 42.0°C and −19.7°C, respectively. The annual average precipitation is about 561.8 mm, and the precipitation in these three months of July, August, and September accounts for more than 50% of the year. However, from November to February, the precipitation only accounts for 5–8%. And the rainfall is mostly heavy, which is easy to cause flooding, soil erosion, and other disasters.

2.2.2 Engineering geology

The project area is located in the loess plateau area and the alluvial plain area north of Weihe River in Xi'an City, with terraces developing on both banks of the Sichuan River and clear terraces at the first and second levels. The soil quality above and below the two levels is different. The terrain is relatively flat and generally high in the north and low in the south. The ground elevation is between 360 and 410 m. The Sichuan River crosses the Yanliang Guanshan area from west to east, and first and second-class terraces are developed on both sides of the river. Dikes and revetment facilities are planned to be built on both sides of the Sichuan River, and the landform units are mainly the first-level terrace and floodplain.

2.3 Existing problems

The dam built in the upper reaches of the river has been holding up the upstream water. Since it has not been systematically managed for many years, it is cut off all year round. The river is covered with sand pits, construction, domestic garbage, and illegal construction occupying beach land, blocking the river. The natural ecology of the river valley and wetland water environment is seriously damaged. There is no flood control project for the river. The current situation of the river cannot be synchronized with the rapid development of the regional economy and high flood control requirements. The lack of ecological landscape function and the backward flood control and rescue facilities are the main problems facing the river. Therefore, it is of great significance to carry out the ecological management project in the urban section of the Sichuan River.

3 ENGINEERING LAYOUT

3.1 Agency flood

In view of the current situation that there is no flood control project in the urban section of Sichuan River, improving the flood control standard in this area and ensuring the safety of many important enterprises, institutions, facilities, and people's life and property in the urban area have become the top priority of river control. To ensure the construction safety, progress, and rationality of the project layout, the calculation of flood has become an essential part of the urban section of the Sichuan River project layout.

Based on the analysis of the runoff production situation between the Agency and Juhe sink entrance, the existing reservoir, and the flood encounter, the design flood calculation of the Engineering Department was transformed into the design flood calculation of Juhe sink entrance, and the flood calculation of Juhe sink entrance was transformed into the design flood calculation of Yaoxian Station of Qishui River and the discharge flood of Taoqupo Reservoir of Juhe River.

According to the measured data of Yaoxian Station of Qishui River from 1959 to 2010, the parameters of the frequency curve were estimated by the moment method according to the method recommended by relevant codes.

The mean is calculated as follows:

$$\overline{Q} = \frac{1}{n} \sum_{i=1}^{n} Q_i \qquad (1)$$

Cv of variation coefficient is calculated as follows:

$$C_V = \frac{\sigma}{\overline{Q}} = \sqrt{\frac{\sum_{i=1}^{n} (K_i - 1)^2}{n - 1}} \qquad (2)$$

where σ is the mean square deviation and Ki is the modulus ratio coefficient, $Ki = Qi/Q$.

Similarly, the peak flood discharge of Taoqupo Reservoir is calculated, the historical survey flood is added, and the mean value is calculated according to the following equation:

$$\overline{Q}_N = \frac{1}{N} \left(\sum_{i=1}^{a+l} Q_{Ni} + \frac{N - a - l}{n - l} \sum_{i=1}^{n-l} Q_i \right) \qquad (3)$$

where Q_N is the mean value of the N year series, including the super flood; n is the number of years of actual floods; N is the number of years since the farthest survey and research year; l is the number of floods extracted from the measured series for extra-large value processing; a is the number of consecutive major floods in N years; Q_{Ni} is the peak flow of the extraordinary flood; Q_i is the peak flow of the general flood.

Cv of variation coefficient is calculated as follows:

$$C_V = \frac{1}{\overline{Q}_N} \sqrt{\frac{1}{N - l} \left[\sum_{i=1}^{a+l} (Q_{Ni} - \overline{Q}_N)^2 + \frac{N - a - l}{n - l} \sum_{i=1}^{n-l} (Q_i - \overline{Q}_N)^2 \right]} \qquad (4)$$

where the letter meaning is the same as before. Then, the discharge flood of Taoqupo reservoir can be obtained by flood regulating calculation.

3.2 *Embankment project*

The total length of the project dike is 15.5 km, of which the left and right bank dikes are 7.73 km and 7.77 km, respectively.

3.2.1 *Design embankment lines*

Under the premise of following the principle of levee layout, according to the characteristics of the river, combined with flood control planning, comprehensive consideration of topography, geological conditions, the location of existing and proposed buildings, and other factors, overall consideration of the upstream and downstream and the left and right banks, section layout of the levee. Two schemes are considered for this embankment layout:

Scheme 1: Bank on both sides of the river needs to be made full use of, bank protection works under the current bank need to be arranged, and adjust the local convex and concave sections on the plane, so as to make the design shoreline as smooth as possible. The advantages of the scheme are the full use of natural terrain and nodes, a small project footprint, and less investment. The disadvantage is that the dike twists and turns, the distance between the dike suddenly enlarges or narrows, the overall very irregular, and not conducive to the flood smooth discharge.

Scheme 2: As far as possible, the existing banks, railway Bridges, and other nodes on both sides of the river should be used to arrange bank protection projects smoothly; the dike distance in the wide area of the river is appropriately relaxed, and the dike project is arranged according to the dike distance of not less than 200 m. The scheme combines the terrain conditions and nodes, considering that the river flood flow is smooth, according to the situation, conducive to the stability of the river, and leaves room for flood control, the implementation difficulty is small, and the wide area of the river has created conditions for the future urban river ecological landscape project construction.

According to the terrain and geological conditions of the river and the advantages and disadvantages of the above scheme, the final choice is scheme two.

3.2.2 *Determine the elevation of the embankment top*
The design dike top elevation is 1.37 m above the design flood level, and the waterward side is masoned to 0.5 m above the design flood level. The overheight of the embankment top is calculated according to the formula $Y = R + e + A$, and the calculation results are shown in Table 1.

Table 1. Calculation results of embankment top superelevation.

Ultra-high Y(m)	Safety of heightening A(m)	Wind obstruct water elevation e(m)	Foreshore R(m)
$Y = R + e + A$	Safety elevation value of the second-level embankment is 0.8 m	$e = \frac{kV^2F}{2gd} \cos \beta$	$R_P = \frac{K_\Delta K_V K_P}{\sqrt{1+m^2}} \sqrt{h_m L_m}$
1.37	0.8	0.0039	0.566

3.2.3 *Design of embankment cross-section*
The section of the new dike is a trapezoidal structure, and the existing loess on the beach surface of the Sichuan River is used as the filling material of the dike. The ratio of the side slope near the backwater was 1:3.0. The width of the top of the embankment is 20 m. The two-way two-lane road is 7 m long, and the electric car and bicycle and pedestrian road is 5 m long. 2 m isolation zone is set between the two lanes, and a 3 m green belt is reserved on both sides.

3.2.4 *Dike type selection*
The main function of the project layout is flood control and disaster reduction, restoration of the ecological environment of the Sichuan River system, and laying the foundation for the establishment of the ecological landscape belt of the Sichuan River (Zhu 2008). Combined with the concept of ecological management, the appropriate structure types of ecological embankments are selected, and the embankment types usually adopted are vertical, slope, compound, and so on. When choosing the type of embankment, consideration should cover the nature of the land around the region, the construction environment, topography, and other conditions selected according to local conditions. Different cross-section types are described in Table 2.

The above analysis and comparison show that the slope dike is similar to the compound dike, and the slope dike is finally selected considering the requirements of occupation space, engineering cost, and hydrophilic ecology. The specific layout of the left and right embankments of the river is shown in Table 3.

Table 2. Brief introduction of dike type.

Dike type	Scope of application
Upright	It is suitable for the terrain, the existing building limitation, and the river section with a large relocation volume.
Sloping	The river reaches high ecological requirements, and large land use space has simple sectional structures, which can meet the flood discharge requirements and ecological diversity of the river.
Compound	The river section has insufficient space but needs to meet the ecological and safety of the river section. The slope of the upper and lower parts of the river section is different, the upper part meets the ecological and landscape needs, and the lower part focuses on flood control.

Table 3. Scale of dike project.

Engineering project	Engineering scale			Total (m)	Engineering scope	Remarks
Revetment (km)	Cross section I	Left bank	2165.2	5407.2	0 + 000~1 + 900 5 + 784.8~5 + 900 5 + 784.8~5 + 900	Fill the gabion by laying bricks and mortar
		Right bank	3242		0 + 000~2 + 200 5 + 708~6 + 750	
	Cross section II	Left bank	1809.2	3824.8	2 + 400~3 + 364.8 4 + 940~5 + 784.8	Hinged slope protection
		Right bank	2015.6		2 + 800~3 + 521.5 4 + 413.9~5 + 708	
	Cross section III	Left bank	1590	2190	1 + 900~2 + 400 4 + 350~4 + 940 5 + 900~ 6 + 400	Hinged slope protection, combined with self-inlaid retaining wall
		Right bank	600		2 + 200~2 + 800	
	Cross section IV	Left bank	2163.4	4077.3	3 + 364.8~4 + 200 6 + 400~7 + 728.2	Cut slope gabion by laying bricks or stones
		Right bank	1913.9		3 + 521.5~4 + 413.9 6 + 750~7 + 771.5	

3.3 *Slope protection engineering*

Sturdy and durable, simple process, and strong adaptability are the three advantages of the traditional berm. It should not be ignored that traditional berm is more destructive to urban river ecosystems, the exchange between urban river water systems will be hindered to some extent, and the urban landscape effect will even be weakened (Liu *et al.* 2007). Therefore, according to the current topographic conditions, the design selects the form of masonry protection according to the local conditions, is combined with the ecological management concept, and compares various aspects such as reliability, applicability, and economy (Luo *et al.* 2022). It selects the gabion retaining mat in the wide and flat section of the river, the articulated slope combined with self-embedded vegetation retaining wall in the wetland park and artificial water section, and the slurry masonry retaining wall only when the river is narrow and the flooding capacity is insufficient.

3.4 Barrage project

As the main means of urban water system planning and construction, the project of river blocking and lake building is widely used in water system construction across the country (Zhang *et al.* 2013). The wide channel of the Sichuan River can reach 650 m, and the narrow channel is less than 100 m. With multiple low-head overflow dams or overflow weirs, the well-arranged river surface can be created to enhance the city image of Yanliang District.

4 CONSTRUCTION ORGANIZATION

4.1 Main project construction

4.1.1 Construction of new embankments

This project can basically meet the balance of earthwork excavation and filling because the embankment filling material can be directly taken from the floodplain excavated by the river bank, and the excavated earthwork can meet the requirement of earthwork filling.

In order to be reasonably equipped with manpower and machinery and ensure the quality of construction and progress of the project, the main process of the embankment—earthwork excavation, backfill, stone cage preparation, steel engineering required period, artificial, mechanical simple calculation—this paper takes the layout of gabion slope protection section earthwork excavation as an example. According to "Dike Engineering Quantity Calculation," the calculation formula for earth excavation of the dike body is as follows:

$$V = \frac{1}{2}(A_1 + A_2) \times (L_2 - L_1) \tag{5}$$

where L_1 and L_2 are the front and rear mileage (m); A_1 and A_2 are the excavation area of the section corresponding to the front and rear mileage (m^2); and V is the corresponding square root between the front and rear mileage (m^3).

4.1.2 Slope protection project

From the reliability, applicability, economy, and other aspects of the comparison, the project slope revetment recommended using a gabion pad and hinged revetment alternately combined, wall revetment using the self-embedded reinforced retaining wall.

1. Gabine pads. The gabion pad section is based on mechanism alloy steel wire gabion. The foundation is a buried depth of 2 m, with two rows of gabion stacked, the lower two and the upper one. The gabion stone cage pad thickness is 50 cm, the stone cage under the shop 300 g/m^2 geotextile, geotextile, and gabion stone cage between filling 10 cm coarse sand cushion, geotextile pad soil compaction. The compaction coefficient is not less than 0.92, and the compaction thickness is not less than 30 cm. To prevent rain erosion, ensure safety during a flood, and meet the need for environmental beautification, the gabion pad is filled with 30 cm thick planting soil, planting grass greening.
2. Hinged slope protection. The foundation of the hinged slope protection section adopts an M 7.5 slurry stone footguard with a buried depth of 2 m and top width of 0.5 m. The hinged slope protection section is paved with 300 g/m^2 geotextile padded with solid soil compaction. The compaction coefficient is not less than 0.92, the compaction thickness is not less than 30 cm, and the surface is covered with soil and grass.
3. Self-inlaid reinforced retaining wall. A newly built self-embedded reinforced retaining wall, 50 cm thick C20 concrete foundation is set. The buried depth of the foundation is 2 m. 50 cm thick graded gravel is laid from the self-embedded block to the front backfill soil, and a 5 m thick clay layer is set after the gravel. The two retaining wall blocks are laid in the middle of the geogrid and fixed with the rubber rod between the blocks. 300 g/m^2 geotextile is laid between the two layers of the grid.

4.1.3 Barrage project

The floor elevation of the rubber dam is generally 0.5 m higher than the average elevation of the riverbed topography at the dam site. In front of the dam, a 15–20 m long concrete cover is laid to increase the permeability diameter, a reinforced concrete seepage wall is set under the bottom of the dam, a silencing pool is set behind the dam, and a flood and anti-flood groove are set behind the pool. The slope protection is connected with the downstream riverbed.

The spillway dam adopts a concrete face to prevent seepage. The main structure of the spillway dam body is M7.5 cement mortar block stone, and the water-facing surface is C20 concrete anti-seepage panel. The settlement joint of the spillway dam is generally 2 cm wide and is filled with polyethylene foam board. The anti-seepage panel is equipped with a rubber water stop belt. Natural landscape stones are set at the top of the overflow dam, allowing pedestrians to pass through when the overflowing head is lower than 0.2 m.

4.2 General arrangement of construction

4.2.1 Temporary construction road

Embankment body filling is obtained from bank cutting slope and beach excavation. The temporary road on the embankment should be set as consistent as possible with the road over the embankment. The construction temporary road will be laid along both sides of the new embankment. The width of the temporary road is 6.0 m. A total of 12 km of temporary roads has been built, covering an area of 72,000 square meters, or 108 mu.

4.2.2 Temporary storage and yard

According to the requirements of storage materials, closed warehouses, and open warehouses are set on the beach, which is convenient for transportation and meets the elevation of the design flood once in 50 years. The closed warehouse is used to store cement, hardware, electrical power, and other valuable materials and easily damaged or lost materials. Open warehouses are used to store materials that are not damaged by the natural climate. A temporary warehouse of 800 m^2 is set up this time.

4.2.3 Temporary housing

Temporary housing is a building built for on-site management and construction personnel. To facilitate production and life, temporary housing should be located close to the construction area, covering a total area of 1000 m^2. The specific construction layout is shown in Figure 1.

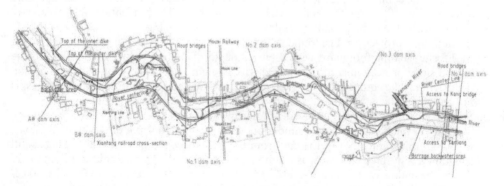

Figure 1. General layout of construction.

4.3 *Overall construction schedule*

Referring to the relevant construction experience of similar projects, according to the characteristics of the project, terrain and geology, and flood traffic conditions, the construction of the project is mainly mechanical and manual assistance. To avoid the main flood season in the engineering area, the construction period is divided into three stages. No construction is performed in May, June, July, or August. Staggered construction is performed in adjacent stages, so as to optimize the configuration of people, materials, and machines and ensure the construction progress.

5 CONCLUSIONS

Based on the Sichuan river section of urban ecological governance on the basis of existing engineering data, the key period of Sichuan river city river embankment, slope protection engineering, a dam engineering are analyzed in general layout and construction organization design through calculation works agency floods, reasonable decorate the project involves the hydraulic structures. At the same time, there is a related calculation to optimize configure each division engineering construction personnel, machinery, and materials, to ensure the progress and quality of the project, and strive to maximize the benefits of the project, and draw the following conclusions:

1. As the construction of an urban river ecological control project is greatly affected by the natural environment, the construction should be based on the regional flood control planning and construction conditions, fully estimate the geological conditions of the construction area, and determine the project layout plan after a comprehensive analysis of technology and economy to ensure the safety of construction.
2. River control involves a wide range of construction projects, which are numerous and scattered. To improve efficiency and reduce construction interference, it is necessary to make comprehensive and reasonable construction organization design and arrangement in advance and formulate standardized, safe, economical, and effective construction schemes and processes to control the construction cost and make the project proceed smoothly.
3. Due to the particularity of the scope of the urban river ecological control project, the construction area is located in the city. There are many surrounding enterprises and residents, which will inevitably cause environmental pollution during the control process. Therefore, various effective protection measures should be taken before, after, and during construction to minimize the construction impact.

ACKNOWLEDGMENTS

This work was supported by the Shaanxi Provincial Department of Education Key Scientific Research Project (number 21JT029).

REFERENCES

He B., Gao H.Q and Xia X.D (2006) Study on the Urban River and its Ecological Management Planning. *J. Soil and Water Conservation in China*, (12): 23–25.

Liu L.M., Qiu W.M., Xu W.N., Sun C. and Guo P. (2007) Comparison and Analysis of Traditional Slope Protection and Ecological Slope Protection. *J. Journal of China Three Gorges University (Natural Science Edition)*, (06): 528–532.

Luan R.Z (2010) *Research on Construction Organization Design of Shuhe River (Shandong Section) Control Project. Shandong University.*

Luo Z.G., Xie Q.M. and Tan B. (2022) Application of Ecological Slope Protection in Urban River Comprehensive Regulation. *J. Urban Architecture*, 19(06): 62–64+68.

Ning H.J., Zhou H.B. and Liu Y. (2012) Planning, Design and Construction Management of Small and Medium-Sized Rivers in Pengshui County, Chongqing. *J. Yangtze river*, 43 (18): 24–27.

Yang S.D., Cai X. and Hua C.L (2012) Discussion on Practice and Methods of Small and Medium-Sized River Management in Yunnan province. *J. Yangtze river*, 43(18): 12–14+21.

Zhang P.L., Fu W., Liu T and Wang Y.X (2013) Construction Organization Design of Longkou Hydropower Station. *J. Hydroelectric power*, 39(02): 72–74.

Zhang Y.C (2012) Cost Control of Urban River Comprehensive Regulation Project. *J. China Rural Water and Hydropower*, (08): 32–33+38.

Zhou X.M. and Gao M.Z (2022) Construction Organization Design of Downstream River Control Project of Shaping First Class Hydropower Station. *J. Yangtze River*, 53 (S1): 68–72.

Zhu G.Z (2008) Ecological Treatment of Urban Sewage River Based on Harmony Between Human and Water. *J. Research of Soil and Water Conservation*, (04): 261–263.

Civil Engineering and Energy-Environment – Gao & Duan (Eds)
© 2023 the Author(s), ISBN 978-1-032-56059-5

Saihanba ecological protection construction based on GIS and its impact on the environment

Bing Xia* & YuLong Lei*
Department of Electrical Engineering, Liuzhou Institute of Technology, LiuZhou, China

ABSTRACT: China insists that green water and green mountains are priceless treasures. It respects, harmonizes, and protects nature and focuses on resource conservation, environmental protection, and natural recovery. It implements the sustainable development strategy, improves the overall coordination mechanism of ecological civilization, builds an ecological civilization system, promotes the transformation of economic and social development to comprehensive green development, and builds a beautiful country. With the help of the Chinese government, China's Saihanba Forest Farm has recovered from the desert and has now become an eco-friendly green farm with a stable sand-control function. By establishing an assessment model of the impact of Saihan Dam on the ecological environment, and comparing this assessment model with the environmental indicators of other cities, provinces, or countries, it is clear whether the reconstruction of Saihan Dam has played a positive role in other ecological health environments, and feasible plans and suggestions are put forward for the construction of ecological protection areas. It will be applied to other parts of the Asia Pacific region to improve people's living environment.

1 INTRODUCTION

Saihanba Mechanical Forest Farm is the largest artificial forest farm in Hebei Province. In 2007, a nature reserve was established on the basis of the original forest farm and Saihanba National Forest Park, which is responsible for building and protecting the forest ecosystem, protecting and restoring the wetland ecosystem, harnessing and improving the desert ecosystem, and maintaining biodiversity. Saihan Dam is located in the ecotone of forest grassland desert sandy land in the plateau, mainly covering natural vegetation communities, important water sources of Luanhe River and Liaohe River, and rare and endangered wildlife resources. It is a forest ecosystem-type nature reserve (Department of Natural Ecology Protection, June 19, 2015). At the same time, we should attach equal importance to resource development and protection, coordinate with the sustainable development strategy, and do a good job in managing and protecting forest ecological resources, to give full play to the ecological public welfare function of forests. (Shao Lixin, 11.05. 2020)

Therefore, index data such as forest coverage, coverage area, forest storage, urban air quality standard (better than Level 2), and carbon dioxide absorption are selected to conduct a natural ecological quality assessment of Saihan Dam, including rarity assessment, naturalness assessment, and vulnerability assessment. The monitoring results show that the Saihanba Mechanical Forest Farm has a total operating area of 1.41 million mu and 1.1 million mu of forest land, including 860,000 mu of planted forest and 240,000 mu of natural forest; The main tree species are larch, birch, spruce and other forest coverage rate of

---

*Corresponding Authors: 823607107@qq.com and 1617684433@qq.com

DOI: 10.1201/9781003433651-40

80%, the total forest stock is 5.025 million cubic meters. There are 618 species of vascular plants belonging to 303 genera and 81 families in the forest area; 22 families, 51 genera, and 79 species of macrofungi; 152 species and subspecies of terrestrial wildlife. (Du Xinglan, September 14, 2017) Research shows that Saihanba plays an important role in resisting sandstorms, protecting the environment, maintaining ecological balance and stability, and the restoration of Saihanba Forest Farm has played a decisive role in Beijing's resistance to sandstorms.

In developing countries, many ecological and environmental problems exist, such as land degradation, desertification, water and soil loss, forest reduction, and water shortage. These problems, which occur in a wide range and affect a large number of people, have seriously restricted the economic development of developing countries and profoundly affected the world's sustainable development. It should be noted that the ecological deterioration of developing countries is related to the long-term imbalance of international economic and technological development. Therefore, the causes and effects of ecological deterioration in developing countries are global. The international community should pay equal attention to this and take concerted action to solve global environmental and development problems (Li Xue 1991)

More and more studies focus on social and economic factors affecting human health and introduced some conceptual models, including Pressure State Response (PSR) (Relationship and Friends 1979). However, to build a comprehensive assessment system for forest farm health, considering the importance of landscape and social factors, there is an urgent need for an integrated ecosystem to evaluate the ecosystem. Many studies have realized the comprehensive evaluation of forest farm health with similar methods, but the evaluation of different urban applications is not comparable, which is difficult to be directly applied to different types and regions and cannot meet the actual requirements

Therefore, in view of the above problems, this paper adopts the pressure state response (PRS) framework system. The linear ecological health evaluation index system adopts the functional model and the comprehensive evaluation method ecological health comprehensive evaluation model and then establishes a set of the ecological health evaluation index system, which is used as the standard or basic evaluation index model for later evaluation. Arctic software is used to sample the index mask, then AHP is used to construct the weight calculation logic of the judgment matrix (T.L. Saaty 1970), which aims to correct that each indicator can be accurately compared and to calculate the indicator weight of the relevant ecological construction health evaluation indicators. Through this process, Saihanba has built an ecological health evaluation index system. The process is to combine actual data and aggregation algorithms with building an evaluation model for other regions and comparing the evaluation index model with Saihanba's ecological health. The relevant area and weight coefficient of Saihanba's ecological health environment are calculated using the judgment matrix and entropy method. Whether the reconstruction of Saihan Dam has a positive effect on other ecological health and the environment relies on the change of weight coefficient.

2 MODEL ARCHITECTURE

According to the impact of the environmental transformation of Saihan Dam on China's environment, an ecological health environment index evaluation system is determined first, and then data are collected according to the established ecological health environment index evaluation system. The collected data are subject to mathematical statistics and classification, and the Arctic software is used for mask sampling and kriging data fitting to obtain the required regional index data (such as the forest coverage of Saihanba from 1962 to 2021). The indicator weight grade is further constructed through the established ecological health and environmental indicator evaluation system and mathematical statistics. The index weight grade is used to establish the evaluation model of the ecological health and environment index. The regional weight grade is divided through the AHP weight calculation logic to obtain a set of the weight index system. Then, the ecological health environment index evaluation coefficient is obtained through the established ecological health environment index evaluation model and entropy weight method to calculate the data. The judgment matrix established by AHP is used to compare each index coefficient, and the correlation coefficient ratio is obtained. Divide the weight of the obtained data according to the value of the ratio, and then obtain the index coefficients of forest coverage, coverage area, forest storage, and water conservation of Saihan Dam through the clustering algorithm. The comprehensive index coefficients of Saihan Dam are provided. The relationship between each index coefficient and the comprehensive index coefficient is obtained through clustering algorithm analysis and calculation. Combined with the ratio between the index coefficient of the Saihanba forest coverage rate and the difference between the index coefficient of the Korean forest coverage rate and the comprehensive index coefficient, the ratio analysis is conducted with the difference between the other three indexes and the comprehensive index coefficient

$$\left(\triangle H_1 = \frac{H_1}{H_1 - R_{1i}}\right)$$

2.1 *Reconstruction model and weight index model of Saihan Dam*

The ecological health evaluation index system is built through the "pressure state response" framework system to evaluate the ecological environment. After the established presentation evaluation index, if the original value of an index falls in a certain score range, its membership degree is 1, and the membership degree of other ranges is 0, and the corresponding score is given

In this paper, the linear weighted function model and comprehensive evaluation model are used for the comprehensive evaluation of ecological health. The expression is:

$$M = \sum_{h=1}^{p}(\sum_{j=1}^{m} \alpha_i \omega_j)\delta_h \tag{1}$$

where M is the comprehensive evaluation score, α_i is the assignment of the evaluation index i; ω_i is the comprehensive weight of evaluation index i relative to the jth subsystem, δ_h is the subsystem weight; p is the number of subsystems, and m is the number of evaluation indicators in the subsystem.

After the ecological health evaluation index system is obtained, the Arctic software and AHP weight calculation logic are used to conduct index mask sampling and analysis for Saihan Dam, and then the four evaluation indexes of forest coverage, coverage area, forest stock, and water conservation are selected for weight grading.

Table 1. Saihanba data weight analysis.

	Land area covered with trees	Coverage area/ten thousand mu	Forest stock /10,000 cubic meters	Water conservation / 100 million cubic meters
Weight index level	8.9235	1.6892	1.4092	1.7208

Combined with the weight of Saihanba forest cover, the entropy weight method and formula (1) were used to analyze the forest cover and other indices after the transformation of Saihanba from 1962 to 2021, as shown in the figure below:

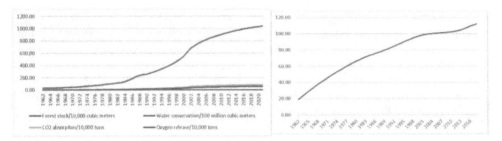

Figure 1. Index analysis chart of Saihan Dam after reconstruction.

3 USE THE MODEL TO ANALYZE OTHER REGIONS

A new weight model coefficient is obtained by comparing the weight model of Saihan Dam with that of North Korea. At the same time, the index coefficients of the reconstructed Saihan Dam are compared with those of a country in the Asia Pacific region.

North Korea was selected for index weight analysis. The selected indexes were the forest coverage rate and biological carbon content of North Korea compared with the forest coverage rate and carbon dioxide absorption rate of Saihanba, and the comprehensive evaluation model formula after adding the ecosystem was used:

$$H_2 = \frac{1}{4} M \sum_{i=1}^{21} \eta_{ki}$$

The comprehensive evaluation weight indicators after joining the ecosystem are constructed as follows:

Table 2. Saihanba and North Korea, South Korea ecosystem weight analysis.

	Forest coverage rate (Saihanba)	Carbon-2 absorption rate (Saihanba)	Forest coverage rate (North Korea)	Biocarbon content (North Korea)
Weight	$\frac{\Delta H_{1i}}{\Delta H_{1j}}$	$\frac{\Delta H_{2i}}{\Delta H_{2j}}$	$\frac{\Delta H_{3i}}{\Delta H_{3j}}$	$\frac{\Delta H_{4i}}{\Delta H_{4j}}$

According to the vegetation analysis in North Korea, where coniferous forests account for the largest proportion of vegetation, a weight index system is built again based on this vegetation, as shown in the following Table 3:

Table 3. Saihanba with North Korea, South Korea vegetation weight analysis.

	Forest coverage rate (Saihanba)	Carbon-2 absorption rate (Saihanba)	Forest coverage rate (North Korea)	Biocarbon content (North Korea)
Weight	$\frac{\Delta H_{1i}}{\Delta H_{1j}} M$	$\frac{\Delta H_{2i}}{\Delta H_{2j}} M$	$\frac{\Delta H_{3i}}{\Delta H_{3j}} M$	$\frac{\Delta H_{4i}}{\Delta H_{4j}} M$

Through the calculation of the clustering algorithm and the relative ratio of forest coverage and carbon dioxide absorption of Saihanba, the following trend chart is obtained.

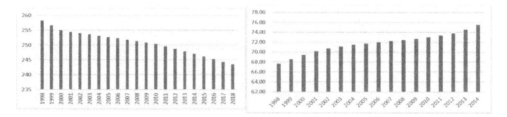

Figure 2. Trend chart of the relative ratio of forest coverage and carbon dioxide absorption.

By formula

$$\zeta_{ki} = \frac{\min_k \min_t |S_{k0} - S_{kt}| + p\max_k \max_t |S_{k0} - S_{kt}|}{|S_{k0} - S_{kt}| + p\max_k \max_t |S_{k0} - S_{kt}|} \quad (4)$$

$$R_i = \frac{1}{4} \sum_{k=1}^{4} \zeta_{ki} \quad (5)$$

The comprehensive evaluation weight indicators of the health environment of North Korea after joining the Saihanba ecosystem are obtained:

ΔH_1 = 4.0754, ΔH_2 = 4.0084, ΔH_3 = 5.9916, ΔH_1 = 5.9246. Combined with the trend chart of forest coverage and carbon dioxide absorption in Saihanba from 1998 to 2018, the comprehensive assessment weight of the Korean conifer health environment was further calculated, and the following results were obtained:

$$\eta_{ki} = \frac{\min_k \min_i |S_{k0} - S_{ki}| + p\max_k \max_i |S_{k0} - S_{ki}|}{|S_{k0} - S_{ki}| + p\max_k \max_i |S_{k0} - S_{ki}|} * \frac{p_{ij} - p_{\min}}{p_{\max} - p_{\min}}$$

$$H_1 = \frac{1}{4} \sum_{i=1}^{21} \eta_{ki}, \quad H_2 = \frac{1}{4} M \sum_{i=1}^{21} \eta_{ki}, \quad \triangle H = H_1 - H_2, \quad \triangle H_5 = \frac{\Delta H_1}{\Delta H_2}$$

If the weight ratio of an item in North Korea is 0-1, and the ratio between the index coefficient of North Korea's forest coverage rate and the index coefficient of Saihanba's forest coverage rate, as well as the difference between the comprehensive index coefficient, it needs to focus on governance according to the specific situation of the region. For example, the ratio of some regions in North Korea is between 0 and 1, but it is not suitable for the growth of trees due to geographical location. Therefore, focusing on its governance cannot improve the ecological environment. If it is in between, it needs to be governed. If it is, governance is not required. In the case of major disasters, priority should be given to regions with large disaster coefficients and low-weight scores. The second is to govern regions with a large coefficient of major disasters and high-weight scores. The last is to govern regions with a small coefficient of major disasters and low weight score.

4 CONCLUSION

According to the model description, the establishment of ecological protection area can make the regional environment recover well. Thus, scientific research can be carried out smoothly, providing a detailed and accurate basis for the rational development and utilization of natural resources in the future.

1. On the premise of not damaging the existing ecology of the Reserve, explore reasonable development and operation activities to improve the self-construction ability of the Reserve, so as to maintain the sustainable development of the ecological reserve, ensure the normal progress of the ecosystem and its natural succession process, and realize the sustainable use of resources and multiple ways of utilization; Only by achieving sustainable development can the Reserve effectively provide the necessary material basis for scientific research, production experiment teaching, and tourism.
2. National linkage to build an ecological protection model to maximize benefits. Environmental protection departments at all levels should further strengthen the management of the ecological environment, strictly carry out environmental impact assessment and implement the "three simultaneous" system of environmental protection.
3. Strengthen supervision and inspection to ensure that all environmental protection measures proposed in the environmental impact report are implemented. The construction shall be carried out according to the model index grade to reasonably change the overall environment.

REFERENCES

Editorial and Editing Department, Liu Yuan. The 2016 National Prairie Monitoring Report. *China Animal Husbandry*, 2017 (8): 18.

High Spring Breeze. Ecological and Environmental Planning Technical Specification and Analysis. *Liaoning Urban and Rural Environmental Technology*, 2004, 024 (002): 45–48.

J. Ke, C. Li Study on Coordinated Development of Resources, Environment and Economy in China Based on DEA Cluster Analysis. *China oft Science*, 2005 (2): 5.

J. Liu, T. Yue, H. Ju *et al.* Comprehensive Ecosystem Assessment in Western. China. *Expert literature Library*, 2006.

Civil Engineering and Energy-Environment – Gao & Duan (Eds)
© 2023 the Author(s), ISBN 978-1-032-56059-5

Research progress of rural wastewater treatment technology

Cong Xiao, Ning Tian*, Hongtao Wang, Xiaoning Qu, Lizhen Ma, Bin Shuai,
Mengshan Yu & Weichong Xu
The First Company of China Eighth Engineering Bureau Ltd, Jinan, China

ABSTRACT: Under the backdrop of carbon peaking and carbon neutralization integration into the overall layout of ecological civilization construction, rural sewage treatment and resource utilization technology will surely move in the direction of "green and low carbon." This poses a severe challenge to the development of rural sewage treatment technology and is also an important opportunity for the next iteration of technology. This paper summarizes the source and treatment status of rural domestic sewage. At the same time, the technical characteristics and applications of rural domestic sewage treatment technology such as earthworm eco-filters, efficient algal ponds, buried micro-power processes, biological contact oxidation, membrane bioreactors, and constructed wetland processes are discussed in depth in order to provide certain technical support for rural sewage treatment problems.

1 INTRODUCTION

In recent years, with the rapid development of the social economy, the level of the rural economy has continued to improve, and the resulting environmental problems have become increasingly prominent. At present, rural domestic sewage has become an important source of non-point source pollution in rural areas. Domestic sewage in rural areas is distinguished by its small quantity, dispersion, distance from the sewage pipe network, large body of water, small water environment capacity, low sewage treatment rate, and low management level (Liang 2007). Rural domestic sewage is generally discharged into nearby rivers and lakes without any treatment, which aggravates the pollution of the water environment. In particular, rural domestic sewage is rich in nitrogen and phosphorus, which is an important reason for water eutrophication (Hong 2013). Under the background of carbon peaking and carbon neutralization being included in the overall layout of ecological civilization construction, rural sewage treatment and resource technology will surely move in the direction of "green low-carbon." As a result, expanding research into rural domestic sewage treatment technology has emerged as a current hot topic.

   At present, the rural sewage treatment methods are all based on the treatment measures of the first region and use the successful treatment cases as a reference. Among them, the treatment measures are only simple imitations, not combined with the actual situation to improve, which makes it difficult to meet the treatment needs, so the treatment effect of rural domestic sewage is not obvious (Cao 2009). As a result, expanding research into rural domestic sewage treatment technology has emerged as a current hot topic. This paper summarizes the technical characteristics and application cases of rural sewage treatment to provide some technical support for rural sewage treatment.

_____
*Corresponding Author: tianning868@163.com

DOI: 10.1201/9781003433651-41

2 CONVENTIONAL PROCESS OF RURAL DOMESTIC SEWAGE TREATMENT

2.1 *Vermibiofilter technology*

Earthworms have amazing phagocytic capacity, and their digestive tract can secrete protease, lipase, cellulase, and other enzymes, which strongly decompose most organic wastes (Fang 2009). The vermibiofilter (VBF) is designed based on the function of earthworms to improve the aeration and permeability of the soil and promote the decomposition and transformation of organic materials. It is a new concept sewage treatment process that can not only remove pollutants in urban sewage with high efficiency and low energy consumption, but also greatly reduce the cost of excess sludge treatment and disposal. The VBF treatment system integrates a primary sedimentation tank, an aeration tank, a secondary sedimentation tank, a sludge return facility, and an oxygen supply facility, greatly simplifying the sewage treatment process. The VBF sewage treatment technology was first researched and developed in France and Chile and has begun to be industrially applied abroad.

Anusha (2015) used VBF to treat rural domestic sewage; the results showed that domestic wastewater treatment using earthworms was found to be more effective than treatment without worms. Chemical Oxygen Demand (COD) and chlorides in domestic wastewater were found to decrease by about 57–66% and 70–77%, respectively, after 2 hours, 4 hours, and 6 hours.

Yang (2000) conducted a pilot test, and the results showed that the eco-filter had a COD removal rate of 83–88%, a BOD_5 removal rate of 91–96%, a SS removal rate of 85–92%, an ammonia nitrogen removal rate of 55–65%, a total phosphorus removal rate of 35–65%, and a total sludge yield of 0–2 mg/L.

2.2 *High-Rate Algal Pond*

The High-Rate Algal Pond (HRAP) was proposed and developed by Oswald of the University of California (Nurdogan 1996). Compared with traditional stabilization ponds, its main characteristics are mainly in the following four aspects: 1) the depth of the pond is relatively shallow; 2) there is a continuous mixing device perpendicular to the pond corridor; 3) the residence time is relatively short, usually less than 12 days; and 4) the width is narrower (Canovas 1996). The continuous mixing device in an HRAP can promote the complete mixing of sewage, adjust the concentration of oxygen and carbon dioxide in the pond, balance the water temperature in the pond, and promote the blowing off of ammonia nitrogen. These characteristics enable the formation of an environment conducive to the growth and reproduction of algae and bacteria in HRAP and strengthen the interaction between algae and bacteria. As a result, the HRAP has more biofacies than the general stabilization pond, resulting in good removal of organic matter, ammonia nitrogen, and phosphorus (Picot 1992).

García (2000) compared the nitrogen removal efficiency and the changes of TN, total organic nitrogen, NH_3-N, and nitrogen oxide in the high-efficiency algal pond under different hydraulic retention times. The conclusion of this study was that HRT determined both the nitrogen removal efficiency and the distribution of nitrogen forms in the effluent of an HRAP. The nitrogen removal level could be controlled through suitable HRT operating strategies. By operating at an HRT of 4 days in spring and summer and ten days in autumn and winter, the nitrogen concentration in the effluent of an HRAP system could be reduced to less than 15 mg/L.

Xu (2001) conducted a pilot study on the HRAP, which showed that the average removal rates of COD, BOD, NH_3-N, and TN in the HRAP were 75%, 60%, 91.6%, and 50%, respectively.

2.3 Anaerobic/ Anoxic/ Oxic treatment technology

The core of the anaerobic, anoxic, and oxic (AAO) process is activated sludge treatment, which removes organic pollutants, nitrogen, and phosphorus from water through various combinations of anaerobic zones, anoxic zones, and aerobic zones, as well as different sludge reflux methods (Huang 2015). The process is mainly composed of anaerobic, anoxic, and aerobic stages. Raw water mixed with reflux sludge from the settling tank enters the anaerobic unit, which releases phosphorus and degrades some pollutants such as COD and nitrogen simultaneously. The effluent from the anaerobic unit enters the anoxic tank, with the mixed nitrification liquid returned from the aerobic tank for denitrification. Nitrification and phosphorus absorption are mainly carried out in the aerobic tank. The AAO process has the following main features: 1) It can remove organics, nitrogen, and phosphorus simultaneously under three different environmental conditions: anaerobic, anoxic, and aerobic, and with the organic combination of different microbial flora; 2) in the process of simultaneous deoxidization, phosphorus removal, and organics removal; 3) the process flow is the simplest, and the total HRT is also less than that of other similar processes; 4) under the alternate operation of anaerobic-anoxic-aerobic, filamentous bacteria will not reproduce in large numbers, and sludge bulking will not occur.

Wang (2018) proposed to use of the AAO process to treat rural domestic sewage. The results showed that the pilot plant was constructed to treat domestic sewage, and the effluent COD, NH_3-N, TN, and TP were 19.79 mg/L, 2.66 mg/L, 8.82 mg/L, and 0.47 mg/L, respectively. The final effluent met the *Discharge Standard of Pollutants for Municipal Wastewater Treatment Plant* (GB 18918-2002) class 1-A criterion.

Langwen (2017) studied the low influent C/N rate and poor nitrogen removal efficiency of rural domestic sewage. The results showed that, under low-carbon conditions, the denitrification efficiency would increase significantly when the internal reflux ratio was 150%, and the DO concentration at the aerobic terminal was 1.0 mg/L. In addition, the increase in sludge caused by the change in water inflow could also increase the denitrification efficiency.

2.4 Biofilm method

The biofilm method is a method that uses microorganisms to oxidize and decompose organic substances and transform them into inorganic substances (Sabaeifard 2014). Biofilm has biochemical activity, which can purify sewage by adsorbing and decomposing pollutants in a suspended, colloidal, or dissolved state. Among them, the most representative biofilm method is bio-contact oxidation (BCO). In this method, a filler is arranged in the reactor, and the wastewater is in contact with the filler after being oxygenated. Under the double action of biofilm on the filler and activated sludge between the filler spaces, the wastewater is purified. Like other biofilm methods, activated sludge is cultivated and domesticated to form a layer of biofilm on the surface of its filler. When dissolved oxygen and organic matter are sufficient, the biofilm gradually thickens. When the biofilm reaches a certain thickness, an anaerobic layer will form inside, and denitrification will occur. Under the washing of anaerobic metabolic gas and water flow, the biofilm will fall off, and a new biofilm will be formed on the carrier after the old biofilm falls off. In rural domestic sewage treatment, contact oxidation combined with front-end hydrolysis acidification can achieve better treatment effects (Gagnon 1999).

Zhu (2010) used the BCO process to treat rural domestic sewage. The research showed that when the concentration of raw water did not change, the removal rates of COD, BOD_5, NH_3-N, and TP by biological contact oxidation were 76.3%, 78.4%, 74.9%, and 58.8%, respectively.

Gao's (2012) study showed that the annual average concentrations of COD, NH_3-N, and TP in the final effluent were 22.40 mg/L, 8.58 mg/L, and 2.09 mg/L, respectively, by the BCO method.

The Japanese Rural Sewage Treatment Association (Li 2016) adopted the BCO method, which could reduce the BOD_5 in sewage to below 20 mg/L, the SS to below 50 mg/L, and the TN content to below 20 mg/L.

3 NEW TECHNOLOGY FOR RURAL DOMESTIC SEWAGE TREATMENT

3.1 *Unpowered Anaerobic Reactor*

The Unpowered Anaerobic Reactor (UAR) mainly relies on anaerobic biology to treat rural domestic sewage. Anaerobic biological treatment is also known as anaerobic digestion or anaerobic fermentation (Shen 2005). It refers to the process in which a variety of anaerobic or facultative microorganisms work together under anaerobic conditions to decompose organic matter and produce CH_4 and CO_2. The UAR is suitable for villages with less drainage and a lower effluent discharge standard.

Leju (2014) studied the UAR for rural domestic sewage treatment at different operational temperatures and HRT with different loading rates. The results revealed efficient removal of COD by the reactor (36%, 48%, and 68%) at an HRT of 1, 2, and 3 days. Besides, the temperature had a significant influence on the reactor's performance, accounting for a COD removal rate of 44% and 68%, respectively.

3.2 *Buried micro-power process*

This wastewater treatment unit utilizes a combination of treatment methods such as hydrolysis, contact oxidation, and stabilization ponds. The sewage enters the hydrolysis pond after hydrolysis and acidification, and the concentration of organic matter is reduced. The sewage in the hydrolysis pond is lifted to the aeration filter pond using a lift pump (Yen 2004). The aeration filtration pond adopts the jet oxygenation method, and after aeration treatment, the sewage then enters the stabilization pond. The air in the stabilization pond is provided by the natural suction of wind-pulling pipes distributed along the trench. Suitable aquatic plants, such as phytoplankton, can also be planted in the stabilization pond to absorb pollutants for treatment. This process is suitable for villages with tight land resources, a high degree of agglomeration, relatively good economic conditions, and the foundation of a rural tourism industry.

Weijun (2020) tracked the design and operation effects of micropower domestic sewage treatment facilities. The results show that the removal rates of SS, COD, TN, NH_3-N, and TP in rural domestic sewage are 89.3%, 82.2%, 58.2%, 90.4%, and 56.1%, respectively, and the effluent quality indicators meet the design requirements.

3.3 *Artificial wetland treatment process*

Artificial wetland wastewater treatment technology is a typical land treatment technology, mainly through the use of natural wetland systems, using physical, biological, and chemical synergies to remove pollutants from water, such as adsorption, filtration, ion exchange, and plant absorption, etc. (Salmon 1998). This technology is a good solution for domestic wastewater treatment problems and can be applied to production wastewater, such as heavy metal ion-contaminated water, mining wastewater, and waste leachate. Artificial wetland wastewater treatment technology has a very wide range of application prospects.

Du (2015) constructed a new type of artificial tidal flow wetland in the laboratory by simulating the characteristics of intermittent discharges of rural domestic sewage. The effects of COD loading on the oxygen environment and pollutant removal efficiency in the bed were studied. The results show that the concentration of organic pollutants was the main factor limiting the COD removal efficiency of the tidal flow; the COD removal efficiency of the system may reach as high as 95.6%; the ammonia nitrogen (NH_3-N) removal efficiency increased with rising organic pollutant loading rates from 85.2% to 98.7%; and higher organic pollutant loading favored assimilation by heterotrophic bacteria and enhanced denitrification.

Sheng-Long (2012) used an anaerobic tank and a composite constructed wetland to treat rural domestic sewage. The results showed that the removal rate of the anaerobic tank of SS was 80.6%, and the removal rates of the combined artificial wetland of COD, TN, and TP were 61%, 61.4%, and 57%–66%, respectively.

3.4 *Membrane bioreactor process*

Membrane bioreactor (MBR) technology is a new technology that uses membrane separation technology and integrates biological treatment technology, making this technology very efficient for wastewater treatment. This technology was first applied in the late 1960s, and it organically combines the activated sludge method and ultrafiltration membrane components, replacing the secondary sedimentation tank used in the traditional activated sludge method with membrane separation technology to achieve the relevant functions (Pollice 2008). It is well known that membrane separation technology can effectively separate different substances, enabling the bioreactor to operate continuously at a low F/M ratio, further accelerating the decomposition of organic matter. With the advancement of technology, various types of membrane technologies have been developed in leaps and bounds, which has led to further breakthroughs in the development of MBR technology (Azimi 2019). It was first researched by Japanese scientists. Due to the characteristics of the Japanese region, which has a large population but a small area, a smaller and more efficient wastewater treatment system was needed, and MBR technology was born, which has the advantages of stable treatment, a simple process, and a small area and can solve the above problems very well.

Jiang (2010) used the combined process of MBR and a constructed wetland to treat rural domestic sewage. A good treatment effect was reached through reasonable parameter selection and elaborate design. The effluent quality met the *Discharge Standard of Pollutants for Municipal Wastewater Treatment Plants* (GB 18918-2002) first-level A criteria. This system showed higher adaptability to influent changes in quality and quantity, and it had the characteristics of a good treatment effect, a small floor area, and convenient maintenance.

Hao (2011) investigated the technical feasibility of an intermittent operation of an integrated MBR to treat rural domestic sewage. The test results showed that the average COD value of effluent is 46 mg/L and the average removal rate was 68.3%; the average concentration of NH_3-N in the effluent was 3.23 mg/L and the average removal rate was 65.3%; the average concentration of TN in the effluent was 7.12 mg/L, and the average removal rate was 65.8%.

4 CONCLUSION

In this paper, the main processes and technical characteristics of rural sewage treatment and their application cases are studied in detail. Among them, the main characteristics of the traditional process are that it has been applied for a long time and is mature and reliable; the main characteristics of the emerging process are that it is applied to rural domestic sewage treatment in a short time with few successful cases or high operating costs. Given that carbon has reached its peak and that carbon neutralization is part of the overall design of ecological civilization construction, the rural domestic sewage treatment process should integrate the characteristics of each treatment process and develop in a green and low-carbon direction. The combination of multiple processes, such as AAO and artificial wetland, can not only meet the effluent quality requirements but also reduce the operation cost and achieve low-carbon operation of rural sewage treatment by optimizing design parameters.

REFERENCES

Anusha V. and Sundar K (2015). Application of Vermifiltration in Domestic Wastewater Treatment. *International Journal of Innovative Research in Science*, Engineering and Technology.

Azimi S.S. and Mousavi S.T (2019). *Study of Urban Wastewater Treatment Systems, A^2/O, MLE, IFAS, MBR, and AOPs and Sewage Outlets for use in Agriculture and Industry by GPS-X 7.0 Software.*

Canovas S. and Picot B (1996). Casellas C et al. Seasonal Development of Phytoplankton and Zooplankton in a High-rate Algal Pond. *Water Science & Technology*, 33(7):199–206.

Cao Q. and She J (2009). Treatment Technologies for Rural Domestic Sewage. *Environmental Science and Management.*

Chunhua Xu and Qi Zhou (2001). Research and Application of HARP. *Environmental Protection*, (8): 3.

Fang C.X. and Luo X.Z (2009). Zheng Z et al. Study on Modified Vermibiofilter for Treatment of Domestic Sewage. *China Water & Wastewater.*

Gagnon G.A. and Slawson R.M (1999). An Efficient Biofilm Removal Method for Bacterial Cells Exposed to Drinking Water. *Journal of Microbiological Methods*, 34(3):203–214.

Gao D.W (2012). Bio-contact Oxidation and Greenhouse-Structured Wetland System for Rural Sewage Recycling in Cold Regions: A Full-Scale Study. *Ecol Eng*, 2012,49.

García J. and Mujeriego R (2000). High-rate Algal Pond Operating Strategies for Urban Wastewater Nitrogen Removal. *Journal of Applied Phycology*, 12(3–5):331–339.

Hao G. and Weifang M (2011). Pilot Scale Study on Intermittent MBR Process for Rural Domestic Sewage Treatment. *Water Treatment Technology*, 37 (10): 4.

Hong C. Zhang M (2013). Occurrence and Removal of Antibiotic Resistance Genes in Municipal Wastewater and Rural Domestic Sewage Treatment Systems in Eastern China. *Environment International*, 55(May):9–14.

Huang M.H. and Zhang W (2015), Liu C et al. Fate of Trace Tetracycline with Resistant Bacteria and Resistance Genes in an Improved AAO Wastewater Treatment Plant. *Process Safety and Environmental Protection.*

Jian Yang. and Yongsen L (2000). Pilot Scale Study on Urban Sewage Treatment by Green Ecological filter. *Jiangsu Environmental Science and Technology*, 13 (4): 3.

Jiang L.L. and Liu J (2010). Treatment of Rural Domestic Sewage by MBR/Constructed Wetland Process. *China Water & Wastewater*, 2010.

Langwen K.E. and Wang T (2017). Factors Influencing Denitrification Efficiency of AAO Process for Domestic Sewage Treatment. *Guizhou Science.*

Leju J. and Ladu C (2014). Efficiency Assessment of the Anaerobic Filter Reactor Treating Rural Domestic Sewage at Different Operational Temperature and HRT. *Ijetae Com*, 73(Suppl 2):458–458.

Li Z (2016). Enlightenment and Application of Japanese Decentralized Sewage Treatment Technologies. *Modern Agricultural Science and Technology.*

Liang Z. and Jin-Ren.N (2007). Treatment Technologies and Approaches for Rural Domestic Sewage. *Journal of China University of Geosciences* (Social Sciences Edition).

Nurdogan Y. and Oswald W.J (1996). Tube Settling of High-Rate Pond Algae. *Water Science & Technology*, 33(7):229–241.

Picot B (1992). Comparison of the Purifying Efficiency of High Rate Algal Pond with Stabilization Pond. *Water Science and Technology*, 25(12):197–206.

Pollice A. and Laera G (2008). Optimal Sludge Retention Time for a Bench Scale MBR Treating Municipal Sewage. *Water Science & Technology A Journal of the International Association on Water Pollution Research*, 57(3):319.

Sabaeifard P. and Abdi-Ali A. (2014). Optimization of Tetrazolium Salt Assay for Pseudomonas Aeruginosa Biofilm Using Microtiter Plate method. *Journal of Microbiological Methods*, 105:134–140.

Salmon C. and Crabos J.L (1998). Artificial Wetland Performances in the Purification Efficiency of Hydrocarbon Wastewater. *Water Air & Soil Pollution*, 104(3–4):313–329.

Shen D. and He Y (2005). Underground Inpowered Anaerobic Reactor for Rural Domestic Sewage Treatment. *Transactions of the Chinese Society of Agricultural Engineering.*

Sheng-Long K. and Cheng Y (2012). Treatment of Rural Domestic Sewage by Anaerobic and Combined Artificial Wetland Method. *Journal of Anhui Radio & TV University.*

Wang T.T. and Zhou W (2018). Nitrogen and Phosphorus Removal from Rural Domestic Sewage Using Modified AAO Integrated Process. *China Water & Wastewater.*

Weijun F (2020). Design and Operation Effect of Buried Micro Power Facilities for Rural Domestic Sewage Treatment *Shanxi Architecture*, 46 (6): 2.

Xin D.U. Shi. C.H. (2015). Impact of Organic Pollutant Loading on Effect of Artificial Tidal Flow Wetland Purifying Rural Domestic Sewage. *Journal of Ecology and Rural Environment.*

Yen and Chih-Jen (2004). Micro-Power Low-Offset Instrumentation Amplifier IC Design for Biomedical System Applications. *IEEE Transactions on Circuits & Systems.*

Zhu L. and Gao R (2010). Research on Combined Packing Treating Domestic Wastewater in Bio-Contact Oxidation Technology[C]// *International Conference on Bioinformatics & Biomedical Engineering. IEEE.*

Civil Engineering and Energy-Environment – Gao & Duan (Eds)
© 2023 the Author(s), ISBN 978-1-032-56059-5

Research progress on the use of wastewater to cultivate oil-producing microalgae

Yuqi Huang* & Wenhao Jian
School of Municipal and Environmental Engineering, Shenyang Jianzhu University, Shenyang, China

ABSTRACT: With the rapid development of industry, carbon emissions are increasing, and China has put forward the goal of "carbon peak and carbon neutrality" to achieve sustainable development. The wastewater treatment industry has responded to the country's "dual carbon" call by adopting low-carbon processes and recycling resources to achieve carbon reduction. The development and use of oil-producing microalgae in wastewater treatment are one of the most effective ways to cope with the energy crisis and improve environmental problems. This paper reviews the latest research progress on the cultivation technology of Chlorella in wastewater, the factors affecting the growth condition and oil enrichment of Chlorella during cultivation, the recovery technology, and the oil enrichment technology, taking into account the research progress at home and abroad. It is explained that the water quality of different types of wastewater is quite different, and the wastewater with a high nutrient utilization rate is more suitable for the cultivation of oil-producing microalgae. After recovery, the energy consumption control of the extraction process by conventional methods can effectively improve the comprehensive economic benefits of microbial fuel.

1 INTRODUCTION

In the context of global efforts to promote low-carbon technologies and energy efficiency, the wastewater treatment industry is also constantly looking for effective ways to cope with the energy crisis. Microalgae, with their great advantages in their own right, are fast coming into the limelight and becoming a major research target for achieving low energy consumption in wastewater treatment.

The advantages of microalgae mainly include wide ecological distribution, simple growth conditions, high environmental tolerance, a fast growth rate, and a high application value. Electron microscopy imaging of microalgae is shown in Figure 1. Microalgae grow by using CO_2 in the air, the organic carbon in the water, and nutrients such as nitrogen and phosphorus, accumulating their own biomass while reducing the nutrient content of the water. More studies have shown that it is feasible to use wastewater culture of oil-producing microalgae to realize the resourcefulness of microalgae, and the related studies are broadly divided into two directions: first, to provide microalgae production and oil yield by optimizing the effect of wastewater culture of microalgae to compensate the cost investment in wastewater treatment processes to the maximum extent; second, to improve the treatment efficiency of oil-producing microalgae on wastewater to realize economic benefits by separating and obtaining the treated microalgae while controlling the treatment cost

---

*Corresponding Author: service@sjzu.edu.cn

316   DOI: 10.1201/9781003433651-42

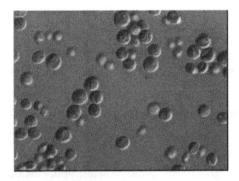

Figure 1. Electron microscopy imaging of microalgae.

The main problems faced in the application of microalgae-related technologies are high cost, low efficiency, low oil production efficiency, and weak growth capacity in actual sewage culture (Shen 2015). In this paper, we will discuss the research related to the cultivation of microalgae in wastewater and oil enrichment at home and abroad, as well as how to efficiently utilize the nitrogen and phosphorus in wastewater for the cultivation and extraction technologies of energetic microalgae, and provide suggestions for realizing the industrialization of related technology.

2 OIL-PRODUCING MICROALGAE CULTURED IN WASTEWATER

When growing in wastewater, microalgae absorb carbon, nitrogen, phosphorus, and other nutrients and convert them into oil, amino acids, and protein, storing them in stem cells (Lai 2013). The metabolic process of microalgae in wastewater is shown in Figure 2. Because the water quality of different kinds of wastewater varies greatly, not all wastewater can be cultured with oil-producing microalgae. At present, the main wastewater sources used for microalgae culture are domestic sewage, breeding wastewater, and food and fermentation industrial wastewater.

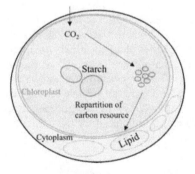

Figure 2. Metabolic process of microalgae in wastewater.

2.1 *Domestic sewage*

The pollution level of domestic sewage is low, and the concentration content of nitrogen and phosphorus in it basically meets the growing demand for microalgae. Microalgae culture using domestic wastewater requires only small amounts of added nitrogen, phosphorus, ferric ammonium citrate, and trace elements (Tao 2020). Lv et al. used municipal domestic wastewater as the medium to cultivate highly productive oleaginous algae in a foam column reactor, and the concentration of microalgae cells in the culture was 1.6 times higher than that in the BG11 medium culture, and no nitrogen and phosphorus residues were detected in the culture water, which showed a very good removal effect on nitrogen and phosphorus (Lv 2011). Yi cultivated two microalgae (C. vulgaris and C. solokiniana) on artificially prepared domestic wastewater, and the highest oil yield of both microalgae was above 150 mg/L. The removal rate of COD by C. vulgaris was 81.99%, which was about 20% higher than that of C. solokiniana, and the removal rates of TN and TP by both were above 80%. Both were above 80% (Yi 2013). Fu cultivated Chlorella using municipal domestic sewage as the culture medium and discovered that sterilization treatment and aeration during the cultivation process were beneficial to Chlorella growth and biomass accumulation, with biomass accumulation reaching 0.2413 g/L. Meanwhile, Chlorella had a better effect in removing pollutants from domestic sewage, with the removal rates of TN being 83.6%, TP being 47.5%, NH^{4+}-N and COD removal rates being 80%, and the removal rate of COD being 55% (Fu 2015). Ling et al. have successfully screened and cultured the slash-and-burn algae and four-tailed algae in the influent water of wastewater treatment plants, and the results showed that their growth rates were basically stable between 0.20–0.28 g/L while they showed better nitrogen and phosphorus removal ability during the culture process (Ling 2016).

2.2 Aquaculture wastewater

The content of COD, TN, and TP in farming wastewater is high, and if it is not treated in time and discharged directly into the environment, it will cause eutrophication of water bodies. Microalgae can absorb and transform a large number of pollutants in culture wastewater to meet discharge standards.

Related research has focused on livestock farming wastewater and aquaculture wastewater. Li et al. (2012) selected sterilized pretreated anaerobic fermentation wastewater from pig manure for culturing Chlorella, and the algal biomass of Chlorella increased by 800% throughout the culturing process, and COD and NH_3-N in the wastewater decreased by 85% and 55%, respectively. Liu et al. (2014) showed that as many as 15 species of microalgae could effectively use nutrients from pig wastewater and degrade pollutants in the wastewater.

Feng used Nile tilapia culture wastewater as a medium to culture Chlorella vulgaris and compared it with BG11 medium in parallel. The experimental results showed that the growth rate of Chlorella in this medium reached 2.73 × 105 cells/h, the oil content was 32.14%, and the removal rates of TN and TP from the wastewater medium were 98.50% and 99.35%, respectively, all of which were better than the BG11 medium (Feng 2011). I.S. Jung and other scholars in the UK found that microalgae could synthesize long-chain polyunsaturated fatty acids by using nutrients from mariculture wastewater, and although the oil yield reached 34%, there was a certain degree of decline in oil yield at a later stage due to the influence of environmental pH (Liu 2017). In addition, aquaculture wastewater from shrimp and turtles in South America can also be used for the cultivation of microalgae (Zhao 2018; Zhou 2012).

2.3 Food and fermentation industry wastewater

A large amount of organic matter remains in food fermentation wastewater, which can be used to cultivate microalgae after a simple pretreatment. Zhou et al. found that only small amounts of magnesium, nitrogen, and phosphorus in soy product wastewater can be used to cultivate Chlorella, which not only has good performance in growth and oil accumulation but also has a good treatment effect on soy product wastewater (72.9% COD removal, 64.2% TN, and 84.9% TP) (Cao 2012). Cao et al. used modified beer wastewater to cultivate

Chlorella (optimizing the nutrient salt composition in it), and the results showed that the growth rate of Chlorella was 1.69 times higher than the blank control group, and the oil yield was 0.97 times higher, which can effectively use the organic wastewater and reduce the culture cost (Miao 2007). Miao et al. investigated the co-culture of sticky red yeast and Spirulina obtusifolia based on MSG wastewater. The COD degradation rate was 70.3%, and the oil yield of Spirulina obtusifolia was 216 mg/L after five days of co-culture. The COD degradation rate and oil yield were 1.36 and 6.54 times higher than those of individual cultures, respectively (Yang 2013).

3 OIL-PRODUCING MICROALGAE RECOVERY

The effective recovery of oil-producing microalgae is the key to the industrialization of biodiesel. The advantages and disadvantages of the recycling process are compared in Table 1. Microalgae recovery is typically divided into two steps: pretreatment and enrichment separation.

Table 1. Comparison of advantages and disadvantages of the recycling process.

Process	Technology	Merit	Defect
Pre-processing	Pre-oxidation	Oxidants can promote algae aggregation	High doses of oxidants can damage algae and even kill them
	Chemical flocculation	Algae cells can complete precipitation in a short time	Narrow application range, not suitable for large-scale recycling
	Physical flocculation	Green clean, will not pollute the algae liquid	High energy consumption
Enrichment and separation	Method of precipitation	Wide application range, simple operation	Long settling time, high cost
	Air flotation method	Low energy consumption, wide application	Complex operations requiring pretreatment
	Centrifugal method	High separation efficiency and high recovery	High operating costs, narrow scope of application
	Filtration method	High efficiency for low-concentration fragile microalgae	Slow filtration, easy to block

3.1 *Pre-processing*

Pre-oxidation, chemical flocculation, physical flocculation, and other processes are commonly used to pretreat algal solutions (Ren 2020).

3.1.1 *Pre-oxidation*
Pre-oxidation is the addition of an oxidizing agent to the algal solution, resulting in its enrichment and separation. Oxidants can promote algal aggregation, but high doses of oxidants can cause damage and even death to the algae. The pre-oxidation process can be divided into three types according to the type of oxidant: ozone oxidation, pre-chlorination, and salt pre-oxidation.

319

(1) 3.1.2 Chemical flocculation
Chemical flocculation flocculates and precipitates microalgae through adsorption bridges and adsorption electro-neutralization (Ren 2020). Zhang et al. added chitosan to the supernatant when collecting the supernatant. The results showed that chitosan not only enhanced the flocculation effect but also significantly promoted the flocculation rate (Zhang 2007). Wang independently developed a new flocculant with calcium, magnesium, and silicon as powder and added it to Dianchi Lake's algae-laden water, and the algae cells could complete the precipitation in a short time through the action of the flocculant (Wang 2014). Chemical flocculants have a narrow application range and are unsuitable for large-scale recycling.

(2) 3.1.3 Physical flocculation
The two main physical flocculation methods are the electric field and ultrasound. Electric field flocculation is to use the electric field applied in the algae solution to make the cells move in the direction of the positive electrode, and when the algae cells reach the positive electrode, electric neutralization will occur, causing them to be aggregate to form larger flocs (Li 2020). Ultrasound treatment of microalgae causes them to polymerize and precipitate with each other under ultrasound.

3.2 Enrichment and separation

After pretreatment, the algal solution is ready for enrichment and separation. Currently, the commonly used enrichment and separation methods are precipitation, air flotation, centrifugation, filtration, etc.

3.2.1 Method of precipitation
Precipitation is the most commonly used separation technology for collecting microalgae. It is mainly used for algae cells with high density and easy sedimentation. The disadvantage of this method is that the precipitation time is long, the cost is high, and the separation effect of general algae cells is poor.

3.2.2 Air flotation method
The air flotation separation method mainly produces bubbles through the air flotation device, which adsorb on the algae cells, thereby reducing the total density of the flocs so that they can be collected onto the water's surface by buoyancy. Zeng et al. studied the effects of different process conditions on the rate of collecting spirulina by air flotation. Compared with the conventional separation method, although the water content of the algae after harvest is the same, the energy consumption of the air flotation method is only 40%–65% of that of the centrifugal method (Zeng 2003).

3.2.3 Centrifugal method
In 1991, Chisti et al. first proposed a centrifugal separation method for microalgae. Microalgae have a 95% recovery rate, but due to their high operating costs and limited application scope, they are only suitable for developing high-value-added products (Chisti 2013).

3.2.4 Filtration method
Filtration is the use of a suction filtration device to filter the environment and impose a negative pressure. The water is separated from the algae solution while retaining the algae biomass filter membrane. Because the filter membrane's pore size is small and microalgae cells easily block it, the filter membrane must be washed, the membrane changed, and other processes performed (Gong 2004).

4 OIL-PRODUCING MICROALGAE OIL ENRICHMENT

4.1 *Influencing factors of oil enrichment*

Environmental factors regulate the growth and fat content of microalgae during plant growth. Under certain conditions, microalgae light, CO_2 ventilation, temperature, nitrogen, and phosphorus concentration will increase the lipid content of microalgae cells. In addition, environmental factors can also change the fatty acid composition of microorganisms.

4.1.1 *Environmental factor*

(1) Light intensity is directly related to the photosynthesis efficiency of microalgae cells. Studies have found that light greatly influences the lipid content of microalgae cells and can regulate their fatty acid composition. In some filamentous green algae, when the light intensity increased from 60 m to 300 m, the oil content in the cells increased significantly; the main growth component was triglycerides.
(2) Temperature regulates the physiological metabolic rate of microalgae by affecting the activity of enzymes and membrane fluidity (growth and oil accumulation of microalgae).
(3) Effect of pH on the photosynthesis of microalgae Microalgae can survive between 6.5 and 9.0. A serious deviation from the suitable acid and alkali environment for its growth will inhibit the growth of microalgae and adversely affect the growth and lipid accumulation of microalgae (Wang 2013).
(4) In the process of aeration culture, with the increase of aeration or CO_2, the total lipid content in microalgae cells decreased; on the contrary, the fatty acid content increased. But if ventilation or CO_2 is excessive, it will reduce the fat content in the cells.

4.1.2 *Nutritional ingredient*

In addition to the influence of environmental factors, whether it is nitrogen, phosphorus, carbon, and other major elements, or iron, manganese, silicon, and other trace elements, these factors have a great influence on the growth of microalgae, oil content, and fatty acid composition (Dou 2004; Shen 2015).

In the absence of nitrogen sources, the synthesis of nitrogen-containing substances such as intracellular proteins and nucleic acids is limited, while the stored fat and most of the membrane can still be continuously synthesized, and the remaining carbon in the cells will continue to enter the secondary metabolic channel to form long-chain fatty acids, thereby increasing the lipid content of the cells. Phosphorus is an important component of DNA, RNA, ATP, and the cell membrane in microalgae. Behrenfeld et al.'s 12-year study confirmed that in high nitrogen, low chlorophyll, and oligotrophic waters, microalgae cells mainly absorb organically complexed iron for cell growth (Behrenfeld 1999). Silicon is not only involved in the formation of the cell wall in diatom cells but also plays an important role in photosynthetic pigments, proteins, and cell division (Yuan 2011). Manganese, which also has a direct role in photosynthesis, is an important material for the synthesis of chlorophyll. Silicon can promote the activity of acetyl-CoA in cells, thereby promoting fat synthesis. Microamounts in nutrient solutions also affect the composition and content of microalgae oil. For example, magnesium is a cofactor of many enzymes and a major component of chloroplasts; although vitamins are not essential, they are coenzymes that enhance the activity of related enzymes and promote the growth of microalgae (Chen 2015).

4.2 *Oil recovery technology*

Microalgae oil extraction techniques include organic solvent extraction, supercritical fluid extraction, thermal cracking, etc. (Zhu 2018). These methods require that algae liquid be concentrated and dried before extraction. Subcritical water extraction, direct ester, in situ extraction, and the promotion of microalgal oil discharge outside the cell are a class of new technologies to avoid concentration. As the core technology, the drying process is currently in the experimental stage and has not yet achieved industrialization.

4.2.1 *Solvent extraction*

4.2.1.1 *Extraction method of double solvent system*
Dual solvent system extraction technology is a single-phase system with polar and nonpolar solvents as raw materials to extract algae oil. The methanol-chloroform system proposed by Bligh and Dyer et al. in 1959 is still widely used to extract microalgae oil (Tian 2016). According to the theory of "similar miscible," the collected algae powder was dissolved in a methanol-chloroform solvent, and the polar solvent methanol was combined with the major polar lipid of algae cells, thus destroying the hydrogen bond and electrostatic force between lipid and protein. Hydrophobic neutral lipid was dissolved by the nonpolar solvent chloroform and fully extracted, and water was injected into the system. Methanol and water were miscible and stratified with chloroform containing oil, and then chloroform was volatilized by heating to obtain crude lipids (Zhang 2012).

To improve oil collection rates, appropriate solvent systems and adding sequences must be chosen (Guo 2016). In the extraction, the order of adding solvent is chloroform, methanol, and water. The reverse addition of solvent will lead to a decrease in the extraction rate because the water in the system will first form a protective film on the oil film so that it cannot be directly contacted with chloroform and is, therefore, difficult to dissolve in nonpolar solvents such as chloroform. The principles of selecting a dual solvent system are: polar solvents can effectively reduce the adhesion of the cell membrane so that the cell membrane is loose and porous; nonpolar solvents should be as similar as possible to the lipids in algae cells; High temperature, high pressure, and mechanical methods for increasing extraction rate (Zhang 2011).

4.2.1.2 *Accelerated solvent extraction*
In 1996, Richter et al. proposed a method of solvent extraction of solid or semi-solid samples at high temperature (50–200°C) and high pressure (10.3–20.6 mPa), namely the rapid solvent extraction method. High temperature and high pressure increase the mass transfer rate, allowing the solvent to penetrate rapidly into algal cells. At the same time, they reduce the dielectric constant of the medium, making it closer to the polarity of the oil and thereby improving the extraction efficiency.

4.2.1.3 *In situ transesterification method*
In situ esterification refers to the conversion of lyophilized algae powder to alcohols to form fatty acid methyl esters under acidic catalyst conditions. The preparation of biodiesel by in situ esterification avoids the traditional fatty acid extraction process and simplifies the biodiesel production process.

4.2.2 *Supercritical fluid extraction*
A supercritical fluid is a gas-liquid two-phase fluid produced when the temperature and pressure in the environment exceed their critical values (D.L 1985). Supercritical fluid has strong diffusivity and solubility, and its extraction effect on algae oil is better than that of common organic solvents. CO_2 is widely used because of its mild critical conditions (7.4 MPa at 31.1°C), non-toxicity, and chemical inertness. Microalgae oil was extracted by supercritical CO_2. During the recovery process, CO_2 was reduced to gas by adjusting the temperature and pressure, and the oil could be separated (Zheng 2014).

4.3 *Subcritical water or ethanol extraction*

Subcritical water extraction (SWE) means that when the temperature of the water is slightly lower than the critical value, it has similar properties to organic solvents due to the decrease in polarity, which greatly improves its solubility in oil. At the same time, the water is kept in a liquid state by high pressure. At high temperatures, water can quickly penetrate into the algae cells and extract the lipid from the cells; at room temperature, water polarity increases, allowing the oil dissolved in water to be quickly separated, which is conducive to the collection.

5 CONCLUSION AND FORESIGHT

(1) The quality of different kinds of wastewater varies greatly, so not all of them are suitable for culturing oil-producing microalgae. At present, the main types of wastewater used for microalgae cultivation are domestic wastewater, aquaculture wastewater, and food and fermentation industry wastewater. Although there are many reports on the cultivation of microalgae from wastewater, a more in-depth research is needed before it can be expected to be applied on a large scale. The main reasons for restricting its large-scale development are that the nutrient content of wastewater is poor, the nutrient content that microalgae can use is low, most of the components are not easy to degrade, the tolerance of microorganisms to sewage is poor, and it is difficult to survive in high concentrations of sewage.
(2) The recovery of oleaginous microalgae is divided into two steps: pretreatment and enrichment separation. Flocculation is the most widely used method (mainly used in laboratories), and its process is relatively simple; Centrifugal separation is commonly used in actual microalgae recovery. Due to many problems, such as energy consumption, air flotation separation technology is not applicable in industry. In addition, the variety of microalgae harvesting processes is cumbersome and difficult to automate, which also seriously affects the development of related industries.
(3) There are a lot of energy-consuming processes in the oil extraction process. Microalgae oil is extracted using two raw materials: dry algae powder and wet algae body. The total energy balance is evaluated by the FER index. When FER > 1, it indicates positive energy, and FER < 1 indicates that it is negative energy. The FER of the dry route of microalgae is 1.23, and the wet route is 1.16. Using conventional methods to control energy consumption can effectively improve the comprehensive economic benefits of microbial fuels.

In summary, the use of wastewater to cultivate oleaginous microalgae for wastewater recycling still has great potential for development. But to achieve large-scale applications, we also need to solve key technical problems. Many scholars have devoted themselves to research in this direction at this stage, accumulating basic data and experience for future large-scale applications.

To solve the above problems, we think we should start with the following aspects:

(1) The comprehensive extraction technology of microalgae oil can integrate the concentration and extraction of algae with the extraction of oil, thus avoiding the process of dehydration and drying.
(2) Microalgae culture technology optimization to investigate low-cost wastewater pretreatment methods to increase the feasibility of microalgae culture in wastewater tolerance
(3) Optimization of microalgae culture technology to investigate low-cost wastewater pretreatment methods in order to increase the feasibility of microalgae culture in wastewater tolerance.

REFERENCES

Behrenfeld M.J. (1999) Widespread Iron Limitation of Phytoplankton in the South Pacific Ocean. *Science*, 283 (5403): 840–843.
Cao H., Zhang Xinyun, Kong Weibao, *et al.* (2012) Cultivation of Chlorella Vulgaris in Brewery Wastewater to Produce Microalgae Biomass and Lipids. *Oils and Fats of China*, 37 (9): 65–69.

Chen X.Y. (2015) *Studies on Cultivation and Application of Microalgae Rich in Functional Components* Fuzhou: Fuzhou University.

Chisti C. and Yusuf P.E. (2013) Constraints to Commercialization of Algal Fuels. *Journal of Biotechnology*, 167(3): 201–214.

Dou.Q.L., Chen. J.F., Wang. Ji. *et al.* (2004) Research Progress and Prospect of Microbial Oil Recovery Technology. *Natural Gas Earth Sciences*, 15 (5).

Feng T.Y. (2011) Cultivation of oil-producing Microalgae Using Aquaculture Wastewater. *Nanjing: Nanjing Agricultural University*.

Fu Y. (2015) Optimization of Efficient Cultivation of Microalgae Using Municipal Sewage. *Jingzhou: Yangtze University*.

Gong Q.L., Cui J.Z., Pan K.H. *et al.* (2004) Application of Ultrafiltration Technology in Concentration of Unicellular Algae. *Marine Science*, 28 (1): 5–7.

Guo Y. and Wang S.X. (2016) Energy Microalgae Harvesting and Oil Extraction Technology. *Research on Urban Construction Theory*, (15): 2224–2224.

Lai Y.Q., Zhao H.Y., Zhou Z.J. *et al.* (2013) Research Progress on Resource Utilization of Wastewater Cultured Microalgae. *Water Treatment Technology*,39 (10): 6.

Li S.L. (2020) Study on Ultrafiltration Flux Attenuation Process and Influencing Factors of Poplar Hot Water Prehydrolysate. *Jinan: Qilu University of Technology*.

Li Y., Zhang Y.J., Ma. H.F. *et al.* (2012) Study on Sequencing Batch Cycle Treatment and Reuse System of Land Aquaculture Wastewater. *Jiangsu Agricultural Sciences*, 40 (9): 297–302.

Ling Y., Chen S., Shi. W.X. *et al.* (2016) Study on Cultivation of Scenedesmus sp. by Municipal Wastewater for Biodiesel Production. *Environmental Science and Technology*, 39 (1): 8.

Liu L.L., Huang. X.X., Wei L.K. *et al.* (2014) Purification of Nitrogen and Phosphorus in Pig Farm Wastewater by 15 Strains of Microalgae and their Cell Nutrition Analysis. *Journal of Environmental Sciences*, 34 (8): 1986–1994.

Liu M., Yuan J.L., He H. S. *et al.* (2017) Study on the Growth and Purification Effect of Microalgae in Shrimp Aquaculture Wastewater. *Abstracts of the 12th Zhejiang Fishery Science and Technology Forum.*

Lv S.J. (2011) Municipal Wastewater for Oil-producing Microalgae Culture. *Journal of Bioengineering*, 27 (3): 445–452.

Miao J.X., Xue F.Y., Zhang. X. *et al.* (2007) Oil Production by Mixed Culture of Red Yeast and Spirulina in Monosodium Glutamate Wastewater. *Journal of Beijing University of Chemical Technology*, Natural Science Edition (S2): 4.

Pavia D.L. (1985) Introduction to Modern Organic Chemistry Experiment Technology. *Science Press.*

Ren J.J. (2020) *Growth, Biochemical Composition and Transcriptome Analysis of Oocystis Borgei Preserved Under Different Illumination.* Guangzhou: Guangdong Ocean University.

Shen S.G., Zhang. M.Z., Fang C.L. *et al.* (2015) Screening and Culture of a Methanol-degrading Microorganism. *Anhui Agricultural Bulletin*, 21 (23): 2.

Shen. Q.H. (2015) *Advanced Wastewater Treatment and Process Control Based on Oleaginous Microalgae.* Hangzhou: Zhejiang University.

Tao. H.Y., Hu. Y., Wu J.Z. *et al.* (2020) Screening, Cultivation of Oleaginous Microalgae and its Research Progress in Wastewater Treatment. *Industrial Safety and Environmental Protection*, 46 (11): 5.

Tian X., Du W., Liu D.H. (2016) Research Status and Prospect of Upstream Process for Preparation of Biodiesel from Microalgae Oil. *Biological Industry Technology*, (2): 8.

Wang Q., Fu J.Y., Ying. Y. *et al.* (2013) Research Progress of High-oil Microalgae Cultured in Wastewater. *Modern Food Technology*, 29 (6): 6.

Wang S.B., Ma X.X., Wang Y A. *et al.* (2014) Preliminary Study on Turbidity and Algae Removal Effect of a New Flocculant in Algae-rich Water of Dianchi lake. *Fudan Journal: Natural Science Edition* (2): 6.

Yang L.B. (2013) Research Progress on Key Technologies of Cultivation and Recovery of Oleaginous Microalgae. *Environmental Pollution and Control*, 35 (9): 6.

Yi C. (2013) *Screening of Oil-producing Microalgae Using Domestic Sewage as Substrate and Optimization of Culture Conditions.* Harbin: Harbin Institute of Technology.

Yuan C. (2011) *Evaluation of Biodiesel Production by Microalgae and Optimization of its High-yield Conditions.* Shijiazhuang: Hebei Agricultural University.

Zeng W.L., Li B.H., Cai Z.L. *et al.* (2003) Continuous Air Flotation Harvesting of Microalgae Cells. *Aquatic Biology*, 27 (5): 5.

Zhang F. (2011) *Study on the Regulation of Microalgae Oil Synthesis and Membrane Dispersion in Situ Extraction.* Hangzhou: Zhejiang University.

Zhang F., Cheng L. H., Xu. X H. *et al.* (2012) Energy Microalgae Harvesting and Oil Extraction Technology. *Chemical Advances*, 24 (10): 2062–2072.

Zhang P., Wang C.L., Li Y. *et al.* (2007) Research Progress of Microbial Flocculant. *Environmental Science and Management*, 32 (10): 4.

Zhao X.X., Yang. K., Fang. T. *et al.* (2018) Growth and Nitrogen and Phosphorus Removal Characteristics of Three Microalgae in Turtle Culture Wastewater. *Water Resources Protection*, 34 (1): 6.

Zheng C.X., Lu W.Z., Li H T. *et al.* (2014) Ionic Liquids and Methods of Oil Extraction Using Ionic Liquids. CN103864692A.

Zhou L.N., Bo W., Wu S.L. *et al.* (2012) Study on the Cultivation of Chlorella Using Soybean Wastewater. *Guangdong Agricultural Sciences*, 39 (19): 3.

Zhu S.N., Liu. F., Fan J.H. *et al.* (2018) Research Status and Prospect of microalgae Bioenergy. *New Energy Progress*, 6 (6): 8.

Civil Engineering and Energy-Environment – Gao & Duan (Eds)
© 2023 the Author(s), ISBN 978-1-032-56059-5

Effective mitigation strategies for improving the outdoor thermal environment of university campuses using the ENVI-met Program

Lina Yang
Shandong Xiandai University, Jinan, China

Jiying Liu*
Shandong Jianzhu University, Jinan, China

ABSTRACT: The outdoor thermal environment in universities directly affects the daily activities of teachers and students. This study investigated thermal comfort during summer in outdoor educational open spaces. The ENVI-met program was used to simulate the outdoor area and verify the accuracy of the model. The different greening rates and building shading cases were compared in the model. The outdoor thermal comfort levels were evaluated to analyze the mitigation strategies for outdoor thermal environment improvement. The results showed that the outdoor thermal environment was closely related to the shading level. Greening can significantly improve the outdoor thermal environment. After greening increased from 25% to 45%, air temperature (T_a), the mean radiation temperature (T_{mrt}), and physiological equivalent temperature (PET) of the research area will decrease by 3.2°C, 14.4°C, and 6.9°C, respectively. For the area with increased building shading, T_a, T_{mrt}, and PET decreased by 3°C, 8.73°C, and 4.3°C, respectively. When the building height is greater than 24 m, the improvement effect on thermal comfort by adding height is not obvious.

1 INTRODUCTION

Outdoor spaces on campus provide places for recreation, socializing, and learning for a growing number of students. Researchers have found that the utilization rate of outdoor spaces on campus is strongly correlated with thermal comfort (Lai *et al.* 2014; Xu *et al.* 2018). The continuous development of universities and the lack of attention to the planning of outdoor open spaces have seriously affected the outdoor thermal environment on campus. The hot and dry summer climate and the thermal stress caused by solar radiation affect students' thermal sensations and indirectly affect their physical and mental health.

An increasing number of studies have focused on the role of urban design in improving urban climate and the methods to alleviate the thermal stress of campus in summer, which include changing the urban geometric, increasing vegetation, changing the surface material albedo, and adding water body (Deng *et al.* 2021; Li *et al.* 2019). Previous studies have found that the high albedo surface can reduce the surface temperature effectively by reducing the long-wave radiation in the sky, but it is not conducive to the improvement of the near-surface thermal environment (Taleghani & Mohammad 2018). Increasing the water body can achieve the cooling effect through evaporative cooling, but the cooling effect of a single water body is not obvious, and it is limited to the area near the water body and usually works with plants (Chatzidimitriou & Yannas 2015).

---

*Corresponding Author: jxl83@sdjzu.edu.cn

The geometric structure of a building mainly affects the outdoor thermal environment by reducing solar radiation. The high-density layout of the building can effectively block solar radiation, which creates greater comfort in the hot summer. Charalampopoulos et al. (2013) conducted field tests at six sites within a Greek university, and the study showed that a low Sky View Factor (*SVF*) spatial layout reduced summer solar radiation and improved outdoor thermal comfort. Similarly, Qaid et al. (2016) conducted experiments in Malaysia and concluded that a high aspect ratio and shading in urban planning could reduce *PET* in summer environments.

On the one hand, vegetation can improve the outdoor thermal environment by reflecting and absorbing solar radiation. On the other hand, photosynthesis and transpiration can also reduce the heat of the outdoor environment. Many scholars have studied the best strategies to relieve thermal stress in summer from the aspects of vegetation coverage, planting arrangement, and tree species (Chengping et al. 2020; Ming et al. 2016). Wang & Akbari (2016) found that a 10% increase in urban vegetation coverage could reduce T_{mrt} by up to 8.3 K. Abdi et al. (2020) found that rectangular planting patterns, the combination of evergreen and deciduous trees, perpendicular to the prevailing wind direction, are most beneficial to improve the outdoor thermal environment of the campus. Through literature research, it is found that the influence of campus vegetation on thermal comfort is more obvious in summer than in winter.

In hot summer, outdoor comfort mainly depends on solar radiation; therefore, reducing solar radiation through vegetation and building shading is an effective way to alleviate thermal stress on campus in summer. Taking a university in Jinan as an example, field measurements of outdoor spaces were completed. The different greening areas and building shading cases were then simulated in the model, and the scenarios with better outdoor thermal comfort were analyzed.

2 STUDY REGIONAL AND CLIMATIC CONDITIONS

Jinan is located in Shandong province, with a latitude of 36°37'N and a longitude of 117°14'E. The study area is located in the teaching building of Shandong Jianzhu University, including the surrounding environment, as shown in Figure 1. Jinan is a cold region, which is hot and rainy in summer as well as cold and dry in winter. According to the meteorological data in recent ten years, Jinan has the highest temperature in July and August every year, with the average daily maximum temperature between 30°C and 35°C. The solar radiation

Figure 1.   Satellite image of the case study location.

was intense during this period, which made students feel uncomfortable in the outdoor environment.

3   METHODS

The methods are summarized as follows: (1) testing the thermal environment of the teaching building and its surroundings on the spot; (2) simulating the campus environment under different cases. ENVI-met is used to simulate the mitigation strategies of different shading schemes in the outdoor thermal environment. The field test is used to evaluate the outdoor thermal environment of the teaching area. Meanwhile, the simulation model is verified by the field measurement.

3.1   *The description of the field test*

The measured time is the hottest two days of the year, that is, August 17th and August 18th. Considering the diversity of the thermal environment in public space and the building orientation, eight measurement points were arranged in the area, as shown in Figure 2, and the field test pictures of each measuring point are shown in Table 1. In this study, T_{mrt} and *PET* were used as indices for evaluating human thermal comfort. T_{mrt} is equal to the radiant heat exchange between an ideal isothermal enclosure surface and the human body. *PET*

Table 1.   Surrounding environment of eight measurement points.

Testing points	Site environment	Fisheye photos	*SVF*	Testing points	Site environment	Fisheye photos	*SVF*
Point 1			0.799	Point 2			0.526
Point 3			0.157	Point 4			0.531
Point 5			0.531	Point 6			0.728
Point 7			0.778	Point 8			0.855

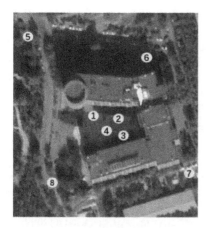
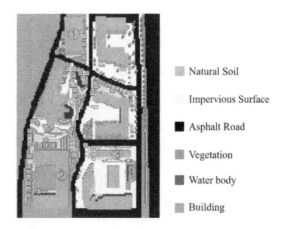

Figure 2. Location of measurement points in the teaching building.

Figure 3. Three-dimensional view of the simulated area.

refers to the standard space temperature when the human body is equal to the core temperature and the skin temperature of the standard space.

The measurement points 1, 2, 3, and 4 were arranged in the square of the teaching building, and the measurement point height was 1.5 m. Points 1 and 2 were near the teaching building on the north side, with no tall trees around and less shade all day. Point 4 was located in the center of a U-shaped square with less shade. The air temperature and relative humidity were recorded by using the iButton, the sun was shielded from solar radiation by using light louvers, and the wind speed and black ball temperature were recorded by using the JA-IAQ-50 multi-functional tester, the solar radiation was recorded by HOBO solar radiation sensor, and the thermal environment testing device was fixed by the tripod. Meanwhile, the infrared thermometer was used to record the thermal image manually every hour, and the fisheye images of sites were taken with a fisheye camera. The four azimuth angles around the teaching building were arranged with one measured point, and the measured height was 4 m. The L99-FSFX anemometer was used to record the wind speed and direction of the four measured points.

3.2 *Simulation methodology*

3.2.1 *Description of the simulation model*
ENVI-met is a microclimate simulation software based on fluid mechanics, thermodynamics, and urban ecology. It takes into account the interaction of surfaces, vegetation, buildings, and air in Urban 3D to simulate the process of numerical dynamic change of urban atmosphere. It includes airflow, turbulence, heat and water exchange between plants and the surrounding environment, particle diffusion, heat exchange, and water vapor exchange between ground and building surface.

3.2.2 *ENVI-met simulation setting*
This study focuses on the improvement of the outdoor thermal environment by shading and greening. Taking the teaching building of the university campus as an example, the orthophoto map of the research area was taken by UAV as the base map of the ENVI-met model and was transformed into BMP format and imported into the submodule SPACES. The buildings, plants, and grounds in the study area were set up according to the real situation. The computational domain for the present study covers a horizontal area of 336×486 m^2

and a vertical height of 30 m. The grid size (L × W × H) was set to be $3 \times 3 \times 3 \text{ m}^3$. In order to better cover the building on the grid, the model was rotated 14°clockwise (Figure 3). Boundary forcing meteorological conditions to use self-measured data (including air temperature, relative humidity, and solar radiation). ENVI-met cannot simulate the time-varying wind field, so the wind speed adopted the average value of test days, the wind speed was 1.5 m/s at 10 m after the correction of formula (1), and the frequency of the wind direction per minute was counted, and its most frequent direction was 225°.

$$V_{a10} = V_{a0} \left(\frac{z_{10}}{z_0} \right)^{\alpha} \tag{1}$$

where V_{a10} is the wind speed at 10 m above the ground, V_{a0} is the wind speed at 1.5 m above the test height, z_{10} means the distance from the ground is 10 m, z_0 is the height of the anemometer installed on the ground at the height of 1.5 m, and α was set as 0.36.

In order to improve the comfort of students in outdoor activities, A0 was used as the base case, and six improvement schemes were simulated by two methods. When vegetation grows luxuriantly in summer, trees use shade to reduce a lot of solar radiation, which in turn reduces air temperature and surface temperature. Therefore, greening is added to block solar radiation, so as to improve students' outdoor thermal comfort. The greening rate is adjusted to 25%, 35%, and 45%, which are operating conditions A1, A2, and A3, respectively, as shown in Table 2.

Table 2. Optimized cases for vegetation greening.

Conditions	A1	A2	A3
Greening rate	25%	35%	45%

SVF is one of the important indexes to evaluate the space form of buildings, which is closely related to the height-width ratio of urban buildings. Therefore, outdoor thermal comfort could be improved by increasing the number of buildings and changing the height of buildings. The study area is relatively open and consists of mostly low-rise teaching buildings. Considering the actual situation of the study area, additional buildings should be added to open spaces and low-rise temporary buildings to reduce SVF if the building density and floor area ratio is controlled within a reasonable range. As shown in Figure 3, buildings 1, 2, 3, and 4 were new. According to the principle of height-width ratio and a reasonable range of floor area ratio proposed by Oke (1988), three building height models were proposed, and the optimized cases are shown in Table 3. The width difference between the new building and the original building is 30 m. The new building broke through the height limit of 4–6 floors of the original building and was set at 24 m, 36 m, and 48 m.

Table 3. Optimized cases for building layouts.

Conditions	A4	A5	A6
Building height	24 m	36 m	48 m
Aspect ratio	0.8	1	1.2
SVF	0.53	0.457	0.414

3.2.3 Validation of the simulation

The accuracy of the ENVI-met model was verified by selecting measurement points 2 and 4 as representative measurement points. Figure 4 shows the difference in the meteorological parameters between the measured and simulated results in different observation points during the 24-hour continuous observation. It is found that the air temperature and relative humidity measured at the two points are similar to those simulated, and the trend of change is the same.

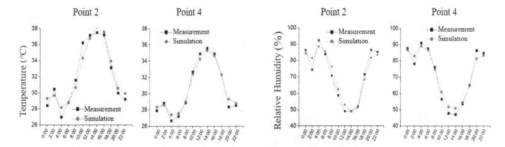

Figure 4. Comparison of measurement results and ENVI-met simulation outputs at points 2 and 4.

Root-mean-square error (RMSE) is often used to evaluate the deviation between ENVI-met simulated and measured values. The RMSE values of air temperature at measured points 2 and 4 were all less than 1 (0.64 and 0.96, respectively), and the RMSE values of relative humidity were higher than those of air temperature (3.73 and 3.34, respectively). The ENVI-met model was valid in this study.

4 RESULTS

4.1 Air temperature (T_a)

Figure 5 shows the impact of greening modifications on reducing air temperature at the study site. The temperature ranged from 34.4 to 37.6°C, which indicated that the occupant stayed in an uncomfortable condition. It is observed that the air temperature in the greening area is obviously lower than that in the non-greening area. As there is more greening on the south side of the teaching building in the actual case, the air temperature in the greening area is 25% lower than that in the non-greening area. The outdoor thermal environment on the south side of the teaching building has hardly been improved, but the activity area on the north side of the teaching building and the square has been significantly improved. When the greening rate reaches 45%, the overall temperature of the teaching area is significantly reduced, and the maximum temperature difference is 3.2°C. Because of the shading of the building, the temperature around the building decreases obviously, and the maximum temperature decreases by 3°C. As the building rises, the temperature drops significantly near the building, but not far from it. On the whole, the cooling effect of trees is better than that of buildings.

4.2 The mean radiation temperature (T_{mrt})

In order to quantitatively compare the influence of the shading type of each measurement point on the outdoor thermal environment, the T_{mrt} was compared and analyzed at 8:00 at

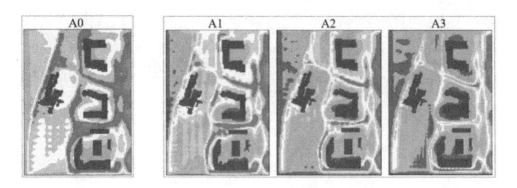

(a) Base Case (b) Vegetation Shading

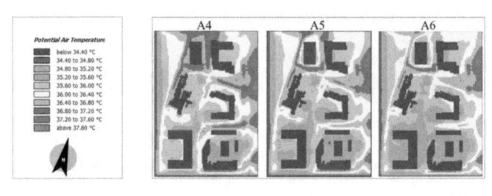

(c) Building Shading

Figure 5. Comparison of Simulated T_a in the seven different working at a height of 1.5 m at 15:00. (a) Base Case (b) Vegetation Shading (c) Building Shading.

the beginning of the working day, 12:00 for the maximum time of direct sunlight, and 15:00 for the highest temperature. Figure 6 shows the T_{mrt} value at the pedestrian height of 1.5 m under the base case and different shading cases. There are obvious differences between different cases during the day. At 8:00, a small difference was shown, which was between 0.98°C and 2°C. For tree shading, T_{mrt} gradually decreased with the increase in the greening rate, and there is a big difference at noon. For instance, in the case of A3, T_{mrt} decreased by 14.4°C because of the most greening. For the case of building shading, when a 24 m building was added, the T_{mrt} decreased obviously, and it decreased by 8.7°C at 12:00. With the building increasing, the T_{mrt} still decreased, but it was not obvious. Therefore, increasing greening can effectively reduce the average radiation temperature, thus improving the outdoor thermal environment. Increasing the number of buildings and reducing the SVF value results in a larger shadow area, thus reducing T_{mrt}. However, with the increase in building height, the effect of reducing T_{mrt} was not obvious.

4.3 Physiological Equivalent Temperature (PET)

Figure 7 shows the PET values at the 1.5 m pedestrian height under the current operating conditions and different shading conditions. At the same time of 8:00, there is a little difference between different cases, which was 0.1–1.3°C. There was a significant difference between 12:00 and 15:00, and the difference in PET between 15:00 was the most obvious,

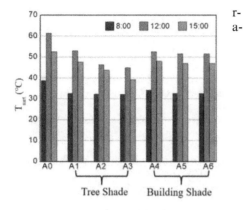
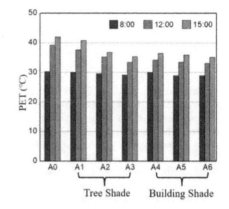

Figure 6. T_{mrt} Values of the Base Case Compared with Different Cases at 8:00, 12:00, and 15:00.

Figure 7. PET Values of the Base Case Compared with Different Cases at 8:00, 12:00, and 15:00.

nging from 1.2°C to 6.9°C. Compared with A1, A2, and A3 cases, PET decreased gradually with the increase in greening rate, and A3 showed the lowest PET value of 35.3°C. Compared with A4, A5, and A6 cases, the PET values at the three moments were not much different. Compared with the A0 case, it is reduced by 5.52–6.69°C. It is stated that adding buildings can reduce the PET value, but increasing the height of the building has no obvious effect on the improvement of the outdoor thermal environment.

5 CONCLUSIONS

In this study, outdoor thermal comfort was investigated during summer in a teaching area, and different optimization strategies were simulated by ENVI-met. The main findings are summarized as follows.

The outdoor thermal environment was closely related to the shading level. The overall thermal environment in the increased greening area has been improved, and the improvement effect of the green area was greater than that of the non-green area. With the increase in the greening rate, the improvement effect was more obvious. After greening increased from 25% to 45%, T_a, T_{mrt}, and PET decreased by 3.2°C, 14.4°C, and 6.9°C, respectively. Increasing building density in open areas can improve the outdoor thermal environment. T_a, $T_{mrt,}$ and PET decreased by 3°C, 8.73°C, and 4.3°C, respectively. The results found that increasing the height of buildings would have a slight effect on improving the outdoor thermal environment of the campus. Therefore, for the sake of the outdoor thermal environment, it is not recommended to construct high buildings. In conclusion of this study, it is suggested to increase tree density to reduce heat stress and obtain students' thermal comfort in the teaching area. Moreover, increasing greening produces more comfort than building shading.

ACKNOWLEDGMENTS

This work was financially supported by the Natural Science Foundation of Shandong Province (ZR2021ME199).

REFERENCES

Abdi B., Hami A. and Zarehaghi D. (2020) Impact of Small-Scale Tree Planting Patterns on Outdoor Cooling and Thermal Comfort. *Sustainable Cities and Society* 56, 102085.

Charalampopoulos I., Tsiros I., Chronopoulou-Sereli A. and Matzarakis A. (2013) Analysis of Thermal Bioclimate in Various Urban Configurations in Athens, Greece. *Urban Ecosystems* 16 (2), 217–233.

Chatzidimitriou A. and Yannas S. (2015) Microclimate Development in Open Urban Spaces: The Influence of Form and Materials. *Energy and Buildings* 108, 156–174.

Chengping X., Chunhui Y., Jing R. and Jiying L. (2020) Numerical Analysis of the Effect of Trees on the Outdoor Thermal Environment and the Building Energy Consumption in a Residential Neighborhood. *IOP Conference Series: Earth and Environmental Science* 546 (3), 032007.

Deng J.Y., Wong N.H. and Zheng X. (2021) Effects of Street Geometries on Building Cooling Demand in Nanjing, China. *Renewable and Sustainable Energy Reviews* 142, 110862.

Lai D., Guo D., Hou Y., Lin C. and Chen Q. (2014) Studies of Outdoor Thermal Comfort in Northern China. *Build. Environ.* 77 (3), 110–118.

Li J., Liu J., Srebric J., Hu Y., Liu M., Su L. and Wang S. (2019) The Effect of Tree-Planting Patterns on the Microclimate Within a Courtyard. *Sustainability* 11, 1665.

Ming J., Jiying L. and Linhua Z. (2016) *Numerical Evaluation of the Impact of Green Wall on the Outdoor Thermal Environment*, pp. 1023–1029, Atlantis Press.

Oke T.R. (1988) Street Design and Urban Canopy Layer Climate. *Energy Buildings* 11 (1–3), 103–113.

Qaid A., Bin Lamit H., Ossen D.R. and Shahminan R.N.R. (2016) *Urban Heat Island and Thermal Comfort Conditions at Micro-climate Scale in a Tropical Planned City. Energy Buildings* 133 (12), 577–595.

Taleghani and Mohammad (2018) The Impact of Increasing Urban Surface Albedo on Outdoor Summer Thermal Comfort Within a University Campus. *Urban Climate* 24 (24), 175–184.

Wang Y. and Akbari H. (2016) The Effects of Street Tree Planting on Urban Heat Island Mitigation in Montreal. *Sustainable Cities and Society* 27, 122–128.

Xu M., Hong B., Mi J. and Yan S. (2018) Outdoor Thermal Comfort in an Urban Park During Winter in Cold Regions of China. *Sustainable Cities and Society* 43 (8), 208–220.

Civil Engineering and Energy-Environment – Gao & Duan (Eds)
© 2023 the Author(s), ISBN 978-1-032-56059-5

Problems and countermeasures found in the safety supervision of water projects based on river basin comprehensive management

Qian Fu*
River and Lake Protection, Construction and Operation Safety Center, Haihe River Water Conservancy Commission, MWR, Tianjin, China

Meng Ting Huang
CHN Energy Dadu Rive Repair & Installation Co., Ltd, Sichuan, China

Hui Tan
Center of Construction Management & Quality & Safety Supervision, Ministry of Water Resources, P.R.C. Beijing, China

Le Kang
China Water Conservancy Engineering Association, Beijing, China

ABSTRACT: From 2016 to 2021, production safety accidents in the field of water project construction accounted for 82.5% of the industry accidents on average, which is the key to the safety production supervision of the water project industry and the difficulty in the supervision of river basin agencies. This study analyzes the erroneous data found in the safety supervision of construction in progress carried out by the river basin agencies to discover the focus of supervision, improves the regulatory standards, make recommendations to improve the supervision information level, and provides a reference for solving the supervision problems, further implementing the application of results and improving the effectiveness of comprehensive management of safety products in the river basin. In addition, it fills the research gap at the academic level.

1 INTRODUCTION

As an important form of safety supervision in the water project industry, safety supervision and special inspections (hereinafter referred to as safety inspection) of the Ministry of Water Resources will mainly focus on the field of engineering construction. The Ministry of Water Resources is responsible for organizing, coordinating, and guiding the national work. River basin management agencies are dispatched by the Ministry of Water Resources in accordance with laws and administrative regulations, and the Ministry of Water Resources is authorized to carry out safety inspections. Based on the guide of supervision and inspection problem list and major accidents hidden danger list in safety production of the water project, river basin agencies should supervise and inspect water project construction projects within the jurisdiction, which includes the project legal person, survey and design units, supervision and construction units, to fulfill the production safety regulations and standards, and then issue "a province with a list" to the project found problems (violations) and urge them to rectify. It can be seen that the basin agencies play an important role in this work.

---

*Corresponding Author: fq400@sohu.com

DOI: 10.1201/9781003433651-44

Weiping Liu stressed in the meeting that water supervision departments should understand the situation and tasks faced in the current supervision of water projects, and clarify what to supervise, study how to supervise, then implement the use of results. The water supervision departments should use supervision work to improve the overall national water safety capacity and play the role of river basin management agencies' supervision, with high-quality supervision to protect the new stage of high-quality water development (Liu 2022). In the process of water project construction, strengthening the management mechanism reform is only one way to achieve the optimization of the quality of China's water project construction. On the other hand, safety supervision of the relevant regulations and systems needs to be updated and optimized, including industry standards and related systems that must be scientifically improved (Dong 2021). The author conducts a network search with the keywords of research on the problems and countermeasures found in the safety supervision of water projects based on river basin comprehensive management. There is no relevant information, indicating that the research is in a blank state.

The difficulty of high-quality safety supervision of construction in progress is that the 2021 version of the problem list has four types of units, 26 categories, and 257 specific problems. The difficulty of high-quality safety supervision of construction in progress is that there are four types of units, 26 categories, and 257 specific problems in the problem list of the 2021 version. In the case of limited investment in supervision resources, finding the focus of supervision and precisely treating it is the urgent need for high-quality supervision work. This study analyzes the data on the problems found in the safety supervision of water projects carried out by the basin agencies, then identifies the commonalities of the problems to discover the key points and weaknesses of supervision, and gives suggestions in terms of improving the regulations and standards. It improves the level of supervision information to solve the supervision problems, further implements the application of the results, and provides a reference for improving the effectiveness of comprehensive management of basin safety production.

2 STUDY SAMPLE SELECTION

In 2019–2021, according to the Ministry of Water Resources notice of special inspection on the water project construction, a basin institution carried out safety inspections on 32 water projects under construction, such as reservoir construction, reinforcement of diseased reservoirs, river management, embankments, water supply (water replacement), pumping stations projects (electric irrigation stations upgrading). Among them, most of the projects are of large scale, high safety risk, difficult construction, strict technical requirements, and many cross operations, while involving tunnel works, high side slopes, deep excavation engineering, underground excavation engineering, high big template engineering, and other hazardous projects. This study takes a sample of 733 problems found during typical inspections.

3 DATA PROCESSING AND ANALYSIS TO FIND PROBLEMS

3.1 Classification analysis of the problem found in the safety main responsibility

Water project inspection of the key responsibility is the project legal person, survey and design units, supervision units, and construction units. The problems found by the units were classified through statistical analysis, as shown in Table 1 and Figure 1. The construction unit found the most problems, accounting for 41.12%, indicating that the construction unit is the most prominent problem of the main responsibility.

Table 1. Classification analysis table of main responsibility units' categories of safety problems found during inspections.

No.	Safety Problems Units	Number	Percentage	Cumulative Percentage
1	Construction Units	302	41.12%	41.12%
2	Project Legal Person	222	30.29%	71.41%
3	Supervision Units	121	16.51%	87.92%
4	Survey and Design units	88	12.08%	100%
Total		733		

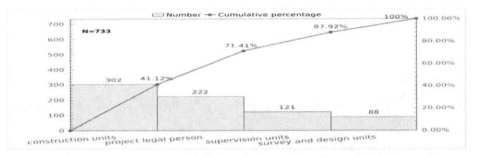

Figure 1. Statistical analysis chart of the main responsibility units' categories of safety problems found during inspections.

3.2 *Deviation classification analysis of the problem found in the safety main responsibility*

The percentage of problems found in the safety main responsibility units and the percentage of the total 257 problems in the list of the main responsibility units were classified and analyzed for deviation, as shown in Table 2 and Figure 2. The results show that the project's legal person is the most prominent problem, and is the focus of safety supervision and

Table 2. Classification statistics table of the percentage of problems found in safety main responsibility and the percentage of problems in the list.

Projects Units	Problems in List	Percentage of Listed Problems	Percentage of Problems Found by Inspection	Deviation of the Percentage of Found Problems and the Percentage of Listed Problems
	①	② = Number of units list questions/257	③ (Table 1)	④ = (③-②)/②
Project legal person	57	22.18%	30.29%	36.57%
Survey and design units	23	8.95%	12.08%	34.98%
Supervision units	41	15.95%	16.51%	3.49%
Construction units	136	52.92%	41.12%	−22.30%
Total	257	100.00%	100.00%	

violation inspection and rectification. Its problem list accounted for 22.18%, and the problems found by the inspection accounted for 30.29%, with a deviation up to 36.57% ((30.29% − 22.18%)/22.18%). Deviation of survey and design units reaches 34.98%, ranks second, and in contrast to the proportion of 12.08% of the total number of safety issues (Table 1), they are the sub-focus of safety supervision and inspection and rectification. The deviation of the supervisory unit is 3.49%, which is basically as expected. The deviation of the construction unit is −22.30%, indicating that the situation is better than expected. The following study is the problems of the project legal person with the highest deviation.

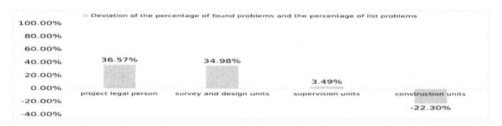

Figure 2. Deviation statistical chart between the percentage of problems found in the safety main responsibility and the percentage of problems in the list.

3.3 Classification analysis of project legal person problem

The list of supervision and inspection problems of the project legal person lists (Luan City Water Project Government 2022) eight categories, including safety management system, safety technology management, safety process control, hidden danger investigation and management, hazard source management, safety accident handling, flood control, and emergency management and others. A total of 222 problems were found. The sorted percentage from the highest to lowest is shown in Figure 3. The results show that four types of problems (safety management system, safety process control, safety technology management, flood control, and emergency management) account for 81.53%, which is the most concentrated area of the project legal person problems, and need to focus on supervision.

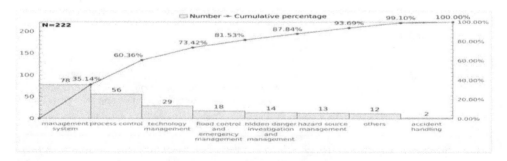

Figure 3. Statistical chart of the percentage of project legal person problem categories in water project safety inspection.

3.4 Major problems in the safety management system of the project legal person

The inspection found that the 78 problems in the safety management system of the project legal person were classified and sorted according to the proportion of problems from high to low, as shown in Figure 4. The results show that the two types of problems-the target management standardization of project safety production and the establishment of a safety production responsibility system account for nearly 70%, which are the most concentrated problems in the safety management system and require enhanced supervision.

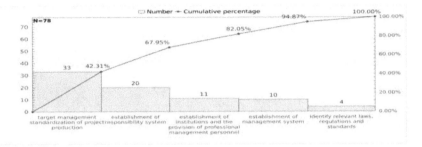

Figure 4. Statistical chart of the safety management system problems of the project legal person.

3.5 The 56 problems of safety process control of the project legal person were classified and counted in descending order of percentage, as shown in Figure 5. The results show that the three types of problems—the supervision and inspection of the implementation of system measures of participating units, the checking of safety expenses payment and usage, and the content and frequency of safety meetings—account for 73.21%, which is the most concentrated area of safety process control problems, and needs focus on supervision.

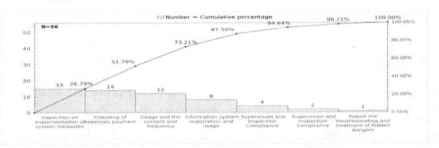

Figure 5. Statistics chart of safety process control issues of project legal person.

3.6 The 29 issues of the project legal person on safety technology management were sorted and categorized from the highest to lowest percentage, as shown in Figure 6. The results show

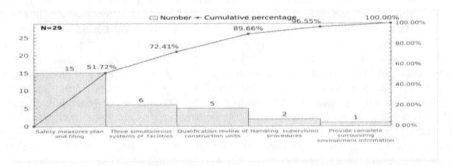

Figure 6. Statistics chart of safety technology management issues of project legal person.

that the two types of problems-safety measures plan and filing and three simultaneous systems of safety facilities account for 72.41%, which are the most concentrated problem areas in safety technology management, and need focus on supervision.

3.7 The 18 issues of flood control and emergency management are categorized and counted in descending order of percentage, as shown in Figure 7. The results show that the two types of problems—audit of the flood emergency plan when exceeding the standard and the drill of the operable target-oriented emergency plan—account for nearly 70%, which are the most concentrated problems in flood control and emergency management, and require key supervision.

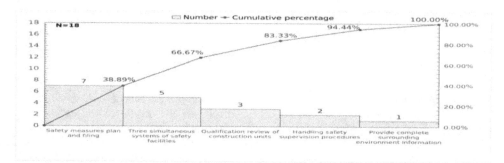

Figure 7. Statistics chart of flood control and emergency management issues of project legal person.

4 CAUSE ANALYSIS

The nine problems frequently occurring in the project legal person are analyzed above, including the target management standardization of project safety production, the establishment of the safety production responsibility system, the supervision and inspection of the implementation of system measures of participating units, the checking of safety expenses payment and usage, the content and frequency of safety meetings, and safety measures plan and filing, three simultaneous systems of safety facilities, audit of the flood emergency plan when exceeding the standard and the drill of the operable target-oriented emergency plan. It should be paid attention to during the inspection.

The common cause of the above-mentioned problems is the lack of pre-construction safety technical planning, resulting in non-compliance with the provisions of national standard 4.0.2 (Uniform Code for Technique for Constructional Safety (GB 50870 – 2013)). Before the start of a project, the relevant units should compile a construction safety technical plan based on the characteristics of the project, and determine the construction safety goals. The planning content should cover the whole process of construction and production. Safety technology planning is according to the requirements of laws and regulations, standards, and norms related to safety production in water conservancy and hydropower engineering construction and the overall safety production target of the project. It is based on the control of hazard sources and combined with the characteristics of the project to systematically identify and analyze risks, develop comprehensive risk control requirements and methods, and plan the required resources to achieve the project safety production target activities. It can be seen that the absence of safety technology planning led by the project legal person makes the management chain incomplete and lacks management foresight, systematization, and relevance, which does not meet the requirements of high-quality development of project safety management.

5 RECOMMENDATIONS

By analyzing the causes of the problems found in the safety supervision of the river basin agencies, it is recommended that the relevant water administrative departments in the policy regulations and industry standards can incorporate safety technology planning into the construction organization design and specify its working principles, objects, basis, main content, and establish the system of safety technology planning which led by the project legal person. The relevant departments should improve the engineering construction safety production system from the perspective of high-quality development.

The relevant departments can use data mining techniques (Han 2010) to identify regular problems in a larger range and longer time, draw a visual picture, carry out digital mapping, and provide data support for basin safety supervision and inspection, city and county selection, project selection, supervision of key parts, and selection of key problems to improve the accuracy and efficiency of river basin agency supervision and achieve the goal of requiring key supervision of water projects to strengthen the foundation of safe production (Yang 2020).

6 CONCLUSION

Through the analysis of the safety supervision data of river basin agencies, this study finds that the project legal person is the focus of current safety supervision from the perspective of the main responsibility. Among them, four types of issues—safety management system, safety process control, safety technology management, and flood control and emergency management during flood seasons—require key supervision, and nine issues also require key supervision. The common reason is related to the lack of safety technical planning work led by the project legal person before the start of construction. Therefore, this paper suggests that water administrative departments clarify the provisions of safety technology planning in industry policy standards, and river basin agencies use data mining technology to improve the accuracy and efficiency of supervision work, to further implement the application of achievement and reach the goal of high-quality supervision.

REFERENCES

Gaosheng Yang. (2020) Evaluation of Supervision Efficiency of Water Conservancy Projects Based on C-OWA Operator and Grey Clustering. *Journal Of Economics Of Water Resources.* 11(38), pp.31.

Hongqi Han. (2010) Research on Data Mining Model of Water Conservancy Project Management. *Yellow River.* 1, pp.73–74.

Key Points of Quality Supervision of Water Projects in Lu'an City in 2021. (2022) Luan City Water Project Government. http://slj.luan.gov.cn/ztzl/slgczlaqgl/gflc/4851184.html.

WeiHong Dong. (2021) Problem Analysis and Countermeasure Research of Safety and Quality Supervision and Management System in Water Project. China Plant Engineering. pp. 259.

Weiping Liu. (2022) The 2022 National Water Resources Work Conference. http://www.gov.cn/xinwen/2022-03/23/content_5680753.htm

Civil Engineering and Energy-Environment – Gao & Duan (Eds)
© 2023 the Author(s), ISBN 978-1-032-56059-5

Research on operation mechanism of distributed electro-hydrogen coupling system

He Wang*, Caixia Tan & Yida Du
North China Electric Power University, Beijing, China

Leiqi Zhang & Qiliang Wu
Electric Power Research Institute of State Grid Zhejiang Electric Power Co., Ltd., Zhejiang, China

Zhongfu Tan
North China Electric Power University, Beijing, China

ABSTRACT: With the gradual improvement of environmental requirements and the development of hydrogen energy, the advantages of wind and solar power generation by the electric-hydrogen coupling system are highlighted. This paper studies the operation mechanism of distributed electro-hydrogen coupling system. Firstly, the structure of distributed electric-hydrogen coupling system is designed, which includes electricity, heat, cold, gas, hydrogen, and other load requirements. Then, the energy structure of five subsystems is analyzed from five subsystems: electric energy subsystem, hydrogen energy subsystem, thermal energy subsystem, cold energy subsystem, and gas energy subsystem. Finally, combined with the energy interaction among the five energy subsystems, the coupling and energy flow of various energy sources in the distributed electro-hydrogen coupling system are analyzed, and the operation mechanism of the distributed electro-hydrogen coupling system is put forward.

1 INTRODUCTION

As a non-renewable resource, the reserves of coal are decreasing year by year. At the same time, the combustion of coal releases a lot of polluting gases, which makes the air pollution problem more serious. On September 23, 2020, China put forward that "carbon dioxide emissions should reach the peak before 2030, and strive to achieve carbon neutrality before 2060", which made the environmental protection issue attract much attention. Hydrogen energy is a clean, efficient, safe, and sustainable secondary energy recognized as the carrier of clean energy and can be obtained in various ways. Its development has become an important direction of the global energy technology revolution.

Wind power generation and photovoltaic power generation are the main utilization forms of renewable energy, and their rapid and large-scale development makes it difficult for the power grid to absorb the scenery. As a clean energy source, hydrogen has the characteristics of high energy density, large capacity, long life, convenient storage, transmission, etc., and has become one of the preferred schemes for large-scale comprehensive green development, storage, and utilization of wind power and photovoltaic. Therefore, the development of the electro-hydrogen coupling system is unstoppable, and the development of electro-hydrogen

---

*Corresponding Author: 2492618731@qq.com

DOI: 10.1201/9781003433651-45

coupling is a new energy development situation in the future. To support the development of the electro-hydrogen coupling system, it is very important to clarify the energy flow mechanism of the electro-hydrogen coupling system. Therefore, this paper studies the operation mechanism of distributed electro-hydrogen coupling system.

(Liu & Han 2022) Hydrogen energy is introduced into renewable energy generation, which solves the problem of poor predictability of energy supply caused by the intermittent nature of renewable energy, such as wind energy and solar energy, and improves the matching of the power grid. In reference (Shao & Zhang 2021), the capacity allocation solution model of hydrogen storage coupled microgrid based on particle swarm optimization (PSO) is constructed considering the operation cost and system reliability. Zhu *et al.* (2022) propose a long-term capacity optimization method for the independent microgrid operated by wind-light-hydrogen storage-supercapacitor. Yang *et al.* (2021) comprehensively consider the characteristics of photovoltaic output, transmission demand, and operation control strategy of the hydrogen system in the hydrogen production process under different hydrogen storage configurations, and proposes a method of power redistribution and capacity allocation for the grid-connected hydrogen production system of photovoltaic power stations. Li *et al.* (2022; Jin *et al.* 2021) propose a multi-microgrid operation control strategy considering the real-time energy supply and demand state and energy storage state of the system, aiming at the electric-hydrogen hybrid energy storage type multi-microgrid system. Fan *et al.*'s work (2022) aims at the characteristics of the small capacity of power terminal users and different energy consumption behaviors. It is difficult for traditional dispatching methods to cope with their dynamic changes, a coordinated dispatching method of multi-buildings in smart parks considering electricity and hydrogen complementation is proposed. Reference (Gregoratti & Matamoros 2015) is based on Lagrange decomposition combined with the subgradient method to solve the problem of building coordinated scheduling. Li *et al.* (2022) put forward a hybrid system of electricity-hydrogen co-production, including solar energy and fossil fuels, which has high efficiency and carbon neutrality. In Li *et al.*'s work, (2022), aiming at the optimal allocation of hydrogen storage capacity of the new energy wind and solar power station on the power generation side, a multi-objective optimal allocation model of hydrogen storage is constructed with the minimum investment cost of hydrogen storage, the minimum error of the system's cumulative tracking plan and the maximum increment of carbon dioxide emission reduction as the objectives, and the power abandonment rate of the station and the actual site area as the constraints.

On the basis of the above literature research, the operation mechanism of the distributed electro-hydrogen coupling system is studied in this paper. The other parts of this paper are as follows. The second part designs the structure of the distributed electro-hydrogen coupling system. In the third part, five subsystems are designed, including the electric energy subsystem, hydrogen energy subsystem, thermal energy subsystem, cold energy subsystem, and gas energy subsystem. The fourth part puts forward the operation mechanism of the coupling system from the aspects of the coupling of the system and the multi-energy flow energy flow.

2 DISTRIBUTED ELECTRO-HYDROGEN COUPLING SYSTEM STRUCTURE

The distributed electric-hydrogen coupling system takes electric energy and hydrogen energy as the leading energy sources and involves a variety of energy conversion forms, so as to meet various load requirements of electricity, heat, cold, gas, hydrogen, and the like on the user side. The typical structure diagram of the distributed electric-hydrogen coupling system is shown in Figure 1.

As can be seen from Figure 1, the distributed electro-hydrogen coupling system is mainly composed of three parts: distributed power generation equipment, electric energy conversion equipment and energy storage equipment, as follows:

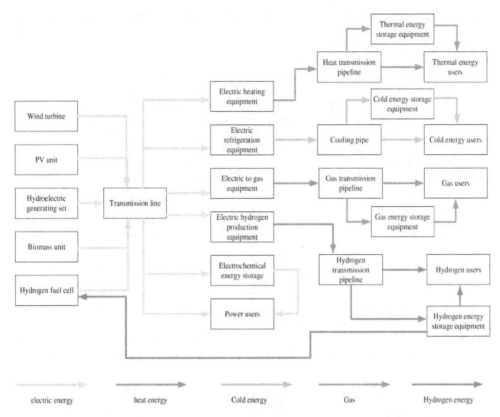

Figure 1. Typical structure of distributed electro-hydrogen coupling system.

Distributed power generation equipment includes wind turbines, photovoltaic units, hydropower units, biomass units, and hydrogen fuel cells. Wind turbines, photovoltaic units, and hydropower units respectively convert wind energy, solar energy, and water energy into electric energy. Biomass units use the biomass energy of biomass to generate electricity, and hydrogen fuel cells directly convert the chemical energy of hydrogen and oxygen into electric energy.

Electric energy conversion equipment includes electric hydrogen production equipment, electric heating equipment, electric refrigeration equipment and electric gas conversion equipment, which respectively converts electric energy into hydrogen energy, heat energy, cold energy and natural gas energy, and delivers them to users through hydrogen pipeline, heat pipeline, cold supply pipeline and gas pipeline, respectively, to meet all kinds of energy needs of users.

Energy storage devices are mainly electrochemical energy storage devices and hydrogen energy storage devices, including thermal energy storage devices, cold energy storage devices, and gas energy storage devices, which charge and discharge electricity, hydrogen, heat, cold, and gas, respectively, and play an important role in energy buffering.

3 STRUCTURAL ANALYSIS OF EACH ENERGY SUBSYSTEM

In addition, classifying the distributed electric-hydrogen coupling system equipment is according to three parts: distributed power generation equipment, electric energy conversion equipment, and energy storage equipment. According to the different energy properties, the distributed electro-hydrogen coupling system can be divided into five systems: electric energy subsystem, hydrogen energy subsystem, thermal energy subsystem, cold energy subsystem, and gas energy subsystem.

3.1 *Electric energy subsystem*

According to the typical structure of distributed electro-hydrogen coupling system, the electric energy subsystem has an energy conversion relationship with the external energy market, hydrogen energy subsystem, gas energy subsystem, cold energy subsystem and thermal energy subsystem, and the energy interaction relationship between the electric energy subsystem and other subsystems of electro-hydrogen coupling system is shown in Figure 2.

As can be seen from Figure 2, the energy interaction between the electric energy subsystem and the external and other subsystems is embodied in six aspects: first, it is connected with the external power grid, and the distributed electric-hydrogen coupling system can interact with the external power grid. On the one hand, it can supply electricity to the outside. On the other hand, it can purchase electricity from the external power grid to enhance the stability of the grid-connected renewable energy. Second, the electric energy subsystem can realize its energy coordination through electrochemical energy storage devices, and at the same time, it can meet the energy demand of external electrical equipment. Third, the electric energy subsystem realizes the conversion of electric energy to gas energy through electric-to-gas energy conversion equipment. Generally, the typical technology is P2G technology, which uses the generated surplus electric energy to electrolyze water to generate hydrogen, which is combined with carbon dioxide in the atmosphere to produce methane to provide gas (i.e., natural gas). Fourth, the electric energy subsystem can convert electric energy into hydrogen energy through electric hydrogen production equipment, and hydrogen energy can also convert hydrogen energy into electric energy through hydrogen fuel cells. Fifth, the electric energy subsystem converts electric energy into heat energy through electric heating equipment. Sixth, the electric energy subsystem realizes the conversion of electric energy to cold energy through electric refrigeration equipment.

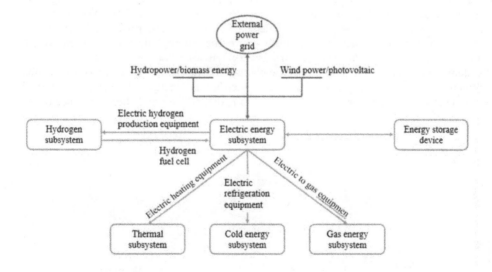

Figure 2. Unit energy supply prices for various types of energy in the park.

3.2 Hydrogen energy subsystem

For the hydrogen energy subsystem, compared with the cold, hot, and gas subsystems, the energy interaction with the electric energy subsystem is more frequent, as shown in Figure 3.

As can be seen from Figure 3, the energy interaction between the hydrogen energy subsystem and the external and other subsystems is mainly manifested in three aspects: first, it can interact with the external hydrogen network; Second, electric energy can be converted into hydrogen energy by an electric hydrogen generator, and hydrogen energy can also be converted into electric energy by a hydrogen fuel cell, so as to realize the interaction between the hydrogen energy subsystem and the electric energy subsystem; Third, hydrogen energy can balance the energy flow of the hydrogen energy subsystem through the hydrogen energy storage device to meet the external demand for hydrogen.

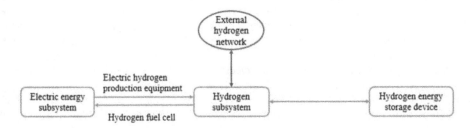

Figure 3. Energy interaction diagram of hydrogen energy subsystem.

3.3 Thermal energy subsystem

The thermal energy subsystem can realize the transformation between the electrical energy subsystem and the thermal energy subsystem. The specific relationship is shown in Figure 4.

As can be seen from Figure 4, the interaction between the thermal subsystem and the external and other subsystems is mainly reflected in three aspects: First, the thermal subsystem has a certain interactive relationship with the external heating network, which ensures the thermal load demand of various terminal users. Second, the electric energy subsystem realizes the conversion of electric energy to heat energy through electric heating equipment. Third, the energy storage device in the thermal energy subsystem is similar to that in the electrical energy subsystem, which can realize the internal energy balance of the system.

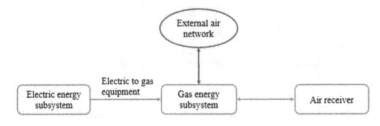

Figure 4. Energy interaction diagram of thermal energy subsystem.

3.4 Cold energy subsystem

The cold energy subsystem is similar to the heat energy subsystem, which realizes the energy conversion between the electric and cold energy subsystems. The specific interaction relationship is shown in Figure 5.

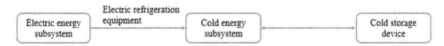

Figure 5. Energy interaction diagram of cold energy subsystem.

As can be seen from Figure 5, the cold energy subsystem and the electric energy subsystem realize the conversion of electric energy to hydrogen energy through the electric refrigeration device. At the same time, energy storage and utilization can be realized through cold storage.

3.5 Gas energy subsystem

The gas-energy subsystem can realize the conversion between the electric energy subsystem and the gas-energy subsystem. The specific relationship is shown in Figure 6.

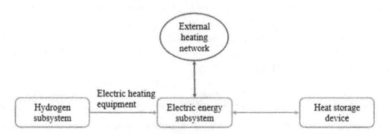

Figure 6 Energy interaction diagram of the gas-energy subsystem.

As can be seen from Figure 6, the energy interaction relationship of the gas-energy subsystem is similar to that of the heat-energy subsystem, which is mainly reflected in three aspects: First, the gas-energy subsystem has a certain interaction relationship with the external gas network, which ensures the gas supply demand of various end-users; Second, the electric energy subsystem realizes the conversion of electric energy to gas energy through electric-to-gas equipment; Third, the gas storage tank in the gas-energy subsystem is similar to the energy storage devices in other subsystems, which can be inflated and deflated, effectively maintaining the energy balance in the system.

4 SYSTEM COUPLING AND ENERGY FLOW ANALYSIS

4.1 System coupling analysis

Based on the structural analysis of distributed electro-hydrogen coupling system, the energy interaction relationship among electricity, heat, cold, gas and hydrogen is further analyzed, and the energy coupling relationship of distributed electro-hydrogen coupling system is shown in Table 1.

Table 1. Energy coupling relationship of distributed electro-hydrogen coupling system.

Dominant Energy	Energy Input Equipment	Energy Output Equipment				
		Electric Energy	Heat Energy	Lengneng	Gas Can	Hydrogen Energy
Electric Energy	Wind turbines, photovoltaic units, hydropower units, biomass units, hydrogen fuel cells, and electrochemical energy storage.	Electrochemical energy storage	Electric heating equipment	Electric refrigeration equipment	Electric gas conversion equipment	Electric hydrogen production equipment
Hydrogen Energy	Electric hydrogen production equipment, hydrogen energy storage	Hydrogen fuel cell	/	/	/	Hydrogen energy storage

As can be seen from Table 1, the distributed electro-hydrogen coupling system takes electric energy and hydrogen energy as the leading energy sources and involves various energy conversion relationships, especially the conversion between electric energy and all other energy sources. The distributed electro-hydrogen coupling system, which integrates multiple energy sources, shows strong coupling and can maximize the coordination and complementary benefits among various energies.

4.2 *Analysis of energy flow of multi-energy flow*

The distributed electric-hydrogen coupling system is dominated by electric energy and hydrogen energy and involves various energy flows such as electricity, hydrogen, cold, heat, and gas.

For the electric energy flow, the distributed electric-hydrogen coupling system firstly converts wind energy, solar energy, water energy, biomass energy, and hydrogen energy into electric energy through distributed power generation equipment such as wind turbines, photovoltaic units, hydropower units, biomass units, and hydrogen fuel cells, and generates electric energy flow. Then, the electric energy flows through electric energy conversion equipment such as electric heating equipment, electric refrigeration equipment, electric gas conversion equipment, and electric hydrogen production equipment, and is converted into heat energy, cold energy, natural gas energy, and hydrogen energy, respectively, forming heat energy flow, cold energy flow, gas flow, and hydrogen energy flow. In addition to energy conversion, the surplus electric energy flow can also be stored by electrochemical energy storage equipment and released when needed.

For the hydrogen energy flow, the distributed electric-hydrogen coupling system converts electric energy into hydrogen energy through electric hydrogen production equipment and generates hydrogen energy flow. In addition to meeting the needs of users, the hydrogen energy flow can be stored by hydrogen energy storage equipment, or converted into electric energy flow by hydrogen fuel cell, so as to complete the cycle of electricity-hydrogen energy conversion.

As for cold and hot energy flow and gas flow, they are respectively generated by electric energy flowing through electric heating equipment, electric refrigeration equipment, and electric-to-gas equipment, and are delivered to corresponding users through pipelines, and can also be stored by cold energy storage, hot energy storage, and gas energy storage equipment.

5 CONCLUSION

In this paper, the structure of distributed electro-hydrogen coupling system and its energy subsystems are analyzed from the whole to the local, and the coupling mechanism of electricity, hydrogen, cold, heat, and gas in distributed electro-hydrogen coupling system is revealed. The distributed electric-hydrogen coupling system is dominated by electric energy and hydrogen energy, involving various energy conversion relationships. The distributed electric-hydrogen coupling system, which integrates multiple energy sources, shows strong coupling and can maximize the coordination and complementary benefits among various energies.

In the future, based on the operation mechanism of this paper and from the perspective of model quantification, the operation strategy of the distributed electro-hydrogen coupling system in different time scales can be proposed.

ACKNOWLEDGMENTS

This research was funded by [State Grid Corporation of China Science and Technology Project, "Research on multi-time scale coordinated control and operation optimization technology of distributed electro-hydrogen coupling system"][B311DS221003].

REFERENCES

Fan H., Yu W., Liu L. & Dou Z. Multi-building Coordinated Dispatching Method in the Smart Park Considering Electricity and Hydrogen Complementation Under Dual-carbon Target *[J/OL]. Power System Automation*: 1–13 [October 08, 2022].

Gregoratti D. and Matamoros J. Distributed Energy Trading: The Multiple-microgrid Case. *IEEE Transactions on Industrial Electronics*,2015,62 (4): 2551–2559.

Jin S., Wang S. and Fang F. Game Theoretical Analysis on Capacity Configuration for Microgrid based on the Multi-agent System. *International Journal of Electrical Power and Energy Systems*, 2021, 125: 106485.

Li L., Han Y., Li Q., Pu Y., Sun C. and Chen W. Event-triggered Decentralized Coordinated Control Method for Economic Operation of an Islanded Electric-hydrogen Hybrid DC Microgrid. *Journal of Energy Storage*,2022,45.

Li R., Li Q., Pu Y., Li S., Sun C. and Chen W. Optimal Capacity Allocation of Multi-microgrid System With Hybrid Energy Storage Including Electricity and Hydrogen Considering Power Interaction Constraints. *Power System Protection and Control*, 2022,50(14):53–64.

Li W., Wang Y., Xu L., Tang Y., Wu X. and Liu J. Thermodynamic Evaluation of Electricity and Hydrogen Cogeneration From Solar Energy and Fossil Fuels. *Energy Conversion and Management*,2022,256.

Liu P. and Han X. Comparative Analysis on Similarities and Differences of Hydrogen Energy Development in the World's Top 4 Largest Economies: A Novel Framework. *International Journal of Hydrogen Energy*,2022,47(16).

Shao Z. and Dongqiang Z. Optimal Allocation of Multi-energy Complementary Power System Capacity based on Contract Load Curve. *Power Grid Technology*, 2021,45(05):1757–1767.

Yang Y. et al. Optimal Power Reallocation of Large-scale Grid-connected Photovoltaic Power Station Integrated with Hydrogen Production. *Journal of Cleaner Production*, 2021, 298.

Zhu X., Hu X., Shi N., Yao Z. and Zhong J. Long-term Capacity Optimization of Electric Hydrogen Coupled Microgrid Considering Dynamic Efficiency of Hydrogen Storage [J/OL]. *High Voltage Technology*: 1–13 [October 08, 2022]. doi: 10.13336/J.1003-6520.HVE.2003.

Civil Engineering and Energy-Environment – Gao & Duan (Eds)
© 2023 the Author(s), ISBN 978-1-032-56059-5

Changing trend and influencing factors of ambient air quality in Gansu Province: Based on grey correlation analysis

Jia Liang
Gansu Ecological Environmental Emergency and Accident Investigation Center, Lanzhou, China

Shan Wang
School of Petrochemical Engineering, Lanzhou University of Technology, Lanzhou, China

Longlong Wang, Guohua Chang, Panliang Liu & Jinxiang Wang*
College of Urban Environment, Lanzhou City University, Lanzhou, China

ABSTRACT: In recent years, with the significant improvement in people's living standards, people have started to require a high-quality ecological environment. In this context, environmental pollution has gradually become the focus of attention. The atmospheric environment has the most extensive impact on human life. The strategic positioning of Gansu Province as a national ecological environment safety barrier plays an important role in protecting the ecological environment of northwest China. To study the changing trend and spatial distribution characteristics of ambient air quality and associated influencing factors in Gansu Province, the grey correlation degree is used to determine the contribution of energy type or industry type to pollution factors and air quality, and statistical analysis method is adopted to study the changing trend of ambient air quality. The real-time monitoring data of ambient air quality and additional data, including those of industrial energy, industrial development, and civil vehicle ownership, were collected from 14 cities (prefectures) in Gansu Province from 2015 to 2019. The results showed that the overall concentration of SO_2, NO_2, PM_{10}, $PM_{2.5}$, and CO atmospheric pollutants in the study area dropped steadily from 2015 to 2019, whereas that of O_3 began to decline after a peak in 2017 and generally showed an upward trend. Moreover, the concentrations of SO_2, NO_2, PM_{10}, $PM_{2.5}$, and CO were "high in winter and low in summer," whereas that of O_3 showed an opposite trend. PM_{10}, $PM_{2.5}$, and O_3 were the primary pollutants in the province. The ambient air quality was better in the eastern region than in the central and western regions of the province, and the air quality in Lanzhou, the provincial capital, was worse than that in other regions. The correlation between O_3 and NO_2, $PM_{2.5}$, and PM_{10} concentrations was the closest. Additionally, an evident correlation was observed between the ambient air quality data and coal and oil consumption, the output value of secondary industries, and the number of trucks.

1 INTRODUCTION

Gansu Province, an important old industrial base in China, has witnessed the rapid development of the power sector under the background of "Western Development." Lanzhou, its provincial capital and the largest petrochemical base in western China (Jia 2018), has long

---

*Corresponding Author: wangjx8541@163.com

been experiencing the problem of air pollution. As early as the 1970s, the photochemical smog phenomenon was reported in the Xingu District of Lanzhou City, making it one of the first areas where photochemical smog pollution was discovered in China (Wang *et al.* 2017). To prevent and control air pollution, the State Council has successively promulgated and implemented the Action Plan for Air Pollution Prevention and Control (hereinafter referred to as the "Ten Atmospheres") and the Three Year Action Plan for Winning the Battle of Blue Sky. The People's Government of Gansu Province has also issued Opinions on the Implementation of the Action Plan for Air Pollution Prevention and Control in Gansu Province and the Three Year Action Plan for Winning the Battle of Blue Sky in Gansu Province (2018–2020) to promote the prevention and control of air pollution throughout the province.

In recent years, there have been many studies on the change rules and causes of ambient air quality in Lanzhou, but there is a lack of studies on the change rules, spatial distribution characteristics and influencing factors of ambient air quality in the whole province. Therefore, this paper takes 14 cities (prefectures) in Gansu Province as the research object, based on the real-time monitoring data (SO_2, NO_2, PM_{10}, $PM_{2.5}$, CO, O_3) of ambient air quality in each city (prefecture) from 2015 to 2019, analyzes the temporal change law and spatial distribution characteristics of air quality in Gansu Province, and uses the gray correlation analysis method to analyze the impact law of industrial energy, industrial development, civil vehicle ownership, etc. on ambient air quality.

2 DATA AND METHODS

2.1 *Regional overview*

Gansu Province is located in the northwest of China, between $92°\sim109°$E and $32°–43°$N, at the intersection of the Qinghai Tibet Plateau, the Loess Plateau, and Inner Mongolia. It includes 14 cities (prefectures), namely Lanzhou, Jiayuguan, Jinchang, Baiyin, Wuwei, Tianshui, Pingliang, Zhangye, Jiuquan, Dingxi, Longnan, Qingyang, Linxia, and Gannan.

2.2 *Data source*

In this study, the annual mean concentrations of SO_2, NO_2, PM_{10}, $PM_{2.5}$, CO, and O_3 in the study area, the number of days with good air quality in 14 cities, and annual mean values of six indicator concentrations were derived from the Development Yearbook of Gansu Province and Bulletin of Ecological Environment in Gansu Province from 2015 to 2019. The monthly mean values of six indicators were derived from China's online air quality monitoring and analysis platform. The daily and hourly concentrations of the six indicators were derived from China's Environmental Monitoring Station. Notably, the data from 2016 to 2019 were obtained after removing sand and dust. Moreover, industrial energy, industrial development, and automobile ownership data of Gansu Province were obtained from the Development Yearbook of Gansu Province during 2015–2019.

2.3 *Research methods*

In this study, the primary factors affecting the ambient air quality in Gansu Province were analyzed using the grey correlation method, a quantitative method that describes the correlation degree between grey systems or sequences affected by various elements. The measure of the size between two systems or factors is called the correlation degree. If the changing trend of the two elements in the system development process is consistent, the correlation degree of the two elements is considered large; otherwise, it is considered low (Zhang *et al.* 2012). The correlation analysis is based on the grey process of the grey system. It is a dynamic process that compares the time series of elements to determine the dominant factors with substantial influence (Wang & He 2018). A correlation degree of 0–0.3 indicates that the grade of correlation is low, that is, the coupling between the two factors is weak; values of 0.3–0.6 and 0.6–0.8 indicate medium and high correlation, implying that the coupling between the two factors is moderate and strong, respectively. Values ranging between 0.8 and 1 indicate a very high correlation, that is, the relative changes between the two factors are almost consistent, indicating that the coupling effect is the strongest (Luo et al. 2003).

3 RESULTS AND DISCUSSION

3.1 *Changing the law of ambient air quality*

3.1.1 *Annual variation law of ambient air quality*

Figure 1 shows the changing trend of the concentrations of the six assessment indicators of ambient air quality in Gansu Province from 2015 to 2019. The concentrations of SO_2, NO_2, CO, $PM_{2.5}$, and PM_{10} all show a downward trend, indicating that the measures implemented under the action plans such as "Ten Rules for the Atmosphere" and "Blue Sky Defense" achieved good results and that the pollutant emissions decreased considerably. Contrary to those of the other five indicators, the concentration of O_3 began to decline after its peak in 2017. However, it generally showed an upward trend, which could be related to the formation mechanism of O_3. As a secondary pollutant generated from the photochemical reaction, O_3 is affected by various factors such as natural light, temperature, volatile organic compounds (VOCs), and nitrogen oxides (NOx). The "Ten Atmospheres" plan and other key controls were focused on the emission of primary pollutants, and the control measures for the VOCs of O_3 precursors were relatively few (Jia et al. 2016).

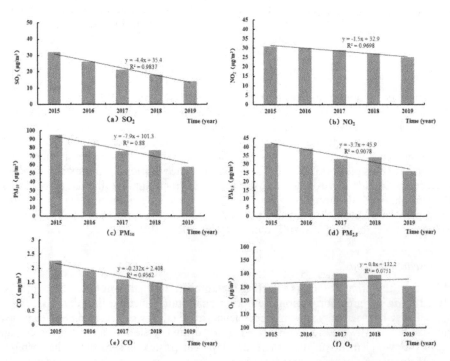

Figure 1. Annual trends of six pollutants in Gansu Province: (a) SO_2, (b) NO_2, (c) PM_{10}, (d) $PM_{2.5}$, (e) CO, and (f) O_3.

3.1.2 Analysis of seasonal variation in ambient air quality

Figure 2 shows the average concentrations of six indicators in 14 cities from January to December 2015. The highest concentrations of SO_2, NO_2, PM_{10}, $PM_{2.5}$, and CO appeared in January, February, and December, with the lowest concentrations in June, July, and August, showing a "U" trend, that is, "high in winter and low in summer." These low concentrations could be attributed to the abundant rainfall in summer, which effectively washes and dissolves pollutants in the atmosphere. Simultaneously, as a result of monsoon, the airflow capacity becomes strong, and the air pollutants spread easily, thereby effectively reducing the concentration of pollutants. In contrast, in winter, coal-fired heating generates a large number of pollutants, including particulate matter, NO_x, and SO_2, whereas some areas without centralized heating are directly discharged with coal-fired flue gas, resulting in a high concentration of atmospheric pollutants being emitted into the air. In addition, owing to low precipitation, temperature inversion, and other limitations in winter, atmospheric pollutants are not easily spread, resulting in a high concentration of pollutants in the air (Daoo et al. 2004).

Our findings revealed that the concentration of O_3 was high in June, July, and August and low in January, February, and December, showing an "inverted U" changing trend, i.e., "high in summer and low in winter." The primary reason for the high O_3 concentration in summer is closely related to the atmospheric photochemical reaction near the earth's surface. In summer, owing to the high light intensity, increased temperature, and long sunshine duration, the photochemical reaction increases the O_3 concentration, whereas the opposite is true in winter (Jia et al. 2016).

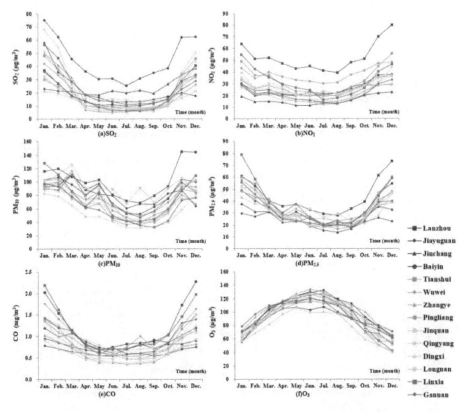

Figure 2. Monthly change in pollutant concentrations in different cities of Gansu Province: (a) SO_2, (b) NO_2, (c) PM_{10}, (d) $PM_{2.5}$, (e) CO, and (f) O_3.

3.1.3 Analysis of daily variation in ambient air quality

Lanzhou is the capital city of Gansu Province, exhibiting the most serious environmental air pollution in the province. Therefore, considering Lanzhou as an example, we selected the hourly data of 2019 as the research object in this study. The hourly average values indicated high concentrations of SO_2, NO_2, PM_{10}, $PM_{2.5}$, and CO in winter (December, January, and February) and of O_3 in summer (June, July, and August). The results are shown in Figure 3. The main conclusions are as follows:

1. SO_2 concentration rose slowly from 0:00 to 9:00, after which it began to rise rapidly, reaching the peak at 12:00 h, then declining gradually, and finally reaching the minimum at approximately 21:00. These changes could be mainly attributed to the combined effects of coal-fired heating, industrial fossil fuel use, and vehicle exhaust emissions (Bao et al. 2017).
2. CO concentration presented a "double peak" feature; the first and second peaks appeared at approximately 11:00 and 20:00 h, respectively. CO is produced under the condition of incomplete combustion of carbon-containing fuels, with vehicle exhaust contributing substantially to CO in the urban atmosphere. The primary reason for the "double peak" feature is the CO vehicular emissions in the morning and evening when people travel to and from work, respectively, and their cumulative effects (Wang et al. 2012).
3. The change rule of PM_{10} and $PM_{2.5}$ concentrations is essentially the same, showing a "double peak" feature. The first and second peaks appeared at approximately 12:00 and 21:00, respectively. Anthropogenic activities such as road cleaning and traveling through motor vehicles in the morning produce a large number of particulate matter particles.

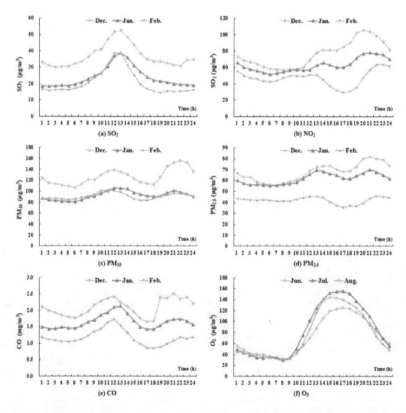

Figure 3. Daily variation of pollutants: (a) SO_2, (b) NO_2, (c) PM_{10}, (d) $PM_{2.5}$, (e) CO, and (f) O_3.

Notably, meteorological conditions such as temperature inversion are not conducive to the diffusion of pollutants. Particles discharged into the atmosphere accumulate, leading to a maximum concentration around noon. Post noon, the inversion layer was destroyed, the particles gradually diffused, and the concentration began to decline, reaching the lowest value around 17:00–18:00. The motor vehicles led to an increase in particulate matter emissions since work began at 00:00. After sunset, the decline in solar radiation led to a temperature inversion, resulting in a peak cumulative concentration of particulate matter at night (Viana et al. 2006).

4. NO_2 concentration began to rise at 8:00, slightly decreased after a small peak at approximately 12:00, rose again at approximately 15:00, and reached the highest value at approximately 19:00. The concentration of NO_2 is greatly affected by the exhaust emissions of motor vehicles. Since 8:00, the peak work hours begin such that the emission of motor vehicles increases. After 12:00, the solar radiation increases, and the NO_2 accumulated in the air is gradually consumed in the photochemical reaction. Moreover, in the afternoon, the NO_2 emission increases again during the off-duty period (Ryerson et al. 2003).

5. The concentration of O_3 was the lowest from 0:00 to 8:00 and began to rise after 8:00, reached the peak at 14:00, and began to decline rapidly after 17:00. The highest concentration of O_3 was observed at 14:00–17:00, which is the period with the strongest solar radiation. After 17:00, because the solar radiation gradually decreases, the concentration of O_3 decreases rapidly.

3.2 Analysis of pollution characteristics

3.2.1 Comparison and analysis of air quality at and better than Class 2

Statistical analysis was conducted on the air quality at and better than Class 2 in 14 cities in Gansu Province. Five cities in Hedong (Lanzhou, Tianshui, Baiyin, Qingyang, and Linxia) and two cities in Hexi (Jinchang and Jiayuguan) were selected to plot the changing trend (Figure 4). From 2015 to 2019, the number of days with air quality at and better than Class 2 in 14 cities in the province, except Lanzhou, remained between 270 and 350 days, and the number of days with air quality at and better than Class 2 in Lanzhou in each year were below 300. Specifically, the number of days with air quality at and better than Class 2 in all cities except Jinchang was the lowest in 2018, and that in most cities was below 300. In 2019, the number of days with air quality at and better than Class 2 in all cities across the province was relatively high, with 296 days in Lanzhou and more than 330 days in other cities. The number of days with air quality at and better than Class 2 in Lanzhou dropped to 213 in 2018 and then rose rapidly in 2019.

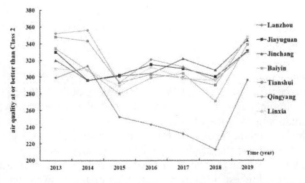

Figure 4. Trend of good days in some cities of Gansu province from 2015 to 2019.

Lanzhou is the capital city of Gansu Province, and its air quality is worse than that of other cities. This may be attributed to the reason that Lanzhou is an industrial city dominated by petroleum and chemical industries in northwest China, and the industrial waste gas emissions from which aggravate air pollution. Additionally, with the increase in population, the number of cars has been increasing, and vehicle exhaust has a great negative impact on air quality (Liu *et al.* 2014).

3.2.2 *Comparative analysis of primary pollutants*

The primary pollutant refers to the air pollutant with the largest individual air quality index (IAQI) when the air quality index (AQI) is greater than 50. Figure 5 shows the distribution of the days concerning the primary pollutant in light pollution weather in 14 cities (states) from 2017 to 2019. The top three pollutants in light pollution weather in Gansu Province were PM_{10}, $PM_{2.5}$, and O_3. Lanzhou City had the maximum number of days it experienced light pollution, followed by Jiuquan, Wuwei, Baiyin, and Jiayuguan. NO_2 was the primary pollutant only in Lanzhou, Longnan, and Gannan, with Lanzhou exhibiting the maximum number of days (44 days) against the pollutant and Longnan and Gannan exhibiting only one day. SO_2 was the primary pollutant only in Baiyin and Pingliang, with ten days and one day, respectively. Specifically, the primary pollutants were PM_{10}, O_3, $PM_{2.5}$, and NO_2 in Lanzhou; PM_{10}, $PM_{2.5}$, SO_2, and O_3 in Baiyin; PM_{10} and O_3 in Jiayuguan and Jinchang; PM_{10}, $PM_{2.5}$, and O_3 in Wuwei; Qingyang, Dingxi, Pingliang, and Gannan; $PM_{2.5}$, PM_{10}, and O_3 in Tianshui, Linxia, and Longnan, and PM_{10}, O_3, and $PM_{2.5}$ in Jiuquan and Zhangye. The number of days of PM_{10}, $PM_{2.5}$, and O_3, the major pollutants in the province, decreased rapidly in 2019 compared to that in 2017 and 2018, indicating that particulate matter is the primary pollutant that affects the ambient air quality in Gansu Province.

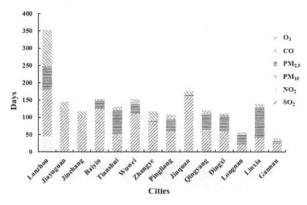

Figure 5. Distribution of total primary pollutants in light pollution weather in cities (states) from 2017 to 2019.

3.2.3 *Spatial comparison analysis*

We selected the concentration data of 2019 of the primary pollutants PM_{10}, $PM_{2.5}$, and O_3, which affected the ambient air quality of each city (state), and used the ArcMap module in ArcGIS10.7 for mapping (Figure 6). Figure 6 shows that the particle concentration in Lanzhou is the highest in the province, followed by that in the western region, and the southeast region has the lowest concentration. The concentration of O_3 is the highest in the middle (Lanzhou City), followed by that in the west, and the lowest in the east. The PM_{10} concentration was the highest in Lanzhou (79 μg/m^3), followed by Jiuquan, Baiyin, Jinchang, Wuwei, Dingxi, Tianshui, Pingliang, Qingyang, Linxia, Longnan, and Gannan.

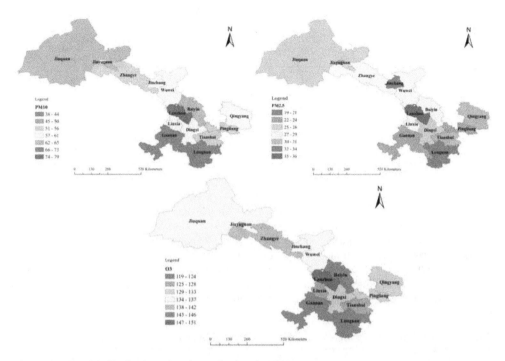

Figure 6. Spatial distribution of major pollutants in 2019.

The highest concentration of PM$_{2.5}$ was 36 μg/m^3 in Lanzhou, followed by Wuwei, Qingyang, Tianshui, Linxia, Zhangye, Baiyin, Longnan, Jinchang, and Gannan. The concentration of O$_3$ was the highest in Lanzhou. The high pollutant concentration in the western region could be attributed to the layout of heavy industrial enterprises in Gansu Province. The heavy industrial regions, mainly located in the central and western regions, discharge a relatively large number of pollutants.

3.3 *Analysis of influencing factors of pollutant concentration in ambient air*

3.3.1 *Correlation between key pollutants*

The average annual concentrations of six pollutants in Gansu Province from 2015 to 2019 were selected, with O$_3$, PM$_{2.5}$, and SO$_2$ as the parent sequence X0 = (x0 (1), x0 (2), ... , x0 (5)) and other five pollutants as the subsequence Xi = (xi (1), xi (2), ... , xi (5)), (i = 1, 2, ... , 5). The original data was initialized and transformed into a sequence that can be easily compared. The grey correlation method was used for calculation, the results of which are shown in Table 1. The degree of correlation between O$_3$ and other pollutants showed the following trend: NO$_2$ > PM$_{2.5}$ > PM$_{10}$ > CO > SO$_2$. O$_3$ had a high correlation with NO$_2$, PM$_{2.5}$, and PM$_{10}$. However, it showed the highest correlation with NO$_2$, with a correlation degree of 0.7364, indicating that the concentration of O$_3$ is most affected by the concentration of NO$_2$. The correlation degree between PM$_{2.5}$ and various pollutants showed the following trend: PM$_{10}$ > CO > NO$_2$ > SO$_2$ > O$_3$. It indicates that PM$_{2.5}$ is highly correlated with PM$_{10}$, with a correlation degree of 0.9265, indicating that PM$_{2.5}$ and PM$_{10}$ are relatively close. Additionally, PM$_{2.5}$ exhibited correlation degrees of 0.7580, 0.7366, and 0.6367 concerning CO, NO$_2$, and SO$_2$, indicating that CO, NO$_2$, and SO$_2$ also have certain effects on PM$_{2.5}$ concentration, respectively. Moreover, the degree of correlation between SO$_2$ and pollutants showed the following trend: CO > PM$_{10}$ > PM$_{2.5}$ > NO$_2$ > O$_3$.

Table 1. Correlation degree between O_3, $PM_{2.5}$, SO_2, and pollutants.

Pollutants \ Pollutants	O_3	$PM_{2.5}$	SO_2
SO_2	0.5345	0.6367	1
NO_2	0.7364	0.7332	0.6138
PM_{10}	0.6180	0.9265	0.7342
$PM_{2.5}$	0.6384	1	0.7090
CO	0.7364	0.7580	0.8337
O_3	1	0.5669	0.5345

3.3.2 Impact of industrial energy on air quality

Based on the data on ambient air quality in Gansu Province from 2015 to 2019, the comprehensive index of ambient air quality and seven indicators of SO_2, NO_2, PM_{10}, $PM_{2.5}$, CO, and O_3 were selected as the parent series X0 = (x0 (1), x0 (2), ... , x0 (5)) and the consumption of coal, oil, natural gas, primary power, and other energy as the sub-series Xi = (xi (1), xi (2), ... , xi (5)), (i = 1, 2, ... , 4) so as to conduct the initial value processing of the original data. It was converted into a sequence that can be easily compared and calculated using the grey correlation method. Table 2 shows the calculation results of the grey correlation.

Table 2. Correlation between provincial industrial energy and environmental quality indicators.

Energy \ Pollutants	Coal	Oil	Natural Gas	Primary Power and Other Energy Sources
SO_2	0.7063	0.6471	0.6188	0.5975
NO_2	0.9019	0.7618	0.6963	0.6596
PM_{10}	0.7887	0.6967	0.6524	0.6212
$PM_{2.5}$	0.8089	0.7104	0.6678	0.6390
CO	0.7299	0.6573	0.6219	0.5969
O_3	0.7403	0.9517	0.8050	0.6806

The correlation degree between SO_2, NO_2, PM_{10}, $PM_{2.5}$, CO, and industrial energy shows the following trend: coal>oil>natural gas>primary power and other energy, in which NO_2, $PM_{2.5}$, PM_{10}, and coal exhibit the largest correlation degree, indicating that coal has a great impact on the concentration of NO_2, $PM_{2.5}$, and PM_{10} is the primary source of pollution. The correlation degree between O_3 concentration and industrial energy is in the order of oil > natural gas > coal > primary power and other energy sources. The correlation degree between O_3 and oil is the largest, indicating that the use of petroleum products has a substantial impact on O_3 concentration. O_3 formation has an essential relationship with VOCs, the main source of which is oil volatilization and leakage (Carter 1994).

3.3.3 Impact of industrial development on air quality

Based on the annual average concentrations of SO_2, NO_2, PM_{10}, $PM_{2.5}$, CO, and O_3 in Gansu Province from 2015 to 2019 and the gross domestic product of all industries in the province and regions, SO_2, NO_2, PM_{10}, $PM_{2.5}$, CO, and O_3 were considered as the parent sequence X0 = (x0 (1), x0 (2), ... , x0 (5)) and the primary, secondary, and tertiary industries in the province and cities as sub-sequences Xi = (xi (1), xi (2), ... , xi (5)), (i = 1, 2, 3). The original data was initialized and transformed into a sequence that can be easily compared. The grey correlation method was adopted for calculation, and the calculation results are shown in Table 3.

The results show that the degree of correlation between each industry and six pollutants (SO_2, NO_2, PM_{10}, $PM_{2.5}$, CO, and O_3) is in the order of secondary industry > primary industry > tertiary industry, with secondary industry exhibiting the largest degree of correlation with each pollutant, indicating that it is the primary factor affecting ambient air quality, followed by the primary and tertiary industries.

Table 3. Correlation between industrial development and pollutants in the Province.

Industrial / Pollutants	Primary Industry	Secondary Industry	Tertiary Industry
SO_2	0.5790	0.6464	0.5717
NO_2	0.6172	0.7651	0.6038
PM_{10}	0.5990	0.6937	0.5882
$PM_{2.5}$	0.6101	0.7104	0.5994
CO	0.5757	0.6546	0.5673
O_3	0.6668	0.8231	0.6409

3.3.4 *Impact of the number of private-owned automobiles on air quality*

The annual average concentrations of six pollutants (SO_2, NO_2, PM_{10}, $PM_{2.5}$, CO, and O_3) and the number of private-owned automobiles in Gansu Province from 2015 to 2019 were selected for analysis. The parent sequence of SO2, NO2, PM10, PM2.5, CO, and O3 was considered as X0 = (x0 (1), x0 (2), ... , x0 (5)), and the ownership of passenger vehicles, cargo vehicles, and other vehicles was considered as the subsequences of Xi = (xi (1), xi (2), ... , xi (5)), (i = 1, 2, 3), respectively, to convert the data to a sequence that can be easily compared. The grey correlation method was adopted for calculation, and the calculation results are shown in Table 4. Table 4 shows that the influence of the number of private-owned automobiles on pollutants varied from 2015 to 2019 and that the influence of cargo and other vehicles on pollutants was greater than that of passenger vehicles. Among all studied vehicle types, other vehicles have the greatest impact on SO_2 and CO and have a greater impact on $PM_{2.5}$ and PM_{10}. Cargo vehicles have the greatest impact on O_3, followed by NO_2, $PM_{2.5}$, and PM_{10}.

Table 4. Correlation of civil vehicle ownership to pollutants.

Car Ownership / Pollutants	Passenger Vehicles	Cargo Vehicles	Other Vehicles
SO_2	0.5586	0.6340	0.8951
NO_2	0.5731	0.7018	0.7072
PM_{10}	0.5683	0.6661	0.7709
$PM_{2.5}$	0.5757	0.6770	0.7918
CO	0.5540	0.6389	0.8240
O_3	0.5988	0.7974	0.6004

4 CONCLUSION

This paper adopted statistical analysis and grey correlation degree methods to study the changing trend and influencing factors of ambient air quality in Gansu Province. The main conclusions can be summarized as follows: (1) From 2015 to 2019, the concentrations of SO_2, PM_{10}, $PM_{2.5}$, and CO in each city showed a downward trend. The concentration of NO_2 in Gannan and Baiyin increased slightly, whereas that in other cities decreased slightly. The concentration of O_3 increased in all cities except Gannan, Linxia, Zhangye, and

Tianshui. (2) From 2015 to 2019, the monthly mean concentrations of SO_2, NO_2, PM_{10}, $PM_{2.5}$, and CO showed a "U" trend; that is, the concentrations were higher in winter and spring than in summer. The monthly mean concentration of O_3 showed an inverted "U" trend, which indicated that the concentrations were higher in summer than in winter and spring. (3) The number of days with air quality at and better than Class 2 in Lanzhou was observed to be minimum in Gansu Province, with the number in Gannan being the maximum. In 2019, the number of days with air quality at and better than Class 2 was more than 300 days in 14 cities, except Lanzhou. The concentrations of SO_2, NO_2, PM_{10}, $PM_{2.5}$, CO, and O_3 reached the national secondary standard in 2019, except those of PM_{10}, $PM_{2.5}$, and NO_2 in Lanzhou. The primary pollutants affecting the ambient air quality of the province were concluded to be PM_{10}, $PM_{2.5}$, and O_3. (4) The concentrations of primary pollutants in the central and western regions of Gansu Province were high, whereas those in the eastern regions were low. The concentrations of NO_2, $PM_{2.5}$, PM_{10}, CO, and O_3 in Lanzhou, the provincial capital, were the highest. (5) Among the six indicators, O_3–NO_2 and $PM_{2.5}$–PM_{10} had the highest correlation, and the correlation between ambient air quality data and coal and oil consumption, secondary industry output value, and the number of trucks was high. In terms of future work, the longer time scale air quality data and energy data should be used to analyze the different sources of atmospheric pollutants in different regions.

ACKNOWLEDGMENTS

This work was financially supported by the Foundation of Key Laboratory for Environmental Pollution Prediction and Control, Gansu Province, China (kleppc-2019-02) and the Natural Science Foundation of Gansu Province, China (20JR10RA286).

REFERENCES

Bao M.Y., Cao F. and Liu S.D., et al. Analysis of the Characteristics and Influencing Factors of Major Air Pollutants at a Rural Site in Suzhou. *Ecology and Environmental Sciences*, 2017, 6(1): 119–128. (in Chinese)

Carter W.P.L. Development of Ozone Reactivity Scales for Volatile Organic Compounds. *Air & Waste*, 1994, 44, 881–899.

Daoo V.J., Panchal N.S., Sunny F., et al. Scintillometric Measurements of Daytime Atmospheric Turbulent Heat and Momentum Fluxes and Their Application to Atmospheric Stability Evaluation. *Exp. Therm Fluid Sci.*, 2004, 28, 337–345.

Jia C., Mao X., Huang T., et al. Non-methane Hydrocarbons (NMHCs) and their Contribution to Ozone Formation Potential in a Petrochemical Industrialized City, Northwest China. *Atmos. Res.*, 2016, 169, Part A: 225–236.

Jia C.H. *Characteristics and Chemical Behaviors of Atmospheric Non-methane Hydrocarbons in Lanzhou Valley, Western China*. Lanzhou: Lanzhou University, 2018. (in Chinese)

Liu H.L., Shi P.J., Liu H.M., et al. Northwest Territories Heavy Industrial City Industrial Carbon Emissions Analysis: A Case Study of Lanzhou City. *Ecological Economy*, 2014, 30(05): 41–45.

Luo S.H., Ma W.C., Wang X.R., et al. A Case Study on Indicator System of Urban Environmental Protection and Ecological Construction. *Acta Ecologica Sinica*, 2003, (01):45–55. (in Chinese)

Ryerson T.B., Trainer M., Angevine W.M., et al. Effect of Petrochemical Industrial Emissions of Reactive Alkenes and NOx on Tropospheric Ozone Formation in Houston, Texas. *J. Geophys. Res.*, 2003, 108, 4249.

Viana M., Chi X., Maenhaut W., et al. Organic and elemental carbon concentrations in carbonaceous aerosols during summer and winter sampling campaigns in Barcelona, Spain. *Atmos. Environ.*, 2006, 40: 2180–2193.

Wang J., Mo J., Li J., et al. OMI-measured SO_2 in a large-scale national energy industrial base and its effect on the capital city of Xinjiang, Northwest China. *Atmospheric Environment*, 2017, 167, 159–169.

Wang O. and He B.Y. Relationships between $PM_{2.5}$ and six elements for air quality in Urumqi based on Grey correlation. *Journal of Arid Resources and Environment*, 2018, 32(06):176–181. (in Chinese)

Wang R., Tao S., Wang W., et al. Black carbon emissions in China from 1949 to 2050. *Environ. Sci. Technol.*, 2012, 46: 7595–7603.

Zhang Y.P., Xu J.L., Zhao X.P., et al. Grey correlation analysis of air quality and its affection factors in Luoyang city. Journal of Henan University of Science and Technology: *Natural Science*, 2012, 33(01):100–104 + 10. (in Chinese)

Civil Engineering and Energy-Environment – Gao & Duan (Eds)
© 2023 the Author(s), ISBN 978-1-032-56059-5

Progress of electrocatalytic technology in treating organic chemical wastewater

Jianguang Wang*, Haifeng Fang*, Shiyi Li*, Shengjie Fu* & Xiaohu Lin*
Huadong Engineering Corporation Limited of Power China, Hangzhou, China
Huadong Eco-environmental Engineering Research Institute of Zhejiang Province, China

ABSTRACT: Electrocatalysis is a green and efficient wastewater treatment technology that plays an important role in treating organic chemical wastewater. This paper introduced the mechanism of electrochemical wastewater treatment, preparation of electrodes, and research progress in the electrocatalytic degradation of organic chemical wastewater. The application limit of electrocatalytic technology in organic chemical wastewater treatment was also analyzed, and the corresponding development direction in the future had been prospected. This paper's systematic and comprehensive review of electrochemical wastewater treatment will be an essential reference for applying electrochemical technology in engineering.

1 INTRODUCTION

With the acceleration of industrialization, a large amount of organic wastewater was generated from industries such as petrochemical, printing and dyeing, papermaking, and pharmaceuticals. This organic wastewater was characterized by difficult biodegradation, high chemical oxygen demand (COD), and high toxicity, which was challenging to degrade quickly and thoroughly by using traditional water treatment methods.

Electrocatalysis was a water treatment technology without secondary pollution and high efficiency. Under the applied electric field, organic pollutants' direct or indirect redox reactions are on the electrode surface or in the water to realize the organic pollutant decomposition. In recent years, much research has been conducted around preparing high catalytic performance electrodes and developing electrocatalytic water treatment processes to realize the organic chemical wastewater degradation. In this paper, an overview was given in terms of electrode material preparation, electrocatalytic redox methods, and electrocatalytic co-process development, and the existing problems were analyzed.

2 BASIC PRINCIPLES OF ELECTROCHEMICAL WATER TREATMENT

2.1 *Electrochemical oxidation*

Electrochemical oxidation refers to generating various free radicals under the action of the electric field. And free radicals on the electrode's surface oxidize and decompose the organic pollutants in the water. The electrochemical oxidation mechanism of organic matter can also

---

*Corresponding Authors: wjg.emails@qq.com; fang_hf@hdec.com; li_sy5@hdec.com; fu_sj@hdec.com and lin_xh@hdec.com

DOI: 10.1201/9781003433651-47

be divided into direct and indirect oxidation. Direct electrochemical oxidation was the process of organic matter decomposition through the direct transfer of strong oxidants and electrons generated on the electrode surface. This process includes electrochemical combustion and conversion. Comninellis proposed a reaction mechanism for electrochemical conversion and electrochemical combustion of organic pollutants on the metal oxide (MOx) electrode surface (Comninellis 1994). Hydroxide ions (OH) in H_2O or H_2O lose electrons on the anode surface and generate active hydroxyl radicals ($\cdot$OH), which are adsorbed on the electrode surface and react with organic matter (R) in an oxidation reaction, decomposing the organic matter gradually until it was utterly inorganic.

$$MO_x + H_2O \rightarrow MO_X(\cdot OH) + e^- \tag{1}$$

$$MO_x + OH^- \rightarrow MO_x(\cdot OH) + e^- \tag{2}$$

$$R + MO_x(\cdot OH) \rightarrow CO_x + zH^+ + ze^- + MO_x \tag{3}$$

If the $\cdot$OH adsorbed on the electrode surface can rapidly react with the MO_x electrode, the oxygen in the $\cdot$OH will be transferred to the MO_x lattice to form the high-valent oxide MO_{x+1}; MO_{x+1} oxidated with organic matter called electrochemical transformation.

$$MO_X(\cdot OH) \rightarrow MO_{x+1} + H^+ + e^- \tag{4}$$

$$R + MO_{x+1} \rightarrow RO + MO_x \tag{5}$$

Indirect electrochemical oxidation is a process in which solid oxidizing substances are produced during an electrochemical reaction, diffuse into the aqueous solution properly, and react with organic matter to cause it to be degraded. This process is divided into irreversible and reversible processes. Irreversible processes are reactions that produce vital oxidizing substances involved in the electrochemical reaction process. The vital oxidizing substances include solventized electrons, $\cdot$OH, O_3, O_2^-, $HO_2\cdot$ and H_2O_2, and if Cl^- is present in the solution, $HClO$, ClO^-, and other substances may be present. Once the applied electric field disappears, these substances are no longer present. The reversible indirect electrochemical oxidation process is that some low-valence stable metal ions or media in the anode surface are oxidized into oxidizing high-valence ions. These high-valence ions can directly oxidize organic matter in the aqueous solution by generating $\cdot$OH, and the ion itself is reduced, and then the organic matter is continuously oxidized on the anode surface.

2.2 *Electrochemical reduction*

Electrochemical reduction is the partial or complete hydrogenation reduction of organic substances by cathodic reduction, converting highly toxic organic substances into less toxic ones. This process can be used as pretreatment for oxidation and biological methods, which are currently primarily used for the halogenated hydrocarbons and dyes degradation. Electrochemical reduction includes direct electrochemical reduction and indirect electrochemical reduction.

Direct electrochemical reduction is a process in which organic substances are reduced by gaining electrons directly on the cathode surface, and the metal recovery in solution is an immediate electrocatalytic reduction process.

Indirect electrochemical reduction is a process that uses redox substances generated during electrochemical reactions or active hydrogen atoms generated in aqueous solutions to reduce organic substances. This method converts a wide range of organic substances containing halogens (X) to less toxic ones in water treatment processes (Chen *et al.* 2006).

$$2H_2O + 2e^- + M \rightarrow 2(H)_{ads}M + 2OH^- \tag{6}$$

$$R - X + M \rightarrow (R - X)_{ads}M \tag{7}$$

$$(R - X)_{ads}M + 2(H)_{ads}M \rightarrow (R - H)_{ads}M + HX \tag{8}$$

$$(R - H)_{ads}M \rightarrow R - H + M \tag{9}$$

Water is electrochemically generated as active hydrogen and adsorbed on the electrode surface. Organic matter (R-X) is adsorbed on the electrode surface and undergoes a hydrogenation reduction reaction with active hydrogen to remove halogen ions and produce a new substance (R-H).

3 CURRENT STATUS OF RESEARCH ON ELECTROCATALYTIC DEGRADATION OF ORGANIC CHEMICAL WASTEWATER

The current research on electrocatalytic degradation of organic chemical wastewater is mainly carried out in electrode materials preparation and development of the electrocatalytic degradation process.

3.1 *Electrode material preparation*

The electrode is the core of electrocatalytic water treatment; the electrode material's performance impacts wastewater treatment. Currently, the commonly used electrode substrate materials are metal and carbon materials. The carbon materials can be further modified to obtain electrode materials with higher catalytic performance.

3.1.1 *Metal substrate electrode*
Metal materials have been used as electrodes for a long time. However, the traditional single-metal materials have problems such as low degradation efficiency and poor chemical stability, so the metal matrix materials with high chemical stability are selected and loaded with catalysts on the surface to improve the electrode's ability to treat organic chemical wastewater. The modification and modification of metal matrix electrodes mainly include noble metal loading and metal oxide modification.

Electrodes prepared by using precious metal catalyst loading are applied to the reductive degradation of organic chemical wastewater, which can decompose highly toxic and difficult-to-biodegrade organic matter into more biochemically viable organic matter for subsequent degradation treatment. The commonly used precious metal catalysts are Au, Ag, Pt, and Pd, which have the characteristics of high catalytic activity and are not easy to passivate. Au/Ti, Ag/Ti, Pd/Ti, and Pt/Ti electrodes have been prepared more maturely and applied to organic wastewater treatment (Yi & Yu 2009; Yoon *et al.* 2012). In recent years, researchers have further improved the catalytic effect of the electrode by composite modification. Sun *et al.* (2014) prepared a composite electrode by using nickel foam (Foam-Ni) as the substrate, loaded metal Pd on the surface, doped and modified it with TiN nanoparticles, and used it for the degradation of 2,4-dichlorophenoxyacetic acid (2,4-D) wastewater. The removal rate of 2,4-D from the composite electrode increased dramatically after doping with TiN nanoparticles, and it could be completely degraded within 120 min (Sun *et al.* 2014). Liu *et al.* (2015) used Foam-Ni as the base material, modified the surface by using graphene (RGO), and loaded metal Pd on the graphene surface by electrodeposition to prepare Pd/RGO/Foam-Ni electrodes used for the degradation of 4-chlorophenol, which could be degraded entirely within 1 h under acidic conditions. Sun *et al.* (2017) prepared a

Pd/CNTs-nafionfilm/Ti electrode by modifying nafion and carbon nanotubes (CNTs) as the substrate material to increase the specific surface area and catalyst active sites of the electrode and then electrodeposited loaded Pd nanoparticles on the surface. This electrode had high catalytic activity and stability and was able to completely dechlorinate 2,3,5-trichlorophenol under acidic conditions and decompose it to the more biochemically available phenol.

Coating metal oxides on the surface of metal materials to prepare dimensionally stable anode (DSA) and doping by mono-or multi-element can improve the oxidation effect of the electrode, reduce the energy consumption in the wastewater treatment process, and can improve the service life of the electrode. In recent years, more research has been conducted on the preparation and modification of Ti-based PbO_2, SnO_2, and IrO_2 electrodes. Pure SnO_2 is an n-type semiconductor with low conductivity at room temperature and a band gap of about 3.5 eV. Doping can significantly improve the conductivity of SnO_2 by swelling the lattice and increasing the defects. Sun *et al.* (2015a; 2015b) prepared Sb-Ni-Ce and Sb-Ni-Nd doped Ti/SnO_2 electrodes by the sol-gel method, and the service life of the doped composite electrodes was greatly improved. The Sb-Ni-Nd co-doped composite electrode had a longer service life than the Sb-Ni-Ce co-doped composite electrode. The modified electrodes were used to degrade phenol-containing wastewater, and phenol could be completely degraded within 2.5 h. The removal rates of total organic carbon (TOC) by the two electrodes reached 75.9% and 90.8%, respectively, and the energy consumption during the degradation process decreased by more than 50%. PbO_2 electrode has good conductivity, chemical stability, high oxygen precipitation overpotential, and high oxidation capacity when electrolyzed in an aqueous solution. Zhuo *et al.* (Zhuo *et al.* 2017) prepared Ti/PbO_2 electrodes and modified Ti/SnO_2-Sb_2O_5/PbO_2 and Ti/SnO_2-Sb_2O_5/PbO_2-PVDF composite electrodes for the electrocatalytic oxidative degradation of perfluorooctanoic acid wastewater. The composite electrode modified with polytetrafluoroethylene (PVDF) had a high oxygen precipitation overpotential with a lifetime 17.67 times higher than the Ti/PbO_2 electrode, allowing the removal of PFOA up to 92.1%.

The reversible adsorption of iridium on oxygen makes IrO_2 electrodes have a long service life, but their chemical stability is poor, so inert oxides such as SnO_2, TiO_2, and ZrO_2 are often added to the coating (Jiang *et al.* 2015). Araujo *et al.* (2017) used CeO_2, SnO_2, and Sb_2O_3 for the composite modification of Ti/RO_2-IrO_2 electrodes and the degradation of wastewater containing naphthalene and benzene, respectively. Compared to the composite modification with CeO_2 and SnO_2, the electrodes prepared by composite modification with Sb_2O_3 showed higher removal rates of naphthalene and benzene, reaching 93.4% and 99.8% within 2 h, respectively.

3.1.2 *Carbon material substrate electrode*

Carbon material matrix electrode is the most widely used non-metallic electrode, mainly including carbon fiber, carbon felt, carbon cloth, graphite felt, graphene, and glass carbon electrodes. Carbon materials can be used alone or as carriers for electrode catalysts. Carbon material electrodes have the characteristics of higher electrical conductivity, better chemical stability, easy preparation, and lower price. Yu *et al.* (2016) prepared a composite electrode for electro-Fenton degradation of dimethyl phthalate (DMP) wastewater by modifying polyaniline on the surface of a graphite mat using electropolymerization, and the degradation rate of the modified electrode for DMP was 5 times higher than that of graphite mat, and DMP could be completely degraded within 1 h. Xu *et al.* (2016) modified the glassy carbon electrode with a polysulfide cordial interlayer and loaded the outer layer with copper oxide nanoparticles to prepare a copper oxide nanoparticle/cordial sulfide/glassy carbon modified electrode (CuO/PTH/GCE), which showed sound electrochemical oxidative degradation of nitrophenol. Ganiyu *et al.* (2017) prepared a cobalt-iron bilayer hydroxide-loaded carbon felt electrode for the degradation of acid orange II wastewater by the hydrothermal method. the TOC removal rate was up to 97% after 8 h of reaction, and the modified electrode had good reusability and maintained the TOC removal rate above 60% after 7 times of use. Boron-doped diamond (BDD) electrode is a new carbon material matrix electrode with high oxygen precipitation overpotential and electrochemical stability and low adsorption capacity (Zhang *et al.* 2017). Labiadh *et al.* (2016) compared methyl orange's electrocatalytic oxidative degradation process at the Ti-Ru-Sn ternary oxide electrode (TiRuSnO$_2$), PbO$_2$ electrode, and BDD electrode, respectively. At 0.5 A current, for an initial concentration of 100 mg/L of methyl orange, both the PbO2 electrode and BDD electrode could completely degrade it. However, the removal rate of methyl orange by the TiRuSnO2 electrode was only 69%, and only BDD electrode could completely mineralize it, indicating BDD's more robust catalytic oxidation performance.

3.2 *The electrocatalytic degradation process of organic chemical wastewater*

3.2.1 *The electrocatalytic oxidative degradation process*

Currently, degradation processes for organic chemical wastewater have been developed based on electrocatalytic oxidation technology, which uses the generated reactive radicals with the solid oxidizing capacity to achieve efficient degradation of organic wastewater and pollutants by controlling the process conditions. He *et al.* (2015) designed an electrocatalytic-activated carbon composite reactor by using Ru-Ir/TiO2 electrode as the anode and Ti plate as the cathode by placing powdered activated carbon between the cathode and the anode, which was able to be used for the degradation of ammonia-containing wastewater, and further studied the degradation process of ammonia, NH4 + , NO3-, NO2-, total nitrogen (TN), and COD. It was found that direct and indirect oxidation occurred simultaneously in the electrocatalytic oxidation of ammonia. The pollutant removal efficiency was influenced by the influent water flow rate, salinity, current density, and chloride ion content. The ammonia removal ratio was up to 80% under the condition of pH of 6.5, 0.9 A current, 2% Na2SO4, 1500 mg/L of Cl- and 0.8 L/h influent flow. Zhang *et al.* (Gao *et al.* 2011) studied the electrocatalytic oxidative degradation of activated orange X-GN dye wastewater by using a BDD electrode as the anode and stainless steel sheet as the cathode and investigated the effects of current density, electrolyte solution concentration, and initial pH on the degradation. When the current density was 100 mA/cm2, the electrolyte concentration was 0.05 mol/L, the initial pH was 3.78, the chromaticity removal efficiency was 99%, the TOC removal ratio was 56.95%, and the energy consumption was 65.4 kWh/m^3 after 5 h of reaction. Ajeel *et al.* (2017) studied the electrocatalytic oxidative degradation of 2-chlorophenol and phenol in wastewater by using the prepared carbon black-diamond electrode and Pt sheet as anodes, respectively. It was found that under the conditions of the electrolyte solution of 0.25 mol/L Na$_2$SO$_4$, the solution with a pH of 3, and a current density of 30 mA/cm^2, the removal ratio of 2-chlorophenol could reach 94% after 6 h of reaction at room temperature, while the removal of phenol was only 20%, with the removal effect of carbon black-diamond electrode and Pt sheet electrode on the mixture of 2-chlorophenol and phenol being consistent.

3.2.2 *The electrocatalytic reduction degradation process*

The electrocatalytic reduction process of organic wastewater is mostly used to decompose chlorinated organics. The subsequent reaction is realized by converting the bio-degradation resistant organics into easily biodegradable organic compounds through hydrogenation reduction with removing chlorine atoms from organic pollutants.

Zhang *et al.* (2015) prepared a Pd/Ti electrode used for the electrocatalytic reduction of 2,4,5-trichlorobiphenyl, and the electrocatalytic removal rate could all reach more than 95% after 5 h of reaction when the pollutant concentration was 60, 80, and 100 mg/L. Ma *et al.* constructed catalytic reaction systems for the reductive degradation of 3,6-dichloropyr-idinecarboxylic acid (3,6-D) by using nickel foam, silver mesh, copper foam, Pd/Ni, Pd/Ag, and Pd/Cu electrodes as cathodes and graphite flakes as anodes, respectively, and investigated the effects of electrode materials, solution pH, initial concentration of 3,6-D and current density on the degradation process. The Pd/Ni electrode exhibited a better catalytic effect and maintained a high current efficiency during the reaction. When the initial concentration of 3,6-D was 250 mmol/L, the electrolyte solution was 1.25 mol/L NaOH solution, the current density was 208 A/m2, the conversion of 3,6-D could reach 99% and the current efficiency was 76.3% after 12 h reaction (Ma *et al.* 2016). Sun *et al.* (2017) prepared carbon nanotube-polypyrrole (CNTs-PPy)-modified Ti mesh-loaded Pd cathodes for the reductive degradation of 2,3-dichlorophenol (2,3-DCP). At the condition of pH of 2.5 and 5 mA current, 2,3-DCP with the initial concentration of 100 mg/L was completely degraded within 70 min, and the addition of impurity ions (NO_3^-, CO_3^{2-}, HCO_3^-, Mg^{2+}, K^+, and Ca^{2+}) in the electrolyte solution did not affect the electrocatalytic reduction of 2,3-DCP.

3.2.3 *Electrocatalytic co-processing*

For some degradation-resistant organic chemical wastewater, the use of single electrocatalytic degradation technology has the disadvantages of long degradation time and high energy consumption. Therefore, in order to further improve the degradation efficiency of organic chemical wastewater, the combination of electrocatalysis with other treatment methods to develop new combined processes has been paid more and more attention. The commonly adopted coupled technologies include bio-electrochemical technology, ultrasound-electrochemical technology, electro-Fenton technology, photoelectrochemical technology, etc. Using Ti/RuO2 as the anode and stainless steel as the cathode, Cui *et al.* (2017) combined electrocatalytic oxidation, ascending flow aeration biofilter and electrodialysis methods together to degrade the discharge effluent with triazole biocide, achieving a final removal rate of over 90% and reducing the overall operating cost compared to a single electrocatalytic oxidation process. Zhang *et al.* (2015) constructed an electro-Fenton system for the degradation of 2,4-dichlorophenol by using a gas diffusion electrode as the cathode and a Ti/IrO2-RuO2 electrode as the anode, and Fe-C particles were added to the reaction device, and the composite electro-Fenton system showed good performance in chlorophenol degradation and TOC removal. Guzman-Duque et al. ((2016) used the BDD electrode and Ti/IrO2 electrode as anode and Pt sheet as cathode, respectively, to treat crystalline violet dye wastewater by combining ultrasonic technology with electrochemical oxidation. And compared with single ultrasonic decomposition and electrocatalytic oxidation, the ultrasonic-electrochemical coupling technology improved the removal of crystalline violet dye and TOC. Aravind *et al.* (2016) established a combined bio-optical-electrocatalytic oxidation process with Ti/IrO2-RuO2-TiO2 electrode as anode for the degradation of dyeing wastewater with up to 95% decolorization and 68% COD removal.

4 LIMIT OF THE APPLICATIONS

Electrocatalytic water treatment technology has gradually advanced with the continuous development of electrocatalytic theory, while it has not been widely used in organic chemical wastewater treatment. The main limitations may include the following two aspects.

(1) High cost of electrode materials, short stability and service life of electrodes, and high energy consumption. Many new electrodes use expensive materials such as rare earth elements, precious metals, carbon nanotubes and graphene in the process of preparation, and the preparation methods are also more complicated, which limits the popularization and application of electrodes and makes the development of electrode preparation methods mostly limited to the laboratory research stage.

(2) Electrocatalytic degradation mechanism is not fully investigated and understood. Electrocatalytic degradation of organic pollutants in water is a complex process, and the actual organic chemical wastewater contains organic substances, heavy metal ions, inorganic salts and other substances, which makes the electrocatalytic process more

complex. The processes of generation and conversion of reactive radicals during the reaction as well as the reaction activities of organic chemical wastewater on the electrode surface and in solution are still unclear, which all limit the application of electrocatalytic technology in the degradation of organic chemical wastewater.

5 CONCLUSION

Electrocatalysis is a green and efficient wastewater treatment technology, which plays an important role in the treatment of organic chemical wastewater and has a broad development prospect due to its good removal effect on the removal of degradation-resistant organic compounds with high concentration. At present, the development of new electrode materials, electrocatalytic reaction equipment and electrocatalytic process, and the research on the mechanism of electrocatalytic degradation of organic pollutants are the trends of studies on electrocatalytic degradation of organic chemical wastewater. Through the continuous improvement of catalytic reaction theory and electrode preparation technology, electrode catalysis technology will play a greater role in organic chemical wastewater degradation and water environmental protection.

ACKNOWLEDGMENTS

This work was financially supported by the Drainage system comprehensive efficiency evaluation and collaborative operation technology research (KY2021-HS-02-11) and the Project contract for the study on the construction technology of urban plain river network ecosystem based on the improvement of self-purification capacity of slow-flowing rivers (KY2018-SHJ-02).

REFERENCES

Ajeel M.A, Aroua M.K, Daud W.M.A.W et al. Effect of Adsorption and Passivation Phenomena on the Electrochemical Oxidation of Phenol and 2-Chlorophenol at Carbon Black Diamond Composite Electrode. *Industrial & Engineering Chemistry Research*. 2017, 56(6): 1652–1660.

Araújo D.T, de A. Gomes M., Silva R.S., et al. Ternary Dimensionally Stable Anodes Composed of RuO2 and IrO2 with CeO2, SnO2, or Sb2O3 for Efficient Naphthalene and Benzene Electrochemical Removal. *Journal of Applied Electrochemistry*. 2017, 47(4): 547–561.

Aravind P., Subramanyan V., Ferro S. et al. Eco-friendly and Facile Integrated Biological-cum-photo Assisted Electrooxidation Process for Degradation of Textile Wastewater. *Water Research*. 2016, 93: 230–241.

Chen G., Wang Z., Yang T. et al. Electrocatalytic Hydrogenation of 4-Chlorophenol on the Glassy Carbon Electrode Modified by Composite Polypyrrole/Palladium Film. *The Journal of Physical Chemistry B*. 2006, 110(10): 4863–4868.

Comninellis C. Electrocatalysis in the Electrochemical Conversion/Combustion of Organic Pollutants for Waste-water Treatment. *Electrochimica Acta*. 1994, 39(11–12): 1857–1862.

Cui T., Zhang Y.H., Han W.Q. et al. Advanced Treatment of Triazole Fungicides Discharged Water in Pilot Scale by Integrated System: Enhanced Electrochemical Oxidation, Up-flow Biological Aerated Filter and Electrodialysis. *Chemical Engineering Journal*. 2017, 315: 335–344.

Ganiyu S.O, Le T., Bechelany M. et al. A Hierarchical CoFe-layered Double Hydroxide Modified Carbon-felt Cathode for the Heterogeneous Electro-Fenton Process. *Journal of Materials Chemistry A*. 2017, 5(7): 3655–3666.

Gao Y., Chang M., Li X. et al. Boron Doped Diamond Electrode and its Application in Electroanalysis(in Chinese). *Chemical Progress*. 2011, 23(05): 951–962.

Guzman-Duque F.L., Grupo De Diagnóstico Y Control De La Contaminación F D I U. Synergistic Coupling Between Electrochemical and Ultrasound Treatments for Organic Pollutant Degradation as a Function of the Electrode Material (IrO2 and BDD) and the Ultrasonic Frequency (20 and 800 kHz)[J]. *International Journal of Electrochemical Science*. 2016: 7380–7394.

He S., Huang Q., Zhang Y. et al. Investigation on Direct and Indirect Electrochemical Oxidation of Ammonia over Ru-Ir/TiO2 Anode. *Industrial & Engineering Chemistry Research*. 2015, 54(5): 1447–1451.

Jiang Y., Lei Y., Cheng H. et al. Review of Titanium-based Iridium Dioxide Coated Electrodes (in Chinese). *Metal Functional Materials*. 2015, 22(06): 55–61.

Labiadh L., Barbucci A., Carpanese M.P. et al. Comparative Depollution of Methyl Orange Aqueous Solutions by Electrochemical Incineration Using TiRuSnO2, BDD and PbO2 as High Oxidation Power Anodes. *Journal of Electroanalytical Chemistry*. 2016, 766: 94–99.

Liu Y., Liu L., Shan J. et al. Electrodeposition of Palladium and Reduced Graphene Oxide Nanocomposites on the Foam-nickel Electrode for Electrocatalytic Hydrodechlorination of 4-chlorophenol. *Journal of Hazardous Materials*. 2015, 290: 1–8.

Ma H.X., Xu Y.H., Ding X.F. et al. Electrocatalytic Dechlorination of Chloropicolinic Acid Mixtures by Using Palladium-modified Metal Cathodes in Aqueous Solutions. *Electrochimica Acta*. 2016, 210: 762–772.

Sun C., Baig S.A., Lou Z. et al. Electrocatalytic Dechlorination of 2,4-dichlorophenoxyacetic Acid Using Nanosized Titanium Nitride Doped Palladium/nickel Foam Electrodes in Aqueous Solutions. *Applied Catalysis B: Environmental*. 2014, 158–159: 38–47.

Sun Z., Ma X., Hu X. Electrocatalytic Dechlorination of 2,3,5-trichlorophenol on Palladium/carbon Nanotubes-nafion film/titanium Mesh Electrode. *Environmental Science and Pollution Research*. 2017, 24 (16): 14355–14364.

Sun Z., Song G., Du R. et al. Modification of a Pd-loaded Electrode with Carbon Nanotubes–Polypyrrole Interlayer and its Dechlorination Performance for 2,3-dichlorophenol. *RSC Advances*. 2017, 7(36): 22054–22062.

Sun Z., Zhang H., Wei X. et al. Fabrication and Electrochemical Properties of a SnO2-Sb Anode Doped with Ni-Nd for Phenol Oxidation. *Journal of the Electrochemical Society*. 2015, 162(9): H590–H596.

Sun Z., Zhang H., Wei X. et al. Preparation and Electrochemical Properties of SnO2-Sb-Ni-Ce Oxide Anode for Phenol Oxidation. *Journal of Solid State Electrochemistry*. 2015, 19(8): 2445–2456.

Xu B., Zhang H., Yang Z. Electrochemical Oxidation and Degradation of Nitrophenol by Nanometer Copper Oxide (in Chinese). *Journal of Anhui Normal University (Natural Science Edition)*. 2016, 39(04): 355–358.

Yi Q., Yu W. Nanoporous Gold Particles Modified Titanium Electrode for Hydrazine Oxidation. *Journal of Electroanalytical Chemistry*. 2009, 633(1): 159–164.

Yoon J., Shim Y., Lee B. et al. Electrochemical Degradation of Phenol and 2-Chlorophenol Using Pt/Ti and Boron-Doped Diamond Electrodes. *Bulletin of the Korean Chemical Society*. 2012, 33(7): 2274–2278.

Yu J., Liu T. and Liu H. Electro-polymerization Fabrication of PANI@GF Electrode and its Energy-effective Electrocatalytic Performance in the Electro-Fenton Process. *Chinese Journal of Catalysis*. 2016.

Zhang C., Zhou M.H., Ren G.B. et al. Heterogeneous Electro-Fenton Using Modified Iron-carbon as the Catalyst for 2,4-dichlorophenol Degradation: Influence Factors, Mechanism and Degradation Pathway. *Water Research*. 2015, 70: 414–424.

Zhang J.L., Cao Z.P., Chi Y.Z. et al. Preparation and Electrocatalytic Dechlorination Performance of Pd/Ti Electrode. *Desalination and Water Treatment*. 2015, 54(10): 2692–2699.

Zhang M., Wang Y., Zeng S. et al. Study on Degradation of Reactive Orange X-GN Azo Dye Wastewater by Boron Doped Diamond Electrode(in Chinese). *Surface Technology*. 2017, 46(07): 128–133.

Zhuo Q.F., Xiang Q., Yi H. et al. Electrochemical Oxidation of PFOA in Aqueous Solution Using Highly Hydrophobic Modified PbO2 Electrodes. *Journal of Electroanalytical Chemistry*. 2017, 801: 235–243.

Civil Engineering and Energy-Environment – Gao & Duan (Eds)
© 2023 the Author(s), ISBN 978-1-032-56059-5

An evolutionary transfer optimization framework on well placement optimization via kernelized autoencoding

Ji Qi, Kai Zhang* & Xingyu Zhou
School of Petroleum Engineering, China University of Petroleum (East China), Qingdao, China

ABSTRACT: Well placement optimization is an important step in oil and gas development engineering. The rapid selection of suitable well locations can be considered to determine the success or failure of oil and gas development because well placement is costly and well locations have a significant impact on the ultimate recovery factor. At present, the methods of oil and gas well selection are mainly to build the numerical model of oil and gas reservoirs, evaluate the candidate well locations by numerical simulation technology, and search the final well locations with the highest economic benefit by an evolutionary algorithm. This approach has achieved amazing accomplishments, but its main shortcoming is that it has to search from scratch for each solution, completely failing to take advantage of the experience of many tasks that have already been optimized. Taking this cue, in this paper, an evolutionary transfer optimization framework is proposed, which can extract experience and knowledge from past completed tasks and add them to the solution process of new tasks to accelerate the search for the optimal well locations. In the proposed framework, kernelized autoencoding is employed to construct the mapping in a Reproducing Kernel Hilbert Space, and the experience and knowledge can be easily transferred across the past and current well placement optimization tasks. In addition, the proposed framework holds a closed-form solution, and it will not bring much computational burden in the search, and the proposed framework can be set as a transfer operator to be easily combined with all the common population-based optimization algorithms. To validate the efficacy, comprehensive experiments on the well placement optimization tasks are presented.

1 INTRODUCTION

Well placement Optimization (WPO) is one of the most important issues in oil and gas development. WPO can significantly reduce development risk and cost, and improve recovery. With the development of various optimization algorithms, numerical simulation combined with optimization algorithms has gradually become the most effective method to solve WPO problems (Bukhamsin *et al.* 2010; Harb *et al.* 2020; Onwunalu *et al.* 2010), and this kind of methods have achieved great success in both theory and practical application. However, each individual WPO task is currently searched from scratch, without using the solving experience of previous tasks, forming an "island of data."

Inspired by human learning processes that use past learning experiences to accelerate new ones, researchers are trying to enhance the performance of evolutionary algorithms with past experiences and knowledge. An interesting evolutionary search paradigm has been proposed in the field of evolutionary computing, namely evolutionary transfer optimization (ETO) (Tan *et al.* 2021). Current ETO approaches include storing and reusing optimal solutions of

---

*Corresponding Author: reservoirs@163.com

DOI: 10.1201/9781003433651-48

past tasks (Cunningham *et al.* 1997; Louis *et al.* 2004), the reuse of implicit search routes and knowledge in the population of multi-objective problems (Jiang *et al.* 2017), and the exploitation of the information contained in tasks and populations (Sultana *et al.* 2022). This approach works because tasks do not exist in isolation in nature, and almost any single task has multiple similar tasks, especially in real-world engineering problems.

The ETO framework proposed in this paper is designed according to this motivation. It combines strategies for storing and reusing optimal solutions of past tasks with strategies for extracting information contained in populations of past tasks. Populations in a fixed generation from the past and the current task are projected to a Reproducing Kernel Hilbert Space (RKHS). The nonlinearity between two tasks can be captured. Then, the mapping is conducted via a kernelized single-layer denoising autoencoder (Zhou *et al.* 2021). The optimal solution of the past task will be explicitly transformed into a learned solution, and the search performance can be improved by injecting the learned solution into the current population. In order to verify the performance of the proposed framework, we build fluvial reservoir models for experiments, and the results show that the proposed framework has a significant performance improvement over the ordinary evolutionary algorithm.

2 THE KERNELIZED AUTOENCODING FOR KNOWLEDGE TRANSFER ON WELL PLACEMENT OPTIMIZATION

2.1 *Autoencoding in evolutionary transfer optimization*

Denoising autoencoder is one type of neural network, which reconstructs the corrupted input to its clean version, i.e., what it does is it builds a mapping from input to output, with the exception that the input is the output that adds noise. It has been successfully applied to solving many challenging learning problems, such as domain adaptation (Chen *et al.* 2015; Glorot *et al.* 2011), natural language processing (Ishii *et al.* 2013; Lu *et al.* 2013), and sentiment analysis (Sagha *et al.* 2017; Zhai & Zhang 2016).

Since the source and target of transfer learning are similar, we can consider the source as a corrupted version of the target, so the denoising autoencoder can be used to construct a cross-task mapping to transfer knowledge.

2.2 *The kernelized autoencoding for knowledge transfer*

In generation g, the population of the current WPO task is defined as P_{tg}, and the population of the past task is defined as P_{sg}, and the size of the population is N. The mapping matrix M can be built through a denoising autoencoder by setting P_{sg} as the input and P_{tg} as the output. After that, we suppose P_{sg} is projected the RKHS by a nonlinear mapping function $\varnothing$, which minimizes the squared reconstruction loss as follows:

$$L_{sq} = \frac{1}{2N} \|P_{tg} - M\varnothing(P_{sg})\|^2 = \frac{1}{2N} tr\left[\left(P_{tg} - \varnothing(P_{sg})\right)^T \left(P_{tg} - \varnothing(P_{sg})\right)\right] \quad (1)$$

where $tr(\ )$ denotes the trace of a matrix. According to [49], the linear mapping M in RKHS can be represented as a linear combination of the data points $\varnothing(X)$ in RKHS, i.e., $M = M_k\varnothing(X)^T$, Therefore $M\varnothing(P_{sg})$ can be rewritten as $M_k\varnothing(P_{sg})^T\varnothing(P_{sg})$. Then the kernel matrix is denoted as:

$$K\left(P_{sg}, P_{sg}\right) = \varnothing\left(P_{sg}\right)^T\varnothing\left(P_{sg}\right). \quad (2)$$

Then the loss function can be rewritten as:

$$L_{sq}(M_k) = \frac{1}{2N} tr\left[\left(P_{tg} - M_k K(P_{sg}, P_{sg})\right)^T \left(P_{tg} - M_k K(P_{sg}, P_{sg})\right)\right]. \tag{3}$$

Finally, the mapping matrix can be deduced into a closed-form solution:

$$M_k = P_{tg} K(P_{sg}, P_{sg})^T \left(K(P_{sg}, P_{sg}) K(P_{sg}, P_{sg})^T\right)^{-1}. \tag{4}$$

The optimal solution S of the past task can be explicitly transferred to a learned solution LS in the current task:

$$LS = M\emptyset(S) = M_k \, K(P_{sg}, S). \tag{5}$$

LS is injected into the g-generation population of the current task to speed up the optimization.

3 THE FLUVIAL RESERVOIR MODELS FOR EXPERIMENTS

3.1 Source task

As a task to store knowledge and experience, the past task is called the "source task." Figure 1 shows the distribution of the reservoir porosity, oil saturation, reservoir thickness, and the optimal injector locations in the source task. The size of the source task is 100×100, the number of injectors to be optimized is four, the reservoir thickness ranges from 5 to 25 ft, the oil saturation ranges from 0.3 to 0.4, and the porosity of each grid is proportional to the permeability and ranges from 0.2 to 0.35. The injection rate is 1700 STB/d, the bottom hole pressure is 2200 psia, and the production control step is 1800 days. The optimal solution is set as LS, and it will be transferred to the target task.

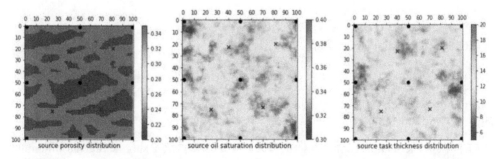

Figure 1. The distribution of the reservoir porosity, oil saturation, reservoir thickness, and the optimal injector locations (blue cross) in the source task.

3.2 The target task

As a task to receive knowledge and experience, the current task is called the "target task." Figure 2 shows the distribution of the reservoir porosity, oil saturation, and reservoir thickness in the source task. The size of the source task is 100×100, the number of injectors to be optimized is four, the reservoir thickness ranges from 10 to 25 ft, the oil saturation

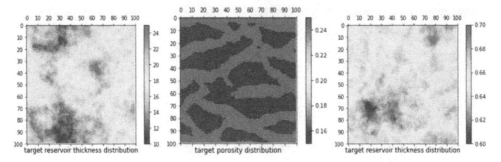

Figure 2. The distribution of the reservoir porosity, oil saturation, and reservoir thickness in the source task.

ranges from 0.6 to 0.7, and the porosity of each grid is proportional to the permeability and ranges from 0.15 to 0.25. The injection rate is 1000 STB/d, the bottom hole pressure is 1800 psia, and the production control step is 5400 days.

In order to verify the performance of the proposed framework, we use a differential evolution algorithm (DE) and DE combined with the proposed framework (labeled as DE + KM) to optimize well placement in the target task. The configurations of evolutionary operators and parameters in DE and DE + TCM are kept the same, and the only difference among them is the injection of solutions during the search process. The specific experimental settings are outlined as follows. The population size is 20, the maximum iteration is 100, the independent number of runs is 5 (without the configuration of any specific random seed), and the evolutionary operators in DE are: scaling factor F = 0.5, crossing probability CR = 0.8. The interval of knowledge transfer occurs is 5, the unit price of oil $c_1 = 80\$/STB$, the unit treatment cost of water production $c_2 = 5\$/STB$, and the unit cost of water injection $c_3 = 5\$/STB$. $-NPV$ (Net Present Value) is the objective value of the minimization optimization algorithm:

$$NPV = c_1 * oil\ production - c_2 * water\ production - c_3 * water\ injection. \qquad (6)$$

4 RESULTS AND DISCUSSIONS

According to the convergence curve of objective value shown in Figure 3, we can see that the proposed framework effectively enhances the search performance of the DE. With the proposed framework, the NPV corresponding to the searched well locations at 47 generations is already higher than the optimal well locations searched by the DE algorithm over 100 generations, and the search speed increased by 53%. And it gets a lower objective value (higher NPV) when the computational resources are exhausted. After subtracting the minimum NPV found by the DE algorithm from all NPV values, the highest NPV increased by 5.49% when using the proposed framework compared with that not using it.

Table 1 lists the NPV, final oil production (total), final water production (total), and final water injection (total) corresponding to the optimal well locations searched by DE and DE + KM. After 100 iterations, the optimal well locations searched by DE combined with the transfer optimization framework resulted in a $\$10^6$ increase in NPV and a 10^4 STB increase in oil production compared to the basic DE algorithm. From what has been discussed above, the proposed framework can boost the search for better well locations in the well placement optimization problems.

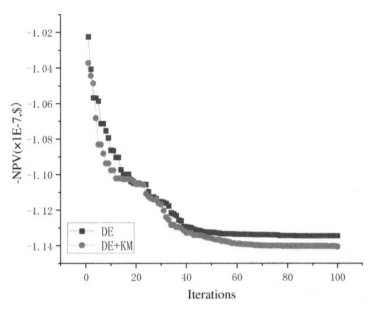

Figure 3. Average convergence curves of the DE and DE + KM over five independent runs on the target task.

Table 1. The NPV, Final Oil Production (Total), Final Water Production (Total), and Final Water Injection (Total) Corresponding to the Optimal Well Locations Searched by DE and DE+KM.

Well Placement	NPV, 10^8 dollars	Final Oil Production (Total), 10^6 STB	Final Water Production (Total), 10^7 STB	Final Water Injection (Total), 10^7 STB
DE Optimal	1.13	1.58	2.01	2.16
DE + KM Optimal	1.14	1.60	2.01	2.16

5 CONCLUSION

In this paper, we propose a transfer optimization framework for well placement optimization problems, which is a new method that can be applied to the oil and gas development industry.

(1) Such a framework can capture the nonlinear relationships contained in the population of optimization tasks, and it uses a kernelized denoising autoencoder to build cross-task mappings.
(2) Through the mapping, the optimal solutions of past tasks can be transformed into learned solutions and added to the population of target tasks to improve the search performance of the algorithm.
(3) The experimental results on well placement optimization problem show that the proposed framework has a significant performance improvement over the conventional evolutionary algorithms, the search speed increased by 53%, and the highest NPV increased by 5.49%.

ACKNOWLEDGMENTS

This work is supported by the National Natural Science Foundation of China under Grants 52074340, 51874335, 12131014 and 51722406, the Shandong Provincial Natural Science Foundation under Grants JQ201808 and ZR2019MEE101, the Fundamental Research Funds for the Central Universities under Grant 18CX02097A, the Major Scientific and Technological Projects of CNPC under Grant ZD2019-183-008, the Science and Technology Support Plan for Youth Innovation of University in Shandong Province under Grant 2019KJH002, the National Science and Technology Major Project of China under Grant 2016ZX05025001-006, and the 111 Project under Grant B08028.

REFERENCES

Bukhamsin Ahmed Y, Farshi Mohammad Moravvej, and Aziz Khalid 2010. Optimization of Multilateral Well Design and Location in a Real Field Using a Continuous Genetic Algorithm. *Proc., SPE/DGS Saudi Arabia Section Technical Symposium and Exhibition.*

Chen M., Weinberger K.Q., Xu Z., and Sha F. "Marginalizing Stacked Linear Denoising Autoencoders," *The Journal of Machine Learning Research*, vol. 16, no. 1, pp. 3849–3875, 2015.

Cunningham Pádraig and Smyth Barry 1997. Case-based Reasoning in Scheduling: Reusing Solution Components. *International Journal of Production Research* 35 (11): 2947–2962.

Glorot X., Bordes A., and Bengio Y. "Domain Adaptation for Large-scale Sentiment Classification: A Deep Learning Approach," in *Proceedings of the 28th International Conference On Machine Learning*, 2011, pp. 513–520.

Harb Ahmad, Kassem Hussein, and Ghorayeb Kassem 2020. Black Hole Particle Swarm Optimization for Well Placement Optimization. *Computational Geosciences* 24 (6): 1979–2000.

Ishii T., Komiyama H., Shinozaki T., Horiuchi Y., and Kuroiwa S., "Reverberant Speech Recognition Based on Denoising Autoencoder." *In Interspeech*, 2013, pp. 3512–3516.

Jiang Min, Huang Zhongqiang, Qiu Liming et al. 2017. Transfer Learning-based Dynamic Multiobjective Optimization Algorithms. *IEEE Transactions on Evolutionary Computation* 22 (4): 501–514.

Louis Sushil J and McDonnell John 2004. Learning with Case-injected Genetic Algorithms. *IEEE Transactions on Evolutionary Computation* 8 (4): 316–328.

Lu X., Tsao Y., Matsuda S., and Hori C. "Speech Enhancement Based on Deep Denoising Autoencoder." in Interspeech, 2013, pp. 436–440.

Onwunalu, Jérôme E and Durlofsky, Louis J. 2010. Application of a Particle Swarm Optimization Algorithm for Determining Optimum Well Location and Type. *Computational Geosciences* 14 (1): 183–198.

Sagha H., Cummins N., and Schuller B. "Stacked Denoising Autoencoders for Sentiment Analysis: A Review," *Wiley Interdisciplinary Reviews: Data Mining and Knowledge Discovery*, vol. 7, no. 5, p. e1212, 2017.

Sultana Nasrin, Chan Jeffrey, Sarwar Tabinda et al. 2022. Learning to Optimize General TSP Instances. *International Journal of Machine Learning and Cybernetics*: 1–16.

Tan, Kay Chen, Feng Liang, and Jiang, Min 2021. Evolutionary Transfer Optimization-a New Frontier in Evolutionary Computation Research. *IEEE Computational Intelligence Magazine* 16 (1): 22–33.

Zhai S. and Zhang Z.M. "Semisupervised Autoencoder for Sentiment Analysis," in *Thirtieth AAAI Conference on Artificial Intelligence*, 2016.

Zhou Lei, Feng Liang, Gupta Abhishek et al. 2021. Learnable Evolutionary Search Across Heterogeneous Problems via Kernelized Autoencoding. *IEEE Transactions on Evolutionary Computation* 25 (3): 567–581.

Civil Engineering and Energy-Environment – Gao & Duan (Eds)
© 2023 the Author(s), ISBN 978-1-032-56059-5

Review of research progress on pharmaceutical wastewater treatment technology

Jianguang Wang*, Shengjie Fu*, Haifeng Fang*, Shiyi Li* & Xiaohu Lin*
Huadong Engineering Corporation Limited of Power China, Hangzhou, ChinaHuadong Eco-environmental Engineering Research Institute of Zhejiang Province

ABSTRACT: Pharmaceutical wastewater is a type of highly concentrated and difficult-to-treat wastewater, and its discharge in large quantities can impose a heavy burden on the environment. This paper reviews the characteristics of current pharmaceutical wastewater, including notable features such as biodegradability, high concentration, and complex composition. At the same time, the current treatment technologies of pharmaceutical wastewater are introduced in detail, including physical, chemical, and biological methods. Compared with a single technology, the combined treatment technology can effectively treat pharmaceutical wastewater, and the characteristics of combined treatment technologies are introduced. This paper systematically composes the characteristics of pharmaceutical wastewater and the treatment technologies, which can provide a research basis for further development of new technologies or engineering applications.

1 INTRODUCTION

With the rapid development of the pharmaceutical chemical and healthcare products manufacturing industry, a large amount of toxic and harmful wastewater discharged in the pharmaceutical process seriously endangers people's health [1]. The quality and quantity of wastewater generated by different drug varieties and production processes in the drug production process also differ significantly. Pharmaceutical industry wastewater can be divided into four categories according to its product characteristics and water quality characteristics [2]: synthetic pharmaceutical production wastewater; biopharmaceutical production fermentation wastewater; proprietary Chinese medicine production wastewater; washing water and rinse water in the production process of various types of preparations. Various preparation production processes and washing water and rinse water generally have a small degree of pollution, mainly from raw material washing water, raw material decoction residue, and ground rinse water.

The direct discharge of pharmaceutical wastewater can seriously damage the environment and endanger human health. Therefore, to treat pharmaceutical wastewater more effectively and develop new treatment technologies, this review introduces the characteristics of pharmaceutical wastewater and the currently available treatment technologies in detail.

---

*Corresponding Authors: wang_jg7@hdec.com; fu_sj@hdec.com; fang_hf@hdec.com; li_sy5@hdec.com and lin_xh@hdec.com

DOI: 10.1201/9781003433651-49

375

2 PHARMACEUTICAL WASTEWATER CHARACTERISTICS

Pharmaceutical wastewater has complex water quality components, is difficult to treat, and generally has the following characteristics.

(1) Complex water quality components
Pharmaceutical reactions are complex with many by-products, and the reaction raw materials are often solvent-like substances or compounds with cyclic structures. Therefore, the pollutant components in wastewater are numerous and complex, which increases the difficulty of wastewater treatment.

(2) High concentration
The pharmaceutical process needs to use various chemical raw materials, involving multi-step reactions and a low utilization rate of raw materials [1]. Most raw materials will be discharged, resulting in high pollutants concentration in the wastewater.

(3) High COD value
The chemical oxygen demand (COD) of pharmaceutical wastewater is tens of thousands or even higher, which is caused by the incomplete reaction, causing a large number of by-products, raw materials, and solvent media to enter the wastewater.

(4) High toxic and harmful substance concentration
Many organic pollutants in pharmaceutical wastewater are toxic and harmful to microorganisms, such as halogen, nitro, organic nitrogen, dispersants, and surfactants.

(5) Many un-biodegradable substances
Most pollutants in pharmaceutical wastewater are un-biodegradable substances, such as halogen, ether, nitro, thioether, alum, and some heterocyclic compounds.

(6) High salinity
Excessive salinity in wastewater has an apparent inhibitory effect on microorganisms. When the chlorine ion in wastewater exceeds a specific concentration, the microorganism activity will be inhibited, and the removal rate of COD will be significantly reduced; the concentration of chlorine ion in wastewater is too large, which is very likely to cause sludge expansion, making the process operation difficult and causing the death of the microorganism.

(7) High chromaticity
Pharmaceutical wastewater contains raw materials and by-products expressed in various colors. The colored wastewater intercepts light transmission in water, thus affecting the growth of aquatic organisms and inhibiting the natural purification ability of organic substances decomposed by daylight catalysis.

3. PHARMACEUTICAL WASTEWATER TREATMENT TECHNOLOGY

The commonly used treatment methods for pharmaceutical industry wastewater are physical-chemical, biological, combined physical-chemical, and biological processes.

3.1 *Physical and chemical technology*

3.1.1 *Coagulation and sedimentation method*
The physical and chemical methods are preferred to the more economical coagulation and sedimentation methods. Usually, coagulation treatment can effectively reduce pollutant concentration and improve the biodegradation performance of wastewater. However, the flocculation and sedimentation process has problems such as generating much chemical sludge, low effluent pH, and low ammonia nitrogen removal efficiency. Therefore, even if there is a better treatment effect, it is still necessary to consider it carefully.

376

3.1.2 *Air flotation method*

The air flotation method usually includes various forms, such as gas flotation, dissolved air flotation, chemical air flotation, and electrolytic air flotation [3] [4]. For example, gentamicin wastewater is often treated by chemical air flotation. After chemical air flotation treatment, the COD removal efficiency of gentamicin can reach more than 50%, and the solids suspension removal rate can reach more than 70%.

3.1.3 *Adsorption method*

The adsorption method uses porous solids to adsorb one or several pollutants to purify the wastewater [5]. Coal ash or activated carbon adsorption is commonly used in wastewater treatment in the pharmaceutical industry to pretreat wastewater generated from proprietary Chinese medicines, vitamin B6.

3.1.4 *Air stripping*

When the concentration of ammonia nitrogen exceeds the concentration that microorganisms are allowed to degrade, the microorganisms will be inhibited by NH_3-N, resulting in poor removal. In pharmaceutical wastewater treatment, air stripping is commonly used to reduce the ammonia nitrogen content, for example, acetaminophen iodofuranone wastewater can be used for ammonia nitrogen removal [6].

3.2 *Biological treatment technology*

Currently, the more advanced treatment method for pharmaceutical wastewater is the activated sludge method. The enhanced pretreatment and aeration method make the activated sludge method more stable, making it the primary method for pharmaceutical wastewater treatment. However, the activated sludge method still has some disadvantages, including the need for significant dilution of the influent, the occurrence of sludge expansion, the high yield of residual sludge, and the necessity of secondary or multi-stage treatment. Therefore, the treatment effect of the activated sludge method can be improved by improving the aeration method and adopting immobilization technology, and they have become an essential part of the research conducted on the activated sludge method.

3.2.1 *Sequential batch reactor*

The sequential batch reactor (SBR) method is more suitable for treating wastewater with intermittent discharge and significant water quantity and quality fluctuations. SBR biological treatment technology has been widely used in the treatment of municipal wastewater and food industry wastewater. The SBR method has also been successfully applied in treating many types of pharmaceutical industrial production wastewater [7], but the SBR method has the disadvantages of sludge settling and long mud-water separation time. It needs to maintain high sludge concentration when treating high-concentration wastewater. High sludge concentration is highly susceptible to viscous sludge swelling. Therefore, powdered activated carbon (PAC) is often considered to be added to the activated sludge system so that the aeration tank foam can be reduced, sludge settling performance, liquid-solid separation performance, and sludge dewatering performance can be improved, and higher removal rates can be obtained.

3.2.2 *The biological contact oxidation*

The biological contact oxidation has the characteristics of both the activated sludge and biofilm and a high treatment load, making it possible to treat organic wastewater that easily causes sludge expansion [8]. In pharmaceutical wastewater treatment, the biological contact oxidation method can be directly used as a pretreatment process. If the concentration of influent wastewater is too high, a large amount of foam also tends to appear in the tank, and prevention and countermeasures must be taken during operation.

3.2.3 Upflow Anaerobic Sludge Bed (UASB)

UASB has the advantages of high anaerobic digestion efficiency and simple structure [9]. The key to the efficient and stable operation of UASB is the formation of granular sludge with high methanogenic activity and good settling performance. However, when using the UASB to treat pharmaceutical production wastewater, it is usually required that the SS content should not be too high to ensure that the COD removal rate can be above 85% to 90%. Upflow anaerobic sludge bed filter (UASB + AF) is a new type of composite anaerobic reactor developed in recent years, which combines the advantages of UASB and anaerobic filter (AF) to improve the performance of the reactor [9]. This composite reactor can effectively trap sludge and accelerate sludge granulation during start-up operation, and it has a better ability to withstand fluctuations in volumetric load, temperature, and pH. This composite anaerobic reactor has been used to treat pharmaceutical wastewater.

3.2.4 Photosynthetic bacteria treatment (PSB)

Many strains of the genus Pseudomonas erythropoiesis in PSB can decompose and remove organic matter using small molecules as hydrogen donors and carbon sources. PSB can metabolize organic matter under aerobic, microaerobic, and anaerobic conditions, and anaerobic acidification pretreatment is often used to improve the treatment effect of PSB [10]. For certain non-antimicrobial biochemicals, PSB can be considered for wastewater treatment in combination with other physical or biological treatment technologies.

3.3 Combined treatment process

3.3.1 Flocculation sedimentation + hydrolysis acidification + SBR (FHS)

The FHS process is a proven method for treating pharmaceutical wastewater, and it is an economical and reasonable treatment process.

Anaerobic hydrolysis is used as pretreatment for various biochemical treatments, which significantly reduces engineering investment and operating costs because aeration is not required. Anaerobic hydrolysis can improve the biochemical properties, reduce the load of subsequent biological treatment, and reduce the aeration of the subsequent aerobic process by a large amount; thus, it is widely used in the treatment of high-concentration organic wastewater of pharmaceuticals which is difficult to biodegrade.

A better treatment effect can be achieved if the hydrolysis temperature is maintained above 10°C. Thus, using the hydrolysis acidification pretreatment process in cold regions has significant advantages in treating pharmaceutical wastewater with high concentration and complex and variable composition. However, in the process of sludge cultivation, the bacteria in aerobic and anoxic sludge are susceptible to the environment. If they are in overload operation conditions for a long time, the nitrification reaction will become slow, resulting in high NO_2^--N accumulation and making the system operation stay in the nitrosation stage.

3.3.2 Electrolysis + SBR

The biochemical process is the most common method to treat pharmaceutical wastewater. However, with the strengthening of environmental awareness and the improvement of environmental standards, the traditional biochemical method is difficult to achieve the treatment goal. The combination of electrolysis and SBR is more feasible for treating pharmaceutical wastewater.

When pretreating pharmaceutical wastewater by electrolysis, the higher the electrolysis voltage, the faster the removal of COD and color. After pretreatment by electrolysis, the biochemical properties of wastewater are greatly improved, but too much electrolysis time can decrease wastewater's biochemical properties. The removal rate of COD from electrolytic pretreatment is only 37%~47%, and then the removal rate can reach 80%~86% in SBR biochemical treatment system. The effect of pH on the electrolysis effect exists; if the pH is too high or too low, it is unsuitable for removing COD. When the pH is 7, the electrolysis effect is relatively good; the effect of pH on chromaticity removal is relatively tiny.

3.3.3 Anaerobic/aerobic process (A/O)

Developed based on a conventional secondary biochemical treatment system, the A/O process is a new process with the simultaneous removal of pollutants such as organic matter and nitrogen. The A/O process can improve the removal efficiency of the difficult-to-degrade organic in wastewater so that the system has a high COD removal rate (96.2%) while maintaining a high BOD_5 removal rate (98.6%).

The A/O combines the advantages of both biofilm and SBR, and it can effectively utilize the reactor volume and improve the treatment efficiency. After the anaerobic treatment of pharmaceutical wastewater, the biochemical properties are improved. Experiments showed that with the A/O, the effluent COD was less than 250 mg/L at an average influent COD of 5832.9 mg/L, and the average COD removal rate reached 96.2%. However, when the temperature fluctuates sharply (greater than 2-3 °C), it significantly decreases gas production and the treatment effect of the anaerobic. Therefore, the temperature should be kept as stable as possible in the design and operation of wastewater treatment projects.

3.3.4 Air flotation+hydrolysis+ aerobic process (AHA)

The AHA combines the advantages of physical, anaerobic, and aerobic treatment, and it is suitable for treating pharmaceutical wastewater that is difficult to biodegrade. The pre-treatment by air flotation can effectively reduce the organic and COD_{Cr} of wastewater, which is beneficial to the subsequent biological treatment. Hydrolysis acidification improves wastewater's biochemical properties and provides conditions for subsequent aerobic treatment. In the hydrolysis stage, large-molecule organic matter is degraded into small-molecule substances, and substances that are difficult to be biodegraded are transformed into easily biodegraded substances so that the wastewater can be treated in the subsequent aerobic treatment unit with less residence time. After hydrolysis and acidification, wastewater directly enters the contact oxidation tank for aerobic treatment. In addition, for pharmaceutical wastewater containing hard-to-degrade organic matter, the addition of domestic wastewater for co-treatment can improve the wastewater treatment effect by co-substrate conditions.

3.4 Advanced oxidation treatment technology

3.4.1 Fenton method

The Fenton method uses iron salts (Fe^{3+} or Fe^{2+}) as catalysts and produces strong oxidizing [·OH] in the presence of H_2O_2 [11] [12]. [·OH] can oxidize many organic molecules, and the reaction process does not require high temperature and pressure. Fenton method reaction conditions are mild, the equipment is relatively simple, and the scope of application is also relatively wide. The disadvantage of this method is that the oxidation capacity is relatively weak, and the effluent water contains many iron ions.

3.4.2 Wet-Air-oxidation process

WAO is a method that uses oxygen or air as an oxidant at high temperature (125-320°C) and high pressure (0.5-10 MPa) to oxidize organic in the aqueous state or inorganic in the reduced state to produce CO_2 and H_2O [13] [14]. When the wet oxidation method was used to treat VC pharmaceutical wastewater, it was found that the COD removal rate of the wastewater could be improved by about 23% after adding the catalyst Ti-Ce-Bi at a reaction temperature of 200°C, partial pressure of oxygen of 315 MPa, a total pressure of 5.5 MPa, and a reaction time of 60 minutes. Meanwhile, it was found that the BOD_5/COD increased from 0.17 to more than 0.6, and the biochemical properties were significantly improved.

3.4.3 *Photocatalytic oxidation*

Photocatalytic oxidation is a simple, efficient, and promising technology [15]. It can oxidize almost all reducing substances in a specific time and has the advantages of high energy utilization, good decolorization effect, no residual sludge, and no secondary pollution [16] [17]. Photocatalytic oxidation is a catalytic process using n-type semiconductors (such as TiO_2, SrO_2, WO_3, and SnO_2) as catalysts. When these catalysts are exposed to near-ultraviolet light radiation, electron-hole pairs (h^+-e^-) are formed. Since holes have a strong oxidizing ability, when these electrons and holes migrate to the surface of the particles, water forms hydroxyl radicals with a powerful oxidizing ability on the surface of the semiconductor. Using [·OH] can then oxidize and completely mineralize various organic substances. The water is made to form hydroxyl radicals with powerful oxidizing ability on the surface of the semiconductor and using [·OH]; it is possible to oxidize various organic substances and make them completely mineralized.

3.4.4 *Ozone oxidation method*

Ozone is considered an effective oxidant and disinfectant with a solid oxidizing ability. The treatment of organic wastewater by ozone oxidation technology has the advantages of fast reaction speed and no secondary pollution [18]. During the oxidation reaction of ozone, the oxidative decomposition reaction of ozone is a radical reaction in which O_3 and OH^- undergo a series of reactions to produce O_2 and the radical [·OH], which has a more substantial oxidative capacity than O_3 and can oxidatively decompose more organic matter [19].

4   CONCLUSION

Pharmaceutical wastewater is highly concentrated and difficult to treat. Researchers have developed various treatments based on these characteristics, including physical, chemical, biological, and coupled treatment technologies. However, the development of pharmaceutical wastewater treatment technologies is not mature. Furthermore, the problems of insufficient stability effluent, poor water quality, high cost, and low resource utilization are still prominent. Therefore, developing new and efficient pharmaceutical wastewater treatment technologies is also important for future research.

ACKNOWLEDGMENTS

This work was financially supported by the Drainage System Comprehensive Efficiency Evaluation and Collaborative Operation Technology Research (KY2021-HS-02-11).

REFERENCES

[1] Tiwari B., Sellamuthu B., Ouarda Y., et al. Review on Fate and Mechanism of Removal of Pharmaceutical Pollutants From Wastewater Using Biological Approach. *Bioresource Technology*. 2017, 224: 1–12.

[2] Guo Y., Qi P.S., Liu Y.Z. A Review on Advanced Treatment of Pharmaceutical Wastewater: 2017 International Conference on Environmental and Energy Engineering (Ic3e 2017). *Binlin D. International Conference on Environmental and Energy Engineering (IC3E)*: 2017: 63.

[3] Suarez S., Lerna J.M. and Omil F. Pre-treatment of Hospital Wastewater by Coagulation-flocculation and Flotation. *Bioresource Technology*. 2009, 100(7): 2138–2146.

[4] Choi M., Choi D.W., Lee J.Y., et al. Removal of Pharmaceutical Residue in Municipal Wastewater by DAF (Dissolved Air Flotation)-MBR (Membrane Bioreactor) and Ozone Oxidation. *Water Science and Technology*. 2012, 66(12): 2546–2555.

[5] Patel M., Kumar R., Kishor K., et al. Pharmaceuticals of Emerging Concern in Aquatic Systems: Chemistry, Occurrence, Effects, and Removal Methods. *Chemical Reviews*. 2019, 119(6): 3510–3673.

[6] Wei S.P., van Rossum F., van de Pol G.J., et al. Recovery of Phosphorus and Nitrogen from Human Urine by Struvite Precipitation, Air Stripping and Acid Scrubbing: A Pilot Study. *Chemosphere*. 2018, 212: 1030–1037.

[7] Mir-Tutusaus J.A., Sarra M. and Caminal G. Continuous Treatment of Non-sterile Hospital Wastewater by Trametes Versicolor: How to Increase Fungal Viability by Means of Operational Strategies and Pretreatments. *Journal of Hazardous Materials*. 2016, 318: 561–570.

[8] Zhai W., Yang F., Mao D., et al. Fate and Removal of Various Antibiotic Resistance Genes in Typical Pharmaceutical Wastewater Treatment Systems. *Environmental Science And Pollution Research*. 2016, 23(12): 12030–12038.

[9] Wang Y., Feng M., Liu Y., et al. Comparison of Three Types of Anaerobic Granular Sludge for Treating Pharmaceutical Wastewater. *Journal Of Water Reuse And Desalination*. 2018, 8(4): 532–543.

[10] Madukasi E.I., Dai X., He C., et al. Potentials of Phototrophic Bacteria in Treating Pharmaceutical Wastewater. *International Journal Of Environmental Science And Technology*. 2010, 7(1): 165–174.

[11] Khan N.A., Khan A.H., Tiwari P., et al. New Insights Into the Integrated Application of Fenton-based Oxidation Processes for the Treatment of Pharmaceutical Wastewater. *Journal Of Water Process Engineering*. 2021, 44.

[12] Liu X., Wang C., Ji M., et al. Pretreatment of Ultra-High Concentration Pharmaceutical Wastewater by a Combined Fenton and Electrolytic Oxidation Technologies: COD Reduction, Biodegradability Improvement, and Biotoxicity Removal. *Environmental Progress & Sustainable Energy*. 2016, 35(3): 772–778.

[13] Segura Y., Del Alamo A.C., Munoz M., et al. A Comparative Study Among Catalytic Wet Air Oxidation, Fenton, and Photo-Fenton Technologies for the On-site Treatment of Hospital Wastewater. *Journal of Environmental Management*. 2021, 290: 112624.

[14] Zhan W., Wang X., Li D., et al. Catalytic Wet Air Oxidation of High Concentration Pharmaceutical Wastewater. *Water Science And Technology*. 2013, 67(10): 2281–2286.

[15] An C., Zhang D., Liu T., et al. Study on Photocatalytic Degradation of Pharmaceutical Wastewater by Heteropoly Acid Silver Photocatalytic Oxidation: 2018 4th international Conference on Environmental Science and Material Application. *4th International Conference On Environmental Science And Material Application (ESMA): 2019: 252*.

[16] You G. Pharmaceutical Wastewater Treatment by cuo/tio2 Photocatalytic Oxidation. *Oxidation Communications*. 2016, 39(4): 3206–3211.

[17] Pablos C., van Grieken R., Marugan J., et al. Simultaneous Photocatalytic Oxidation of Pharmaceuticals and Inactivation of Escherichia Coli in Wastewater Treatment Plant Effluents with Suspended and Immobilised TiO2. *Water Science and Technology*. 2012, 65(11): 2016–2023.

[18] Almomani F.A., Shawaqfah M., Bhosale R.R., et al. Removal of Emerging Pharmaceuticals from Wastewater by Ozone-based Advanced Oxidation Processes. *Environmental Progress & Sustainable Energy*. 2016, 35(4): 982–995.

[19] Buffle M., Schumacher J., Salhi E., et al. Measurement of the Initial Phase of Ozone Decomposition in Water and Wastewater by Means of a Continuous Quench-flow System: Application to Disinfection and Pharmaceutical Oxidation. *Water Research*. 2006, 40(9): 1884–1894.

Civil Engineering and Energy-Environment – Gao & Duan (Eds)
© 2023 the Author(s), ISBN 978-1-032-56059-5

Research on the impact factors of urban terminal industrial carbon emissions based on the Kaya-LMDI method

Hao Chen*
State Grid (Suzhou) City Energy Research Institute, Suzhou, China

Yewei Tao & Peng Wang
State Grid Suzhou Power Supply Company, Suzhou, China

Yahui Ma & Wenbo Shi
State Grid (Suzhou) City Energy Research Institute, Suzhou, China

ABSTRACT: Carbon reduction in the industrial sector is an important basis for cities to achieve carbon peak and carbon neutrality, especially for such super industrial cities as Suzhou. In order to study the main impact factors of terminal industrial carbon emissions, this paper proposes an extended Kaya equation that takes into account such impact factors as scale growth, industrial structure, energy consumption intensity, and energy structure, decomposes the impact factors based on LMDI method, and then performs an empirical study based on the actual data of Suzhou from 2015 to 2019. The research shows that scale growth is the main driving factor for the growth of industrial carbon emissions; the improvement of the industrial structure significantly inhibits industrial carbon emissions; energy efficiency improvement in key industries is an important measure to reduce industrial carbon emissions; considering the clean production of electricity and heat, the improvement of energy structure will help conspicuously reduce industrial carbon emissions.

1 INTRODUCTION

Since China put forward the strategy of carbon peak and carbon neutrality, the country and local governments have successively introduced a series of low-carbon policies to accelerate the implementation of pollution and carbon reduction actions in key domains represented by industry. Low-carbon development in the industrial sector is of great significance for mitigating climate change. China is currently in the stage of in-depth industrialization and urbanization. In 2021, China's industrial-added value accounts for 32.6% of GDP, higher than 17.65% in the United States and 19.6% in the United Kingdom [1]. The proportion of traditional industries is still high, and strategic emerging industries and high-tech industries have not yet become the leading force of economic growth. As the representative industrial city of China, Suzhou shoulders heavy responsibilities to go through the green and low-carbon transformation. Currently, the six traditional pillar industries in Suzhou include metallurgy, chemical industry, textile, and other industries, which are not only important support for the economy but also key industries of carbon emissions. In 2019, the total carbon emissions of the six most energy-consuming industries represented by the steel industry of Suzhou reached 66 million tons (excluding indirect emissions), accounting for

---

*Corresponding Author: chenhao201108@126.com

382   DOI: 10.1201/9781003433651-50

94% of the total carbon emissions of the terminal industry and 35% of the total carbon emissions of the whole society; if indirect carbon emissions related to electricity and heat are considered, the six most energy consuming industries account for 60% of the total carbon emissions of the whole society. To sum up, energy conservation and carbon reduction in key industrial sectors will be the key to achieving carbon peak and carbon neutrality in Suzhou.

Factors affecting carbon emissions of urban energy consumption include economic growth, industrial structure, energy structure, energy efficiency, energy price, per capita income, permanent population size, and other factors [2]. The Kaya equation proposed by Japanese scholar Kaya connects urban carbon emissions with human activities such as economy, policy, population, and other factors, which has strong scalability and is widely used [3]. In terms of decomposition methods of influencing factors, common factor decomposition methods mainly include index decomposition analysis (IDA) [4], structural decomposition analysis (SDA) [5], and data fitting method [6]. IDA can be further divided into the Laspeyres method and the Divisia method [7]. The LMDI decomposition method proposed by B.W. Ang [8] belongs to the representative branch of the Divisia decomposition method, which is characterized by full decomposition, unbiased estimation, and low data requirements, and is widely used. Many scholars use the LMDI model to study the influencing factors of regional and industrial carbon emissions, including regional total carbon emissions [9], industrial [10], manufacturing [11], logistics [12], transport sector [13], carbon emissions, etc. Some scholars combined the LMDI model with the STIRPAT model to build a carbon emission prediction model [14]. Some scholars conducted research on the influencing factors of carbon emissions in the whole region [15] and the coal industry [16] based on the Kaya-LMDI method.

Therefore, this paper focuses on the carbon emissions of major high-energy-consuming industries. First, an analysis model of the driving factors of carbon emissions in key industries is constructed. Considering the four impact factors of scale growth, industrial structure, energy consumption intensity, and energy structure, the Kaya equation is extended. Based on the LMDI method, the carbon emission increment caused by each influencing factor is calculated, so as to obtain the impact of different factors on the carbon emissions of key industries and the whole industry sector. Taking Suzhou as an example, an empirical analysis is carried out to verify the effectiveness of the method. The research shows that scale growth is the main driving factor for the growth of industrial carbon emissions; the improvement of the industrial structure significantly inhibits industrial carbon emissions; energy efficiency improvement in key industries is an important measure to reduce industrial carbon emissions; considering the clean production of electricity and heat, the improvement of energy structure will help conspicuously reduce industrial carbon emissions. In the future, with the standardization of carbon emission accounting, the research on influencing factors will be further improved.

2 KAYA-LMDI METHOD

The classical Kaya equation was first proposed by Japanese professor Yoichi Kaya, which is usually expressed as:

$$c_t = p_t \times \frac{g_t}{p_t} \times \frac{e_t}{g_t} \times \frac{c_t}{e_t} \tag{1}$$

where c_t represents the carbon emissions in the year t, and p_t, g_t, and e_t represent the total population, GDP, and energy consumption in the year t, respectively. Considering that the main research object of this paper is terminal industry, the main factors affecting carbon emissions in the industrial sector are output value scale, industrial structure, energy intensity, and energy structure. Therefore, on the basis of the traditional Kaya equation, we eliminate the population influencing factors and extend the impact factors to the energy structure and industrial structure. The extended Kaya equation is constructed as follows:

$$
\begin{aligned}
CI_t &= \sum_{i \in I} \sum_{j \in J} c_{i,j,t} \\
&= \sum_{i \in I} \sum_{j \in J} g_t \times \frac{e_{i,j,t}}{e_{j,t}} \times \frac{g_{j,t}}{g_t} \times \frac{e_{j,t}}{g_{j,t}} \times \frac{c_{i,j,t}}{e_{i,j,t}} \\
&= \sum_{i \in I} \sum_{j \in J} g_t \times A_{i,j,t} \times B_{j,t} \times C_{j,t} \times a_{i,j,t}
\end{aligned}
\tag{2}
$$

where CI_t represents the terminal industrial carbon emissions in the year t, and i, j represents energy type and industry type, respectively. $c_{i,j,t}$ represents the carbon emission corresponding to the energy type i consumed by the industry j in the year t, $e_{i,j,t}$ represents the consumption of the energy type i of the industry j in the year t, g_t, $A_{i,j,t}$, $B_{j,t}$, $C_{j,t}$, respectively, represents the four influencing factors affecting the carbon emissions of the terminal industry: the overall scale growth, the industrial energy structure, the industrial structure of the terminal industry and the industrial energy consumption intensity, and $a_{i,j,t}$ represents the carbon emission coefficient.

Assuming that t_0 is the base year, the carbon emission difference between the year t and the base year ΔCI_t can be decomposed into:

$$
\begin{aligned}
\Delta CI_t &= \sum_{i \in I} \sum_{j \in J} c_{i,j,t} - c_{i,j,0} \\
&= \sum_{i \in I} \sum_{j \in J} \Delta c_{i,j,t}^g + \Delta c_{i,j,t}^A + \Delta c_{i,j,t}^B + \Delta c_{i,j,t}^C + \Delta c_{i,j,t}^a
\end{aligned}
\tag{3}
$$

where ΔCI_t represents the annual carbon emission increment of the terminal industry, $c_{i,j,0}$ and $c_{i,j,t}$ represents the corresponding carbon emissions of the energy type i of the industry j in the base year and the year t, and $\Delta c_{i,j,t}^g$, $\Delta c_{i,j,t}^A$, $\Delta c_{i,j,t}^B$, $\Delta c_{i,j,t}^C$, $\Delta c_{i,j,t}^a$ are, respectively, the carbon emission increment corresponding to the impact factors: scale growth, energy structure, industrial structure, energy consumption intensity, and carbon emission coefficient change. In this paper, the carbon emission coefficient is treated as a constant value, so the carbon emissions caused by the change in the carbon emission coefficient are not considered.

Based on the LMDI method, we decompose Formula (3) [15]:

$$
\begin{aligned}
\Delta c_t^g &= \sum_{i \in I} \sum_{j \in J} \Delta c_{i,j,t}^A \\
&= \sum_{i \in I} \sum_{j \in J} \frac{(c_{i,j,t+1} - c_{i,j,t})(\ln g_{t+1} - \ln g_t)}{\ln c_{i,j,t+1} - \ln c_{i,j,t}}
\end{aligned}
\tag{4}
$$

where Δc_t^g represents the carbon emission increment corresponding to the scale growth in the year t. Similarly, the carbon emission increment corresponding to other impact factors can be obtained.

3 DATA SOURCE DESCRIPTION

This paper takes Suzhou as an example to carry out a case analysis. The data source is mainly the statistical yearbook of Suzhou from 2015 to 2020, including the output value of the Suzhou terminal industry by industry, energy consumption by category, etc. The carbon emission coefficient refers to the IPCC greenhouse gas accounting guidelines and the Provincial Greenhouse Gas Inventory Guidelines of China. The electric power emission factor of the provincial power grid in the same period is used. The coal equivalent conversion coefficient of various types of energy refers to the China Energy Statistics Yearbook.

4 CASE ANALYSIS

4.1 *Decomposition results*

This paper only considers carbon emissions from energy consumption. Taking 2015 as the base year, the Kaya formula and LMDI decomposition method were used to decompose the main factors affecting the total carbon emissions (including indirect emissions) in the industrial sector in Suzhou from 2016 to 2019, and the annual impact effects of various factors were decomposed.

The total contribution of scale growth, industrial structure, energy intensity, and energy structure to Suzhou's terminal industrial carbon emissions in the four years from 2016 to 2019 was 1.32, -9.46, 6.49, and 3.48 million tons, respectively. The largest incremental contribution in order is from scale growth, energy intensity, and energy structure, and the restraining factor is industrial structure. The decomposition results are shown in Table 1 and Figure 1. The decomposition results in some key industries are shown in Figure 2.

Table 1. Decomposition of terminal industrial total CO_2 emissions change of Suzhou.

(Unit: 10000 tons)	2015–2016	2016–2017	2017–2018	2018–2019	Year Cumulative
Scale growth effect	216.12	310.05	728.47	65.06	1319.69
Industrial structure effect	−168.00	-140.80	−707.50	70.74	−945.56
Energy intensity effect	183.04	33.80	35.78	396.16	648.79
Energy structure effect	54.88	126.43	73.34	93.63	348.28
Combined effect	286.04	329.48	130.09	625.60	1371.20

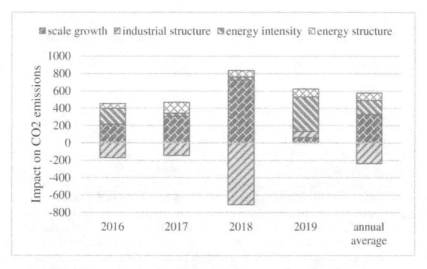

Figure 1. Decomposition of terminal industrial total CO_2 emissions change of Suzhou (unit: 10000 tons).

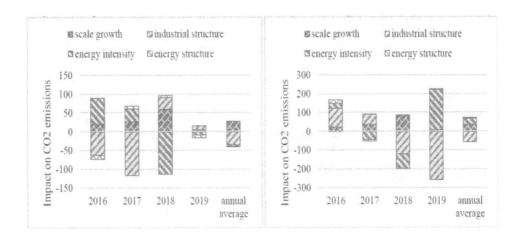

(a) Textile industry   (b) Chemical industry

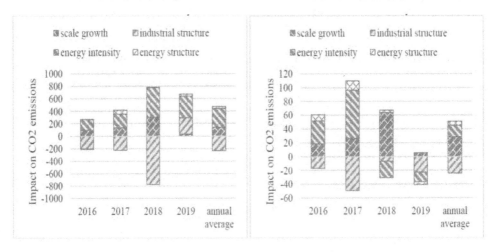

(c) Steel industry   (d) Computer equipment manufacturing industry

Figure 2. Decomposition of total CO_2 emissions change in textile, chemical, steel, and computer equipment manufacturing industries of Suzhou (unit: 10000 tons).

4.2 *Main findings*

It is found that scale growth is the main driving factor for the growth of industrial carbon emissions. During the four years from 2016 to 2019, the carbon emission increment brought by the scale growth accounted for 96% of the total, and the annual carbon emission increment was a large positive value, indicating that the industrial scale growth continued to drive the increase of energy consumption carbon emissions in the industrial field.

The improvement of the industrial structure plays a significant role in restraining industrial carbon emissions. The steel industry was taken as an example. In 2018, the output value of Suzhou's steel industry decreased by 8%, and its proportion in the total output value of the terminal industry dropped by one percentage point. As a result, the adjustment of the industrial structure reduced carbon emissions by 7.79 million tons. In 2019, the output value of the steel industry increased by 5%, and its proportion rose by 0.3 percentage points.

Accordingly, the adjustment of the industrial structure promoted carbon emissions to increase by 0.71 million tons conversely.

Affected by the energy intensity effect of the steel industry, the energy intensity effect plays a role in promoting the carbon emission of the Suzhou terminal industry. From 2016–2019, the energy intensity of the steel industry increased rather than decreased, and the energy intensity of the electrical machinery industry and computer equipment manufacturing industry slightly increased or remained flat, accordingly promoting the continuous increase of carbon emissions in the terminal industry, which have increased by 1.83, 0.34, 0.36, and 3.96 million tons in four years, respectively. Among them, the energy intensity effect of the steel industry drives the carbon emissions of Suzhou's terminal industries to increase by 1.74, 2.27, 4.75, and 3.51 million tons. Excluding the steel industry, the energy intensity effect drives the carbon emissions to increase by 0.09, -1.93, -4.40, and 0.45 million tons instead. It shows that promoting energy efficiency improvement in key industries is an important measure to reduce carbon emissions.

Since the indirect carbon emission factor of electricity and heat are relatively high, the energy structure effect generally promotes the growth of carbon emissions; if the electricity and heat are produced by clean energy, the improvement of the energy structure will help significantly reduce carbon emissions. The energy structure factor is reflected in the change in the average carbon emission effect of energy consumption. When we calculate the total carbon emissions of various industries, the indirect carbon emissions from electricity and heat are considered. The indirect carbon emission factors are generally higher than the coal, oil, and natural gas carbon emissions. Therefore, the higher the proportion of electricity and heat consumed by the industry, the higher the average carbon emissions of energy consumption. If electricity and heat are produced by clean energy, that is, its carbon emission factor is 0, the opposite conclusion will be drawn: the improvement of energy structure has a significant inhibition effect on the growth of carbon emissions (as shown in Figure 3). Therefore, it is necessary to strengthen clean power and heating supply, and further promote industrial electrification, methanol or hydrogen energy.

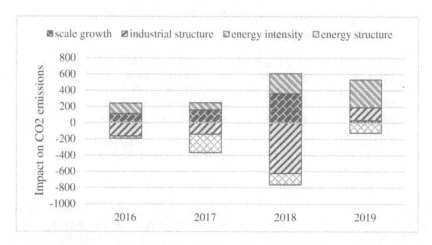

Figure 3. Decomposition of terminal industrial direct CO_2 emissions change of Suzhou (unit: 10000 tons).

5 CONSTRUCTION

This paper proposes a decomposition model of impact factors of urban terminal industrial carbon emissions based on the extended Kaya equation and LMDI decomposition method and studies the impact factors of terminal industrial carbon emissions in Suzhou during the 13th Five Year Plan period. The study found that scale growth is the main driving factor of carbon emission growth, industrial structure upgrading plays a restraining role in industrial carbon emissions, and energy efficiency improvement in key industries is an important measure to reduce carbon emissions. In particular, the industrial structure effect and energy intensity effect of the steel industry have a significant impact on Suzhou's industrial carbon emissions. If electricity and heat are produced by clean energy, the improvement of the energy structure will help significantly reduce total carbon emissions by reducing indirect carbon emissions. Therefore, Suzhou needs to strictly control the development of high-emission industries, promote high-end industrial upgrading, systematically improve industrial energy efficiency, strengthen clean energy substitution, and take multiple measures to promote low-carbon and green industrial development.

ACKNOWLEDGMENTS

This work was financially supported by the project Energy Efficiency Analysis of Key Industries in Suzhou and Research on Energy Efficiency and Carbon Reduction Service Strategy for Power Grid Companies.

REFERENCES

[1] Jiang Huiqin, Li Yixuan, Chen Miaomiao and Shao Xinxiao (2022) Prediction and Realization Strategy of the Carbon Peak of the Industrial Sector in Zhejiang Under the Vision of Carbon Neutrality. *Areal Research and Development*, 41(04):157–161+168.

[2] Xue Yue-xin, Xie Jing-chao, Huai Chao-ping, Li Hang and Liu Jia-ping (2022) Decomposition Analysis of Influencing Factors of Energy Related Carbon Emission in Beijing. *Journal of BEE*, 50 (09):128–132.

[3] Deng Xuankai (2021) Study on the Influencing Factors of Land Use Carbon Emissions in Wuhan City - Based on the Extended Kaya Equation and LMDI Decomposition Method. *Agriculture and Technology*, 41(20):104–109.

[4] Wu Zhenxin, Shi Jia and Wang Shuping (2014) *Analysis of the Driving Factors of Carbon Emissions in Beijing based on LMDI Method. Forum on Science and Technology in China*, 2: 85–91.

[5] Chai Qimin, Chen Yi and Xu Huaqing (2015) Study on the Decomposition Scheme of Regional Indicators Under the Double Control Target of Carbon Intensity and Total Amount - Taking Wenzhou City as an Example. *Energy of China*, 37 (004): 28–32.

[6] Zhao Xianchao, Ding Meng and Yang Ying (2012) Fitting Variation Analysis of Carbon Emissions and Influencing Factors in Changsha City. *Journal of Hunan University of Technology Social Science Edition*, 17 (001): 15–20.

[7] Liang Qidi, Feng Xiangzhao et al. (2020) Study on the Influencing Factors of Carbon Emissions From Energy Consumption Based on LMDI Method: Taking Tangshan City as an Example. *Environment and Sustainable Development*, 45(1):5.

[8] Ang B., Liu F. and Chew E. (2003) Perfect Decomposition Techniques in Energy and Environmental Analysis. *Energy Policy*, 31: 1561–1566.

[9] Chen Feng, Zhang Jing, Ren Jiao, et al. (2022) Spatiotemporal Variations and Influencing Factors of Carbon Emissions in the Yellow River Basin based on LMDI Model. *Journal of Earth Environment*, 13 (04):418–427.

[10] Liu Teng, Dong Hongguang, Gao Lehong and Luo Tao (2022) Factors Influencing Decoupling of Industrial Carbon Emissions and Countermeasures to Reduce Emissions in Ningxia. *Journal of Ningxia University (Natural Science Edition)*, 1–6[2022-10-10].

[11] Liu Xiaoli and Wang Yongli (2022) Analysis of Driving Factors of Carbon Emission in China's Manufacturing Industry based on LMDI Decomposition. *Statistics & Decision*, 38(12):60–63.

[12] Jiang Leping (2022) LMDI-based Carbon Emission Measurement and Influencing Factors of Logistics Industry in Anhui Province. *Logistics & Material Handling*, 27(06):173–177.

[13] Yang Shaohua, Zhang Yuquan and Gen Yong (2022) Examining the Changes in Carbon Emissions of the Transportation Sector in the Yangtze River Economic Belt – A LMDI and Theil index-based approach[J]. *China Environmental Science*, 1–23[2022-09-30].

[14] Liu Maohui, Zhai Huaxin and Liu Shengnan, et al. (2022) Comparative Analysis of Carbon Emissions in Tianjin based on LMDI method and STIRPAT Model. *Journal of Environmental Engineering Technology*, 1–11[2022-09-30].

[15] Wang Sheng, Tan Jian, Ma Yahui and Zhou Fenghua (2022) Impact Factor Analysis and Forecasting of the Carbon Emissions from Industries based on LMDI Method Under Multiple Uncertainties: The Case of Suzhou City. *Integrated Intelligent Energy*, 44(02):1–7.

[16] Chen Li and Li Zhengyan (2021) Analysis of Factors Influencing China's Coal Consumption Demand Based on Kaya-LMDI. *Coal Economic Research*, 41(06):4–9.

[17] Zhao Yixuan. (2019) Decomposition of Influencing Factors of Carbon Emissions in Beijing based on Kaya LMDI model. *Modern Economic Information*, (17):488–489.

Civil Engineering and Energy-Environment – Gao & Duan (Eds)
© 2023 the Author(s), ISBN 978-1-032-56059-5

Study on conventional process optimization after water source switching in the drinking water plant

Shuo Zhang*, Jiajiong Xu, Yan Wang & Ruhua Wang
Shanghai Municipal Engineering Design Institute (Group) Co., Ltd., Shanghai, P.R. China

ABSTRACT: Because of the switching of raw water from high turbidity to low turbidity, an optimization method has been tried on the conventional treatment process and the sludge treatment process. Production practice results showed that the dosage of coagulant was about 21 mg/L, which was reduced by more than 50% compared with the high turbidity raw water. The dosage of ammonium sulfate was increased by more than 60% to control bromate. The disinfection could be free chlorine disinfection or chloramine disinfection. The sludge discharge cycle of the advection sedimentation tank and the filter backwash cycle should be appropriately extended. After the adjustment of the operating parameters of the water plant, the factory water could comply with the national standard, effectively dealing with low-turbidity and algae-containing raw water.

1 INTRODUCTION

In order to improve the quality of raw water in Shanghai, the Qingcaosha Raw Water Project has been implemented. The raw water of the half-existing water plants in the urban area has been gradually switched from the Huangpu River to the Yangtze River. The quality of the raw water has been significantly improved compared with the upper reaches of the Huangpu River. The planned water plants will also use the raw water of the Yangtze River. The water quality of the raw water of the Huangpu River and the raw water of the Yangtze River is quite different. The two types of raw water are significantly different in each interval, and the total amount is the main water quality indicators, such as turbidity, oxygen consumption, and ammonia nitrogen. The concentrations of various pollutants in the Huangpu River are higher than those in Qingcaosha Reservoir. The water quality of the Yangtze River is better than that of ordinary rivers [1]. Turbidity, chromaticity, ammonia nitrogen, COD_{Mn}, and other indicators were significantly lower.

The quality of raw water has an extremely important and decisive role in the process combination, process type, and operating parameters [2]. Water plants sometimes require pretreatment and advanced treatment to strengthen the removal of ammonia nitrogen, and organic matter [3], but conventional processes are an important part of all water plants. For example, coagulation is always an important process for removing natural organic matter. In the past forty years, most research, design, and production were conducted around the raw water of the Huangpu River. However, the research on the relevant characteristics of the raw water in the Yangtze River and, the adaptive strategies for the operation of the water plant were relatively lacking. The water plants urgently need to solve the adjustment and optimization of process and operation parameters in various aspects such as dosing amount, turbidity of post-sinking water, flushing method, and intensity. Therefore, productive

---

*Corresponding Author: zhangshuo@smedi.com

390

DOI: 10.1201/9781003433651-51

research was carried out in the existing water plants, and the method of conventional process optimization after the new water source was systematically analyzed.

2 RAW WATER QUALITY

At present, the commonly used water source of water plants comes from Qingcaosha Reservoir. Qingcaosha Reservoir is located on Changxing Island at the mouth of the Yangtze River, and the water is drawn from the Yangtze River. The raw water entering the plant mainly has the characteristics of low turbidity, low organic matter, low ammonia nitrogen, high pH, and seasonal algae.

3 RESULTS AND DISCUSSIONS

3.1 *Coagulant reduction*

Aluminum sulfate has been long used as a coagulant in water plants, and there were extensive experience and application practices. When the source water of the Huangpu River was used, the dosage was mostly 40 mg/L, and it was as high as 60 mg/L in winter. The turbidity and concentration of organic pollutants in Qingcaosha raw water are significantly lower than those in the upper reaches of the Huangpu River, and thus the consumption of alum in production could be significantly reduced. The stirring test showed that after adding 5 mg/L of poly sulfur aluminum chloride to Qingcaosha raw water, the alum flowers were loose, and the number was small. When the turbidity of the supernatant was below 0.5 NTU, the dosage of aluminum sulfate could be 9 to 15 mg/L. When the turbidity of the water after sedimentation was controlled to be 0.5 to 1.0 NTU, the amount of alum added could be controlled to 10 to 25 mg/L. After the Huangpu River raw water was switched to Qingcaosha raw water, the amount of coagulant used in January decreased by 50%.

3.2 *Coagulant replacement*

After the raw water was switched, the concentration of aluminum ions in the effluent of some water plants increased for the high pH of Qingcaosha raw water, which was close to the national standard limit of 0.2 mg/L. In order to reduce the residual aluminum concentration in the effluent, aluminum sulfate was replaced. After the use of poly sulfur aluminum chloride, the pH of the raw water was reduced by about 0.8, and the aluminum ion in the factory water was effectively controlled below 0.1 mg/L.

3.3 *Sedimentation adjustment*

After the water source was switched, the treatment effect of the advection sedimentation tank was basically good, but the turbidity of the settled water increased slightly for the low turbidity of the raw water. The advective sedimentation tank of a plant in Yangpu was constructed a long time ago, and the designed hydraulic retention time was only one hour. The maximum turbidity of the water after the subsidence in the low-temperature period was no more than 3 NTU, with an average of 1.76 NTU, as shown in Table 1. The design hydraulic retention time of the advective sedimentation tank of a factory in Nanhui is 1.8 h, and the turbidity of the water after sinking was below 1 NTU, as shown in Table 2.

When the raw water from the Huangpu River was used, the sludge discharge cycle of the sedimentation tank was once a day. After switching to Qingcaosha raw water, the amount of sediment in the sedimentation tank was significantly reduced, and the sludge discharge cycle of the sedimentation tank was extended to once every two day.

Table 1. The turbidity of sedimentation effluent in Yangpu during the low-temperature period.

Raw water	Maximum turbidity NTU	Average turbidity NTU	Minimum turbidity NTU
Qingcaosha Reservoir	2.69	1.75	1.30
Huangpu River	2.59	1.43	0.59

Table 2. The effluent water of a plant in Nanhui during the low-temperature period.

Setting	Turbidity NTU	Chroma CU
Old sedimentation tank 1	0.82	<5
Old sedimentation tank 2	0.51	<5
New sedimentation tank 1	0.65	<5
New sedimentation tank 2	0.81	<5

3.4 Filter adjustment

After the water source was switched, the sand filter treatment effect was basically good, meeting the internal control indicators of the enterprise. However, the turbidity of the filtered water increased slightly for the low turbidity of the raw water, with a maximum of 0.25 NTU and an average of 0.11 NTU, as shown in Table 3.

Table 3. The turbidity of sand filter effluent in Yangpu during the low-temperature period.

Raw water	Maximum turbidity NTU	Average turbidity NTU	Minimum turbidity NTU
Qingcaosha Reservoir	0.26	0.12	0.07
Huangpu River	0.23	0.07	0.06

Affected by the addition of powdered activated carbon to the water pump station, black powdered activated carbon was deposited on the surface of the sand filter, but no significant effect was found on the filtration cycle and backwashing, as shown in Figure 1.

Figure 1. Powdered activated carbon deposited on the surface of the sand filter.

3.5 *Disinfection adjustment*

According to the quality of raw water, the sterilization effect of sodium hypochlorite was better than that of chloramine. Chlorine was added at the breakpoint first to generate free chlorine as the main disinfectant. Then ammonia and chlorine were added in sequence, and the factory water was disinfected with monochloramine. In this disinfection method, chlorination at the breaking point could completely consume the ammonia in the raw water, which made the final control in the chloramine stage easier. Due to the low ammonia nitrogen concentration in Qingcaosha raw water, the dosage of ammonium sulfate in the water plant was significantly increased by 60%.

In terms of ammonium sulfate injection, the dosage of liquid ammonium sulfate had a good linear relationship with the increase of ammonia nitrogen in water and had nothing to do with the initial value of ammonia nitrogen in the water. In the production process, the dosage of liquid ammonium sulfate could be determined according to the increase of ammonia nitrogen. When the concentration of bromide ions in the raw water was high and there was a disinfection contact tank, ammonia could be added after the disinfection contact tank and before the clean water tank in order to reduce the risk of bromate exceeding the standard. If there was no disinfection contact tank, ammonia could be added to the effluent of the sand filter. The dosage of ammonium sulfate was about 0.2 to 0.5 mg/L, and the bromate in the factory water was controlled below 5 µg/L.

3.6 *Sludge water treatment*

The structure related to the sludge water treatment system of a water plant in Pudong included a sludge adjustment tank, concentration tank, pre-concentration tank, balance tank, and dehydration machine room. After the filter tank flushing wastewater entered the pre-concentration tank for concentration, the preliminary thickened sludge with a solid content of about 1% entered the sludge discharge adjustment tank and the sedimentation tank to flow into the concentration tank for concentration. There was a sludge water regulation tank, and the sludge discharge water came from the clarifier, and the sludge was evenly discharged within 24 hours. There was one thickening tank, which could meet the sludge treatment requirements when the scale of the water plant was 400,000 m^3/d. The enrichment tank was divided into two compartments that could operate independently. The actual effective area of each grid was 126.7 m^2. The effective pool depth was 3.0 m. The designed solid load was 70 to 90 kg TDS/m^2/d, and the maximum solid load was 150 kg TDS/m^2/d. The organic polymer flocculant was injected into the influent water of the concentration tank. A 2.35-m-long inclined plate was on the upper part of the concentration tank, and the bottom was provided with a central drive mud scraper. The scraping wall of the scraper was equipped with a stirring grid to improve the sludge thickening effect by slow stirring. The solid content of the concentrated sludge at the bottom of the concentration tank was about 3%. The supernatant from the concentration pool was reused or discharged. The amount of dry sludge was calculated using the calculation formula of the British Water Research Center, as shown in Formula 1.

$$DS = SS + 0.2C + 1.53A + 1.9F \tag{1}$$

where DS = dry sludge content mg/L; SS = suspended solids in raw water removed mg/L; C = chroma removed; A = dosage of aluminum salt mg/L Al_2O_3, and F = dosage of iron salt mg/L Fe^{2+}.

Because of the characteristics of low turbidity and low organic matter in the raw water of Qingcaosha Reservoir, the amount of dry mud was significantly reduced, which was only 24% of the design expectation. According to the situation, the sludge discharge of the clarifier, the sludge discharge of the pre-concentration tank, the backwash of the filter, the operation of the centrifugal dehydrator, and the dosing of PAM at each point were optimized in production. Measures were taken such as reducing the time and volume of sludge discharge, reducing the frequency, intensity, and time of backwashing, and shortening the working time of the centrifugal dehydrator to achieve the purpose of saving resources, which could save energy and reduce consumption, as shown in Table 4.

Table 4. Review and comparison of design parameters of sludge drainage.

	Raw water	
Parameter	Huangpu River	Qingcaosha Reservoir
Scale/ m^3/d	400,000	400,000
Self-use water coefficient	1.05	1.05
Raw water turbidity (85%)/NTU	50	5
Original water chromaticity (85%)	23	5
Coagulant (average)/mg/L (Al_2O_3)	20	5
SS and turbidity coefficient	1.0	1.3
Dry mud/T/d	26.8	6.4
Pre-concentration tank sludge discharge/ m^3/d	93	93
Clarifier sludge discharge/ m^3/d	4160	4160
Concentrator Solids Flux/ kg DS/m^2/d	106	25
Concentrated tank liquid level load/ m^3/m^2/h	1.1	1.1
Thickener run time/h	16	16
Dehydrator running	Dual-purpose, work 16 h/d	One use and one preparation, work 7 h/d
Dehydrator Slurry volume/ m^3/h	30	30
Dehydrator dry mud volume/ kg/h	900	900

3.7 Water plant operation

The raw water of a water plant in Yangpu was switched to the water of Qingcaosha Reservoir. The conventional process operation parameters were determined after production debugging and operation, which mainly included the reduction of aluminum sulfate by more than 60%. The ammonia addition to the raw water was adjusted to ammonia addition after filtration, and the ammonia addition amount was increased.

The turbidity of the clarifier effluent was generally controlled at 1 to 2 NTU. The amount of alum added was 15 to 25 mg/L. When the turbidity was low, and the water temperature was high, the dosage should be low, and when the turbidity and algae were high, the high value should be used. The initial dose of PAM was 0.3 mg/L, and after the optimization of operation, only 0.10 to 0.13 mg/L PAM was added. The original design value of the sediment return water in the sedimentation zone was 40 m^3/h, and it was 20 m^3/h after operation optimization.

The normal rinsing cycle for a homogeneous filter tank was about 36 hours. During the period of high algae in the raw water, the flushing cycle was about 18 to 20 hours. When the water load reached the design scale, the flushing cycle was about 24 to 28 hours. The turbidity of the filter effluent was generally less than 0.1 NTU.

Compared with the original design value of 2.5 mg/L, the dosage of ozone was significantly reduced. The dosage of ozone was no more than 1 mg/L. The normal flushing cycle of the activated carbon filter was about 100 to 120 hours. During the period of high water temperature, the flushing cycle was about 48 to 72 hours. The turbidity of the effluent from the activated carbon filter was generally less than 0.1 NTU. The average removal rate of COD_{Mn} could reach 32.9%.

There were two chlorine points. One chlorine addition point was in the raw water inlet pipe of the pre-ozone contact tank, and the dosage was 1.0 mg/L. The other chlorine addition point was in the water inlet pipe of the disinfection contact tank, and the dosage was 1.1 mg/L. The ammonia addition point was in the outlet pipe of the disinfection contact tank, and the dosage was 0.2 to 0.3 mg/L.

The quality of the factory water has been significantly improved. According to the Standard for Drinking Water Quality GB 5749-2006, the qualified rate of the conventional water quality index of the existing conventional process production system and the advanced treatment system of the 7# production system reached 100%. The mean value of turbidity was lower than 0.14 NTU. The mean value of ammonia nitrogen was lower than 0.32 mg/L, and the mean value of COD_{Mn} was lower than 1.4 mg/L.

4 CONCLUSION

Based on the results and discussions presented above, the conclusions were obtained as follows:

(1) It was shown that the dosage of coagulant could be 21 mg/L, which was more than 50% less than that of the source water of the Huangpu River in the adaptive production practice of the water plant.
(2) The disinfection could be a combination of free chlorine disinfection and chloramine disinfection for the new raw water.
(3) It was concluded that the consumption of ammonium sulfate was increased by more than 60% when the source water of Huangpu River was used.
(4) The sludge discharge cycle of the advective sedimentation tank and the filter backwash cycle could be appropriately extended to reduce the running time of the sludge dewatering equipment.

ACKNOWLEDGMENTS

This work was financially supported by the Shanghai Science and Technology Committee Program (No. 22dz1209103).

REFERENCES

[1] Wen W., Xiang R., Jing L., Hao W., Li M., Jun L. and Jia Z. (2019) Sources, Distribution, and Fluxes of Major and Trace Elements in the Yangtze River. *Environ. Sci.*, 40 (11): 4900–4913.
[2] Platikanov S., Baquero D., González S., Martín-Alonso J., Paraira M., Cortina J.L. and Tauler R. (2019) Chemometric Analysis for River Water Quality Assessment at the Intake of Drinking Water Treatment Plants. *Sci. Total. Environ.*, 667: 552–562.
[3] Serajuddin M.d., Chowdhury M.d. and Aktarul I. (2018) Towards a Novel Approach to Improve Drinking Water Quality at Dhaka, Bangladesh. *Environ. Eng. Res.*, 23(2): 136–142.
[4] Du R., Zhou J.J., Wang F.J., Liu J., Tang X.B., Zou J.F., Yuan Y.X. and He J.G. (2015) Control of the Ozonation by-Products by O3/BAC in Shanxi Yellow River Water Treatment. *Adv. Mate. Res.*, 1119: 408–412.

Civil Engineering and Energy-Environment – Gao & Duan (Eds)
© 2023 the Author(s), ISBN 978-1-032-56059-5

Design of on-site wastewater treatment facilities in highway service areas

Bo Fan
Sichuan Jiuma Highway Co.Ltd, Aba Prefecture, China

Jiawei Wang*
Beijing Xinqiao Technology Development Co.Ltd, Beijing, China

Zhong Yan
Sichuan Jiuma Highway Co.Ltd, Aba Prefecture, China

ABSTRACT: Wastewater generated from highway service areas is the main pollutant source and gaining more attention in recent years. The goal of this study is to design and apply on-site wastewater treatment facilities in highway service areas for water conservation and environmental protection. The characteristics of wastewater from highway service areas were investigated. Meanwhile, an integrated wastewater treatment system based on MBBR technology (Moving bed biofilm reactor) was established. It is found that the wastewater in highway service areas shows great uncertainty and fluctuation in water quantity and quality, and suspended solids, COD, ammonium, and oil content are the main substances to be removed. The designed wastewater treatment system is robust and sustainable and has high efficiency in organic matter, nitrogen, and phosphorus removal. The effluent was reclaimed for farm, landscape, and road watering.

1 INTRODUCTION

Highways have become an increasingly vital transportation network in China in recent years. Providing convenience for passengers and drivers, highway service areas are usually constructed with public toilets, shopping malls, restaurants, parking (car wash) yards, maintenance stations, and petrol stations. The highway service area is constructed far away from the city, so the generated sewage is unable to be discharged into the municipal wastewater treatment plant. It has an adverse impact on the surrounding environment if the sewage is discharged in situ without suitable wastewater disposal. Therefore, it is necessary to set up a separate wastewater treatment system in highway service areas.

The generated wastewater quantity in the highway service area is about 30-300 m3/d in China, which shows high variability and is greatly affected by traffic intensity and climate conditions. Moreover, the water quality of the wastewater from highway service areas varies from the typical domestic wastewater, due to its different sources. Nowadays, China is facing a significant challenge of water shortage [1]. Therefore, the appropriate treatment of wastewater in highway service areas and the reclamation of the effluent are meaningful for both water conservation and environmental protection.

---

*Corresponding Author: wangjiawei1109@163.com

396

DOI: 10.1201/9781003433651-52

2   CHARACTERISTICS OF WASTEWATER IN HIGHWAY SERVICE AREAS

Sewage in highway service areas is mainly produced from public toilet flushing, restaurant cleaning, and parked vehicle maintenance, in addition to the petrol station. The composition of wastewater in the highway service areas is similar to the domestic water to some degree, where the main pollutants are COD, BOD_5, suspended solids (SS), ammonia (NH_4+-N), total nitrogen (TP), animal and vegetable oils and petroleum [2]. To evaluate the influent water quality from the on-site wastewater treatment facilities in highway service areas, several water samples were taken from several highway service areas in China [3], and the results are listed in Table 1.

Table 1.   Water quality of raw wastewater from the highway service areas in China (**mg/L**).

Highway	COD	BOD_5	SS	NH_4^+-N	TP
Xian-Baoji Highway	289–497	164–234	143–238	46.5–109.3	1.9–5.8
Yingkou-Songyuan Highway	428–618	82–237	156–273	48.2–82.6	3.4–6.9
Beijing-Hong Kong-Macao Highway	314–736	121–318	217–376	42.3–94.7	5.6–14.3
Lianyungang-Huoerguosi Highway	349–836	164–363	234–417	52.4–142.6	4.7–12.1
Shanghai-Chengdu Highway	352–640	138–256	220–369	34.2–53.7	4.8–15.7
Wuhu-Xuancheng Highway	286–689	168–354	184–357	39.8–61.4	1.9–5.7
Shenyang-Haikou Highway	236–789	168–354	104–257	19.8–57.9	2.9–6.7
domestic wastewater	250–400	100–200	200–220	30–40	1.5–5

As can be seen in Table 1, the concentration range of COD, BOD_5, SS, NH_4+-N, and TP in the sewage in the highway service areas are about 300 - 600 mg/L, 100 - 300 mg/L, 150 - 400 mg/L, 50 - 100 mg/L, and 3 - 15 mg/L, respectively. The measured concentration of COD is up to 830 mg/L, and NH_4^+-N is up to 140 mg/L, which is much higher than the average concentration of typical domestic wastewater. Moreover, the amount of influent fluctuates a lot and is affected by various factors, such as season, weather, and flow of people and vehicles. The water quantity of wastewater in highway service areas always reaches a peak during vacation and is extremely low in winter. Based on the above analysis, the total wastewater treatment scale in the highway service areas is limited, but the water quantity and quality fluctuate greatly. The concentration of organic pollutants, nitrogen, and phosphorus in wastewater is high, and the water quality is worse than that of typical domestic sewage. Hence, it is an essential task to design and apply the optimal treatment process for wastewater in highway service areas.

3   DESIGN OF ON-SITE WASTEWATER TREATMENT FACILITIES IN HIGHWAY SERVICE AREAS

3.1   *Background*

The aim is to design a robust and sustainable wastewater treatment system to treat and reclaim wastewater from highway service areas on Jiuzhi-Maerkang Highway. The project starts at the border between Sichuan and Qinghai provinces in China, where the section of the water environment and ecosystem is sensitive. Moreover, the highway was constructed on the plateau with an average altitude of more than 3,300 m. Under low-temperature and hypoxic conditions, the treatment efficiency of microorganisms will be adversely affected. Therefore, suitable and advanced treatment technology should be selected and applied to ensure effluent quality in this project.

According to the Environment Impact Assessment of the project, the treatment scale of each wastewater treatment system is designed as 50 m³/d. The water quality should meet the level I standard of *Integrated Wastewater Discharge Standard (GB 8978 - 1996)* [4] and the *Standard for Irrigation Water Quality (GB 5084 - 2021)* [5]. The effluent is supposed to be reclaimed for farm, landscape, and road watering instead of being discharged to the surface water. The required water quality of influent and effluent is listed in Table 2.

Table 2. The required water quality of influent and effluent from wastewater facilities in highway service areas (mg/L).

Parameters	BOD_5	COD_{Cr}	ammonium	petroleum products	animal oil content
Influent concentration	≤ 200	≤ 300	≤ 80	≤ 10	≤ 40
Effluent concentration	≤ 20	≤ 100	≤ 15	≤ 5	≤ 10

Based on the above, the designed on-site wastewater treatment facilities in highway service areas should possess the following features: (1) ensured high water quality of the effluent under low temperature and hypoxic conditions. (2) limited energy consumption and operation cost. (3) robust and compact.

3.2 Treatment process

According to the characteristics of wastewater in highway service areas, suspended solids, COD, ammonium, and oil content are the main substances to be removed. In consideration of the specific situation and nature of this project, the designed wastewater treatment system in highway service areas is mainly based on MBBR technology. The MBBR applied system consists of nano-plastic media with a specific surface area of about 900 m²/m³. Microorganisms such as *Bacillus subtilhls, Photosynthetic bacteria, Denitrifying bacteria*, and *Nitrifying bacteria* were cultivated in the bio-film. Compared to activated sludge systems, MBBR has no bulking problems and has the advantages of simple construction and maintenance, low cost, and saving land use. Compared to MBR technology (Membrane bioreactor), MBBR has no concern about membrane maintenance and blockage [6]. Therefore, MBBR is selected as the appropriate wastewater treatment process in highway service areas in this project.

The treatment process is designed and shown in Figure 1, it is able to produce high-quality effluent that meets the standard, and the clean water with low COD, BOD, SS, TN, and TP will be reclaimed for farm, landscape, and road watering. The specific process flow is as follows.

(1) The underground pipes bring the raw wastewater to the system, and the feed water flows through a bar grid to preliminarily achieve the reduction of large particles and partially suspended solids.
(2) The feed water is then temporarily stored in the regulation tank to adjust the unstable water quality and quantity and to reduce the instantaneous impact.
(3) After the primary treatment, the wastewater is pumped into MBBR tanks for biological treatment. It creates anoxic and aerobic conditions for nitrification and denitrification. The organic matter, nitrogen, and phosphorus are effectively removed through this process. Denitrification is an anoxic process to oxidize nitrate to nitrogen gas by nitrifiers, which consume organic compounds such as glucose and acetate as their carbon source and electron donor. Since the uncertainty of water quantity in highway service areas, the nutrient should be dosed when the intake amount is low. The activity of the microbe is constantly monitored, and the microbe will be added or adjusted if needed.

(4) The sludge mixture is refluxed in the secondary sedimentation tank by gravity sedimentation.
(5) The unreacted pollutants in the sludge and sludge mixture are discharged into the sludge tank for gravity concentration and digestion treatment to reduce the volume of sludge and improve the stability of sludge. In the meanwhile, the sanitation manure suction truck is regularly used for external transport and treatment.
(6) UV disinfection is a compact and inexpensive unit that requires less maintenance, the remaining microbe and organic matter will be degraded, and clean water will be produced.
(7) The clean water storage tank is the last unit to store clean water for reuse and works as a short-term buffer in case of maintenance and short extreme period as well.

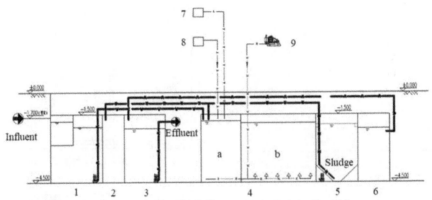

1- Bar grid and regulation tank; 2- Sludge tank; 3- Clean water storage tank
4- MBBR tanks (a- anoxic tank, b- aerobic tank); 5- Secondary sedimentation tank
6- Disinfection tank; 7- Nutrient dose; 8- Microbe activity monitoring system; 9- Air compressor

Figure 1. Schematic of the on-site wastewater treatment facilities.

3.3 *Main structure*

The properties, dimensional size, and operational conditions are elaborated in detail, and the plan and elevation view of wastewater treatment facilities are shown in Figure 2.

- Bar grid and regulation tank. The regulation tank is constructed underground, and made of reinforced concrete with a length, width, and height of 3 m, 3.5 m, and 4.5 m, respectively.
- MBBR tanks. MBBR tanks consist of an anoxic tank and an aerobic tank, 2 in use and 2 on standby. They are constructed underground, and adopt carbon steel anticorrosive material, with a length, width, and height of 6 m, 2.5 m, and 3 m, respectively.
- Sludge tank. The sludge tank is constructed underground, and made of reinforced concrete with a length, width, and height of 3 m, 3.5 m, and 4.5 m, respectively.
- Equipment room. The equipment room is the only aboveground unit in this project, which consists of nutrient-dose equipment, a microbe activity monitoring system, and control systems. It is designed as a brick room with a length, width, and height of 3 m, 5 m, and 3.5 m, respectively. A microbe activity monitoring system and nutrient dose equipment are installed to detect the microbe and nutrient content in the influent. Carbon sources such as glucose will be automatically dosed into the MBBR tanks to maintain the microbe activity.

- Clean water storage tank. The clean water storage tank is constructed underground, and made of reinforced concrete with a length, width, and height of 3 m, 5 m, and 4.5 m, respectively.

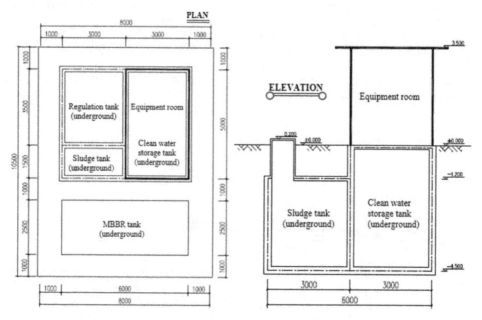

Figure 2. Plan and elevation view of wastewater treatment facilities.

4 CONCLUSION

The following conclusions can be drawn from this paper:

1) The wastewater in highway service areas has great uncertainty and fluctuation in water quantity and quality. Appropriate technology and processes with low cost and good treatment performance should be designed and applied in highway service areas.
2) According to the characteristics of wastewater in highway service areas, suspended solids, COD, ammonium, and oil content are the main substances to be removed.
3) A robust and sustainable wastewater treatment process based on MBBR is feasible for wastewater treatment in highway service areas, with cost-effectiveness and high requirement of COD, nitrogen, and phosphorus removal.
4) An integrated wastewater treatment system is designed and applied in the Jiuzhi-Maerkang Highway service area. The system can produce high-quality effluent that meets the standard under low-temperature and hypoxic conditions. The clean water with low COD, BOD, SS, TN, and TP will be reclaimed for farm, landscape, and road watering.

REFERENCES

[1] Yi Q. Appropriate Process and Technology for Wastewater Treatment and Reclamation in China. *Water Science and Technology* [0273-1223], Published 2000, Volume 42, Issue 12, Pages 107–114.

[2] Dai M., Bai J., and Q. T. Discussion on the Environmental Supervision During the Construction of Communication Projects. *Environmental Protection in Transportation*, 2003, vol.24, no.2, pp.10–12.

[3] Jian L., Yao J., Liu X. et al. Evaluation and Analysis of Wastewater Treatment Technology Applied in Highway Service Area. 2020 *IOP Conf. Ser.: Earth Environ Sci*, 474 072097.

[4] Ministry of Environmental Protection. *Integrated Wastewater Discharge Standard: GB 8978-1996 [S]*. Beijing: China Environmental Press, 1996. (in Chinese).

[5] Ministry of Environmental Protection. *Standard for Irrigation Water Quality: GB 5084-2021 [S]*. Beijing: China Environmental Press, 2021. (in Chinese).

[6] Andreottola G., Foladori R, Ragazzi M. et al. Experimental Comparison Between MBBR and Activated Sludge System for the Treatment of Municipal Wastewater. *Water Science & Technology*. 2000, 41(4).

Civil Engineering and Energy-Environment – Gao & Duan (Eds)
© 2023 the Author(s), ISBN 978-1-032-56059-5

Experimental study on influencing factors of grouting ability and fluidity of cement slurry

Weiquan Zhao*, Jianhua Zhou, Wei Lu* & Zengzeng Ren
China Institute of Water Resources and Hydropower Research, Beijing, China

ABSTRACT: Cement slurry is widely used in rock grouting. Except for fracture aperture and fineness of cement, water-cement ratio, stirring speed, stirring time, slurry temperature and additional admixture have a significant influence on the grouting ability of cement slurry. Funnel viscosity and rheological parameters of cement slurry were systematically tested, and the main influence factors on the grouting ability and fluidity of cement slurry were analyzed. The results reveal that increasing the water-cement ratio can significantly improve the fluidity of cement slurry. For a water-cement ratio of 0.5:1, higher stirring speed and adding admixture could improve the fluidity and grouting ability of the cement slurry, and increasing stirring time and temperature would lead to poor fluidity. The funnel viscosity and rheological parameters of cement slurry were significantly changed under water-cement ratios of 0.5:1 and 0.8:1, and the difference was little under water-cement ratios of 3:1 and 5:1.

1 INTRODUCTION

Cement slurry is a widely used material in rock groutings, such as consolidation grouting and curtain grouting. Except for fracture aperture of rock mass and cement fineness, water-cement ratio, stirring speed, stirring time, slurry temperature, and additional plasticizer also have an important influence on grout grouting ability [1–6]. To avoid abnormal phenomena during the grouting, such as pipe blockage, slurry thickening caused by water separation, and precipitation at the bottom of the grouting pipe, and ensure the filling effect for consolidation and curtain grouting, laboratory tests were necessary to be done. In this paper, Ordinary Portland cement P.O42.5 was used as grouting material, and systematic tests were done to fully understand the fluidity and grouting ability of the cement slurry, and the fluidity of the cement slurry with different proportions, stirring time, admixture and temperature was also analyzed. The results would give reference for in situ consolidation, and curtain grouting works.

2 EXPERIMENTAL MATERIAL AND METHODS

2.1 *Material*

Ordinary Portland cement P.O42.5 produced in the Jidong cement factory was used in the tests. The main composition of the cement is shown in Table 1.

---

*Corresponding Authors: zhaowq@iwhr.com and luwei@iwhr.com

Table 1. Main chemical composition of cement (wt.%).

CaO	SiO$_2$	Al$_2$O$_3$	Fe$_2$O$_3$	MgO	SO$_3$	Na$_2$O+K$_2$O	TiO$_2$	Cl	Total
52~53	26~27	9~10	3~4	1~2	1~2	1.6~1.8	0.4~0.5	<0.1	99.2

2.2 *Experimental methods*

A marsh funnel viscometer could be used to characterize the fluidity of cement slurry. The rheological parameters measured and calculated by the rotational viscometer could be used to characterize the grouting ability of cement slurry, including plastic viscosity η and initial yield strength τ_0 [7]. Funnel viscosity was expressed as the time of 500 mL slurry flowing out of the 700 mL funnel [8]. By using a rotational viscometer, the viscosity values of slurry at 300 r/min and 600 r/min could be measured, and the rheological parameters of slurry could be calculated based on Equation (1) and (2) [9] [10].

$$\eta = \varphi 600 - \varphi 300 mPa \cdot s \tag{1}$$

$$\tau_0 = 0.511 \times (2 \cdot \varphi 300 - \varphi 600) Pa \tag{2}$$

where η is the plastic viscosity; τ_0 is the initial yield strength; φ is the rate of rotational viscometer. The testing equipment and process are shown in Figures 1 and 2.

Figure 1. Funnel viscosity test.

Figure 2. Rheological parameters test.

3 INFLUENCE OF THE WATER-CEMENT RATIO

According to technical specifications for cement grouting construction of hydraulic structures [11], rheological parameters and funnel viscosity of cement slurry of water-cement ratios from 5:1 to 0.5:1 were tested as shown in Figures 3 and 4. The stirring time of the slurry was greater than 30 s under high-speed stirring of 1200 r/min.

The funnel viscosity of cement slurry was lower by increasing of water-cement ratio. If the water-cement ratio of the slurry was less than 1:1, the funnel viscosity of the slurry would rapidly increase with a lower water-cement ratio. For example, for the water-cement ratio from 0.8:1 to 0.5:1, funnel viscosity was increased from 24.1 s to 68.6 s. If the water-cement

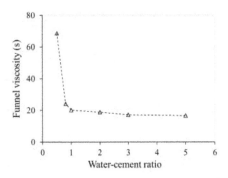
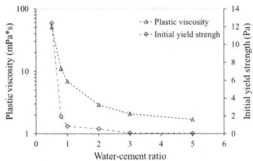

Figure 3. Funnel viscosity of cement slurry with different water-cement ratio.

Figure 4. Rheological parameter of cement slurry with different water-cement ratio.

ratio was greater than 3:1, funnel viscosity would slightly decrease, 17.2 s and 16.6 s under the ratios of 3:1 and 5:1, respectively.

The rheological parameters of cement slurry would also increase with a lower water-cement ratio, especially when the water-cement ratio is less than 1:1. For example, as the water-cement ratio reduced from 1:1 to 0.5:1, the plastic viscosity changed from 6.9 mPa's to 51.1 mPa's, while the initial yield strength increased from 0.84 Pa to 12.42 Pa, respectively.

In fact, there was little difference in funnel viscosity and rheological parameters of cement slurry for water-cement ratio from 3:1 to 5:1, and the grouting ability was nearly the same. Thus, for curtain grouting, the suitable water-cement ratio was suggested as 3:1, 2:1, 1:1, 0.8:1, and 0.5:1, and the water-cement ratio was decreased step by step based on the grouting quantity; for consolidation grouting, the suitable water-cement ratio was suggested as 2:1, 1:1, 0.8:1, and 0.5:1.

4 INFLUENCE OF TEMPERATURE AND STIRRING TIME

During cement grouting work, in order to prevent precipitation and ensure the fluidity of cement slurry, the prepared slurry would be continuously stirred under low stirring speed in the slurry storage barrel. Cement slurry with a water-cement ratio of 0.8:1 and 0.5:1 was selected to study the fluidity of cement slurry under different temperatures and continuous stirring conditions. The low-speed stirring of the slurry was 400 r/min with temperatures of 20°C and 30°C, and the stirring time was from 0 to 3h. The test results are shown in Figure 5. Based on

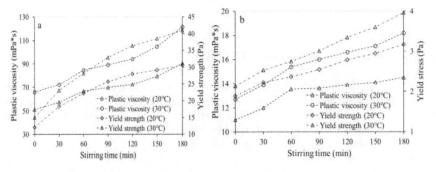

Figure 5. Rheological parameters of cement slurry under different stirring time and temperatures. (a: water-cement ratio 0.5:1; b: water-cement ratio 0.8:1)

the testing results, the rheological parameters of cement slurry were increased with longer stirring time and higher temperature, especially for slurry with a water-cement ratio of 0.5:1.

Under the condition of 20°C temperature and 3 hours of stirring, the plastic viscosity and yield strength of slurry with a water-cement ratio of 0.5:1 were increased from 51.1 mPa's to 90.5 mPa's and from 12.11 Pa to 30.56 Pa, which was 2.5 times and 1.8 times than the initial, respectively. The reason was that with the increasing stirring time, the cement hydration process increased the viscosity of the slurry. So, according to the analysis of grouting ability, for slurry with a water-cement ratio of 0.5:1, fresh grouting slurry should be used in 2 hours and as quickly as possible.

Higher temperatures would increase the hydration process, and lead to poor fluidity for cement slurry. As the temperature increased from 20°C to 30°C, for slurry with a water-cement ratio of 0.5:1, the plastic viscosity and yield strength increased 1.5 times and 1.4 times, respectively. Thus, during the grouting work, the influence of environment temperature and stirring time on the fluidity and grouting ability of cement slurry should be considered.

5 INFLUENCE OF ADMIXTURE

In fact, the setting time of the cement grouting slurry with a water-cement ratio from 0.5:1 to 5:1 satisfied the construction requirement of grouting in deep holes. However, with the increasing stirring time (working time), funnel viscosity, plastic viscosity, and yield strength of the cement slurry were increased, and the fluidity and grouting ability tended to worsen. In order to avoid precipitating and precipitating at the bottom of the grouting pipe under the action of grouting pressure, and ensure the filling effect of grouting, it was necessary to use cement slurry with higher fluidity and stability. Therefore, the addition of suitable admixtures, such as plasticizers in cement slurry, was considered to improve the fluidity and stability of the slurry without affecting the physical strength of consolidation and to improve the grouting ability of the slurry. For cement grouting slurry with a water-cement ratio of greater than 3:1, the basic performance of the slurry was mainly affected by the water-cement ratio, and the effect of adding little amount of admixture on the fluidity and grouting ability of cement slurry was limited. Thus, the influence of adding plasticizer in slurry generally focused on slurry with a water-cement ratio from 0.5:1 to 2:1.

Considering the factors of improving cement slurry performance, compatibility with cement slurry, and recommended dosage, BC- naphthalene series admixture was selected. For a water-cement ratio of 0.5:1, basic parameters that influenced fluidity and grouting ability with additive contenting in the grouting slurry were shown in Figure 6.

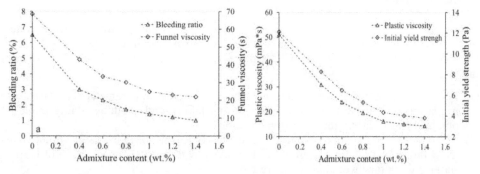

Figure 6. Physical properties of cement slurry with water-cement ratio of 0.5:1 under different contenting of BC- naphthalene. (a: bleeding ratio and funnel viscosity; b: rheological parameters)

Testing results revealed that for cement grouting slurry with the water-cement ratio of 0.5:1, the addition of BC- naphthalene admixture obviously improved the fluidity and grouting ability of the slurry. Bleeding ratio, funnel viscosity, initial yield strength and plastic viscosity of the cement slurry tended to lower with higher containing of admixture, and it would be helpful for cement diffusion during grouting. The suitable amount of BC- naphthalene in cement slurry was about 1%, as the tested parameters of the slurry were with little change under higher contending of admixture.

Physical properties of cement slurry with 1% of BC- naphthalene under water-cement ratios of 0.5:1 and 2:1 were shown in Figure 7. Adding 1% BC- naphthalene admixture had an obvious effect on reducing the bleeding rate of slurry, and was beneficial to improving the stability, fluidity and grouting ability of slurry. Also, the influence of admixture was lower with a higher water-cement ratio.

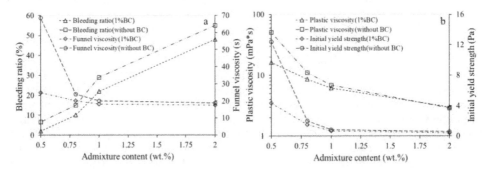

Figure 7. Physical properties of cement slurry with different water-cement ratios contenting 1% of BC- naphthalene. (a: bleeding ratio and funnel viscosity; b: rheological parameters)

Therefore, BC- naphthalene admixture could obviously improve the fluidity and grouting ability of the grouting slurry with a water-cement ratio of 0.5:1, and the influence was little on a water-cement ratio greater than 0.8:1. It was suggested that adding 1% BC- naphthalene admixture in cement grouting slurry with a water-cement ratio lower than 0.8:1 to decrease the rheology parameters and improve the grouting ability.

6 CONCLUSION

This paper was generally dealing with the influence of fluidity and grouting ability of cement slurry and could give reference for material selection of in situ grouting work, and further research, such as controlling the process of cement grouting with different water-cement ratios, was suggested.

(1) For cement grouting slurry, funnel viscosity, and rheological parameters decreased with increasing of water-cement ratio, and better fluidity and grouting ability could be obtained by increasing the water-cement ratio.
(2) The funnel viscosity and rheological parameters of cement slurry increased with a higher temperature and longer stirring time. For slurry with a water-cement ratio of 0.5:1, a suitable using time would be less than 2 hours.
(3) For deep hole grouting, 1% of BC- naphthalene was suggested to be added in cement grouting slurry with a water-cement ratio of lower than 0.8:1.

(4) Suggested water-cement ratios of curtain grouting and consolidation grouting were from 3:1 to 0.5:1 and 2:1 to 0.5:1, respectively, because there was little difference in funnel viscosity and rheological parameters between the water-cement ratio of 3:1 and 5:1.

ACKNOWLEDGMENTS

This work was financially supported by the Special Project of Basic Scientific Research of IWHR (EM0145B412021).

REFERENCES

[1] Zhang J.X., Pei X.J. and Wang W.C., et al. Hydration Process and Rheological Properties of Cementitious Grouting Material. *Construction and Building Materials*, 2017, 139: 221–231.

[2] Xiao F., Zhao Z.Y. and Chen H.M. A Simplified Model for Predicting Grout Flow in Fracture Channels. *Tunnelling and Underground Space Technology*, 2017, 70: 11–18.

[3] Sonebi M. Rheological Properties of Grouts with Viscosity Modifying Agents as Diutan Gum and Welan Gum Incorporating Pulverised Fly Ash. *Cement and Concrete Research*, 2006, 36(9): 1609–1618.

[4] Liu Q.S., Lei G.F., Peng X.X., et al. Rheological Characteristics of Cement Grout and its Effect on Mechanical Properties of a Rock Fracture. *Rock Mechanics and Rock Engineering*, 2018, 51(2): 613–625.

[5] Kremieniewski M. and Rzepka M. The Influence of a Superplasticizer on the Hydrophilicity of Cement Slurries. *Nafta-Gaz*, 2018, 74(10): 745–751.

[6] Kremieniewski M. Improvement of the Early Mechanical Strength of Cement Sheath Formed From Lightweight Cement Slurry. *Nafta-Gaz*, 2018, 74(8): 606–612.

[7] Cristelo N., Soares E., Rosa I., et al. Rheological Properties of Alkaline Activated Fly Ash Used in Jet Grouting Applications. *Construction and Building Materials*, 2013, 48: 925–933.

[8] Celik F. and Canakci H. An Investigation of Rheological Properties of Cement-based Grout Mixed with Rice Husk Ash (RHA). *Construction and Building Materials*, 2015, 91: 187–194.

[9] Xu L., Gao H., Xu J., et al. Experimental Insights into a Novel Over-saturated Brine Cement Slurry Used in Anhydrite Formation. *Bulgarian Chemical Communications*, 2017, 49: 214–219.

[10] Li Z.Y, Liu K.Q, Cheng X.W, et al. The Influence of Sulfomethyl Phenol Formaldehyde Resin (SMP) on Cementing Slurry. *Journal of Adhesion Science and Technology*, 2015, 29(10): 1002–1013.

[11] SL/T 62-2020, *Technical Specification for Cement Grouting of Hydraulic Structure*. China Water & Power Press, 2020.

Civil Engineering and Energy-Environment – Gao & Duan (Eds)
© 2023 the Author(s), ISBN 978-1-032-56059-5

Research on the risk measurement and management methods of power purchase by power grid agents

Nannan Xia*, Bo Dong*, Zhanyuan Feng* & Jian Zhang*
Liaoning Electric Power Trading Center Co., Ltd, Shenyang, Liaoning, China

Xilin Xu*
Economic and Technical Research Institute of State Grid Liaoning Electric Power Co., Ltd, Shenyang, Liaoning, China

ABSTRACT: With the introduction of the new policy of power purchase by agents, grid enterprises will replace a large number of commercial and industrial users to enter the electricity market, and will face the challenge of increased business volume and business difficulty in the future. There will be more risk issues if not handled properly, and these enterprises need to bear the large or small losses brought by the risk. Therefore, how to do a good job of risk management and promote the smooth implementation of the electricity reform is the problem that the grid enterprises have to solve. This paper firstly interprets the new policy, analyzes the possible risks in the process of power purchase by grid agents from four aspects: policy, market, credit, and volume, establishes the risk metric model VaR model to estimate the risk value specifically, introduces two common calculation methods, and finally puts forward targeted suggestions for various types of risk.

1 INTRODUCTION

1.1 *Background and significance of the study*

Since the issuance of the No. 9 document in 2015, the National Development and Reform Commission (NDRC) has deeply promoted the market-oriented reform of electricity, fully completed the task of transmission and distribution price reform, and made phased achievements in the reform of the electricity system. In order to ensure the stable supply of electric energy, the National Development and Reform Commission issued Document 1439 to continue to deepen the market-based formation mechanism of a feed-in tariff for coal-fired power generation, which has strongly promoted the development of full market entry for industrial and commercial users. Considering China's large base of commercial and industrial users, it is difficult to enter the market all at once. In order to ensure the smooth implementation of the electricity price reform policy, on October 23, 2021, the National Development and Reform Commission issued a notice on matters related to the organization and implementation of power purchase by power grid enterprises as an agent and the power purchase by power grid enterprises as an agent for commercial and industrial users who have not yet entered the market directly. Customers can enter the market when they have the conditions to enter the market directly [1]. In this context, small and medium-sized commercial and industrial users can participate in electricity market transactions through indirect means. The role of the power

---

*Corresponding Authors: 110006@13.com; dongbo1@qq.com; ttklfzw2008@163.com; stonezhj@163.com and 15841237101@163.com

408   DOI: 10.1201/9781003433651-54

grid has changed from the originally unified purchase and sale to the agency of commercial and industrial users and residential agricultural users. Commercial and industrial users gradually entering the market for direct transactions will become the future development trend, thus accelerating the realization of the goal of full marketization.

1.2 *Review of research status*

Chen [2] provides an important theoretical basis for grid companies to monitor risks by establishing several models such as risk monitoring of grid power purchase and sale business; Guo [3] analyzes the possible risks of power purchase by grid agents through four aspects: load forecast deviation, power trading method, market settlement, and transmission blockage, and proposes corresponding countermeasures. Liu [4] establishes a multi-dimensional power market under the power purchase and sale risk prevention and control analysis model for power sales companies, which is of great significance for power sales companies to study risk prevention topics. Wang [5] divides risk management into four parts: risk identification, evaluation, prevention and control, and monitoring, and establishes a risk analysis model and evaluation model to study the risk issues of power grid companies. All of the above literature analyzes the possible risks from different perspectives and provides theoretical support by establishing relevant models, in order to deeply explore the countermeasures for enterprises to deal with various risks.

1.3 *Research objectives and ideas*

Under the background of the new electricity reform policy, with a large number of commercial and industrial customers entering the electricity market, the power grid enterprises change from sellers to agents, and the change of role will inevitably increase the business volume and business difficulties for the power grid enterprises. To this end, this paper will conduct an in-depth study based on the following three steps: firstly, the risks of power purchase by power grid enterprises under the full liberalization of commercial and industrial market access are analyzed; by exploring the daily operation of grid enterprises, this paper selects four different perspectives to analyze the possible risks of power purchase by agents of grid enterprises; secondly, a risk measurement model VaR is established and its calculation method is briefly introduced; finally, corresponding risk management countermeasures are formulated for the above four types of risk.

2 RISK ANALYSIS OF POWER PURCHASE BY POWER GRID AGENTS

2.1 *Policy risk*

In terms of policy, grid enterprises are mainly affected by the tariff mechanism and the clearing mechanism. The time span of various projects of grid enterprises is long. During this period, if there is a policy adjustment or new policy introduction, it may bring corresponding risks to grid enterprises. During the transition period, when the tariff mechanism is not yet mature, the tariff is customized by the government with a unified policy. Small changes in the tariff are likely to affect the revenue of the entire grid enterprise. If the stipulated electricity price is lower than the cost of the normal operation of the grid enterprise, then the enterprise will face a huge risk crisis.

2.2 *Market risk*

In terms of market, grid enterprises may suffer from the risk mainly from the market environment and market competition. The electricity sales situation is deeply affected by economic development, so if the economy is in recession, the sales of electricity may also decline. At present, industrial users account for a large proportion of trading users. Grid enterprises can reduce the risk by diversifying the user structure, but this method is more difficult. The second best way to reduce the risk is to screen the information of users. Screening content mainly includes: whether users are in the high-pollution and high-energy-consumption industry, whether the industry the users are in is gradually declining, corporate credibility, etc. If the wrong customer is selected, then the resulting loss may make the whole enterprise's long-term effort come to naught. Another market risk is market competition. Many enterprises mistakenly see all enterprises of the same type as competitors, but in fact, many companies of the same type can become partners and achieve the goal of mutual benefit and a win-win situation.

2.3 *Credit risk*

Credit risk exists in every power financial transaction cooperation. The credit risk of grid enterprises is reflected in both contract execution and contract performance. After a power grid enterprise signs a contract with a customer, the execution strength and completion degree of the contract content will bring certain risks to the power grid enterprise. If a power grid enterprise can not complete the contract in a timely manner, resulting in power shortages, the user will ultimately lose trust in the power grid enterprise, and its credit will be reduced, leading to the continuous loss of customers, which can seriously hinder the sustainable development of the grid enterprise.

2.4 *Volume risk*

The volume risk for the grid enterprise is mainly affected by both the contract power and the actual load of customers. The contract between the grid enterprise and its user needs to be signed according to the actual load of the user, and the more accurate the forecast of the user's load, the greater the contract revenue. Specifically, when the contracted power is less than the actual load of the user, it may cause a shortage of power for the user due to insufficient generation, transmission, and distribution equipment, and the grid enterprise needs to take responsibility and incurs shortage costs. When the contract power is more than the user's load, the excess of generation, transmission, and distribution equipment, and idle and wasted resources will generate power loss, which can, in turn, increase the power purchase cost for grid enterprises. Since electricity is highly susceptible to economic and social factors and fluctuates, the forecast deviation is large. As a result, grid enterprises need to face certain volume risks.

3 RISK METRICS FOR GRID PROXY POWER PURCHASE

3.1 *Power grid agent purchasing risk measurement model – VaR*

VaR (Value at Risk) is a method of evaluating financial risk by measuring the value of return at risk. It is also known as "value at risk", which means that a certain value is at risk of loss, and the probability of loss is quantified by a certain probability [6]. This method was introduced by JP Morgan Bank in 1994, and soon became the VaR method became prevalent internationally and an important benchmark for risk assessment by various financial institutions.

VaR applied to the power industry refers to the maximum value of expected loss of revenue for grid companies at a certain confidence level. Assuming that the probability distribution density of a certain portfolio value of the power grid is $F(w)$, given the confidence level c, the initial value of the asset is set to w_0, and the minimum value at the confidence level is set to $w*$ [7], then:

$$VaR = E(w) - w*$$ (1)

The confidence level is generally taken between 95% to 99%. If the confidence level is too low, the probability of the loss exceeding the VaR value is too large, and the VaR value has no reference meaning [7].

3.2 Solving method of VaR model for power grid agent purchasing risk measurement

3.2.1 Historical simulation method

The historical simulation method is used to simulate the future price changes of the portfolio by using the available historical information. The value of VaR can be found by first calculating the frequency distribution of portfolio risk returns over the past period of time and finding the average return over the last period of time and the lowest return at the corresponding level [8]. Assuming that the risk factor of a certain portfolio is $F(i)$ $(i = 1, 2, ..., n)$, by calculating the historical price situations, it is possible to simulate the direction of the price changes over the next 100 days. Assuming that these 100 changes are equally likely to occur in the future, the current value of each risk factor $F(i)$ plus the vector of observed changes yields the sum of the respective possible future price levels [4], expressed by the following equation.

$$A_F (i)_1 = F(i)_0 + \Delta F(i)_{-1}$$
$$A_F (i)_2 = F(i)_0 + \Delta F(i)_{-2}$$
$$......$$
$$A_F (i)_{100} = F(i)_0 + \Delta F(i)_{-100}$$ (2)

The risk factor and historical time length are determined, the relevant information is collected to calculate the current price and the possible future price, and then the corresponding profit and loss distribution can be found. The profit and loss are ranked in order of size, its way odds mapping is calculated to get the future profit and loss distribution, the quantile at a specific confidence level is selected, and the value of the quantile is the value of VaR.

3.2.2 Parametric method.

VaR is the maximum loss at a certain confidence level, which can reflect the degree of loss and loss probability under various market assumptions. If the given confidence level is σ, the value in the following formula is VaR.

$$Prob(Absolute\ value\ of\ loss > VaR) < \sigma$$ (3)

The value of VaR can be calculated directly if the probability density function of the gain or loss in the selected time horizon is known in advance; if the probability density function obeys a normal distribution, the calculation formula is as follows [3].

$$VaR = \alpha \sigma p$$ (4)

where α is the value corresponding to the selected confidence level. For example, when $\sigma = 95\%$, $\alpha = 1.645$; when $\sigma = 97.5\%$, $\alpha = 1.96$; when $\sigma = 99\%$, $\alpha = 2.33$. p is the current market value of the commodity.

4 RISK MANAGEMENT OF POWER PURCHASE BY POWER GRID AGENTS

Risk management refers to the active and purposeful management of the economic unit or individual to choose the most effective way through the identification, monitoring, and evaluation of risks. Good risk management is conducive to reducing decision-making errors in power grid enterprises, avoiding unnecessary operating costs, and the sustainable development of enterprises. It is one of the important topics of enterprise management.

Common risk management methods are risk transfer and risk avoidance. Risk transfer is a means of risk management by transferring all or part of the risk enterprises face to other parties by contractual or non-contractual means, and insurance is one of them. Risk transfer does not reduce the loss caused by the risk. For the same kind of risk, different risk managers have different tolerance, so something may cause irreversible and significant losses to party A, but may only have a transient impact on party B. Therefore, risks can be effectively addressed and safeguarded through risk transfer.

Risk aversion refers to avoiding the impact of risk by measures such as active abandonment or adjustment of activity arrangements when an activity is assessed to have a high probability of risky losses [9]. It is one of the most effective risk control techniques that can nip risk factors in the bud and avoid harm to project activities. The common risk avoidance methods include termination and change. Once the frequency of risk loss or the severity of the loss is found to be high, the risk is rejected by terminating the project; if the risk impact is within tolerable limits, the risk is avoided by changing the project workplace, workflow, etc.

In response to the above analysis of the risks of power purchase by grid agents, the following solution measures are proposed.

(1) For the policy risk, it can only be reduced, instead of being transferred or completely eliminated. Grid companies should fully cooperate with the policy requirements, coordinate with the government and society, and create a good corporate image. They can also use Internet technology to establish a high-quality information collection platform to learn the latest industry information in advance, so that they can make corresponding adjustments to their own corporate development strategies in a timely manner, and even make full use of this time difference to achieve profitability. Grid companies also need to strengthen their own management to make the environment develop in their favor as much as possible.

(2) For the market risk, decision-makers should have a keen sense of risk and risk management throughout the concept of enterprise management because the market environment is changing rapidly, and government policies are updated, which requires good policy interpretation ability and timely attention to market trends. The market risk should be kept in the bearable range to do risk aversion. If the risk is too big, the risk transfer should be taken to reduce the risk. At the same time, the power grid enterprises should find the positioning of their own enterprises and adopt a rational view of the same type of enterprises, because those who thought they were competitors may also become partners.

(3) For credit risk, grid enterprises should reasonably develop their own agency power purchase strategy, have more communication with customers to get a deep understanding of customer needs and provide customers with the best quality service, conscientiously fulfill the contract content, and improve their credit level, so as to improve customer loyalty and get more user recognition.

(4) For volume risk, power grid enterprises should adopt scientific methods to forecast users' electricity consumption, monitor market data in real-time, and develop power purchase packages suitable for users, so as to reduce the risk of contract electricity. Specifically, users can be divided into high-energy-consumption users, low-electricity-demand users, flexible electricity users, etc. For different users, power grid enterprises should analyze their electricity consumption characteristics and electricity consumption structure, and target load forecasting with the help of traditional load forecasting methods. Then, on the basis of load forecasting, trading contracts should be set up accordingly, and finally, the contract trading mode and trading time sequence should be determined. At the same time, for commercial and industrial users with flexible

characteristics, the power grid enterprises act as an agent for the users to participate in the mechanism and process related to time-sharing transactions, and they can do a good job of connecting with time-sharing transactions.

5 SUMMARY AND OUTLOOK

Based on the introduction of the new policy of power purchase by power grid agents, this paper discusses the possible risks faced by power purchase by power grid enterprises' agents, and the main contents and related conclusions are as follows.

(1) The background, content outline, and significance of the new policy are analyzed. Based on the combining of related literature, four risks are proposed: policy risk, market risk, credit risk, and volume risk. Then a risk measurement model VaR is established to quantify the risk value of power purchase by grid agents, and finally, management suggestions are made for different risks.

(2) The government coordinates and manages the whole market, and grid enterprises inevitably face the risk from government policies. In this regard, grid enterprises should strictly implement policy regulations, coordinate with the government and strengthen their own construction. For the risk from the market environment and market competition, grid enterprises should pay attention to market changes in a timely manner and participate in a rational competition. For credit risk, grid enterprises should strive to provide customers with quality service and improve corporate reputation, so as to improve customer stickiness; grid enterprises should do a good job of load forecasting and develop appropriate customer packages to reduce the losses brought about by volume risk.

The power market is ever-changing, and risk is everywhere. Grid companies can only be comfortable with unknown challenges by improving their risk awareness, continuously refining their risk management capabilities, and taking various measures to avoid risks as much as possible. But the risk cannot be completely eliminated; therefore, the research on the topic of risk analysis of power purchase by power grid enterprises' agents still a long way to go.

ACKNOWLEDGMENTS

Upon the completion of the paper, I would like to express my heartfelt thanks and sincere respect to all the leaders and colleagues of the project study on the impact of deepening the reform of the coal pricing mechanism and fully opening up the market of industry and commerce on the electricity market. Thank you for your patient guidance and for providing me with so many learning materials. The successful completion of this paper cannot be done without your help.

REFERENCES

[1] National Development and Reform Commission Deploys Local Organizations to Carry Out Power Purchases by Power Grid Enterprises' Agents. *Rural Electrician*, 2021, 29(12):2. DOI: 10.16642/j.cnki. ncdg.2021.12.003.

[2] Chen Kun. *Risk Monitoring and Analysis Model of Power Purchase and Sale Business of Power Grid Enterprises*. North China University of Electric Power, 2015.

[3] Guo Jin. *Research on Risk Analysis Model and Management Strategy of Power Purchase and Sale of Electricity for Grid Enterprises*. North China University of Electric Power (Beijing), 2005.

[4] Liu Zhuhui. *Analysis Model of Risk Prevention and Control of Power Purchase and Sale by Power Sales Companies in Multi-dimensional Power Market [D]*. North China University of Electric Power (Beijing), 2019. DOI: 10.27140/d.cnki.ghbbu.2019.000760.

[5] Wang Xiaokang. *Research on Investment Risk Management of Power Grid Construction Projects*. Shandong University of Finance and Economics, 2015.

[6] Zhao Yujing. An Analysis of Risk Cost Measurement Methods. *New Accounting*, 2018(04):19–21.

[7] Chen Shoujun. *Research on the Optimization Model of Risk Measurement, Early Warning, and Prevention and Control of Power Grid Operation*. North China University of Electric Power (Beijing), 2016.

[8] Shao Xinwei. *VaR-based Financial Risk Measurement and Management*. Jilin University, 2004.

[9] Fang Shaofeng, Zhou Renjun, Yanyuan Peng, Li Bin, Xu Fulu, Shi Liang Yuan. A Balanced Market Trading Model for Electricity Retailers with Risk Aversion. *Journal of Power Systems and their Automation*, 2020,32(02): 22–27. DOI: 10.19635/j.cnki.csu-epsa.000237.

Civil Engineering and Energy-Environment – Gao & Duan (Eds)
© 2023 the Author(s), ISBN 978-1-032-56059-5

Prospect of typical post-combustion carbon capture technology in fossil-fuel power station

Guochen Zhao*

Datang Northeast Electric Power Test & Research Institute, China Datang Corporation Science and Technology Research Institute, Changchun, P.R. China

ABSTRACT: To control climate warming, the growth rate of global carbon emissions has dropped significantly since the beginning of the 21st century, and energy conservation and emission reduction have become the main consensus and historical trends. Currently, carbon emissions in 54 countries around the world have peaked. For China, raw coal still dominates energy consumption, followed by crude oil, non-fossil energy, and natural gas. Under the "carbon neutrality" goal, vigorously developing carbon dioxide capture, utilization, and storage (CCUS) technology is not only a strategic choice for China to reduce carbon dioxide emissions and ensure energy security in the future but also an important means to build an ecological civilization and achieve sustainable development. In this paper, several typical post-combustion carbon capture technologies for fossil-fuel power stations are described in detail, including the alcohol amine method, frozen ammonia method, and membrane separation method. The carbon mentioned above capture technologies is introduced in detail from three aspects: technical introduction, domestic and foreign progress, and prospects aiming to provide a detailed analysis of carbon capture technologies for researchers in the fossil-fuel power industry. It also disseminates the relevant knowledge of carbon emission reduction to the public and contributes to the early realization of the Chinese carbon peak and carbon neutrality goals.

1 INTRODUCTION

The fossil-fuel power industry is the focus of the current carbon capture, utilization and storage technologies (CCUS) demonstration in China. It is estimated that by 2025, the emission reduction of coal power CCUS will reach 6 million tons per year. It will peak in 2040 at 200-500 million tons/year and remain unchanged. The deployment of gas electric CCUS will gradually expand and remain unchanged after reaching the peak in 2035, with emission reductions of 0.2 to 100 million tons per year. Adding CCUS to fossil-fuel power stations can capture 90% of carbon emissions, making it a relatively low-carbon power generation technology. In Chinese current installed capacity, about 900 million kilowatts will still be in operation by 2050. The deployment of CCUS technology helps to fully use existing coal power units, properly retain coal power capacity, and avoid resource waste caused by the early retirement of some coal power assets. The realization of low-carbon utilization and transformation of active advanced coal-fired power units combined with CCUS technology is an important way to release the emission reduction potential of CCUS (Wang *et al.* 2022; Zhang *et al.* 2021).

---

*Corresponding Author: 1257593514@qq.com

DOI: 10.1201/9781003433651-55

CCUS technology can achieve near-zero carbon dioxide emissions in fossil fuel utilization. It can not only reduce the total carbon emissions but also significantly control the total emission reduction costs to ensure a safe and stable supply of energy for economic development. In the era of a global joint response to climate change, after coal power has achieved ultra-low emission of pollutants, carbon emissions have become the primary issue affecting the sustainable development of the coal power industry. CCUS technology has been regarded as an important technical route to greatly reduce carbon emissions from fixed emission sources such as coal-fired power plants. The development of CCUS technology is of special significance to China.

Carbon capture technology is mainly divided into pre-combustion carbon capture technology, post-combustion carbon capture technology, oxygen-enriched combustion technology, and chemical chain combustion technology. The pre-combustion carbon capture technology mainly includes the integrated coal gasification combined cycle power generation system (IGCC) to realize the clean technology of CO_2 separation. The physical absorption method is mainly based on the low-temperature methanol washing method/polyethylene glycol dimethyl ether method. Among the many technical routes, the most widely used is post-combustion carbon capture technology, which mainly includes chemical absorption, physical adsorption, membrane separation, and low-temperature distillation. Chemical absorption methods include the alcohol amine absorption method, carbonate circulation absorption method, hot caustic potash absorption method, ionic liquid absorption method, and so on. The absorbents of physical adsorption mainly include activated carbon, molecular sieve, silica gel, activated alumina, activated soil, metal-organic framework, etc. According to the separation principle, it can also be divided into pressure swing adsorption and temperature swing adsorption. Membrane separation methods are mainly divided into gas permeation separation and membrane-based absorption separation, as well as the newly developed supported liquid membrane method (Chen & Lu 2022; Zhang *et al.* 2022).

In this paper, several typical post-combustion carbon capture technologies used in thermal power plants are introduced in detail based on the wide application of various carbon capture technologies. From the three aspects of technology introduction, domestic and foreign progress, and prospects, we will make an objective assessment as possible to provide a reference for policymakers, investment decision-makers, and scientific researchers. This paper also plays an important role in studying the strategic positioning and development path of CCUS under the carbon neutralization goal. It will help policymakers carry out CCUS-related work at the strategic, planning, and policy levels and help researchers determine emission anchors in future periods based on current knowledge of CCUS. At the same time, it will also help the public to understand the relevant knowledge of CCUS, understand the status and role of CCUS, and jointly promote the realization of China's carbon peak carbon neutrality goal.

2 ALCOHOL AMINE CARBON CAPTURE TECHNOLOGY

2.1 *Technical introduction*

The chemical absorption method uses an alkaline absorbent to come into contact with CO_2 in the flue gas and undergo a chemical reaction to form unstable salts. Under certain conditions, salts will reversely decompose to release CO_2, regenerating the absorbent's absorption capacity and separating and enriching CO_2 from the flue gas. It is convenient for utilization or storage, and at the same time, the absorbent can be recycled. Commonly used CO_2 absorbents are alcohol amine solution, strong alkali solution, hot caustic potash solution, etc. Among them, ethanolamine (MEA) has a small molecular weight and a strong ability to absorb acid gases, so it is most advantageous to capture low-concentration CO_2 in flue gas after combustion (Figure 1) (Rao & Rubin 2022; Wang *et al.* 2015).

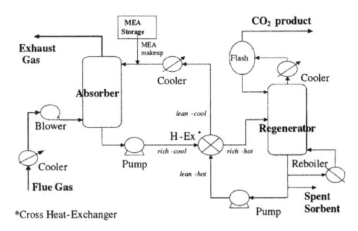

Figure 1. Flowsheet for CO_2 capture from flue gases using an amine-based system (Rao & Rubin 2022).

2.2 Research status

2.2.1 Current status of foreign research

The alkanolamine desulfurization and decarbonization process technology was invented and patented by Bottoms in 1930 after more than 90 years of development. So far, this process is still the most widely used in the desulfurization and decarbonization of natural gas and refinery gas. It is also used in the synthetic ammonia industry to prepare downstream products from syngas. Fluor Corporation of the United States has developed an MEA-based Econamine FG process for CO_2 capture, demonstrated in the Warrior Run coal-fired power plant in Maryland in 2000, with a CO_2 capture capacity of 150t/d. Chapel D proposed Econamine FG for CO_2 capture in 1999, and the study showed that the heat consumption requirement of the regeneration process was 4.2 GJ/t CO_2. Yeh AC et al. measured the absorption capacity of ethanolamine through a semi-continuous reactor, and the results showed that the maximum CO_2 capture rate of ethanolamine was 94%, and the absorption capacity of ethanolamine was 0.4 kg CO_2/kg MEA. Singh D et al. conducted a technical and economic analysis of ethanolamine capture technology, and the results showed that the CO_2 capture cost was 33-55$/t CO_2 (Astaria et al. 1983; IEA 2004).

2.2.2 Current status of Chinese research

The Nanjing Chemical Group Corporation, a subsidiary of Sinopec, uses MEA as the main solvent and preferably adds active amines, antioxidants, and corrosion inhibitors to form a composite absorbent suitable for recovering low partial-pressure CO_2. It is combined with the low partial pressure of Xi'an Thermal power research institute CO., LTD. The CO_2 recovery process has been successively used in Huaneng Beijing Thermal Power Plant, Huaneng Shanghai Shidongkou Power Plant, and Shengli Oilfield Shengli Power Plant. The amine-based carbon capture device independently developed and designed by CPI Yuanda Environmental Protection Engineering CO., LTD. with an annual capture of 10,000 tons of industrial-grade carbon dioxide was also successfully operated in 2010 in CPI Chongqing Hechuan Shuanghuai Power Plant (Huang et al. 2009; Ma et al. 2014).

2.3 Future outlook

The post-combustion carbon emission reduction process based on the chemical absorption method does not require much modification to the existing coal-fired power generation system. It has the characteristics of fast absorption and high absorption efficiency for CO_2

with low partial pressure in the flue gas. Currently, the decarbonization process based on MEA solvent has been demonstrated in many coal-fired power plants at home and abroad. The CO$_2$ treatment capacity of a single project is also gradually increasing. The process flow improvement, solvent modification, and research and development of new solvents, such as the amine complex process and the sterically hindered amine process, are still in progress. Whether it is the MEA method or the ammonia method, the decarbonization of coal-fired power plants can decrease the power generation efficiency of the power plant by 10% or more, resulting in higher power generation costs. The search for high-efficiency and low-energy-consuming CO$_2$ absorbents is still the focus of this field. The CO$_2$ removal process based on chemical absorption is currently the most suitable for stock coal-fired power plants. With the improvement of the absorption solvent performance, the improvement and optimization of the process, and the implementation of the effective resource utilization of CO2. The chemical absorption method will be popularized and applied in more coal-fired power plants, which is significant for decarbonizing coal-fired power plants.

3 FROZEN AMMONIA CARBON CAPTURE TECHNOLOGY

3.1 Technical introduction

The decarburization process using ammonia water as the absorbent often has the problem of a high ammonia escape rate at the top of the absorption tower. Therefore, the Alstom company in the United States proposed the Chilled Ammonia Process (CAP) decarbonization process to control the escape of ammonia gas. The process of absorbing CO$_2$ by frozen ammonia is a process in which a weak base (NH$_3$) and a weak acid (CO$_2$) react reversibly to form soluble salts and release heat. Soluble salts absorb heat at high temperatures and decompose through reverse reactions to form NH$_3$ and CO$_2$. In the NH$_3$-CO$_2$-H$_2$O system, CO$_2$ is dissolved in the solution as carbonate, bicarbonate, etc. The frozen ammonia method has strong pollutant treatment capacity, wide product application prospects, strong CO$_2$ absorption capacity, low regeneration energy consumption, and low supplementary cost (Figure 2) (Philippe 2007; Sherrick et al. 2008).

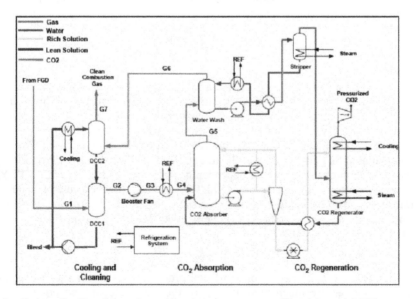

Figure 2 Carbon dioxide refrigeration ammonia capture process (Sherrick et al. 2008).

3.2 Research status

3.2.1 Current status of foreign research

In 2006, Gal E proposed a patent for refrigerated ammonia absorption technology using ammonia water as an absorbent and described the technology in detail in the patent. Alstom established a 5-year plan to develop refrigeration ammonia technology in 2006, aiming to achieve commercial operation by the end of 2011. Alstom has built a pilot plant for capturing 15,000 tons of CO_2 per year in the Karlshamn power plant of Sweden E.ON and the Pleasant Prairie power plant of We Energies in the United States. Commissioning and testing began in 2008. Beginning in 2007, Alstom has cooperated with AEP in the United States and Statoil Hydro in Norway to build commercial demonstration units for capturing 100,000 tons of CO_2/year at the Mountaineer Power Plant and Mongstad Refinery, which are planned to be in 2009 and 2010 respectively year for debugging. In 2008, Alstom signed an agreement with Canada's TransAlta and planned to build a commercial operation device to capture 1 million tons of CO_2 per year in TransAlta's Alberta power plant by 2012. The captured CO_2 will be used to improve the oil field's recovery rate (Telikapalli *et al.* 2011; Valenti *et al.* 2009).

3.2.2 Current status of Chinese research

Jiang Wenmin *et al.* used the Rate Based method of Aspen Plus to establish a process simulation model for the removal of CO_2 from ammonia water in a 300MW coal-fired unit. Bai Yanfei, Wang Chunbo, etc., conducted experimental research on the removal of ammonia water and MEA from flue gas components under oxygen-enriched combustion. The experimental results showed that the removal effect of ammonia water was better. Qi Guojie, Wang Shujuan, etc. analyzed the economics of combined ammonia-water removal of CO_2 and SO_2 from flue gas based on Aspen Plus. Ma Shuangchen *et al.* conducted a comparative study on the MEA method and the ammonia water method to remove CO_2. The results show that the ammonia method is superior to the MEA absorption method in absorbent regeneration, energy consumption, and by-product utilization value. Sun Long studied the law of ammonia escape in the carbon capture process by the amine method and proposed an effective system and method to control ammonia escape (Ma *et al.* 2013).

3.3 Future outlook

The energy consumption of the ammonia regeneration process is much smaller than that of the MEA solution regeneration process. Secondly, the mixture of ammonia water and CO_2 is stable, so the regeneration tower can use relatively high pressure. Once again, ammonia water has a lower cost, no organic products will be produced from degradation, and it has great application prospects. However, the equipment investment of the CAP method is significantly higher than that of the MEA method, and the cost of the CAP method decarbonization system is 21.2% higher than that of the MEA method decarbonization process. Based on the analysis results of technical economics, it can be seen that the power generation cost of the MEA decarbonization process is 0.02 yuan/kWh higher than that of the CAP decarbonization process. The decarbonization cost is 31.21 yuan/t CO_2, higher than that of the MEA method, and the CAP method has obvious advantages for the unit to decarbonize the unit. Compared with the MEA decarburization process, the ammonia decarburization process reduces the heat consumption of the reboiler during the decarburization process but, at the same time, increases the refrigeration work of the refrigeration ammonia process. Since ammonia removal equipment is added to the ammonia water decarbonization system, the investment and energy consumption for controlling ammonia escape are increased.

4 MEMBRANE SEPARATION CARBON CAPTURE TECHNOLOGY

4.1 Technical introduction

Membrane separation is a physical process in which the diffusion coefficients of different components in the gas are different on the polymer material to achieve gas separation. When used for CBM decarbonization, CO_2 gas in CBM is used as a "fast gas" and permeates from the membrane to separate from "slow gas" such as CH_4, O_2, N_2, etc. Membrane separation methods are mainly gas permeation and membrane-based absorption, as well as the newly developed supported liquid membrane method. The membrane separation method has the advantages of simple operation, small footprint, no solution storage, foaming degradation and corrosion of equipment, and other problems. While removing the acid gas can also remove part of the water vapor, reducing the load of the dehydration device. No moving parts, not easily affected and disturbed by the external environment (Figure 3) (Amooghin et al. 2015; Wilcox et al. 2014).

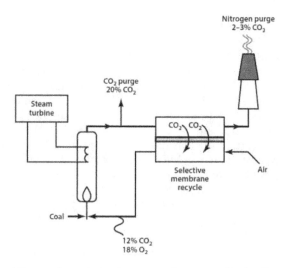

Figure 3 Schematic of the membrane technology research process allowing for enhanced CO_2 flue gas partial pressure at no energy cost (Wilcox et al. 2014).

4.2 Research status

4.2.1 Current status of foreign research

At this stage, the separation process of CO_2 and N_2 based on the gas separation membrane method has not been realized in industry, while the industrialized CO_2 membrane separation process includes CO_2/CH_4 and CO_2/H_2 separation. The gas separation membranes mainly used in industry are polymer membranes, such as polyimide Matrimid®5218, block polyetherimide Pebax®, etc. The earliest commercialized gas membrane separation system was the Prism hollow fiber composite gas separation system developed by Monsanto Company in 1979, using polysulfone as the membrane material. Dow Chemical began supplying the N_2-rich Generon system to the market in 1985 (the system uses poly-4-methyl-1-pentene as a membrane material) for oil and gas production. UOP established the first membrane separation CO_2 removal device in 1996, which can reduce the CO_2 volume fraction from 4.5% to below 1.5%, and the hydrocarbon recovery rate can reach 96.1%. Norway's Kvaerner Oil and Gas Company have used a membrane separation/adsorption method to separate CO_2 from natural gas since 1998. It has successively conducted field tests in Scotland, Norway, and the United States and has finally succeeded in commercialization (Car et al. 2008; Fels 1986).

4.2.2 Current status of Chinese research

The development of gas separation membranes in China started in the 1980s. In 1985, the Dalian Institute of Chemical Physics, Chinese Academy of Sciences successfully developed the hollow fiber nitrogen and hydrogen membrane separator for the first time, and its performance can reach the level of a Prism separator, filling the domestic gap. In the past 30 years, my country's gas membrane separation technology has also been developed by leaps and bounds. Recently, the field pilot plant of the hollow fiber membrane contactor for CO_2 removal from natural gas, jointly developed by the Dalian Institute of Chemical Physics and Petronas Corporation (PETRONAS), was successfully tested at the natural gas purification plant in Malaysia. The device is the world's first hollow fiber membrane contactor system for high-pressure natural gas purification, and the membrane material used is polytetrafluoroethylene (Sun *et al.* 1990; Tao 2014).

4.3 Future outlook

Compared with the traditional chemical absorption method and pressure swing adsorption method, gas membrane separation of CO_2 in flue gas has the advantages of high efficiency, energy saving, environmental protection, wide applicability, and flexible operation. It is a new green separation technology with great development potential. The development and application prospects of this technology mainly include the following aspects. Develop new materials to prepare CO_2 separation membranes and membrane modules with high permeability and selectivity. Develop new film-forming technologies to prepare defect-free films. The gas membrane separation process of CO_2 is modified to improve the separation effect of CO_2 and reduce operating costs and energy consumption.

5 CONCLUSION

The maturity of CO_2 capture technology varies greatly. Currently, the pre-combustion physical absorption method is in the commercial application stage, the post-combustion chemical adsorption method is still in the pilot stage, and most other capture technologies are in the industrial demonstration stage. Post-combustion capture technology is currently the most mature capture technology and can be used to decarbonize most thermal power plants. The 150,000-ton carbon capture and storage demonstration project by Guohua Jinjie Power Plant are under construction. It is currently China's largest post-combustion carbon capture and storage demonstration project. The pre-combustion capture system is relatively complex, and the integrated gasification combined cycle (IGCC) technology is a typical system that can carry out pre-combustion carbon capture. Domestic IGCC projects include the Huaneng Tianjin IGCC project and Lianyungang clean energy power system research facility. Oxygen-enriched combustion technology is one of the most potential large-scale carbon capture technologies for coal-fired power plants. Oxygen-enriched combustion technology has developed rapidly and can be used in new coal-fired and partially retrofitted thermal power plants. The current first-generation carbon capture technologies (post-combustion capture technology, pre-combustion capture technology, and oxygen-enriched combustion technology) are becoming increasingly mature. The main bottlenecks are high cost and energy consumption and lack of extensive large-scale demonstration project experience. The second-generation technologies (such as new membrane separation technology, new absorption technology, new adsorption technology, pressurized oxygen-enriched combustion technology, etc.) are still in the stage of laboratory research and development or small test. After the technology is mature, its energy consumption and cost will be reduced by more than 30% compared with the mature first-generation technology. It is expected to be widely promoted and applied around 2035.

ACKNOWLEDGMENTS

Datang Northeast Electric Power Test & Research Institute financially supported this work.

REFERENCES

Astaria G., Savage D.W. and Bisio A. (1983) Gas Treating with Chemical Solvents. *Gas Treatment with Chemical Solvents.*

Amooghin A.E., Omidkhah M. and Kargari A. (2015) Enhanced CO_2 Transport Properties of Membranes by Embedding Nano-porous Zeolite Particles into Matrimid®5218 matrix. *RSC Advances*, 5(12): 8552–8565.

Chen W.H. and Lu X. (2022) Research on Optimization of CCUS Cluster Deployment in China's Coal-fired Power Plants Under Carbon Neutrality. *Advances in Climate Change Research*, 18(03).

Car A., Stropnik C., Yave W. et al. (2008) Pebax®/Polyethylene Glycol Blend thin Film Composite Membranes for CO_2 Separation: Performance with Mixed Gases. *Separation and Purification Technology*, 62(1): 110–117.

Fels M.F. (1986) PRISM: An Introduction. *Energy and Buildings*, 9(1–2): 5–18.

Huang B., Xu S.S., Gao S.W. et al. (2009) Industrial Test Research on CO_2 Capture in Huaneng Beijing Thermal Power Plant. *Proceedings of the CSEE*, (17): 14–20.

IEA G.H.G (2004) Improvements in Power Generation with Post-Combustion Capture of CO_2, Report PH4/33. IEA Greenhouse Gas R&D Programme.

Ma Y.F., Hao J., Wan Z.P. et al. (2014) Engineering Practice of CO_2 Capture and Purification in Oilfield Coal-Fired Power Plants. *Clean Coal Technology*, (5): 20–23.

Ma S., Song H., Zang B. et al. (2013) Experimental Study on Additives Inhibiting Ammonia Escape in Carbon Capture Process Using Ammonia Method. *Chemical Engineering Research and Design*, 91(12): 2775–2781.

Ma S., Song H., Zang B. et al. (2013) Experimental Study of Co(II) Additive on Ammonia Escape in Carbon Capture Using Renewable Ammonia. *Chemical Engineering Journal*, 234: 430–436.

Philippe P. (2007) *Technologies for CO_2 Capture Roadmaps and Potential Cost Reductions. Paris Alstom Power Systems.*

Rao A.B. and Rubin E.S. (2022) A Technical, Economic, and Environmental Assessment of Amine-based CO_2 Capture Technology for Power Plant Greenhouse Gas Control. *Environmental Science & Technology*, 36, 20, 4467–4475.

Sherrick B., Hammond M., Spitznogle G. et al. (2008) *CCS with Alstom's Chilled Ammonia Process at AEP's Mountaineer Plant.* Baltimore, Maryland, USA.

Sun Z.G., Zhang M.Q., Zheng G.P. et al. (1990) Application of Domestic Hollow Fiber N_2-H_2 Membrane Separation Device. *Chemical World*, 5.

Telikapalli V., Kozak F., Francois J. et al. (2011) CCS with the Alstom Chilled Ammonia Process Development Program-Field Pilot Results. *Energy Procedia*, 4: 273–281.

Tao J. (2014) *The First Hollow Fiber Membrane Contactor for CO_2 Removal from Natural Gas. 2014 National Magnesium Compound Industry Annual Conference and Technical Equipment Exchange Conference Album.*

Valenti G., Bonalumi D. and Macchi E. (2009) Energy and Exergy Analyses for the Carbon Capture with the Chilled Ammonia Process (CAP). *Energy Procedia*, 1(1): 1059–1066.

Wang L.D., Jing R.Q., Wang R.J. et al. (2022) Research on the Technology Path of CO_2 Capture from Coal-fired Flue Gas in China's Thermal Power Industry. *Journal of Beijing Institute of Technology (Social Sciences Edition)*, 24(4): 66–73.

Wang H., Hou F.Z., Shang H. et al. (2015) Analysis of Secondary Pollution of Tail Gas from CO_2 Capture System of Ethanolamine (MEA) Coal-fired Power Plant. *Chemical Industry and Engineering Progress*, 34(09).

Wilcox J., Haghpanah R., Rupp E.C. et al. (2014) Advancing Adsorption and Membrane Separation Processes for the Gigaton Carbon Capture Challenge. *Annual Review of Chemical and Biomolecular Engineering*, 5: 479–505.

Zhang X., Li K., Ma Q. et al. (2021) Development Orientation and the Prospect of CCUS Technology Under the Carbon Neutrality Target. *China Population, Resources, and Environment*, 31(9): 29–33.

Zhang F., Fu A.Y., Liu K.Z. et al. (2022) Research Progress of Carbon Capture, Storage, and Utilization Technology. *Leather Manufacture and Environmental Technology*, 3(01).

Civil Engineering and Energy-Environment – Gao & Duan (Eds)
© 2023 the Author(s), ISBN 978-1-032-56059-5

Growth conditions and growth kinetics of Chlorella Vulgaris cultured in domestic sewage

Xingguan Ma
School of Municipal and Environmental Engineering, Shenyang University of Architecture, Shenyang, Liaoning, China
Liaohe River Basin Water Pollution Control Institute, Liaoning Shenyang, China

Wenhao Jian*
School of Municipal and Environmental Engineering, Shenyang University of Architecture, Shenyang, Liaoning, China

ABSTRACT: This paper assess the feasibility of achieving the dual objectives of domestic wastewater treatment and biomass accumulation. Growth kinetic models were used to analyze the growth pattern of Chlorella in domestic wastewater. The logistic model can simulate the growth trend of Chlorella in domestic wastewater better than the other two models. However, the currently developed model still cannot fully predict the growth of Chlorella. Factors such as nutrient removal, aging, and algae death need to be considered to develop a more accurate model.

1 INTRODUCTION

Due to the similar nutrient composition, domestic wastewater can be used as an alternative nutrient source for microalgae (Chisti 2007; Wang *et al.* 2020). To prevent eutrophication of the stowed water, biological treatment is required to reduce the nitrogen and phosphorus content before discharge (Jun-Rong *et al.* 2012; Lam & Lee 2012). Chlorella is a good option for achieving simultaneous microalgae culture and wastewater treatment, which can absorb elements such as N, P, and C from the environment for the synthesis of cellular components (Kassim *et al.* 2017; Mohd *et al.* 2017). Therefore, using domestic wastewater for microalgae cultivation can achieve biomass enrichment and effective purification of N, P, and organic matter in the water body.

At present, microalgal culture techniques for domestic wastewater are still unable to consistently achieve the goal of efficient culture (Li *et al.* 2021; Zhou *et al.* 2014). Microalgae have a specific growth rate, mainly related to the activity of related intracellular enzymes and the nutrient reserves in the environment. The growth of Chlorella in domestic wastewater and the effectiveness of the growth kinetic model applied in this scenario were evaluated to achieve an efficient culture of microalgae in domestic wastewater. In this study, three growth conditions were selected, namely initial pH, light intensity, and algae addition, and growth kinetics were analyzed and evaluated.

---

*Corresponding Author: kobe_jian@foxmail.com

DOI: 10.1201/9781003433651-56

2 MATERIALS AND METHODS

2.1 *Culture medium culture*

The algal species used for the study were purchased from the Freshwater Algal Species Bank of the Chinese Academy of Sciences: Chlorella Vulgaris (No. FACHB31). Before the experiment, Chlorella Vulgaris was preserved and cultured in BG11. The culture process was carried out in a constant temperature light incubator with a light intensity of 4000lux, light time of 24h, and temperature of 25°C. The composition and content of BG11 are shown in Table 1.

Table 1. BG-11 composition and content.

Chemical composition	Dosage · mg(L^{-1})
$NaNO_3$	1500
K_2HPO_4	40
$MgSO_4 \cdot 7H_2O$	75
$CaCl_2 \cdot 2H_2O$	36
$C_6H_8O_7$	6
$C_6H_8FeNO_7$	6
$EDTANa_2$	1
Na_2CO_3	20
H_3BO_3	2.86
$MnCl_2 \cdot 4H_2O$	1.86
$ZnSO_4 \cdot 7H_2O$	0.22
$Na_2MoO_4 \cdot 2H_2O$	0.39
$CuSO_4 \cdot 5H_2O$	0.08
$Co(NO_3)_2 \cdot 6H_2O$	0.05

2.2 *Domestic wastewater culture*

The domestic sewage used in the study was provided by Guodian North Wastewater Treatment Plant (Shenyang). After the sewage samples were left for 24h, the supernatant was taken and filtered in a 1.2mm GF/C filter unit to prevent the suspended material in the sewage samples from affecting the growth of Chlorella Vulgaris. The algae were inoculated into the filtered domestic wastewater using a photobioreactor built into the medium. The reactor is shown in Figure 1, and the ambient temperature, light intensity, and light time can be controlled stably.

Table 2. Water quality index of domestic wastewater.

Water Quality Indicators	CODcr	NH_3-N	TP	TN	pH
Concentration . mgL^{-1}	195-200	42–46	4.1–4.3	81–86	8.1–8.2

2.3 *Collection of Chlorella*

After Chlorella has reached a stable growth period, aeration to the culture medium is stopped. Chlorella settles naturally for three days. Then, two different layers are formed: the upper layer contains water containing suspended Chlorella cells, and the bottom layer is concentrated Chlorella. The algal cells in the reactor at the end of the culture were enriched using a high-speed centrifuge (5000r/min, 10min), transferred to a culture dish, and dried

Figure 1. Algal bioreactor.

using a blast drying oven with heating for 24h (65°C). The dried algal slurry was ground into powder and stored airtight in a dry environment at about 10°C for backup.

2.4 Measurement of Chlorella biomass

Stem cell weight is commonly used to determine the biological yield of a biological process, and it is obtained by measuring the total suspended solids concentration in the medium. The absorbance of the algal solution at 680 nm was measured using a spectrophotometer, followed by drying the samples in an oven at 65°C for 24 h and then weighing them. In this way, the relationship between the concentration of algal cells and the absorbance of the algal solution was established, and the standard curve was plotted. The correlation between microalgal biomass concentration and optical density was determined by equation (1).

$$N_x = 0.3638 OD_{680} - 0.0294 \left(R^2 = 0.9983 \right) \tag{1}$$

$$\text{Specific growth rate}: \mu = \frac{\ln \frac{N_2}{N_2}}{t_2 - t_1} \tag{2}$$

$$\text{Biomass production}: P = \frac{N_2 - N_1}{t_2 - t_1} \tag{3}$$

The specific growth rate (μ) and biological yield (P) were determined by equations (1) and (2), respectively, where N_1 and N_2 are the biomass (g/L) at t_1 (days) and t_2 (days), respectively.

2.5 Kinetic growth models

Three nonlinear mathematical models were used to predict the growth of Chlorella in different concentrations of domestic wastewater. The experimental data were referenced to Chlorella's biomass production when different wastewater concentrations were added to the culture setup.

Pearl and Reed originally developed the logistic model to describe the growth process of organisms based on the initial amount, growth rate, time, and final amount (Lacerda et al. 2011; Pearl & Reed 1920).

$$y = \frac{A + C}{1 + exp^{-B(t-M)}} \tag{4}$$

The Gompertz model has been widely used in the literature, and most of the kinetic data are described based on the model shown below (Lacerda *et al.* 2011).

$$y = A + Cexp^{-\exp\{-B(t-M)\}} \tag{5}$$

The Richards model is a four-parameter shown below (Richards 1959).

$$y = A\left[1 + vexp\{k(\tau - t)\}\right]^{-\frac{1}{v}} \tag{6}$$

A: $ln\frac{x_t}{x_0}$ (asymptotic value at t→0); C: $ln\frac{x_t}{x_0}$ (asymptotic value at t→∞); B: relative growth rate at time m (day 1); t: residence time (s); M: time to reach maximum growth rate (s); X_t: biomass concentration at t (g·L^{-1}); X_0: initial biomass concentration (g·L^{-1}); τ: lag time of biological decay point; n, k: parameters

2.6 Methods for studying the growth kinetics of Chlorella Vulgaris

The study used a nonlinear regression technique to solve the growth model. The algorithm for the analysis of microalgal growth kinetics was as follows: (1) values were calculated from the experimental data; (2) data were loaded into the IBM SPSS Statistics (hereafter SPSS) program; (3) the corresponding growth models were then inserted into the SPSS program under the nonlinear column; (4) after the respective models were inserted and checked, the program showed the independent and model variables ; (5) initial estimation of the parameter values; and (6) the accuracy of the growth kinetic model predictions was assessed using S^2, R^2, RMSD, prediction plots with residual plots.

$$R^2 = 1 - \frac{\sum_{i=1}^{n}\left(y_{iobs} - y_{cala}\right)^2}{\sum_{i=1}^{n}\left(y_{iobs} - \bar{y}\right)^2} \tag{7}$$

$$\bar{y} = \frac{1}{n}\left(\sum_{i=1}^{n} y_{iobs}\right) \tag{8}$$

$$S^2 = \frac{\sum_{i=1}^{n}\left(y_i - \bar{y}\right)^2}{n - 1} \tag{9}$$

$$RMSD = \frac{1}{n}\left\{\sum_{i=1}^{n}\left(y_{iobs} - y_{calc}\right)^2\right\}^2 \tag{10}$$

In equations (7)–(10), the subscript n refers to the number of detections, obs refers to the actual detection data, and calc refers to the model calculation data.

3 RESULTS AND DISCUSSION

3.1 Effect of light intensity

Microalgal growth is influenced by different factors, among which light intensity is one of the most important. Light is the necessary source for the autotrophic growth of microalgae and the most important element for photosynthetic activity. It contributes to cell proliferation, respiration, and photosynthesis. Microalgae need light to produce ATP and NADPH and to synthesize molecules necessary for growth. The optimal light intensity for growth and biomass production varies mainly from one microalgal species to another. In culture, the biomass of microalgal species usually increases with increasing light intensity due to higher absorption and utilization of the light by the photosynthetic machinery. However, photo-inhibition is observed at high light intensities, beyond the saturation point, due to photo-oxidation reactions occurring within the cells. This saturation point depends on the particular algal species and culture conditions. The reactor is shown in Figures 2 and 3. Light intensity was found to have a large effect on Chlorella's growth and biological yield. In the constant temperature experiment, the speChlorella'sic growth rate and biomass production of Chloed with increasing light intensity. However, when the light intensity reached a certain range (>12000lux), both data of Chlorella showed a decreasing trend, and a precipitous drop occurred when the light intensity exceeded 14000lux.

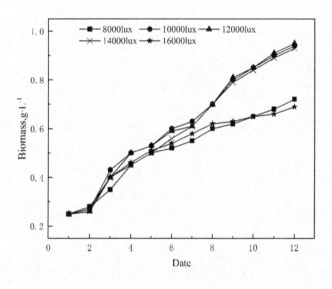

Figure 2. Effect of light intensity on Chlorella growth.

3.2 *Effect of initial pH*

Environmental acidity and alkalinity have important effects on the growth and metabolism of microalgae cells. Neutral or weakly alkaline environments are suitable for the growth of most microalgae (Mostafa et al. 2012). However, some specific microalgal species (e.g., Duchenne) can grow in extremely acidic conditions with pH values as low as 1 (Raven 1990). According to Goldman(1982), pH in biomass media significantly affects the production of green microalgae (e.g., Chlorella) in continuous culture (Goldman et al. 1982). The extent to which cell metabolism is affected by pH determines the pH tolerance limit of microalgae. Goldman claimed that the maximum tolerance pH is not affected by the availability of inorganic carbon (Goldman et al. 1982). However, pH determines the growth and development of carbon species in water and regulates the availability of different carbon sources used for photosynthesis in microalgae. The inconsistency of CO_2 concentration in the ambient gas can lead to pH changes (Tang et al. 2011). Figure 4 shows the effect of pH in the medium on the growth of Chlorella. As seen from the figure, the growth of Chlorella

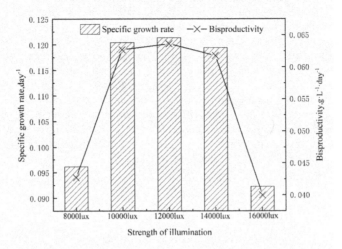

Figure 3.   Effect of light intensity on growth rate and biomass yield.

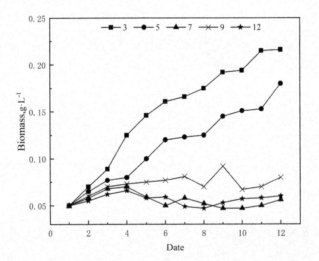

Figure 4.   Effect of initial pH value on Chlorella growth.

Vulgaris cultured at pH 3 and 5 showed an almost linear growth trend, and algal cell concentrations reached a maximum after 12h at about 0.15–0.25 g·L^{-1}. Chlorella did not show a satisfactory growth curve in environments with high pH. At pH 7, 9, and 12, the growth of Chlorella stagnated, and there was no significant increase in microalgae biomass after 12 days of incubation. As can be seen in Figure 5, the specific growth and biological yield of the medium with pH 3 were 0.1366 day^{-1} and 0.0149 g·L^{-1}·day^{-1}, respectively, on day 1. The results indicate that the Chlorella used in this study was well adapted to the low pH medium, which facilitated the suppression of other biological contaminants (e.g., fungi) that may be present in the unsterilized wastewater medium. At pH below 4.5, when CO_2 is dissolved in water, carbon in the medium is dominated by free CO_2 molecules or CO_3^{2-}. Therefore, this study speculates that Chlorella prefers to absorb carbonic acid (H_2CO_3) as a carbon source for growth.

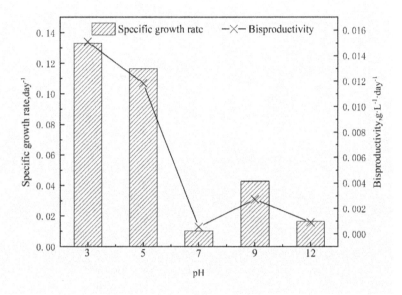

Figure 5. Effect of initial pH value on growth rate and biomass yield.

3.3 *Effect of microalgae inoculum concentration*

During the complete culture process, as the density of microalgae increases with time, the growth rate will be slowed down by nutrient depletion, accompanied by an increase in the density of microalgae resulting in a gradual decrease in the light transmission of the culture system. Therefore, it is important to allow a uniform distribution of light in the microalgae culture system (Qiang *et al.* 1998). A sufficient number of algal species should be added to the culture to increase the initial growth rate of microalgae and enhance their survival rate. Figure 6 shows the growth of microalgae at different inoculation concentrations. All five experimental groups of Chlorella showed similar growth trends throughout the incubation process. As shown in Figures 6 and 7, the highest biological yield was in the 0.02v/v experimental group (0.0229 g·L^{-1}·day^{-1}), followed by 0.15v/v, 0.10v/v, 0.05v/v and 0.03v/v. The highest specific growth rate of 0.1675 d^{-1} was observed in the experimental group, with the lowest inoculum concentration for Chlorella Vulgaris.

Therefore, a certain reduction in algal concentration can increase the specific growth rate and thus facilitate the removal of nutrients from the wastewater.

3.4 *Nutrient removal*

If domestic wastewater is discharged directly into a water body, the high content of N, P, and organic matter in domestic wastewater can cause eutrophication of water bodies by direct discharge (Molazadeh *et al.* 2019, Zeinab *et al.* 2020). Elemental N in domestic wastewater exists mainly as NH_4^+, NO_2^-, NO_3^-, and organically bound nitrogen situation; while

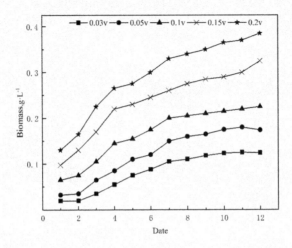

Figure 6. Effect of inoculation concentration on Chlorella growth.

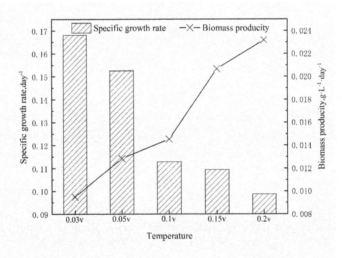

Figure 7. Effect of inoculation concentration on growth rate and biomass yield.

phosphorus is usually present in the form of phosphate (PO_3^{4-}) ions. Therefore, nutrients in domestic wastewater can be used for microalgal growth (Cai *et al.* 2013). Microalgal systems can be used as tertiary treatment units in typical wastewater treatment plants to improve further the efficiency of removing the remaining NO_3^- after the post-denitrification process (Filippino *et al.* 2015). Figures 8 and 9 depict the effect of the conventional incubation process in a laboratory setting (light intensity of 12000lux, light time of 24h, temperature of 25°C, pH of 3, and inoculation ratio of 0.03v/v). The removal efficiencies of T.N., T.P., NH$_3$-N, and COD in the wastewater samples were 94.01%, 90.08%, 97.33%, and 85.37%, respectively, which shows a good removal effect (Aslan & Kapdan 2006, Whitton *et al.* 2016).

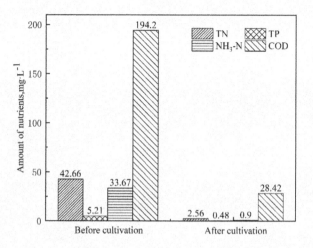

Figure 8. Nutrient removal.

3.5 *Growth kinetic studies of Chlorella Vulgaris*

In recent years, mathematical models have been widely used to predict the growth trends of microorganisms, which are necessary for microbial growth studies and industrial microbiology (Elekl & Yavuzatmaca 2009; Li *et al.* 1999; Mansouri 2017; Zhang 2008). This study selected three growth kinetic models: the Logistic, Gomperz, and Richards models.

As shown in Table 3, the three models fit better ($R^2 > 0.99$) at an algal inoculum concentration of 0.03v/v (hereafter 0.3v/v). However, successive additions of algal species to the

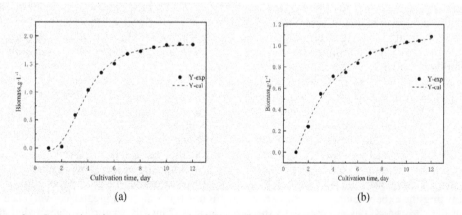

Figure 9. Comparison between the growth of Chlorella and model calculation data (experimental value (exp) and calculated value (cal)) under different inoculation concentrations. (a) A logistic model with algal inoculation concentration of 0.03v/v. (b) Logistic model with algal inoculation concentration of 0.2v/v. (c) Gompertz model with algal inoculation concentration of 0.03v/v. (d) Gompertz model with algal inoculation concentration of 0.2v/v. (e) Richards model with algal inoculation concentration of 0.03v/v. (f) Richards model with algal inoculation concentration of 0.2v/v.

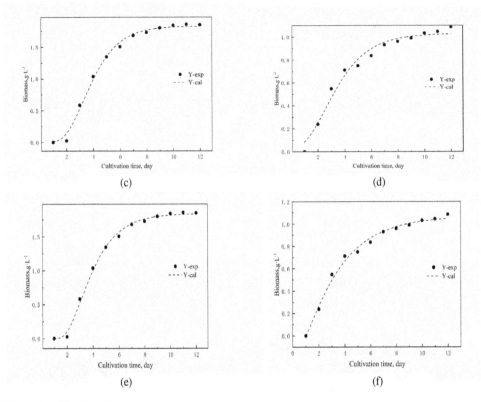

Figure 9. (Continued)

microalgal culture resulted in a decrease in R^2 values ($R^2<0.99$) for some of the models. It may be due to the inaccuracy of the kinetic model for inferring the decaying growth of microalgae. As shown in Figure 9, the growth of Chlorella predicted by the kinetic model coordinated well with the experiments performed in the 0.03v/v group. However, in the experiment at 0.2v/v, the deviation of predicted values increased in the 0.2v/v group experiments. Since all three models did not consider the intra-species competition of Chlorella during growth and the aggregation of algal cells due to insufficient aeration disturbance, it led to an increase in the prediction error when the inoculum concentration increased. Although Gompertz's growth kinetic model also showed higher R^2 ($R^2 = 0.99461$) and lower RMSD values (0.5699) in the 0.02v/v group, the predicted values were more discrete from the measured than the Logistic and Richards models.

Figure 10. reflects the difference between the predicted values and the actual growth of Chlorella. Good residual plots must have y-values close to the x-axis and should not have a random distribution. From Figures 10(c) and (d), the Gompertz model shows a higher-order S-shaped curve distribution along the X-axis, which indicates that the model was inaccurate in fitting the experimental data. In particular, Figure 10(a) shows a greater convergence of the data to the X-axis compared to the other groups. This observation suggests that the logistic model predicts less error than the other growth models and, therefore, can represent the growth of Chlorella Vulgaris, but only in the case of cultures with low inoculum concentrations. At higher inoculum concentrations (e.g., 0.2v/v), it can be seen that the calculated values of all three kinetic models shown in Figures 10(b), (d), and (f) deviate more from the X-axis than at low inoculum concentrations (0.03v).

Table 3. R^2, RMSD and variance values of different growth models of Chlorella under different inoculation concentrations.

Model	0.03v	0.05v	0.1v	0.15v	0.2v
R^2					
Logistic	0.99639	0.9907	0.99482	0.99242	0.99396
Gompertz	0.99461	0.98672	0.99106	0.97707	0.97785
Richards	0.99677	0.99096	0.99454	0.98988	0.99192
RMSD					
Logistic	0.04946	0.6967	0.0365	0.03752	0.03119
Gompertz	0.05699	0.07864	0.0452	0.06152	0.0563
Richards	0.04681	0.06883	0.03748	0.04335	0.03606
Variance					
Logistic	0.01957	0.03897	0.01066	0.01126	0.00778
Gompertz	0.02923	0.05566	0.01839	0.03406	0.02853
Richards	0.01753	0.0379	0.01124	0.01504	0.0104

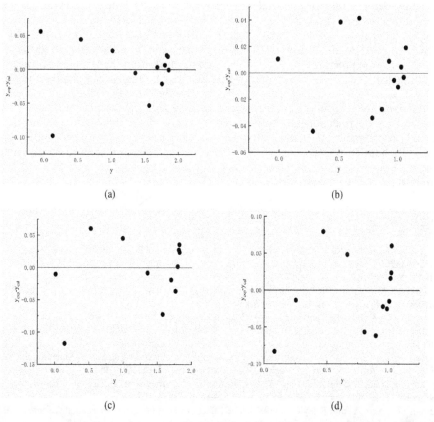

Figure 10. Residual diagram of the growth model. (a) Logistic model with an algal inoculation concentration of 0.03v/v, (b) logistic model with an algal inoculation concentration of 0.2v/v, (c) Gompertz model with algal inoculation concentration of 0.03v/v, (d) Gompertz model with an algal inoculation concentration of 0.2v/v, (E) Richards model with an algal inoculation concentration of 0.03v/v, (f) Richards model with an algal inoculation concentration of 0.2v/v.

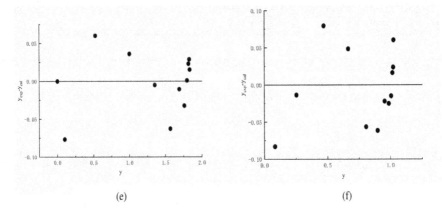

Figure 10. (Continued)

4 CONCLUSION

This study demonstrates the considerable feasibility of culturing Chlorella in domestic wastewater. The fastest growth rate was achieved at an algal inoculum of 0.03 v/v, an initial pH of 3, a temperature of 25°C, and a light intensity of 12000lux, with a biological yield of 0.0229 g·L^{-1}. The nutrient removal effect of Chlorella in domestic wastewater was better during the culture process, and the removal rate reached more than 85% in all cases. After comparing the three types of growth models, the logistic model can simulate the growth trend of Chlorella in domestic wastewater better than the other two models. However, when the algal inoculum concentration increased, the model did not include the simulation calculation of the microalgal death process, which increased the error between the calculated data and the experimental data. Existing growth kinetic models cannot fully and accurately describe the growth process of microalgae. More stable models need to be developed by considering various factors such as changes in nutrient transport, senescence and death of microalgae, and intra-species competition.

REFERENCES

Aslan S. and Kapdan I.K. (2006). Batch Kinetics of Nitrogen and Phosphorus Removal From Synthetic Wastewater by Algae. *J. Ecological Engineering*. 28(1). pp. 64–70.

Chisti Y. (2007). Biodiesel from Microalgae. *J. Biotechnology Advances*. 25(3). pp.294–306.

Cai T., Park S.Y. and Li Y. (2013). Nutrient Recovery From Wastewater Streams by Microalgae: Status and Prospects. *J. Renewable and Sustainable Energy Reviews*. 19(3). pp.360–369.

Devgoswami C., Kalita M., Talukdar J., Bora R. and Sharma P. (2011). Studies on the Growth Behavior of Chlorella Haematococcus and Scenedesmus sp in Culture Media with Different Concentrations of Sodium Bicarbonate and Carbon Dioxide Gas. *J. African Journal of biotechnology*. 10 (61).

Elekli A. and Yavuzatmaca. M. (2009).Predictive Modeling of Biomass Production by Spirulina Platensis as a Function of Nitrate and NaCl Concentrations. *J. Bioresource Technology*. 99(5). pp.1847–1851.

Filippino K.C., Mulholland M.R. and Bott C.B. (2015). Phytoremediation Strategies for Rapid Tertiary Nutrient Removal in a Waste Stream. J. Algal Research. 11. pp.125–133.

Goldman J.C., Azov Y. and Riley C.B. (1982). The Effect of pH in Intensive Microalgal Cultures. I. Biomass Regulation. *J. Journal of Experimental Marine Biology and Ecology*. 57(1). pp .1–13.

Jun-Rong L.I., Shen T. and Lei H.U. (2012). Effect of Chlorella Pyrenoidosa on Wastewater Treatment From Pig Farms. *J. Acta Ecologiae Animalis Domastici*. 33(5). pp.80–82.

Kassim M.A., Rashid M.A. and Halim R. (2017). Towards Biorefinery Production of Microalgal Biofuels and Bioproducts: Production of Acetic Acid from the Fermentation of Chlorellasp. and Tetraselmis Suecicahydrolysates. *J. Green and Sustainable Chemistry*.7(2). pp. 152–171.

Lam M.K. and Lee K.T. (2012). Microalgae Biofuels: A Critical Review of Issues, Problems, and the Way Forward . *J. Biotechnology Advances*. 30(3). pp.673–690.

Li C., Hui X.M. and Pan Z.H. (2021). Study on the Treatment Performance of Simulated Domestic Wastewater by Tetrahymena and Protein Nucleated Chlorella. *Journal of Taiyuan University of Technology*.52(06). pp.880–886.

Lacerda L.M.C.F., Queiroz M.I., Furlan L.T., Lauro M.J., Modenesi K., Jacob-Lopes E. and Franco T.T. (2011). Improving Refinery Wastewater for Microalgal Biomass Production and CO_2 Biofixation: Predictive Modeling and Simulation. *Journal of Petroleum Science and Engineering*. 78(3). pp. 679–686.

Li X.F., Wang C.H. and Ju B. (1999) Microalgae and their Growth Kinetics. *J. Marine Bulletin*.18(6). pp.6.

Lam Man K., Lee K.T., Khoo C.G., Uemura Y. and Lim J.W. (2016). Growth Kinetic Study of Chlorella Vulgaris Using Lab-scale and Pilot Scale Photobioreactor Effect of CO_2 Concentration. *Journal of Engineering Science and Technology*. 11(somche 2015). pp.73–87.

Mohd S., Ainur A., Uemura, Yoshimitsu, Kutty and Shamsul. (2017). Lipid for Biodiesel Production from Attached Growth Chlorella Vulgaris Biomass Cultivating in Fluidized Bed Bioreactor Packed with Polyurethane Foam Material. *J. Bioresource Technology*.239. pp. 127–136.

Mostafa S.S.M., Shalaby E.A. and Mahmoud G.I. (2012). Cultivating Microalgae in Domestic Wastewater for Biodiesel Production. *J. Notulae Scientia Biologicae*. 4(1). pp.

Molazadeh Marzieh Shahnaz D., Hossein A. and Hamid R.(2019) Pourianfar. Influence of CO_2 Concentration and N:P Ratios on Chlorella Vulgaris-assisted Nutrient Bioremediation, CO_2 Biofixation and Biomass Production in a Lagoon Treatment Plant. *Journal of the Taiwan Institute of Chemical Engineers*. 96. pp.114–120.

Mansouri M. (2017). Predictive Modeling of Biomass Production by Chlorella Vulgaris in a Draft-tube Airlift Photobioreactor.*J. Advances in Environmental Technology*. 2(3). pp .119–126.

Pearl R. and Reed L.J. (1920). On the Rate of Growth of the Population of the United States since 1790 and Its Mathematical Representation. *J. Proceedings of the National Academy of Sciences of the United States of America*. 6(6). pp .275–288.

Praveen Kumar, Sudharsanam. A. R., Natarajan M., Senthamil K. (2018). Biochemical Responses from Biomass of Isolated Chlorella sp., Under Different Cultivation Modes: Nonlinear Modeling of Growth Kinetics .*J. Brazilian Journal of Chemical Engineering*. 35(2).pp.489–496.

Qiang H., Zarmi Y. and Richmond A.(1998) Combined Effects of Light Intensity, Light Path, and Culture Density on Output Rate of Spirulina Platensis (Cyanobacteria). *European Journal of Phycology*. 33(2). pp.165–171.

Richards F.J. (1959). A Flexible Growth Function for Empirical Use. *Journal of Experimental Botany*. 10(29). pp.

Raven J.A. (1990). Sensing pH? .*J. Plant, Cell & Environment*. 13(7). pp. 721–729.

Tang D., Han W., Li P., Miao X. and Zhong J. (2011). CO_2 Biofixation and Fatty Acid Composition of Scenedesmus Obliquus and Chlorella Pyrenoidosa in Response to Different CO_2 Levels. *J. Bioresource Technology*. 102(3). pp. 3071–3076.

Wang Y., Shi W. Q. and Guo Y. M. (2020). Environmental Optimization of the Efficient Culture of Chlorella Vulgaris in Domestic Wastewater. *J. Journal of Lanzhou University of Technology*. 46(4). pp.6.

Whitton R., Mével A.L., Pidou M., Ometto F., Villa R. and Jefferson B. (2016). Influence of Microalgal N and P Composition on Wastewater Nutrient Remediation. *J. Water Research*. 91. pp.371–378.

Zhou W., Chen P., Min M., Ma X., Wang J., Griffith R., Hussain F., Peng P., Xie Q., Li Y., Shi J., Meng J. and Ruan R. (2014). Environment-enhancing Algal Biofuel Production Using Wastewaters .*J. Renewable and Sustainable Energy Reviews*. 36: pp. 256–269.

Zwietering M.H. (1990). Modeling of the Bacterial Growth Curve. *J. Applied and Environmental Microbiology*. 56(6).pp .1875–1881.

Zeinab A.S., Fereidun E., Dariush M. (2020). Integrated CO_2 Capture, Nutrients Removal and Biodiesel Production Using Chlorella Vulgaris. *Journal of Environmental Chemical Engineering*. Prepublish .pp.

Zhang Z.B. (2008) *Light Transfer and Growth Kinetics in the Photoautotrophic Culture of Microalgae .D.* Nanchang: Jiangxi Normal University.

Civil Engineering and Energy-Environment – Gao & Duan (Eds)
© 2023 the Author(s), ISBN 978-1-032-56059-5

Coordinated control of Ozone and PM$_{2.5}$ under the perspective of air pollution treatment

Dong Li*

Shandong Environmental Protection Scientific Research and Design Institute Co. LTD Jinan, Shangdong, China

ABSTRACT: There is a bond connection between the two pollutants, Ozone, and PM. With this in hand, people can make a workable strategy for coordinated control over these pollutants. Based on the analysis of the relevance between Ozone and PM as well as the current situation on the pollution of Ozone and PM in China, this essay aims to discuss the issues related to coordinated control over Ozone and PM under the perspective of air pollution treatment and give the corresponding optimizing strategy, to provide some clue to the faculty in this field.

1 INTRODUCTION

PM$_{2.5}$ has complex causes, sources, and diverse chemical compositions, which can affect climate change, atmospheric visibility, and environmental quality and endanger human health to a certain extent. Ozone is also an atmospheric pollutant with strong oxidation. Suppose the concentration of Ozone in the ambient air is high. In that case, it will cause direct harm to human health and damage the physiological function and internal structure of plants, adversely affect the normal growth of plants, and affect ecological security and food security. With the increasing range and concentration of Ozone in ambient air, the pollution of high Ozone concentration will affect more living organisms and populations, so people must pay great attention to this problem. In the field of air pollution control at the present stage, the collaborative control of PM$_{2.5}$ and Ozone has become an important topic for relevant scholars to carry out research. To protect human health and improve air quality, relevant scholars should deeply explore and think about the collaborative control strategy of the two.

2 CORRELATION BETWEEN OZONE AND PM$_{2.5}$

In one place, for example, when the average concentration of PM$_{2.5}$ was 30μg/m3, 92.7% of the days, there was good air quality. Although the achievements of air pollution control are good, there is still bad weather with mild pollution or more. Ozone and PM$_{2.5}$ are the main causes of this problem, and each pollutant has a duration of at least 8d. Compared with

---

*Corresponding Author: 13793105715@163.com

2020, 55.5mm more precipitation was recorded in 2021. Precipitation can effectively purify the air, especially by reducing the concentration of PM_{10} and $PM_{2.5}$, which will affect the corresponding air indexes to a large extent. The concentration of Ozone in the atmosphere will be affected by light intensity and temperature, and the increase in temperature will also increase the concentration of Ozone, which will affect the air quality (Chai 2020). In this area's dynamic change of Ozone for 24 hours, the highest Ozone concentration is reached at noon.

In most cases, when the light reaches the strongest, the maximum Ozone concentration is reached. After that, the Ozone concentration will gradually decrease, so the concentration of Ozone is low at night. Under the influence of different factors, promoting Ozone concentration will also increase $PM_{2.5}$ concentrations. When the air environment reaches more than mild pollution, Ozone will be the longest at 8h. This phenomenon occurs mainly from May to October. Compared with other months, this month, with more light and higher temperature, simultaneously has a high Ozone concentration, then will greatly reduce the air's excellent rate. Similar situations can be found in many Chinese cities. Table 1 shows the changes in O_3 and $PM_{2.5}$ concentrations in 339 prefecture-level and above cities from July to October 2021.

Table 1. Year-on-year changes of O_3 and $PM_{2.5}$ concentrations in 339 prefecture-level and above cities from July to October 2021.

Time	Ozone Concentration, $\mu g/m^3$	The Increased Proportion, %	$PM_{2.5}$, $\mu g/m^3$	The Increased Proportion, %
July 2021	113	0.9	16	−15.8
August 2021	138	3.8	17	Flat
September 2021	140	1.4	19	−9.5
October 2021	139	1.5	26	−13.3

Note: Data from the website of the Ministry of Ecology and Environment.

To sum up, the link between Ozone and $PM_{2.5}$ is very close, and although it makes it possible to coordinate the control between them, it is difficult and complex.

3 CURRENT STATUS OF $PM_{2.5}$ AND OZONE IN CHINA

Through the analysis of the present situation of Ozone pollution in China, the concentration of Ozone has been rising in recent years. Although compared with the relevant standard, the concentration of Ozone in our atmosphere is relatively high, but it is mainly mild pollution (Huang & Zhang 2020). However, the Ozone concentration is increasing to a small extent every year in more than 300 cities in China. Combined with the overall development situation, the Ozone concentration of Chinese air is rising slowly. According to the Ozone concentration in April and May 2022, the Ozone concentration in 339 cities at the prefecture level and above increased year-on-year. As shown in Figure 1, the Ozone concentration increased by 16µg/m3 in April. Through the comprehensive analysis of the key development areas of China, compared with other regions of China, the Ozone concentration in Tianjin, Beijing, and other places is higher.

Therefore, Ozone has become the same as $PM_{2.5}$ as a major factor affecting the atmosphere, which to a certain extent, also increases the difficulty of air pollution control. In

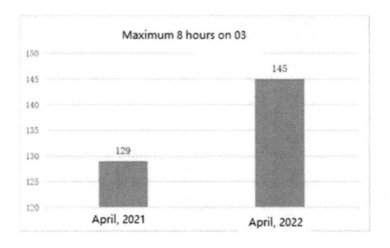

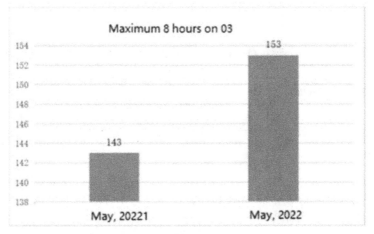

Figure 1. Ozone concentration and year-on-year change in 339 cities at the prefecture level and above in China from April to May 2022.

environmental governance and urban development, many reasons can cause air pollution. First, Ozone pollution has been unable to control VOCs and NO_X effectively. In 2020, our country's atmosphere effectively reduced the emissions of $PM_{2.5}$ and SO_2, but the overall emissions of VOC_S and NO_X still can not effectively be reduced. Especially in the Yangtze River Delta, Tianjin, Beijing, and other places, the VOC_S and NO_X emissions industries are also concentrated in the process of industrial development and, to a large extent, improve the overall emissions of these cities.

Moreover, the continuous development of the Chinese shipping, household, chemical, and pharmaceutical production industries leads to the appearance of many small-scale production enterprises, which leads to the country's difficulty in effectively adjusting the overall industrial layout. Second, our country has a vast territory and large climate differences between various regions. High temperatures and low rainfall areas will lead to the formation of Ozone in a large area. The relevant survey results show that in the past five years. However, the environmental protection department has reduced the VOC_S and NO_X to the greatest extent so that the composition of Ozone has been fully offset, and the rise of natural temperature also, to some extent, offset the above-related substances (Bao & Li 2019). Third, through the measurement and observation of the global Ozone, it is found that with the constant speed up of urban construction, the basic parameter of Ozone material will also be improved. In recent years, there have been increasing Ozone substances in the world, and the increased rate of its production is about 1μg/m3 per year. Because of this situation, we should implement the ecological and environmental protection development strategy effectively. Also, we should pay full attention to the control of $PM_{2.5}$ pollution and take effective collaborative control measures to control Ozone.

4 PROBLEMS EXISTING IN THE COORDINATED CONTROL OF $PM_{2.5}$ AND OZONE AT THE PRESENT STAGE

4.1 *Management control system needs improvement*

Currently, air pollution prevention and control work mainly focuses on individual cities in China, and the relevant regional linkage prevention and control system has not really formed. Even if the work of supervision and prevention and control of Ozone pollution is carried out, the main target is within the city area. For example, many cities apply the summer Ozone control technology means. Although experiences are accumulated in environmental management, there is still a large range of persistent Ozone pollution problems. Take the Ozone pollution situation in 2019 as an example. In September, Ozone pollution concentration exceeded the standard in the Pearl River Delta and North China regions. At the same time, the process of continuous industrial production is also expanding the scope of pollution, eventually affecting the Guangxi Zhuang Autonomous Region, the eastern coastal region, central China, Shandong Peninsula. Among them, the Ozone pollution in North China has reached a moderate level, and nearly one-third of the country's area is covered by it. Given this situation, deep research into the actual situation and pollution characteristics of Ozone pollution, active exploration, and research of Ozone pollution control technology is done. On this basis completed, the development of relevant control documents and systems and the optimization and improvement of laws and regulations of air pollution prevention is promoted to a great extent. However, in the documents mentioned above and in regulations, the basic requirement is to control $PM_{2.5}$ effectively. This situation proves that China can still not build a joint prevention and control mechanism for atmospheric pollution in combination with the actual situation of atmospheric pollution. It is unable to control key industrial development areas finely.

4.2 *Lack of relevant technical means*

China has a short history of studying the coordinated control of $PM_{2.5}$ and Ozone and is still in the exploration stage. There is a relative lack of a feasible management plan and governance system. Although China has formed the correct understanding of the collaborative control of air pollution, there are still some defects and deficiencies in the practical implementation process. First, there is an obvious lack of coordination ability at the present stage, which leads to the inability to effectively control Ozone pollution in some areas when dealing with $PM_{2.5}$ pollution and continuously improve the main body of Ozone pollution. Second, China lacks sufficient synergistic intensity. Pollution prevention and control in China mainly focuses on SO_2 substances, but NO_X substances have been ignored to a certain extent. It is still necessary to continuously improve the intensity of the management of different pollutants (Ren & Wang 2020).

4.3 *Performance of the relevant equipment is not functional*

Combined with the actual situation of industrial and urban development, there are differences in the management and supervision of air pollution in various cities. Based on this, the basic ability of government law enforcement needs to be continuously improved. Although after a long exploration and practice, each city's environmental monitoring technical personnel can fully understand and grasp the content of the common atmospheric pollution monitoring and management. The use of related equipment is relatively insufficient, plus all sorts of atmospheric pollution control equipment appear constantly. However, the relevant standards are not clear. Technical personnel can not effectively identify the performance of various equipment, resulting in difficulty in carrying out supervision and monitoring of the environment effectively.

5 STRATEGIES FOR COORDINATED CONTROL OF OZONE AND $PM_{2.5}$ FROM THE PERSPECTIVE OF AIR POLLUTION CONTROL

5.1 Complete the formulation of cooperative control strategy

With the combination of air pollution in our country, the government should pay great attention to air pollution prevention and control. First, a coordinated control strategy should be established at the national level with Ozone and $PM_{2.5}$ as the core to accelerate the construction of a coordinated control system. And then, according to the transmission and regional characteristics of Ozone pollution, the main joint prevention and control areas of Ozone pollution should be reasonably divided. Moreover, in combination with the energy and industrial structure of each region, the distribution characteristics. The current situation of Ozone pollution, the long-term control strategy, and short-term stage goals of each region are formulated scientifically to promote the formation of effective prevention and control programs for Ozone pollution precursors. Among the closely connected regions, the joint prevention and control mechanism of Ozone pollution should be further improved, and the cooperative control plan should be formulated in combination with the Ozone pollution period to control regional Ozone pollution effectively.

5.2 Make goals of environmental development clearer

When managing environmental development goals and formulating collaborative plans, relevant personnel should improve the management of $PM_{2.5}$ and Ozone pollution according to air pollution control needs. At the same time, combined with the basic emission inventory of air pollution, the air quality improvement structure route plan is reasonably formulated. We can improve the atmospheric quantitative cooperative planning scheme from the following three aspects (Wang 2018). First, to clarify the goals of environmental development, we should effectively plan and coordinate the management of areas and pollutants. Secondly, VOC_S and NO_X substances should be added to the core factors affecting the development of air pollution to make the emission reduction ratio more reasonable. Finally, in developing regional air pollution control programs and collaborative control of different pollution sources, the relevant supervision and management departments and technical personnel should ensure that environmental control quality standards are fully considered simultaneously. It is also necessary to research and analyze all industrial production needs to use governance technology to manage VOC_S and NO_X effectively.

5.3 Strengthen the corresponding infrastructure

To further improve the photochemical monitoring network, monitoring the characteristics of key precursor components can be more accurate. Relevant departments and personnel should further optimize and improve the existing ground-level Ozone observation system to enhance the intensity of observation in forestry and agricultural areas and improve the coverage of the observation network. Effective analysis and research of relevant data can help relevant personnel control Ozone pollution more accurately. Secondly, it is necessary to ensure that the monitoring and early warning of atmospheric Ozone cover the major grain-producing areas and fully grasp the exposure level of crops. It then rationally improves the control policy of air pollutants from the perspective of ensuring the safety of food production. Finally, the establishment of the Ozone sounding observation operation system should be completed to acquire more valuable data, such as the data of calibration satellite products, stratospheric Ozone change data, etc.

5.4 *Strengthen research and development in science and technology*

Only by clarifying the main principles of the formation of Ozone and $PM_{2.5}$ and deeply analyzing the influence of the two pollutants on the atmospheric environment can we control Ozone and $PM_{2.5}$ in a more targeted way. Therefore, it is necessary to strengthen the relevant scientific and technological research and development and dig deeply into the spatial and regional distribution of Ozone pollutants. At the same time, researchers should analyze the factors leading to the formation of Ozone and $PM_{2.5}$ and the source of pollutants, clear the spatial distribution and sources of Ozone and $PM_{2.5}$ in each region. In addition, the regulations on VOCS and NOX emissions should be clarified further. Multi-level and all-round coordinated emission reduction strategies should be formulated according to the situation in key regions and fields to carry out coordinated control of Ozone and $PM_{2.5}$ in a more targeted manner.

5.5 *Strengthen the management and control of mobile pollution sources*

The atmosphere is also heavily influenced by mobile pollution sources, so the concentration of air pollutants can be effectively reduced by strengthening the management and control of mobile pollution sources. In order to take more strict management of vehicles, relevant personnel can set corresponding prompts on the ring road and equip them with effective detection equipment. In particular, it is necessary to capture vehicles with large exhaust emissions accurately. Meanwhile, cooperation should be taken with the transportation department to enhance the punishment of illegal freight vehicles, which can also reduce the emission of pollutants to a certain extent. In addition, the management of non-road mobile machinery should be enhanced to ensure that relevant machinery and equipment can meet energy conservation and emission reduction standards. The pollution caused by relevant factors should be effectively controlled.

5.6 *Guide to public participation*

Relevant authorities can publish the monitored air quality data on their official websites, guide more public supervision of air pollution prevention and control through platforms such as Tiktok and Wechat, and encourage the public to put forward their suggestions and opinions on related work (Shi 2021). Secondly, relevant departments should disseminate knowledge about environmental protection and air pollution to the public through various channels to ensure that the public's role in supervising the prevention and control of air pollution is fully played. This way of public information can play a role in urging the public to protect the environment and promote the innovation of air pollution control work in various regions.

6 CONCLUSION

To sum up, our country has faced a serious air pollution situation nowadays. Although great importance has been attached to air pollution control work, there are still some problems in the coordinated control of $PM_{2.5}$ and Ozone. In carrying out related work, the relevant departments and personnel can improve the quality of air pollution control by formulating a cooperative control strategy and promoting environmental development goals more clearly. Corresponding to strengthening infrastructure construction, strengthening the research and development of science and technology, strengthening the management and control of mobile sources, guide the work involved in such measures as strengthening $PM_{2.5}$ and Ozone coordination control effect.

ABOUT THE AUTHOR

Li Dong, male (1981~), is from Xintai, Shandong province, is a master's student and senior engineer engaged in environmental impact assessment, environmental planning, and air pollution control direction.

REFERENCES

Bao C. & Li J., Generation and Control of Ozone in the air, *Energy and Environment*, 2019(4):2.

Chai F., A Long Time for Work, Blue Sky Forever: Coordinated Control of $PM_{2.5}$ and Ozone. *Environment and Sustainable Development*, 2020, 45(6):2.

Huang C. & Zhang G., *Expert Opinion\Effectiveness and Prospect of Coordinated Control of $PM_{2.5}$ and O_3 in Shanghai*, 2020.

Ren R. & Wang Y., *Study on the Characteristics of Atmospheric Ozone Pollution and Its Control Measures*, 2020.

Shi H., Study on Public Willingness to Participate in Air Pollution Control and Countermeasures, *Science and Technology for Development*, 2021, 17(1):8.

Wang P., *The Bidirectional Coupled WRF-CMAQ Model Used to Study the Effects of Emission Control Strategies on $PM_{2.5}$ and O_3 Concentrations in the Yangtze River Delta Region*, Zhejiang University, 2018.

Civil Engineering and Energy-Environment – Gao & Duan (Eds)
© 2023 the Author(s), ISBN 978-1-032-56059-5

The regional difference and convergence of tourism eco-efficiency in the Yellow River basin

Yu Jie* & Wen Ya*
Business School, Shandong Normal University, Jinan, China

ABSTRACT: An evaluation of the tourism eco-efficiency of nine provinces and regions along the Yellow River was made for 2010-2019, seeking the overall efficiency levels, regional differences, and convergence of development. The Super-efficiency SBM model based on the unexpected output, Dagum gini coefficient decomposition method, and spatial Durbin model is applied to achieve these research objectives. The results show that: (1) the overall tourism eco-efficient of the Yellow River Basin is low and does not usher in the inflection point of the growth rate to good development, but its tourism eco-efficiency has maintained an upward trend. (2)The Yellow River basin presents an unbalanced state due to the large gap in the tourism eco-efficient of the three regions. The main source of the overall regional differences is the regional differences. (3) Considering the spatial effect and the conditional β convergence of the control variables, the development of tourism eco-efficient has significant conditional β convergence, which means that the tourism eco-efficient of provinces and regions along the Yellow River basin will gradually converge to their steady level. Besides, environmental regulation and openness to the outside have a significant positive spatial correlation with tourism eco-efficiency.

1 INTRODUCTION

The Yellow River basin's ecological protection and high-quality development is a major national strategy of great significance to coordinate regional development and reshape China's economic pattern. Sustainable and coordinated development is one of the manifestations of high-quality development in the Yellow River Basin. The best form of sustainable development is ecological civilization, maintaining the harmonious coexistence between human beings and nature (Deng 2020). The coordinated development reflects that the development level and speed of the three regions keep the same pace in the Yellow River Basin's upper, middle and lower reaches. Fortunately, the provinces and regions along the Yellow River basin are rich in our country's historical and cultural heritage and have unique advantages in tourism. Therefore, the tourism industry is more important to economic development, and the ecologically coordinated development of its tourism industry is bound to affect high-quality development (Wang 2020). However, the dual effects of tourism have been a concern.

On the one hand, tourism relies on the natural environment to attract tourists and obtain tourism income. On the other hand, large-scale tourism flows and the development mode inefficient and extensive constantly challenge the environment's carrying capacity (Castilho & Daniela 2021), so how to develop the economy and environment coordinately becomes the core of sustainable development of tourism (Peng 2017). Therefore, by tracking the eco-efficiency of tourism in the last decade, this paper probes into the dynamic changes in the development of tourism ecological civilization in the three regions of the Yellow River basin's

---

*Corresponding Authors: yujiewy@163.com and 17865331145@163.com

DOI: 10.1201/9781003433651-58

upper, middle, and lower reaches. It also finds the regional differences among the three regions and the root causes of the differences to achieve the goal of their coordinated development.

2 LITERATURE REVIEW

Tourism eco-efficiency is one form of sustainable tourism development, combining ecological and economic indicators. The eco-efficiency was first proposed by Schaltegger and Sturm in 1990, who viewed it as getting the maximum economic output with the least resource consumption and environmental cost. Stefan first introduced eco-efficiency into tourism research in 2005, using the ratio of economic returns to environmental costs and tourism income to represent the economic benefits of human production. The environmental impact is expressed in terms of carbon emissions from tourism (Yao 2016).

The study of ecological tourism efficiency provides theoretical support for developing tourism ecological civilization. However, the previous studies mostly started from the macro-scale or the more developed regions (Guo 2021; Jia 2017), and only some scholars studied the spatial scale of the river basin. However, as a special geographical unit, different provinces in the same basin will form different ecological landscapes and human environments. Moreover, the diversity between the upper, middle, and lower reaches of the Yellow River basin is even more pronounced because of its fragile ecology, which has increased the difficulty of tourism development while imperceptibly opening the gap between the provinces (Romano 2020). Therefore, the study on the balanced development of tourism eco-efficiency is of great significance to sustainable and high-quality tourism development in the Yellow River basin.

This paper applied a Super-efficiency SBM model to assess tourism eco-efficiency. The Dagum gini coefficient decomposition method was used to explore the regional differences and the sources, and the conditional β convergence model was used to analyze the convergence.

3 METHODS AND MATERIALS

3.1 *Super-efficiency SBM model based on the unexpected output*

In this paper, we combine Tone's (2003) SBM model with undesired output with Qian Zhenhua's (Qian 2013) global super-efficiency model and calculate each province's tourism eco-efficiency. The decision-making units (DMU) calculated in the non-expected output model are further calculated, and the DMU of multiple $\rho=1$ is rearranged to avoid the incomparable problem. The model can be expressed as follows: Supposing several decision-making units, each DMU contains three elements: input, expected output, and non-expected output, which are represented by (X Y Z), respectively.

$$p = \min \frac{1 + \frac{1}{m}\sum_{i=1}^{m}\frac{s_i^x}{x_{i0}}}{1 - \frac{1}{s_1+s_2}\left[\sum_{k=1}^{s_1}\frac{s_k^y}{y_{k0}} + \sum_{l=1}^{s_2}\frac{s_l^z}{z_{l0}}\right]} \tag{1}$$

$$s.t.\ x_{i0} \geq \sum_{j=1,\ \neq 0}^{n} \lambda_j x_j - s_i^x,\ \forall i;$$

$$y_{k0} \leq \sum_{j=1,\ \neq 0}^{n} \lambda_j y_j + s_k^y,\ \forall k;$$

$$z_{l0} \geq \sum_{j=1, \neq 0}^{T} \lambda_j z_j - s_l^z, \ \forall l;$$

$$1 - \frac{1}{s_1 + s_2} \left(\sum_{k=1}^{s_1} \frac{s_k^y}{y_{k0}} + \sum_{l=1}^{s_2} \frac{s_l^z}{z_{l0}} \right) > 0$$

$$s_i^x \geq 0, \ s_k^y \geq 0, \ s_l^z \geq 0, \ \lambda_j \geq 0, \ \forall i, j, k, l$$

3.2 Decomposition of regional differences

The Dagum gini coefficient and its decomposition method are applied to study the source and evolution of unbalanced tourism ecological development in three regions along the Yellow River basin. Total gini coefficient G is decomposed into intra-regional differential contribution G_w, inter-regional differential contribution G_{nb} and hypervariable density contribution G_t, and the relationship satisfies.
$G = G_w + G_{nb} + G_t$. The formula can be expressed as:

$$G = \sum_{j=1}^{k} \sum_{h=1}^{k} \sum_{i=1}^{n_j} \sum_{r=1}^{n_h} |y_{ji} - y_{hr}| / 2n^2 \bar{y} \tag{2}$$

$$G_w = \sum_{j=1}^{k} p_j s_j \frac{\sum_{i=1}^{n_j} \sum_{r=1}^{n_j} |y_{ji} - y_{jr}|}{2n_j^2 \bar{y}_j} \tag{3}$$

$$G_{nb} = \sum_{j=2}^{k} \sum_{h=1}^{j-1} \frac{\sum_{i=1}^{n_j} \sum_{r=1}^{n_h} |y_{ji} - y_{jr}|}{n_j n_{h(\bar{y}_j} + \bar{y}_h)} (p_j s_h + p_h s_j) D_{jh} \tag{4}$$

$$G_t = \sum_{j=2}^{k} \sum_{h=2}^{j-1} \frac{\sum_{i=1}^{n_j} \sum_{r=1}^{n_h} |y_{ji} - y_{hr}|}{n_j n_h (\bar{y}_j + \bar{y}_h)} (p_j s_h + p_h s_j)(1 - D_{jh}) \tag{5}$$

$$D_{jh} = (d_{jh} - p_{jh}) / (d_{jh} + p_{jh}) \tag{6}$$

3.3 Convergence model

Conditional β convergence means the possibility that each region's eco-efficiency tends to be stable in its own development space under many factors. As the spatial spillover effect of tourism development is obvious, a spatial Dubin model based on a second-order inverse distance matrix is applied. The expression as:

$$y_{t+1} = (\ln F_{1, \ t+1} - \ln F_{it}) / \ln F_{it} \tag{7}$$

$$= \alpha + \beta_1 \ln F_{it} + \beta_2 control + \delta_1 \omega_{ij} \ln F_{it} + \delta_2 \omega_{ij} control + \varepsilon_{it} \tag{8}$$

In constructing the conditional β convergence model, technological innovation (Ti), environmental regulation (Er), green development level of tourism (Gdt), tourism

445

development level (Td), rationalization of tourism structure (Rts), and tourism open to the outside (To) are control variables. Ti, Er, and Gdt are expressed in terms of expenditure on science and technology, investment in pollution control, and carbon emissions per unit of tourism income, respectively, Ts and To are expressed as the proportion of per capita travel expenses and the radio of foreign-invested star-rated hotels' operating income to total operating income of star-rated hotels, and the calculation process of SR is based on Liu Chunji (Liu 2014).

3.4 *Data collection*

Nine provinces along the Yellow River basin were selected as samples to study the level of ecological tourism development and its spatial characteristics and divided into three regions. The upper reaches include Qinghai, Sichuan, Gansu, Ningxia, and Inner Mongolia; the middle includes Shanxi and Shaanxi; and the lower includes Shandong and Henan. The division is based on the cut-off point.

The article assumes real fixed assets of travel agencies and star-rated hotels for the data measuring input and output variables. The number of scenic spots is the capital input, selects the number of the tertiary sector of the economy is the input of the labor force, and selects the tourism income. The number of tourists is the expected output selecting carbon emissions as an undesired output. Fixed assets and tourism income are based on the 2010 price index for conversion (Llanquileo-Melgarejo 2021; Lu 2019).

All the above indexes and data are from the "China Statistical Yearbook," "The Yearbook of China Tourism Statistics," and the statistical bulletin of each province and region from 2010 to 2019. Finally, panel data from 2011 to 2020 were collected for 9 Chinese provinces and regions.

4 THE REGIONAL DIFFERENCE AND ITS SOURCE OF TOURISM ECO-EFFICIENCY

4.1 *The overall description of tourism eco-efficiency*

Table 1 indicates that the mean values of tourism eco-efficiency of the regions along the Yellow River from 2010 to 2019 are less than 1.00, meaning there is still room to improve. However, the average annual growth rate of the composite index is 20.23%, which shows a steady upward trend. In other words, the ecological construction of tourism has achieved

Table 1. A comprehensive index of tourism eco-efficiency in three regions.

Province	2010	2011	2012	2013	2014	2015	2016	2017	2018	2019	AAGR
Qinghai	0.047	0.043	0.049	0.056	0.065	0.074	0.091	0.103	0.104	0.133	12.14%
Sichuan	0.165	0.156	0.190	0.192	0.214	0.243	0.258	0.281	0.277	0.326	7.84%
Gansu	0.055	0.062	0.114	0.138	0.124	0.147	0.156	0.210	0.261	0.557	29.27%
Ningxia	0.063	0.041	0.058	0.070	0.065	0.057	0.065	0.090	0.094	0.267	17.34%
Inner Mongolia	0.128	0.079	0.088	0.095	0.109	0.151	0.144	0.182	0.480	1.065	26.58%
Upstream	0.092	0.076	0.100	0.110	0.115	0.134	0.143	0.173	0.243	0.470	19.89%
Shaanxi	0.179	0.139	0.210	0.226	0.245	0.258	0.268	0.320	0.399	0.557	13.47%
Shanxi	0.204	0.171	0.222	0.282	0.339	0.368	0.432	0.545	0.678	1.556	25.34%
Mid-stream	0.191	0.155	0.216	0.254	0.292	0.313	0.350	0.433	0.538	1.057	20.92%
Henan	0.159	0.174	0.189	0.211	0.270	0.264	0.244	0.286	0.308	1.054	23.35%
Shandong	0.101	0.110	0.124	0.133	0.152	0.154	0.181	0.191	0.219	0.266	11.40%
Downstream	0.130	0.142	0.157	0.172	0.211	0.209	0.213	0.238	0.263	0.660	19.78%
The whole	0.122	0.108	0.138	0.156	0.176	0.191	0.204	0.245	0.313	0.642	20.23%

initial results, but there is still much room for improvement. The main reason is the higher input cost and lower expected output.

Second, the ecological efficiency of the three regions is unbalanced. The average eco-efficiency of the middle reaches more than twice that of the upper reaches and 1.5 times that of the lower reaches. The main reason is that the mean value of tourism eco-efficiency in Shanxi province is the largest, which shows that its input-output ratio is the highest. With the lowest labor and capital costs in exchange for as many tourists and a tourism economy, the resulting carbon emissions are relatively small. The tourism eco-efficiency of Qinghai is the lowest, maybe the restriction of economic development, the coordination between tourism development and eco-development is more difficult.

4.2 *Differential decomposition*

The Dagum gini coefficient decomposition method was used to measure the overall differences between the provinces along the Yellow River basin and the intra-regional differences, inter-regional differences, and sources of differences in the three major regions of the Yellow River during 2010 and 2019.

1) Overall differences. As can be seen from Figure 1, the overall Gini coefficient increase of tourism ecology in the Yellow River basin is 4.59%, indicating that the difference among provinces is rising. From 2010 to 2017, the growth rate was only 1.3%, and the difference in eco-efficiency of tourism between provinces has suddenly increased since 2018. In addition, the overall difference has significant phase characteristics, showing phase fluctuations. 2011–2012 and 2014–2015, the overall difference decreased significantly. 2012–2014 and 2015–2019 have an upward trend.

2) Intra-regional differences (Figure 2). The development of provinces and regions in the upper is quite different. Whose mean within-group difference coefficients were 7.9 and 8.9 times those in the middle and lowered, respectively. The level of tourism development in Ningxia is relatively backward, which may be the reason for the large difference within the group. However, the fluctuation in the upper is not big, indicating that the five regions in the upper are developing synchronously. The middle's intra-group difference fluctuates greatly because Shanxi's tourism eco-efficiency is developing rapidly and gradually diverging from Shaanxi's. The internal difference between the regions with weak tourism ecological development is bigger, but that with good development is smaller.

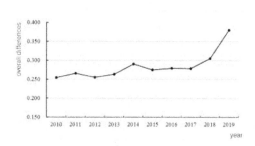

Figure 1. Overall differences in tourism eco-efficiency.

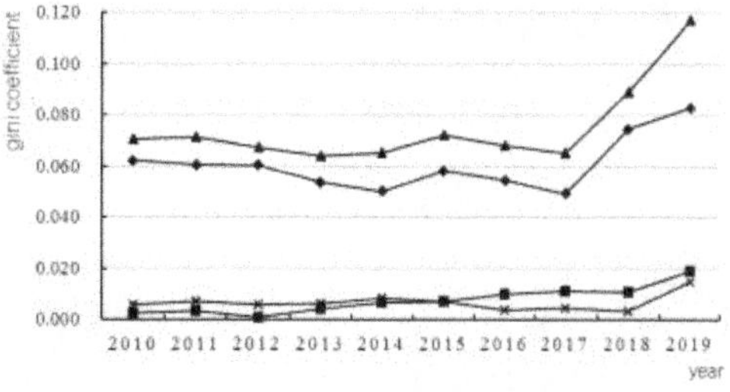

Figure 2. Intra-regional differences.

3) Differences between regions. The fluctuation of inter-regional differences is greater than that of intra-regional differences. Overall, the difference between upstream and middle was the most significant. The regional difference between the middle and lower is below 0.05 every year, and the development level of the middle and lower reached provinces is equal.

The difference is small. The regional difference between the upper and middle was similar to the variation trend of the total difference coefficient between groups. The regional difference was the main part of the total difference.

4) Contribution rate of regional difference sources. Table 2 indicates that intra-regional differences are the main source of the regional difference. There are obvious differences in economic development level and transportation convenience between the Yellow River basin's upper, middle, and lower.

Table 2. The source decomposition of regional differences in three regions.

year	Regional contributions	Contribution rate(%)	Interval contribution	Contribution rate(%)	Super density contribution	Contribution rate(%)
2010	0.071	27.84	0.164	64.43	0.020	7.73
2015	0.072	26.24	0.191	69.28	0.012	4.48
2019	0.117	30.81	0.180	47.29	0.083	21.90
mean	0.075	26.36	0.185	65.12	0.024	8.52

4.3 Convergence analysis

Large-scale tourism flows are common due to the characteristics of tourism across time and space, which makes the significant interaction between regions. Therefore, The spatial convergence model with spatial factors is introduced to study the convergence of its development trend.

(1) Spatial autocorrelation. The spatial correlation of tourism eco-efficiency values in nine provinces and regions from 2010 to 2019 and the local Moran scatter plots are plotted in Table 3 and Figure 3.

Table 3. Global Moran's I Index of tourism eco-efficiency.

year	2010	2011	2012	2013	2014	2015	2016	2017	2018	2019
Moran I	0.253	0.203	0.034	0.116	0.182	0.212	0.47	0.196	0.132	0.157
P	0.014	0.028	0.177	0.256	0.006	0.024	0.000	0.025	0.045	0.038
Z	2.211	1.904	0.927	0.657	1.251	1.984	3.471	1.953	1.693	1.778

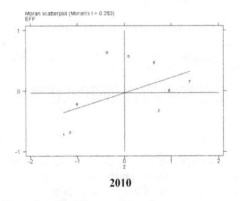 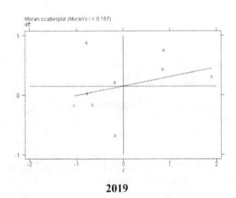

Figure 3. The Moran scatter plots.

The Global Moran's I index was positive for all the years. The other years passed the 5% significance test except for 2012 and 2013, which indicated a positive spatial correlation between the whole tourism ecological development.

The Global Moran's I index only reflects the overall spatial correlation pattern. In order to further reflect the local clustering pattern of the nine provinces, the Moran scatter plots in 2010 and 2019 are drawn (Figure 3). Specifically, the H-H agglomeration, which shows a high level of agglomeration mainly includes Inner Mongolia, Shaanxi, Shanxi, Henan. The pattern of L-L agglomeration mainly includes Qinghai, Sichuan, Gansu, and Ningxia, whose tourism ecology development is weak and has not formed coordinated development with the neighboring provinces.

(2) Conditional β convergence. The results of conditional β convergence are presented in Table 4. The results are divided into two parts: the internal influence of the region and the external influence of other regions on the region, that is, the spatial spillover effect.

Table 4. Conditional β convergence of tourism ecological.

Variable	eff	*ln*Td	Rts	Gtd	*ln*Er	To	ln *rd* Ti
β	−0.412***	−1.370**	−0.060	−1.794**	0.242***	0.039**	−0.524***
	(0.788)	(0.535)	(0.050)	(0.456)	(0.057)	(0.016)	(0.155)
Variable	W eff	W *ln*Td	W Rts	W Gtd	W *ln*Er	W To	W Ti
β	26.277***	−2.885	−0.209**	1.483	0.180**	0.156**	−4.245***
	(6.219)	(3.087)	(0.089)	(1.226)	(0.074)	(0.071)	(0.870)

Note: * * * , * * , * indicate that the estimated coefficients are significant at 1% , 5% , and 10%

The overall convergence coefficient is significantly negative at a 1% level, which indicates a significant conditional β convergence trend in the tourism eco-development of the provinces along the Yellow River basin. Every region has gradually converged to its respective steady-state levels.

From the perspective of self-correlation, environmental regulation and tourism openness have a significant positive impact on the development of tourism ecology. The positive impact of environmental regulation shows that the government's intervention in tourism forced companies to cut their carbon emissions. In addition, the degree of opening to the outside promotes the development of tourism ecology, which shows that foreign investors' advanced technology and management ideas have been applied to the development of China's tourism industry. Tourism development level, green development level, and technological innovation significantly negatively affect ecological tourism development. The negative influence of the tourism development level indicates that the tourism economy can not promote the development of ecological tourism technology.

From the perspective of the spatial spillover effect, the spatial spillover effect of tourism ecology is very strong, which means the ecological progress of the region will lead to the development of neighboring provinces. The rationalization of the tourism industry structure has a negative effect. The reasonable structure of the tourism industry will bring tourists a more comfortable travel experience, reducing the number of local tourists traveling out of the province and attracting more tourists from other places. Both environmental regulation and openness have positive effects, promoting provinces to converge to their steady-state level. The increase in investment in pollution control and strict environmental supervision will not only improve the region's environmental quality but also affect the environmental management of the surrounding areas. The opening-up can still promote the high-quality development of the tourism industry. Technological innovation harms the development of tourism ecology, indicating that technological innovation in tourism can not benefit other regions, but the innovation of eco-technology in tourism is still insufficient.

5 CONCLUSION AND RECOMMENDATIONS

Based on the Super-efficiency SBM model, Dagum gini coefficient decomposition method, and spatial Durbin model, the tourism eco-efficiency of the provinces and regions along the Yellow River basin was calculated, and the regional difference and convergence were studied. The results show that from 2010 to 2019, the overall level of ecological tourism development in the Yellow River basin and all regions increased. However, its efficiency was relatively lower than in other watersheds. Differences between regions are the main source of regional differences. Promoting the ability of environmental regulation and technological innovation can reduce regional differences and realize the balanced development of tourism ecology. The tourism eco-efficiency has a strong spatial correlation, the H-H agglomeration mode is concentrated in the middle, and the L-L plate is concentrated in the upper.

How does the Yellow River basin seek a breakthrough in the tourism ecology, mainly from the following aspects: first, The strengthening of interactions, cooperation, and exchanges between economic belts and provinces, especially with a focus on environmental protection and institutional strengthening. To alleviate the problem of inadequate development of the tourism ecosystem by using the developed tourism resources to deepen and propagandize the culture of the Yellow River. Second, Local governments have strengthened environmental control and established environmental laws and regulations and tourism Enterprises to improve the use of clean energy and reduce carbon emissions in tourism. Tourists enhance their awareness of environmental protection and do not do arbitrary damage to the environment and ecology of tourist attractions. Finally, Enhancing technological innovation and investment in tourism pollution control will contribute to reducing tourism's carbon emissions and increasing clean energy use.

REFERENCES

Castilho & Daniela F.: The Impacts of the Tourism Sector on the Eco-efficiency of the Latin American and Caribbean Countries. *Socio-Economic Planning Sciences* 78. 101–111 (2021).

Deng Zongbing F. Study on Regional Differences and Convergence of Ecological Civilization Development in Eight Comprehensive Economic Areas of China. *The Journal of Quantitative & Technical Economics.* 37 (06):3–25(2020).

Guo Lijia F. Tourism Eco-efficiency at the Provincial Level in China in the Context of Energy Conservation and Emission Reduction. *Progress in Geography* 40(08): 1284–1297 (2021).

Jia Liu F.: Tourism eco-efficiency of Chinese coastal cities e Analysis based on the DEA-Tobit model. *Ocean & Coastal Management* 148.164–170 (2017).

Liu Chunj F. Changes in the Structure of the Tourism Industry and Their Effect on the Growth of the Tourism Economy in China. *Tourism Tribune* 29(08):37–49. (2014).

Lu Xiaojing F. Calculating Green Production Efficiency of Tourism in the Yangtze River Economic Belt and Analysis of Its Spatial and Temporal Evolution. *China Population, Resources and Environment* 29(7). 19–30 (2019).

Llanquileo-Melgarejo P. Evaluation of Economies of Scale in Eco-efficiency of Municipal Waste Management: An Empirical Approach for Chile. *Environmental Science and Pollution Research* 28(22). 28337–28348 (2021).

Peng Hongsong F. Eco-efficiency and its Determinants at a Tourism Destination: A Case Study of Huangshan National Park, China. *Tourism Management* 60. 201–211 (2017).

Qian Zhenhua F. A Two-Stage Approach to Overcoming the Infeasibility of SBM Super-Efficiency Model in Data Envelopment Analysis. *Journal of Mathematics in Practice and Theory* 43(05).171–178 (2013).

Romano G. Affecting Eco-efficiency of Municipal Waste Services in Tuscan Municipalities: An Empirical Investigation of Different Management Models. *Waste Management* 105. 384–394 (2020).

Wang Shengpeng F. The Spatio-Temporal Evolution of Tourism Eco-Efficiency in the Yellow River Basin and Its Interactive Response with Tourism Economy Development Level. *Economic Geography* 40(05).81–89 (2020).

Yao Zhiguo F. A Literature Review of Tourism Eco-Efficiency. *Tourism Science* 30(06):74–91 (2016).

Civil Engineering and Energy-Environment – Gao & Duan (Eds)
© 2023 the Author(s), ISBN 978-1-032-56059-5

Environmental evaluation using the analytic hierarchy process

Yihan Jiang*
Chanjun Highschool International Department, Changsha, China

ABSTRACT: Environmental evaluation has always been difficult to be quantified. This paper provides a solution using the Analytic Hierarchy Process (AHP) as the main methodology for quantification. In this paper, the application of AHP will be shown in a case study where the three major cities, Changsha, Zhuzhou, and Xiangtan in Human Province, China, are ranked based on their environments.

1 INTRODUCTION

Environmental performance is a key component that gauges a city's sustainable development. There are many ways to evaluate the environmental performance of cities mathematically. Different factors will be considered critical criteria for a city's environmental performance. However, some factors may outweigh others, which will complicate quantification. Therefore, there has to be a mathematical way to evaluate all criteria best while considering each criterion's weight. Analytic Hierarchy Process is the most flexible and useful among all mathematical approaches. This paper will demonstrate how Analytic Hierarchy is applied in evaluating environmental performance. By this method, three cities in China will be assessed and ranked for their environmental performance (Cao 2017).

2 CRITERIA SELECTED

People nowadays are increasingly concerned about environmental issues, most of which are side effects of societal development. Global warming, resource shortage, air pollution, and deforestation jeopardize people's everyday lives if not dealt with promptly. However, those environmental issues cannot be dealt with before an environmental evaluation is available. There are many ways to evaluate the environment of a region. One of the user while underrated approaches are the Analytic Hierarchy Process (AHP). All environmental evaluation approaches must be based on a series of carefully selected criteria. The following are three criteria selected to demonstrate the application of AHP for environmental evaluation (Liu 2020).

2.1 *Air pollution*

The World Health Organization has estimated that urban air pollution in developing countries has resulted in more than 2 million deaths per year, along with various respiratory illnesses. (Gulia *et al.* 2015) The air quality of a region can have a huge impact on residents' lives. The AQI will represent air quality in this paper.

---

*Corresponding Author: hunter789654@gmail.com

DOI: 10.1201/9781003433651-59

2.2 Average freshwater resource per capita

Freshwater resources play a fundamental role in human society and are indispensable ingredients for human health. However, globally there are still 1.1 billion people that lack access to water due. (*Water Scarcity | Threats | WWF*, n.d.) Average freshwater resource per capita represents how much water a person can be assigned from the total amount of freshwater resources in a region. Average freshwater resource per capita is also one of the criteria for Chinese Green City Evaluation.

2.3 Vegetation cover

Vegetation cover is the percentage of soil covered by vegetation in a region. Vegetation cover is essential to an ecosystem to prevent soil erosion, increase soil organic matter, and contribute to carbon neutralization.

2.4 Carbon emission intensity

Carbon neutralization is one task that should be prioritized. Carbon emission shows a city's sustainability, which will be a reasonable criterion selected. Low-carbon development is also a criterion for the Chinese Green City Evaluation.

3 EVALUATION METHODOLOGY AHP

Analytic Hierarchy Process (AHP) is a broadly applied multi-criteria decision-making method to determine the weights of criteria and priorities of alternatives in a structured manner based on the pairwise comparison. (Liu *et al.* 2020) It was developed by Thomas L. Saaty in the 1970s and contained goals, criteria, and alternatives. The pairwise comparison of each criterion can help us obtain weight for each criterion. The alternatives will then be compared based on their performance of each weighted criterion (Net Ease 2020).

3.1 Pairwise comparison

All criteria are compared to each other in pairs using the relative importance scale (Table 1):

Table 1. Relative importance scale.

Relative Importance Scale	
Value	Meaning
1	Equally Important
3	Moderately Important
5	Strongly Important
7	Very Important
9	Extremely Important
2,4,6,8	In-between Value

The pairwise comparison is shown in the table below:
Four Criteria are notated:
Air Pollution=x1
Average Freshwater Resource Per Capita=x2
Vegetation Cover=x3

Carbon Emission Intensity=x4

For example (Table 2), the first-row second column means Air Pollution is moderately more important than Freshwater Resource Per Capita. The important comparison is conducted based on data from Environmental Performance Index and the Chinese Green City Index.

Table 2. Pairwise comparison matrix.

	x1	x2	x3	x4
x1	1	2	3	1/2
x2	1/2	1	2	1/3
x3	1/3	1/4	1	1/4
x4	2	3	4	1

3.2 *Weight calculation*

The weight of each criterion can be calculated arithmetically in the following way.

Calculate the sum of each column. Then convert each element into a decimal, the percentage of the original value in the sum of its column to normalize the matrix.

A weight for each criterion will be obtained as the average of each row (Table 3).

Table 3. Normalized pairwise comparison matrix.

	x1	x2	x3	x4	Weighted Criteria
x1	0.26	0.31	0.30	0.24	0.28
x2	0.13	0.15	0.20	0.16	0.16
x3	0.09	0.08	0.10	0.12	0.10
x4	0.52	0.46	0.40	0.48	0.47

3.3 *Consistency proof*

The consistency of pairwise comparison has to be ensured by calculating the consistency ratio, which signifies consistency when smaller than 0.1. Elementwise multiplication of the original pairwise comparison matrix and the vector of criteria weight will generate a new matrix. Then a vector of weighted sum value will be obtained by adding all the elements in each row. Then divide the weighted sum value vector by the criteria weight elementwise vector, obtaining a new vector. Find the average of the elements in the new vector.

The average is called λmax, which is 4.03 in this case.

CI consistency index is then calculated by:

$CI = \frac{\lambda max-}{in-1} = \frac{4.03-4}{4-1} = 0.0103$, where n is the number of criteria.

CR consistency ratio is CI divided by the Random index, which is 0.9 when the number of criteria is 4.

$$CR = \frac{0.0103}{0.9} = 0.0114$$

If the CR is smaller than 0.1, it can be assumed that the matrix is reasonably consistent so that the weighted criteria can be accepted.

4 CASE STUDY

In this paper, the weighted criteria obtained from the Analytic Hierarchy Process will be used to evaluate the environmental performance of three cities in Hunan Province, Changsha, Zhuzhou, and Xiangtan. The three cities have environmental policies and work together to tackle environmental issues. Based on their annual statistics, an evaluation will be made to rank the three cities' environmental performance using weighted criteria.

Figure 1 shows how the Analytic Hierarchy Process is applied to the case study. These three cities are the alternatives, and they all have common criteria.

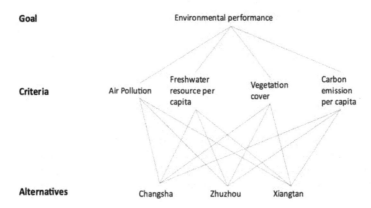

Figure 1. Goal, criteria, and alternatives.

For each criterion, all three cities will be pairwise compared in terms of their performance in this criterion using the relative importance scale, just like above. A matrix can be obtained for each criterion. The matrix is built based on the data from the annual statistics of each city. Notably, if a city is doing better than another one in Air pollution or Carbon emission per capita, namely having less of this problem, the importance value will be higher.

The normalized value reflects the score of each city in this criterion
Air quality:
Data source
Changsha's average AQI from 2021 January to December: 75.5
Zhuzhou's average AQI from 2021 January to December: 71.3
Xiangtan's average AQI from 2021 January to December: 73
(*PM2.5 Past data AQI China Air Quality online monitoring platform*, n.d.)

Table 4. Air quality.

x1	C	Z	X	Normalized
C	1	1/3	1/2	0.16
Z	3	1	2	0.54
X	2	1/2	1	0.30

Freshwater resource per capita (Table 4):
Data source
Changsha's freshwater resource per capita: 1544.7 m3(2017 Changsha Water Resource Report, Biao Cao)

Zhuzhou's freshwater resource per capita: 2680 m3(" Zhuzhou City's Water Resource Report" Hunan Province Platform n.d.)

Xiangtan's freshwater resource per capita: 1345 m3 (Pingxiang Government Website, n.d.) (Table 5)

Table 5.   Freshwater resources per capita.

x_2	C	Z	X	Normalized
C	1	1/3	2	0.24
Z	3	1	4	0.62
X	1/2	1/4	1	0.14

Vegetation cover (Table 6):

Table 6.   Vegetation cover.

x_3	C	Z	X	Normalized
C	1	1/3	3	0.27
Z	3	1	4	0.61
X	1/4	1/4	1	0.11

Changsha's vegetation cover: 55% (Changsha was ranked high among all capital cities in vegetation cover, n.d.)

Zhuzhou's vegetation cover: is 61.96%(NetEase 2020)

Xiangtan's vegetation cover: 45.81(Xiangtan 45.81%, n.d.)

Carbon emission intensity (Table 7):

Table 7.   Carbon emission intensity.

x_4	C	Z	X	Normalized
C	1	2	3	0.54
Z	1/2	1	2	0.30
X	1/3	1/2	1	0.16

A ranking of three cities' carbon emission intensity in tons per 10 thousand Yuan can be obtained. (Yang n.d.)

Then we can obtain a matrix with each score in each criterion and the corresponding weight (Table 8):

Table 8.   Scores and weight.

	Weight	C	Z	X
x_1	0.28	0.16	0.54	0.30
x_2	0.16	0.24	0.62	0.14
x_3	0.10	0.27	0.61	0.11
x_4	0.47	0.54	0.30	0.16

455

Sum all the scores of each criterion times the corresponding weight will obtain the final score for a city's environmental performance:

Changsha: 0.364

Zhuzhou: 0.452

Xiangtan: 0.1926

Therefore, the correct ranking should be Zhuzhou first, Changsha second, and Xiangtan third. We can see that using the Analytic Hierarchy Process can yield a ranking that is not too reliant on precise data.

5 CONCLUSION

In this paper, conducting environmental evaluation using Analytic Hierarchy Process is demonstrated. Four factors, air quality, freshwater resource per capita, vegetation cover, and carbon emission intensity, are chosen as criteria in Analytic Hierarchy Process. Environmental evaluation of three cities in Hunan Province in China, Changsha, Zhuzhou, and Xiangtan, are ranked using AHP. Zhuzhou is ranked first, Changsha second, and Xiangtan third. The more criteria are chosen, the more comprehensive and precise the evaluation will be. A wider range of criteria should be chosen if a city wants to evaluate its environmental performance. This paper provides insight into quantifying evaluation, which can apply not only to environmental evaluation and many other assessments.

REFERENCES

Biao Cao (2017). 2017 *Changsha Water Resource Report.*

Changsha was Ranked High Among All Capital Cities in Vegetation Cover. Retrieved August 29, 2022, from http://hn.people.com.cn/n2/2022/0322/c195194-35185604.html

Gulia S., Shiva Nagendra S.M., Khare M., & Khanna I. (2015). Urban Air Quality Management-A Review. *Atmospheric Pollution Research*, 6(2), 286–304. https://doi.org/10.5094/APR.2015.033

Liu Y., Eckert C.M., & Earl C. (2020). A Review of Fuzzy AHP Methods for Decision-Making with Subjective Judgments. *Expert Systems with Applications*, 161, 113738. https://doi.org/10.1016/j.eswa.2020.113738

Net Ease. (2020, June 4). *Zhuzhou's Vegetation Cover Reached 61.96%.* https://www.163.com/dy/article/FEA0CT630550HHWE.html

PM2.5 Past data AQI China Air Quality Online Monitoring Platform. (n.d.). Retrieved August 29, 2022, from https://www.aqistudy.cn/historydata/

Pingxiang Government Website. (n.d.). Retrieved August 29, 2022, from https://www.pingjiang.gov.cn/35048/35055/35061/content_1800998.html

Water Scarcity | Threats | WWF. (n.d.). *World Wildlife Fund.* Retrieved August 28, 2022, from https://www.worldwildlife.org/threats/water-scarcity

Xiangtan 45.81%.(n.d.). Retrieved August 29, 2022, from http://www.hunan.gov.cn/hnszf/hnyw/20180408_sxhy/sxjjshfzmb/szqx/202003/t20200312_11810844.html

Yang Yifan (n.d.). *Hunan's Carbon Emission.* Retrieved August 29, 2022, from https://hn.rednet.cn/content/2022/06/10/11377440.html

Zhuzhou's Water Situation Report. (n.d.). Retrieved August 29, 2022, from http://www.hunan.gov.cn/hnszf/hnyw/szdt/201212/t20121210_4751244.html

Civil Engineering and Energy-Environment – Gao & Duan (Eds)
© 2023 the Author(s), ISBN 978-1-032-56059-5

Research on the digital twin intelligent management platform in the communication industry existing data center

Wei Wei Kou
China Mobile Group Gansu Co. LTD, Lanzhou GanSu, China

Xue Mei Zhang
GanSu Province Institute of Civil Engineering and Architecture, Lanzhou Gansu, China

Yun Long Fan*
GanSu AnHang Engineering Testing Co., Ltd, Lanzhou GanSu, China

ABSTRACT: Aiming at the problems of the existing and new data centers in the communication industry, such as wide distribution, difficult management, and high energy consumption, based on various management platforms, based on the integration of BIM, big data, artificial intelligence, and digital twinning, this paper explores the construction theory, mode, and application of data center intelligent management platform. Relying on the intelligent management platform project built by a communication industry for the data center clusters distributed in Gansu, the research theory has been applied in actual buildings, and has achieved good results, the application results show that:(1) According to the existing as built drawings, construction drawings and the actual situation on site, the technical implementation scheme can be determined to better present the professional models and functional modules of the data center, so as to achieve the establishment of the digital twin intelligent management platform;(2)It can realize the sharing and intercommunication between various management platforms, various sensor data and models on site, achieve the effect of digital twin intelligent management, give full play to computing power as early as possible, and initially achieve the goal of building energy conservation and emission reduction, cost reduction and efficiency increase, which has certain reference significance for the subsequent construction of the actual digital twin intelligent management platform project.

1 INTRODUCTION

In February 2022, the National Development and Reform Commission, the Central Cyberspace Office, the Ministry of Industry and Information Technology, the National Energy Administration, and other relevant departments agreed to start the construction of national computing hubs in eight places. The construction includes Beijing, Tianjin, Hebei, the Yangtze River Delta, Guangdong, Hong Kong, Macao Bay Area, Chengdu and Chongqing, Inner Mongolia, Guizhou, Gansu, and Ningxia. They set up 10 national data center clusters. The project that channels more computing resources from the eastern areas to the less developed western regions was officially launched and built a collaborative innovation system of national integrated big data centers. Gansu Qingyang has been selected as one of the 10 national data center clusters by the state due to its rich renewable energy,

---

*Corresponding Author: 370168169@qq.com

DOI: 10.1201/9781003433651-60

suitable climate, and large green development potential of the data center. Each cluster is a physically continuous administrative region, which is the construction of large and super-large data centers within the specific computing hub. While making every effort to build large and super large data centers, there are also several problems in the implementation. One is to solve the problems of existing and new data centers that are widely distributed, difficult to manage and control, and have high energy consumption. The other cooperates with existing and new data centers to give early play to their computing power.

Digital Twins is the digitalization of the real world (Grieves 2005, 2006). Since the concept was put forward, it has been continuously integrated with all walks of life. Some scholars have also studied the implementation of the data center. Luo Shanshan (Luo& Leng 2021) built a 3D visualization monitoring system based on the digital twin computer room, taking a single computer room in the data center as the research object, to realize the 3D visualization of the computer room, monitor the system status in real time, display the equipment information dynamically, and improve the management efficiency; Quanwu (2022) proposed a standardized deepening design scheme based on BIM technology according to the characteristics and difficulties of the implementation of the refrigeration room supporting the data center, providing solutions for the efficient implementation and standardized construction of the refrigeration room in the data center; Wang Siyuan (2022) summarized the full life cycle digital twin method of the data center, proposed that BIM technology at this stage is a digital twin in the engineering planning, design and construction stages, and introduced the current application status and application methods in the operation and maintenance stage; Ruan Qian (2022) explored the digital intelligent operation and maintenance of China Mobile IT cloud data center, discussed the digital intelligent operation and maintenance management mode of the data center, and practiced some management mode innovations.

To sum up, it is not difficult to see that the research on the integration of data centers and digital twins mainly focuses on the summary of application methods, the application of a single computer room, 3D visualization, and the exploration of digital intelligent operation and maintenance. There is less research on the digital twin intelligent management of data centers in the communication industry, especially data center clusters. Based on the project of a digital twin intelligent management platform built by a communication industry for cluster data centers distributed in many places in Gansu, this paper applies the research theory to the intelligent management of actually existing and new data centers. It has achieved good results, which also has certain reference significance for the subsequent construction of the actual digital twin intelligent management platform.

2 TECHNICAL IMPLEMENTATION SCHEME OF DATA CENTER IN THE COMMUNICATION INDUSTRY

The communication industry's data center is characterized by many types of equipment, pipelines, systems and monitoring points, complex system operation, and high professionalism and security in external service requirements (Kou et al. 2017). To enable the data center in the communication industry to operate, maintain and manage safely and reliably, specific technical implementation plans need to be determined (Kou & Chen 2018). A digital twin intelligent management platform for the data center should be built to control it, improve management efficiency, and share various information resources to reduce operating costs.

2.1 Determine implementation requirements

According to the actual situation of data center construction and operation and maintenance in the province, there are data centers with various professional systems that have been running for a certain period, and there are also data centers being built. Therefore, before the construction of the management platform, it is necessary to jointly evaluate the overall performance of the existing and new data centers and the performance of various professional systems with various professional construction departments, maintenance departments, and demand departments. See Table 1 for details.

Table 1. Table of contents of professional evaluation.

Assessment content	Concrete content	Remarks
Evaluation of existing buildings	**Assessment structure**: Evaluate the main structure according to the needs	Professional report or maintenance report
	Non-structural evaluation: Evaluate external walls, interior walls, parapets, curtain walls, suspended ceilings, doors and windows, awnings, aprons, equipment supports, and other components as required.	
	Systematic evaluation of each discipline: Evaluate the lighting, air conditioning, fresh air, meeting, power supply, environment, security (monitoring, alarm, access control), and other systems according to the needs and project characteristics	
	Structural and non-structural evaluation: ditto	
New building evaluation	**Systematic evaluation of each discipline:** It is necessary to compare and evaluate the corresponding system of the existing building, consider its compatibility and scalability, and propose targeted transformation measures	The feasibility study report, construction drawing, and system diagram

According to the professional evaluation report, maintenance report, feasibility study report, construction drawing, and system diagram, combined with the demand, planning content, and investment of the demand department, the implementation leading department shall find out the problems existing in the implementation and operation and maintenance of existing and new data centers and the direction for improvement and determine the implementation demand plan.

2.2 Data collection and sorting

According to the preliminarily determined implementation demand plan, the implementation department first collects various engineering data, mainly including as-built drawings, reconstruction drawings, system drawings, and as-built data of all disciplines of the existing data center, and the feasibility study report, construction drawings and system drawings of the new data center. Secondly, after the completion of data collection and sorting, the implementation department needs to cooperate with each professional construction department. It includes the maintenance and demand departments to conduct a full on-site survey of key parts, revise the contents with differences, sort, number, and save according to the standards, and form the final project data.

2.3 Determination of implementation standards

The BIM model implementation standards and platform data standards in the implementation process are determined according to the actual situation of the final project data. The BIM model's executive standard specifies the BIM model's fineness and color standard; Its BIM model implementation standard specifies the BIM model fineness, color standard, and other contents. Because this paper involves the implementation of new and existing data centers, including the stages from project initiation to the full life cycle of operation and maintenance, the model fineness in Gansu provincial and local standards (China Construction Industry Press 2018) is adopted. Other projects can choose the appropriate model fineness according to their stage and specific needs. The color standard of the model shall be unified to ensure model maintenance and data sharing. The architectural and structural disciplines shall be determined according to the as-built drawings and the actual situation on site. The color standard of the components related to the installation discipline shall be implemented according to the relevant provisions in the local standards of Gansu Province (China Construction Industry Press 2018).

2.4 *Determination of data sources*

According to the final implementation demand scheme, the data sharing scheme shall be determined with the data interface protocol types and sensor types of various systems on the site. The system's interface protocol shall be prioritized to realize data transmission and feedback. For the system that needs to be modified or added with sensors, the sensor that can transmit data wirelessly shall be given priority. Then the data can be transferred to the data-sharing platform through the LAN in the building to achieve the integration exchange and sharing of multi-source heterogeneous data between physical entities and virtual models and realize real-time interaction between physical space and information space.

2.5 *Platform construction*

Relying on the existing DICM platform of a data center in Lanzhou New Area of the communication industry, build the digital twin intelligent management platform of the existing data center. First, use the universal IFC standard to connect the three-dimensional visualization model with the management platform. Then set the actual point information of each system in the three-dimensional visualization interface of the operation and main-tenance platform. Then convert the data through the monitoring or sensing equipment at the end of the point. Upload the data to each subsystem of the data storage center of the data sharing platform for data analysis. Finally, upload the data analyzed by each subsystem to the comprehensive operation and maintenance platform through the API interface, and feedback the instructions to the end of each point according to the analysis results to realize the digital twin intelligent management of the existing data center.

3 ENGINEERING EXAMPLES

3.1 *Overview of a data center in Lanzhou*

A data center located in Lanzhou New Area covers an area of 171.3 mu. The total building area of Phase I and Phase II is 79000 square meters, including a 4-floor standardized machine room building, a 2-floor standardized machine room building, a refrigeration sta-tion, an oil machine room, and a maintenance support room, which can provide about 3400 racks. The park has nearly 300 rooms, 14000 monitoring points, and 1500 sets of air con-ditioning terminals.

3.2 *Overview of a data center in Tianshui*

A data center located in Tianshui covers an area of about 20 mu. The total building area of Phase I is about 14000 square meters, including a 4 - story machine room and a 2 - story auxiliary room, which can provide about 1164 racks. There are nearly 50 rooms in the park, about 3000 monitoring points, and about 500 sets of air conditioning terminals.

3.3 Overview of a data center in Qingyang

According to the national, a data center in Qingyang covers an area of about 12 mu, and another 200 mu will be acquired for the subsequent construction of the data center. The project that channels more computing resources from the eastern areas to the less developed western regions was officially launched. The total building area of Phase I is about 10000 square meters, including a 7 - story machine room building and a 2 - story auxiliary room, which can provide about 1007 racks. There are nearly 40 rooms in the park, about 2500 monitoring points, and about 400 sets of air conditioning terminals.

4 DATA COLLECTION

4.1 Overview of a data center in Lanzhou

In the early stage of designing a data center in Lanzhou, BIM models of various specialties were built by Revit software. Based on the BIM models, they were refined according to the requirements of Lod600 in the operation and maintenance stage, as detailed in Figures 1–6.

Figure 1. Architectural model.

Figure 2. Structural model.

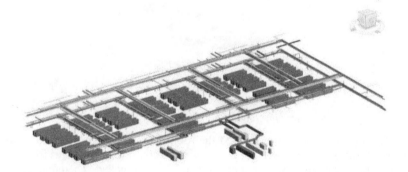

Figure 3. Electromechanical model.

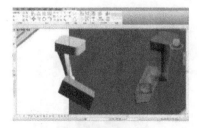

Figure 4. Monitoring model.

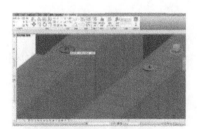

Figure 5. Environmental model.

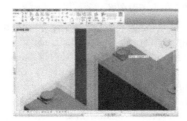

Figure 6. Power supply model.

4.2 *Overview of a data center in Tianshui*

A data center in Tianshui has been completed and is in the final acceptance stage. According to the completion phase model fineness LOD500 requirements re - modeling, see Figures 7–9.

 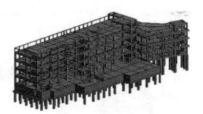

Figure 7. Architectural model.   Figure 8. Structural model.

Figure 9. Electromechanical model.

4.3 *Overview of a data center in Qingyang*

A data center in Qingyang has completed the roof of its main structure and is now in the implementation stage. The project that channels more computing resources from the eastern areas to the less developed western regions was officially launched. Therefore, following the requirements of model fineness Lod600 in the completion stage, fine modeling is carried out, as detailed in Figures 10–15.

Figure 10. Full professional model.   Figure 11. Architectural model.

Figure 12. Structural model.

Figure 13. Water Supply and drainage model.

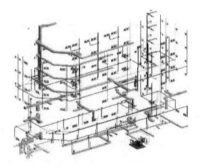
Figure 14. HVAC model.

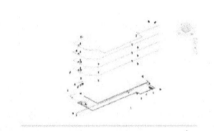
Figure 15. Electrical model.

5 DATA COLLECTION

5.1 *The original system*

According to the actual management requirements of the existing and newly built data centers in the communication industry, each professional system has built its management system, and modules such as data collection and remote control are set up at the end of the equipment. Still, the data of each system is not shared. The data of each system is integrated, exchanged, shared, and interacted with through the management platform to realize the intelligent operation of the data center.

5.2 *Add sensors*

It is necessary to add sensors for equipment without a management system or no data collection module. If structural health safety is analyzed, the related sensors for structural health monitoring can be added.

6 SET UP THE PLATFORM

6.1 *The overall architecture*

This paper presents the overall solution to the park as a unit and other cities' data center management platforms through the reserved interface access platform. According to the characteristics and requirements of data center operation, focus on data sharing, safe operation, and energy analysis of each system, as detailed in Figure 16.

Figure 16. Datacenter overall solution architecture.

6.2 *Platform function*

The engineering modeling, system transformation, platform construction, and so on complete the communication industry data center digital twinning intelligence management platform's construction work mainly includes 6 functional modules such as power supply, cold source, environment, capacity, security, and energy efficiency management, as detailed in Figure 17.

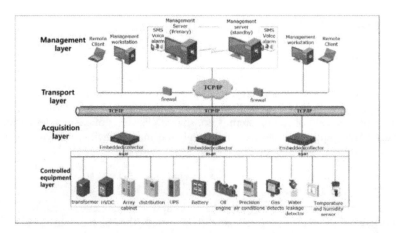

Figure 17. Main interface of the platform.

The power management module can realize the intelligent management of equipment such as distribution cabinets, UPS, ATS, STS, electric generators, storage batteries, etc. It can realize real-time monitoring and analysis of data and can be daily reports, monthly reports, annual reports, user-defined reports, and other forms.

464

In addition to daily data checking and analysis, the cold source management model can also reduce energy loss through digital twinning technology for peak shaving operation of cold storage tanks, low-frequency operation of water pumps and heat recovery of water-source heat pumps, etc., energy saving and emission reduction.

The environmental management module can detect and analyze the temperature, humidity, smoke sensing, gas (carbon dioxide, hydrogen sulfide), and water leakage of the campus and the computer room, and feedback to the corresponding system according to the threshold value to achieve fire alarm, switch exhaust fan, open and close valves, and other intelligent management functions.

The capacity management module can show the capacity and usage of the cabinet, power distribution, and refrigeration in the form of diagrams and provide the installation scheme according to the installation requirements.

The security management module can support the access of video monitoring on campus. It can forecast, warn and prevent security events through big data and video AI analysis; it includes the functions of vehicle violation, personnel crossing the boundary, illegal invasion, fire alarms, perimeter alarms, and environmental protection alarms, etc. It can ensure the safety of the park's personnel, assets, network, and data.

The Energy Efficiency Management module is an important part of the platform through the real-time monitoring and collection of the energy-consuming unit data. It involves air-conditioning, fresh air, strong electricity, power, water supply, drainage, natural gas, and various professional systems in the park and the machine room area. Energy consumption statistics and comparative analysis can be carried out by layer, item, year (season, month, week, day), and other ways to diagnose energy consumption problems and embed solutions into the platform for operation and maintenance control. For example, the platform has realized the linkage between air quality and fresh air systems. The dynamic of personnel and vehicles and intelligent lighting system, the linkage between water immersion system and alarm system, and the linkage between temperature, humidity, and air conditioning system in the computer room, thus, the goal of energy consumption control, cost reduction, efficiency increase and safe operation of the park can be realized.

6.3 *Platform running effect*

1) the operation of the platform minimizes the number of operation and maintenance personnel. The three data centers managed by the platform can reduce the number of owners and property management personnel by about 5 persons/year/place and save about 750,000 yuan/year in labor costs. As the number of data center access increases, the labor costs saved will increase.

2) Through the platform's intelligent management and control of operating equipment, applications such as the peak-shaving operation of ice storage tanks, low-frequency operation of water pumps, heat recovery of water-source heat pumps, and intelligent lighting can be realized. Taking a data center in Lanzhou as an example, the annual energy saving is about 3650793 kW h, and the total cost saving is about 2.3 million yuan/year according to the average electricity price classified by electricity consumption.

7 CONCLUSION

Based on the research of digital twinning intelligence management platforms in data centers of the communication industry, this paper draws the following conclusions:

1) According to the existing as-built drawings, construction drawings, and the actual situation on the spot, the data center's professional models and function modules can be

better presented through the determined technical implementation scheme to realize the construction of a digital twinning intelligence management platform.

2) it can realize the sharing and interworking of all kinds of management platforms and all kinds of sensor data and models on the spot. It can also achieve the effect of digital twinning intelligence management, bring computing ability early into play, achieves energy-saving goals, reduces costs, and increase efficiency initially. It has certain reference significance for constructing the following real digital twinning intelligence management platform project.

ACKNOWLEDGMENTS

This work was supported by the Science and technology planning project from the Department of Housing and Urban-Rural Development of Gansu Province, "Digital Twin City Basic Research-Digital Twin Intelligent Building Management Platform" (JK2020-15).

REFERENCES

Gansu Provincial Department of Housing and Urban-Rural Development, Gansu Provincial Bureau of Quality and Technical Supervision (2018) *DB62/T 3150-2018, Building Information Model (BIM) Application Standard.* Beijing: China Construction Industry Press.

Grieves M. *Product Lifecycle Management - Driving the Next Generation of Lean Thinking.* New York: McGraw-Hill Companies, 2006.

Grieves M. Product Lifecycle Management: The New Paradigm for Enterprises. *International Journal of Product Development* (S1477–9056), 2005, 2(1/2): 71.

Kou W.W., Chen C.I., Zhang K. *et al.* (2017). Comprehensive Application of BIM Technology in China Mobile (Gansu) Data Center Project. Chinese Institute of Cartography. *Proceedings of the Third National BIM Academic Conference*, 155–160.

Kou W.W., Chen C.L.. (2018). Research on the Application of BIM Technology in the Whole Life Cycle of Construction Projects. *Construction Quality*, 36(9), 42–46.

Luo S.H. and Leng J. (2021) The Design and Implementation of a Three-dimensional Visual Monitoring System Based on a Mathematical Twin Computer Room. *Computer Technology and Automation*, 135–139.

Quan W. (2022) The Further Design and Application of Data Center Refrigeration Machine Room Based on BIM. *Electromechanical Installation*, 9:47–49.

Ruan Q. and Liu H. (2022) The Exploration and Practice of Intelligent Operation and Maintenance of China Mobile IT Cloud Data Center. *Internet Plus Technology*, 68–69.

Wang S.Y., Zhao Q. and Liu H.C.H. (2022) Life Cycle Digital Twin Method for Data Center Based on BIM. *Building Intelligence*, 2:132–134.

Civil Engineering and Energy-Environment – Gao & Duan (Eds)
© 2023 the Author(s), ISBN 978-1-032-56059-5

Simulation method of flood routing based on mathematical morphology

Shaobo Wang*, Yulong Xiong* & Wanhua Yuan
Bureau of Hydrology and Water Resources, Pearl River Water Resources Commission of Ministry of Water Resources, Guangzhou, China

ABSTRACT: In this study, we analyzed the flood routing simulation method based on Mathematical Morphology (FRMM). Consulting the physics meaning of hydraulics flood routing model - St. Venant continuity equation discretization, a new simulation method, 'ladder flow,' was proposed on the basis that flux is continuous in time and discrete in space. The method improves the two aspects of FRMM: 1. The expansion operator was modified, which makes the spread velocity of the wave crest can be changed. 2. Cascade method was applied to calculate. Thus, the dynamic distribution of water depth can be found. In the end, combining the actual process of flood routing, combining the actual process of flood routing, we analyzed the rationality of the method, and the result shows that the method corresponds to some extent to the actual flood process.

1 INTRODUCTION

Flood is one of the major natural disasters in human history, which causes more damage than any other natural disaster. Accurate simulation of flood processes is an important basis for producing flood risk maps and a fundamental basis for flood damage assessment. Much research has been done, and fruitful results have been achieved, for example, the empirical river routing model (Huang & He 2006). The Maskingen model, widely used in river flood forecasting (Liu 2004), is the impulse algorithm that assumes a steady flow and storage relationship (Todin & Ciarapica 2001). Water surface curve model for a steady flow of the river, one- and two-dimensional St. Venant equations for unsteady flow (Morris 2000). have been widely used. However, there are also problems with cumbersome and long time-consuming calculations. In recent years, with the development of mathematical morphology in various fields and the strong support of GIS technology, the flood routing simulation method based on mathematical morphology (FRMM for short) has also entered the spotlight, but there are still few related studies. This paper analyzes the FRMM method and proposes a simulation method for flood routing in floodplains.

2 MATHEMATICAL MORPHOLOGY AND ITS APPLICATION IN FLOOD ROUTING SIMULATION

Mathematical morphology is a discipline based on rigorous mathematical theory. G. Matheron's rigorous and detailed arguments for random set theory and integral geometry

---

*Corresponding Authors: 2018183528@qq.com and 874123818@qq.com

DOI: 10.1201/9781003433651-61

laid the theoretical foundation for mathematical morphology. This theoretical foundation also involves a series of mathematical branches such as topology, modern probability theory, modern algebra set theory, and graph theory. At the same time, mathematical morphology is a subject closely integrated with practical matters, has been applied to the study of practical problems, especially in the processing of digital images has achieved a lot of excellent results. With the strong technical support of GIS, the application of this method in flood routing simulation also reveals certain advantages. There are many basic mathematical morphology operations, and we briefly describe the operations related to our study here (Gong 1997; Tang & Lv 1990).

2.1 Simulation of flood dispersion range

Simulation of the flood spreading situation from the geodesic circle principle of mathematical morphology is based on the following assumptions:

$$u_f = c \qquad c \geq 0 \qquad (1)$$

Where: u_f is the forward velocity of the flood front and c is the constant number. The wave crest is assumed to spread at a uniform speed. This spreading on the horizontal ground is the uniform spreading outward with a point as the center of a circle. When it encounters an obstruction such as an upland area, the spreading is explained as follows.

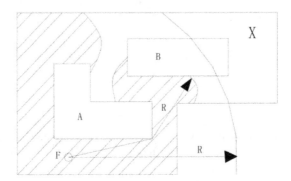

Figure 1. Geodesic circle and Euclidean circle.

Let X be an irregularly shaped submerged connected block, as shown in Figure 1, where several obstructions are A and B. When the floodwater spreads outward from point F, it can be considered as spreading forward in concentric circles when no obstruction is encountered and spreads forward in geodesic circles when an obstruction is encountered. Geodesic circles are defined as follows.

Let point F be the set of central points.

$$B_F(F, R) = \{x | x \in X, d_F(F, x) \leq R\} \qquad (2)$$

is called the "geodesic circle" with the center at F and the radius at R, where R is the farthest geodesic distance from F. The set of all x points whose distance to F is less than and equal to R constitutes the geodesic circle $B_F(F, R)$. The shortest arc of two consecutive points, F and x in X, is called geodesic distance in the continuous plane. The specific implementation of floodwater diffusion by a geodesic circle is based on the expansion algorithm in mathematical morphology.

2.2 Expansion operations in mathematical morphology

Expansion is a basic operation in mathematical morphology and has a very intuitive geometric background. A simple introduction of the expansion algorithm is as follows: let A be the image under study and B be another image, using B as a structural element, and suppose there are three points in B, b1 (0, 0), b2 (1, 0), and b3 (0, 1), respectively, Figure 2. The result of the translation of A by B is obtained as three new graphs, A_{b1}, A_{b2}, and A_{b3}, respectively, where A_{b1} and A overlap, and A_{b2} and A_{b3} are equivalent to the translation of A to the right, and up by one the result of the expansion of A by B is the merging of the three new graphs, denoted as A ⊕ B.

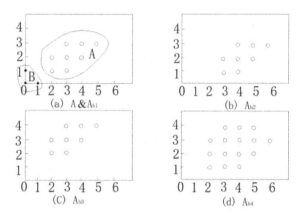

Figure 2. Results of image A being inflated by structural element B.

The result of A being inflated by B is called the Minkowski sum, defined as follows.

$$A \otimes B = \bigcup_{b_i \in B} A_{b_i} \qquad (3)$$

$A \otimes B$ One of the many equivalence forms of

$$A \otimes B = \{p | \exists_{b_i \in B} p \in A_{b_i}\} = \{p | \exists_{b_i \in B} \exists_{a_i \in A} (p = a_j + b_i)\} \qquad (4)$$

Where: ∃ means "exists," i.e. $A \otimes B$, is the set of all points p that satisfy the following conditions: there exists a point a_j in A and a point b_i in B such that $p = a_j + b_i$.

The expansion operation of graph A is based on the structural element (referred to as structural element, also called expansion operator) B. It can be seen that the structural element directly affects the expansion speed and expansion form of graph A. It is reflected in the mathematical image (graph) as the magnification speed of the image (graph) and the shape of the image edge, and in the flood diffusion as the diffusion speed and diffusion form of the wave crest. In raster images, two aspects are considered to construct structural elements: radius r and connectivity (Xiang & Chen 1995). In some simulation methods, a fixed value of radius r is used, and structural element connectivity is used for the eight-way structural element (Li et al. 2005).

Using the above algorithm, the dynamic submerged area map of flood routing is obtained using structural elements from the breach to perform the expansion operation. Then the time when the flood reaches a certain place is found based on the geodesic distance, the diffusion rate, and the number of swellings before the swell map captures a certain place to calculate when the place is flooded.

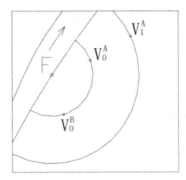

Figure 3. Wave crest diffusion diagram.

It can be seen that the calculation process follows the following two conditions:

$$V_0^A = V_0^B \qquad V_0^A, V_0^B \in V_0 \tag{5}$$

$$V_0^A = V_1^A \qquad V_0^A, V_1^A \in V^A \tag{6}$$

Where: V_0 is the set of all points on the wave crest at the moment T_0, and V_0^A, V_0^B is any two points on the wave crest. In space, the wave crest's velocity is the same regardless of where it is. V^A is the velocity of the flow at the point A on the crest of the wave, and V_0^A, V_1^A is the velocity of the flow at the point A at the time T_0, T_1. In time, the wave crest's velocity is the same at any point in time. As shown in Figure 3, the point F is the breach.

2.3 Simulation method of flood submerged area and water depth

The algorithm of submerged area calculation adopts the water conservation principle $Q = VQ$ for the total flood inflow to the floodplain and V the total volume of water within the submerged area based on GIS calculations. After excluding all "local depressions" using connectivity analysis, the following calculation method was used (Li et al. 2005).

$$Q = \sum_{i=1}^{N}(EW_i - EG_i) \times \Delta A_i \tag{7}$$

Where: EW_i is the water surface elevation of the unit area; EG_i is the ground elevation of the unit area; ΔA is the unit area simplified to the same plane, with the following relationship.

$$M = \sum_{i=1}^{N} \Delta A_i \tag{8}$$

Where: M is the area of the inundation zone; N is the number of cells in the inundation connection zone.

In order to simplify the calculation, the horizontal water level is used to replace the water level of each unit, which is obtained from equations (8) and (9).

$$E_W = \frac{Q}{M} + \frac{\sum_{i=1}^{N} EG_i}{N} \tag{9}$$

Another way to solve E_w under the simplified condition using horizontal water level is to solve (Ding et al. 2004):

$$f(E_W) = Q - \sum_{i=1}^{N}(E_W - EG_i) \times \Delta A \tag{10}$$

The trial algorithm and dichotomous method were used to improve t$E_W$he values to make $f(E_W) \to 0$ and, finally, the corresponding flood submerged area and water depth based on the spatial analysis capability of GIS.

It can be seen that the FRMM method for flood routing simulation starts from a macroscopic point of view to dissect the flooding process and then uses specific computational methods to achieve quantitative simulations. The critical aspects of the simulation are two: 1. The wave velocity during the periodΔT; 2. The water depth distribution within the flood spread at the end of the period.

The above calculation method assumes the constancy of the wave crest in space and time. It simplifies the water level to the horizontal plane for simulating the flood-submerged area. The analysis concludes that:

The assumption of spatial constancy of wave crest is reasonable (Li *et al.* 2005), while the use of constancy in time still needs to be explored because the floodwater spreads outward continuously. The submerged area gradually increases, the water depth near the crest becomes more and more shallow, and the flow velocity also weakens accordingly.

Compared with the complex hydro-hydraulic flood routing model, the water level is simplified to horizontal plane calculation in the simulation of flood-submerged areas. This simplified method applies to the flood routing of lakes, reservoirs, and flood storage areas with elevated water levels. However, floodplains cannot reflect the changes in water depth distribution, flow velocity distribution, and other hydraulic elements in the floodplain at different times. Therefore, this paper tries to propose a simulation method for the flood routing in the floodplain caused by dam failure and other reasons.

3 SIMULATION OF FLOOD ROUTING PROCESS BASED ON MATHEMATICAL MORPHOLOGY

How to determine the rate of flood wave advancement and the distribution of water depth within the submerged area is challenging to find a direct calculation method from a macroscopic point of view, and it is necessary to seek ideas from the flood routing simulation methods of hydrologic and hydrodynamics. St. Venant's continuum equation describing the flow in a channel is

$$\frac{\partial Q}{\partial x} + B\frac{\partial H}{\partial t} = q \tag{11}$$

Its discretization implies a continuity between the variation of water depth in time and the variation of flow in space. Based on this idea, it is considered that the routing of floods based on mathematical morphology can be carried out in the following way, as in Figure 4.

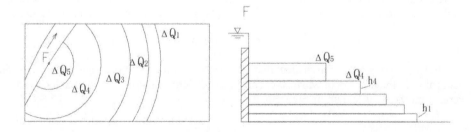

(a) Top view after discretization   (b) Section with F as the center and radius as the direction

Figure 4.   Flood flow discrete in time.

For descriptive convenience, assume the ground is level, ground friction is not considered, and F is at the river breach.

1. The flood flow is discrete in time, i.e., the flow into the floodplain at each timeΔQ is obtained from the flow $\sim$ time function at the breach, as in Figure 4 (a). The right subscript of the following parameter i indicates the inflow to the floodplain for theΔQi period.
2. The flood keeps moving forward in the process. EachΔQ tiled diffuses in the floodplain, forming a stacked wave layer, as in Figure 4 (b).
3. If these wave layers are carried out independently for the next period wave velocity calculation, it is obvious that the wave height of the first wave h_1 will always be determined by the flow of the first periodΔQ_1. In contrast, the wave velocity of this layer V_1 shows a rapidly decreasing trend and appears infinitely close to a certain boundary. The flood will be almost stagnant, which does not correspond to reality, and the reason for this situation is that the spatial continuity of the flow is not considered. Therefore, the effect of pre-flow $\sum_{n=1}^{i-1} \Delta Q_n$ on its wave propagation speed V_i needs to be considered, i.e.

$$V_i = f(h_i, \sum_{n=1}^{i-1} h_n) \tag{12}$$

The flow profile should be considered a form of interlayer connectivity, as shown in Figure 5.

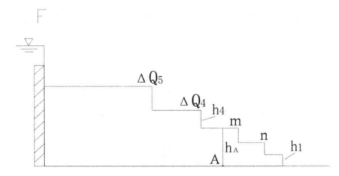

Figure 5. Form of interlayer connectivity.

4. Step 3 wave layer will catch up. The first is the second period ΔQ_2 generated by the wave catching up with the first period ΔQ_1 to prevent the upper layer of traffic overhang due to the catch-up situation. At this time will be ΔQ_2 into ΔQ_1 as the first layer of traffic, that is

$$\Delta Q_1' = \Delta Q_1 + \Delta Q_2 \tag{13}$$

After the above 4 steps are calculated, the water depth can be obtained by superposition according to the number of wave layers and wave heights passing through the location A at a certain moment, as shown in Figure 5. That is

$$H_t = h_1 + \sum_{i=n}^{m} h_i \tag{14}$$

Where: n, m are the lowest and highest layers that are not incorporated into the first layer$\Delta Q_1'$, but pass-through point A.

After finding the velocity of each layer V, the size of the radius of the structural element r is determined, and the advancement range of each wave layer the next Δt time is obtained by the expansion algorithm.

This calculation method has the following two constraints:

1. The wave crest is variable in time, but the wave crest speed is considered uniform within the examined unit of time Δt.
2. This simulation process applies to wave crest in the continuous advance. When the flood submerges the whole floodplain, the depression level will lift, which does not apply to the tiled diffusion, and equations (9) or (10) can be considered for calculation.

An analysis of rationality based on the above method is as follows:

Regarding the actual flood routing process, the water depth at point A should only be related to the discharge in the early period of the dam break, not to the later flow that flows into the floodplain. For example, when the wave crest is about to reach point AΔt, the flow into the floodplainΔQ has just started to spread and is far from reaching point A. It will have no effect on the flood depth at point A. The choice of m in equation (14) conveys the implication that only a part of the flood flow impacts water depth at point A, which shows that the method corresponds to some extent to the actual flood process.

4 CONCLUSIONS

Starting from the algorithm of using mathematical morphology to simulate the flood routing process, combined with the physical significance of the discretization of the St. Venant equation for unsteady flow. The method of simulating flood routing with the flow discrete in time and connected in space is proposed, which can obtain the flood arrival time, water level change process, and flow velocity change process for each submerged unit in the flood spreading process. The method enriches the means of flood routing simulation, but some issues still need further in-depth study.

1. Each periodΔt, each layer of the wave crest speedV_i calculation method still needs further in-depth research.
2. The method is only applicable to the floodplain's spreading process. When the lowest point of the floodplain starts to catch water, the floodwater spreads in the way of water level uplift. Then we have to use equation (9) or (10) for calculation, in which the parameters need to be adjusted, and this issue needs further exploration.

REFERENCES

Ding Z.X., Li J.R., et al. Flood Inundation Analysis Method Based on GIS Grid Model. *JHE*, 6, 56–60, (2004)

Gong W. *Mathematical Morphology in Digital Space: Theory and Applications(Chinese)*. Beijing: Science Press, (1997)

Huang J.C. and He X.Y. Unified 2-D Numerical Model for simulating Dam Break Wave Propagation (Chinese). *JHE*. 37(2): 222–226, (2006)

Li F.W., Zhang X.N., et al. Flood Inundation Research Based on GIS and Mathematical Morphology. *Advances in Science and Technology of Water Resources*. 25(6),14–16, (2005)

Liu Z.Y. Application of GIS-based Distributed Hydrological Model to Flood forecasting (Chinese). *JHE*. 4,70–75,(2004)

Morris M.W. *CADAM: A European Concerted Action Project on Dam Break Modeling, in Proceedings of the British Dam Society*, BritishDamSociety. Thomas Telford, (2000)

Tang C.Q. and Lv. H.B., *Mathematical Morphological Methods, and their Applications (Chinese)*. Beijing: Science Press. (1990)

TodinE C. *Mathematical Models of Large Watershed Hydrology*, WRP, Littleton, Colorado,(2001)

Xiang S.Y., J. Urban Flood Submerging Simulation Analysis Based on GIS. *Journal of the China University of Geosciences*. 20(5), 575–580, (1995)

Civil Engineering and Energy-Environment – Gao & Duan (Eds)
© 2023 the Author(s), ISBN 978-1-032-56059-5

Comparative analysis of structural changes and rationality of water conservancy investment in the five Northwestern Provinces of China

Jia He*, Jiwei Zhu*, Hong Zhao & Jianmei Zhang

State Key Laboratory of Eco-hydraulics in Northwest Arid Region, Xi'an University of Technology, Xi'an, China

Research Center of Eco-hydraulics and Sustainable Development, New Style Think Tank of Shaanxi Universities, Xi'an, China

Department of Engineering Management, School of Civil Engineering and Architecture, Xi'an University of Technology, Xi'an, China

ABSTRACT: The study uses the improved TOPSIS method to analyze the structural rationality of water conservancy investment in the northwestern provinces. It examines the structural changes of water conservancy investment regarding water sources and uses. The results show that all five northwestern provinces have significant water investment funds. Shaanxi's water investment mainly goes to flood control and water supply, while Gansu and Qinghai invest primarily in irrigation and water supply. Water investments in Qinghai and Ningxia are mostly for flood control and irrigation. The reasonable degree of water investment in the five northwestern provinces shows an overall upward trend, with the source structure changing slowly with volatility. The overall direction of growth in the use structure gradually stabilized. The results of this paper help to understand the development trend of the change in the structure and rationality of water resources investment, improve the structure of water resources investment, and improve the efficiency of water conservancy investment in the five northwestern provinces of China.

1 INTRODUCTION

As a controlling factor in the development of water conservancies, investment in water conservancy plays an essential role in ensuring the sustainable use of regional conservancies. With the water conservancy in the five northwestern provinces and cities, there is an imbalance between the supply and demand of water conservancy, which seriously restricts sustainable socio-economic development and the improvement of people's living standards. The unreasonable investment in water conservancy in some regions and sectors has made limited water conservancy even more scarce. Therefore, it is essential to take a deeper perspective into the different periods of the five northwestern provinces. The study explores the characteristic patterns of changes in the structure and rationality of water resources investment to develop a scientific and reasonable investment plan, improve the efficiency of water resources utilization, and achieve sustainable economic and social development in the region.

Many domestic scholars have discussed the structure of water conservancy investment. Zhao *et al.* (2019) remeasured and analyzed the investment structure of water conservancy in

---

*Corresponding Authors: hj22016@163.com and xautzhu@163.com

DOI: 10.1201/9781003433651-62

Jiangsu during the 13th Five-Year Plan and accordingly put forward proposals for financing water conservancy infrastructure in the next five years. Wang (2018) proposed countermeasures to promote the future supply-side structural reform of water conservancy, taking into account the current situation and problems of the supply-side reform of water conservancy. Li (2017) constructed a panel data model based on panel data from 31 Chinese provinces from 1991-2015 to analyze the impact of investment structure on the performance of farmland water conservancy investment in various regions of China. Fang et al. (2012) compared the "Ninth Five-Year Plan" and "Tenth Five-Year Plan" during the water conservancy investment data from the perspective of investment structure, investment policy, investment effect analysis of the "Eleventh Five-Year Plan" period in Jiangsu. Zhuang et al. (2012) used the grey correlation method to evaluate the structure of water conservancy investment in Jiangsu Province from 1996 to 2010. They found that the overall rationality of water conservancy investment in Jiangsu Province significantly improved. Jing (2021) used the improved TOPSIS method to measure and analyze the rationality of the change in the source structure. The researcher used structure in Jiangsu and Shaanxi provinces and compared the differences in the size of fluctuations and growth rate between the East and West in the rationality of water investment structure. As we can see, few articles use mathematical formula models to measure reasonableness, mainly for the eastern provinces. Water conservancy investment has not yet met the actual construction needs of the backward western region. Local financial support for conservation is still needed. In this paper, we study the rationalization of the water conservancy investment structure from two perspectives: source and use, which will help to provide a basis for the adjustment of the water conservancy investment structure in Northwest China.

2 OVERVIEW OF THE STUDY AREA AND DATA SOURCES

2.1 *Study area overview*

The five northwestern provinces are located west of the Great Xing'an Mountains and north of the Great Wall and the Kunlun Mountains - Aljinshan - Yinshan Mountains, specifically containing Shaanxi, Gansu, Qinghai, Ningxia, and Xinjiang. The location is depicted in Figure 1. The area is vast, the population is sparse, the climate is arid, the precipitation is scarce, and evaporation is vigorous. Over the years, the average surface water conservancy is about 146.3 billion cubic meters, underground water conservancy is 99.8 billion cubic

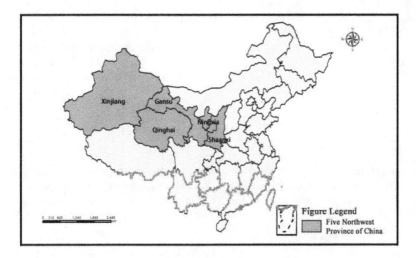

Figure 1. Location map of the five northwestern provinces.

meters, total water conservancy is 167.2 billion cubic meters, and comprehensive water conservancy per capita is 2189 cubic meters. The northwest region is mainly a loess plateau area with small vegetation cover, severe soil erosion, and water resource development and utilization difficulties. Therefore, scientific research and analysis of the rational structure of water resource investment in the five northwestern provinces will help to identify possible inconsistencies in the proportion of water conservancy investment. They would also provide a basis for testing the rationality of the water conservancy investment structure and adjusting the direction of the investment structure.

2.2 *Data sources*

The data in this article are mainly derived from the 2009-2019 China Water Conservancy Yearbook and China Water Conservancy Statistical Yearbook of the five provinces in the northwestern region and their portal websites' public information.

3 EVALUATION METHODOLOGY

Evaluating the rationality of the water investment structure is the basis for the adjustment and optimization of the water investment structure, and the optimization of a water investment structure is the core content of water construction. Although some results have been achieved in studying the signs, objectives, mathematical models, and characteristics of water investment structure optimization (Fang *et al.* 2010; Han *et al.* 2004; Pan *et al.* 2011; Zhang 2004), they are primarily qualitative descriptions and less quantitative calculations. Given this, this paper intends to use the improved TOPSIS method to study the source's rationality and use the structure of water conservancy investment in five northwestern provinces from 2009 to 2019. TOPSIS method belongs to a decision-making method based on ideal points. HwangCL and YoonK (1981) and others first used the TOPSIS method to study multi-attribute decision-making problems. With the depth of research, the TOPSIS method has been improved in terms of the weight coefficient matrix, the optimal and inferior solutions, and the evaluation value formula of the evaluation object (Ren *et al.* 2017). It has been widely used in the comprehensive evaluation of climate and environmental quality, pollution control programs, and evaluation of the rationality of land use structure (Liu *et al.* 2007, 2011). The basic principle of the TOPSIS method is to define the positive and negative ideal solutions of the decision problem and then, find a solution that is closest to the positive ideal solution and farthest from the negative ideal solution from the feasible solutions (Fang & Huang 2011). Its specific calculation process is divided into the following steps:

Step 1: Construct the normalized decision matrix. Let the multi-attribute decision problem have m alternatives $H_1, H_2, \ldots, H_m$ and n measures $D_1, D_2, \ldots, D_n$. The attribute value of the solution H_i under the attribute D_j is y_{ij}. The multi-attribute decision matrix is shown in (1).

$$Y = \left[y_{ij} \right]_{m \times n} = \begin{bmatrix} y_{11} & \cdots & y_{1n} \\ \vdots & \ddots & \vdots \\ y_{m1} & \cdots & y_{mn} \end{bmatrix} \tag{1}$$

In the target decision, to solve the problem of the different magnitudes of the indicator so the evaluation indicators are comparable and more reflective of the changes in the actual situation indicators, the multi-attribute decision matrix is normalized using the coefficients of each hand after calculating the weights. The normalized decision matrix is obtained as shown in Equation (2).

$$Z = \left(Z_{ij} \right)_{m \times n} \tag{2}$$

Benefit-based indicators include central investment, local investment, domestic loans, corporate and private investment, and other investments for the source structure. Flood control, irrigation, flood removal, water supply, hydropower, water conservation and ecology, institutional capacity building, preliminary work, and other investment for the use structure. Benefit-based indicators are standardized according to Equation (3).

$$Z_{ij} = \begin{cases} (y_{ij} - y_{j\,\text{max}})/(y_{j\,\text{max}} - y_{j\,\text{min}}) & y_{j\,\text{max}} \neq y_{j\,\text{min}} \\ 1 & y_{j\,\text{max}} = y_{j\,\text{min}} \end{cases} \tag{3}$$

Cost-type indicators are uncompleted investments. Source structure uncompleted investment is the difference between the planned investment and the investment in place in the current year. It is obtained from the difference between the actual total investment required for the project under construction and the cumulative completed investment in the current year. Cost-based indicators are standardized according to Equation (4).

$$Z_{ij} = \begin{cases} (y_{j\,\text{max}} - y_{ij})/(y_{j\,\text{max}} - y_{j\,\text{min}}) & y_{j\,\text{max}} \neq y_{j\,\text{min}} \\ 1 & y_{j\,\text{max}} = y_{j\,\text{min}} \end{cases} \tag{4}$$

Step 2: Determine the weights. Define $\omega_j (j = 1, 2, \ldots , n)$ as the weight of attribute G_j, then $\sum_{j=1}^{n} \omega_j = 1$. By constructing an objective planning optimization model and making a Lagrangian function, the final expression of weight ω_j is determined as in Equation (5).

$$\omega_j = \mu_j / \sum_{j-1}^{n} \mu_j \tag{5}$$

Where μ_j is calculated by Equation (6).

$$\mu_j = 1 / \sum_{i-1}^{m} \left[(1 - z_{ij})^2 + z_{ij}^2 \right] \tag{6}$$

Step 3: Calculate the weighting matrix.

$$X = [x_{ij}]_{m \times n} = [w_j Z_{ij}]_{m \times n} \tag{7}$$

Step 4: Determine the positive and negative ideal solutions. The equations of the positive ideal solution X^* and negative ideal solution X^- are shown in (8) and (9).

$$X^* = \left\{ \left(\max_{1 \leq i \leq m} x_{ij} | j \in j^* \right), \left(\min_{1 \leq i \leq m} x_{ij} | j \in j^- \right) \right\} = (x_1^*, x_2^*, \ldots x_n^*) \tag{8}$$

$$X^* = \left\{ \left(\max_{1 \leq i \leq m} x_{ij} | j \in j^* \right), \left(\max x_{ij} | j \in j^- \right) \right\} = (x_1^-, x_2^-, \ldots x_n^-) \tag{9}$$

3.1 Among them, j*, j¯ are benefit-type, cost-type indicator sets.

Step 5: Calculate the Euclidean distance. The Euclidean distance S^* of the indicator vector to the positive ideal solution and the Euclidean distance to the negative ideal solution for each year is given by (10) and (11).

$$S_i^* = \left[\sum_{j=1}^{n} \left(x_{ij} - x_j^* \right)^2 \right]^{1/2} \tag{10}$$

$$S_i^- = \left[\sum_{j=1}^{n}\left(x_{ij} - x_j^-\right)^2\right]^{1/2} \tag{11}$$

The closer the item to be evaluated is to the positive ideal solution, and the further it is to the negative ideal solution, the better it is.

Step 6: Reasonableness measurement. The expression of water investment structure reasonableness D is Equation (12). The higher the value of D, the more reasonable the structure of water investment.

$$D = d_i^* = S_i^- / \left(S_i^* + S_i^-\right) \tag{12}$$

4 RESULTS AND ANALYSIS

4.1 *Weighting of evaluation indicators and positive and negative ideal solutions*

From the distribution of the positive and negative ideal solutions of the evaluation scheme, it can be seen that the source structure and use structure of the investment in the five northwestern provinces are all positive ideal solutions of 0.000 for the uncompleted water investment and negative ideal solutions of 0.000 for the rest of the evaluation indicators. The weights of the evaluation indicators determine non-zero positive and negative ideal solutions.

4.1.1 *Source structure of the investment*

According to the water investment source, each indicator's weight is calculated from five aspects: central, local, and domestic loans, corporate and private investments, and other investments. Among them are the central and local investments, including special funds for state bonds, water conservancy construction funds, etc., and other investments, including bonds, the use of foreign investment, etc. The weights of the evaluation indicators by source and the positive and negative ideal solutions of the evaluation scheme are shown in Table 1.

Table 1. Water investment structure weights and positive and negative ideal solutions (by source).

Province	Index	Central Investment	Local Investment	Domestic Loan	Corporate and Private Investment	Others	Unfinished Investment
Shaanxi	ω_j	0.1894	0.1876	0.1584	0.1816	0.1529	0.1301
	X^*	0.1894	0.1876	0.1584	0.1816	0.1529	0.0000
	X^-	0.0000	0.0000	0.0000	0.0000	0.0000	0.1301
Gansu	ω_j	0.1899	0.1697	0.1294	0.1582	0.1810	0.1720
	X^*	0.1899	0.1697	0.1294	0.1582	0.1810	0.0000
	X^-	0.0000	0.0000	0.0000	0.0000	0.0000	0.1720
Qinghai	ω_j	0.1790	0.1827	0.1519	0.1356	0.1769	0.1739
	X^*	0.1790	0.1827	0.1519	0.1356	0.1769	0.0000
	X^-	0.0000	0.0000	0.0000	0.0000	0.0000	0.1739
Ningxia	ω_j	0.1802	0.1852	0.1360	0.1453	0.1553	0.1980
	X^*	0.1802	0.1852	0.1360	0.1453	0.155	0.0000
	X^-	0.0000	0.0000	0.0000	0.0000	0.0000	0.1980
Xinjiang	ω_j	0.1676	0.1762	0.1851	0.1740	0.1479	0.1492
	X^*	0.1676	0.1762	0.1851	0.1740	0.1479	0.0000
	X^-	0.0000	0.0000	0.0000	0.0000	0.0000	0.1492

The most influential evaluation indicators in Shaanxi are central investment and local investment, with central investment having a slightly higher weight than local investment. Corporate and private investments also being important sources of funding for water conservancy investment in Shaanxi. The evaluation indicator with the greatest weight in Gansu is a central investment, followed by other sources, and the smallest is domestic loans. The most influential indicator of the reasonableness of funding sources in Qinghai is local investment, followed by central investment, and the smallest is corporate and private investment. Ningxia has the highest weight of uncompleted investment, followed by local investment and central investment, and the smallest weight is domestic loans. The most influential evaluation indicator in Xinjiang is domestic loans, followed by local investment, and the smallest is other investments.

4.1.2 *Structure of the use of investment*
From flood control, irrigation, flood removal, water supply, hydropower, water conservation and ecology, institutional capacity building, preliminary work, others, and uncompleted investment, the use structure of the weight of each indicator is calculated. Among them, in Shaanxi, Gansu, Qinghai, and Xinjiang, the amount of investment in flood removal and Ningxia hydropower investment is very small, and only a few years exist to improve the accuracy of the measurement results to do away with them. The weights of the evaluation indicators by use and the positive and negative ideal solutions of the evaluation scheme are shown in Table 2.

The indicator with the greatest impact of comprehensive evaluation in Shaanxi is irrigation works, followed by water supply works. The smallest is institutional capacity building. In Gansu Province, the most influential indicator in the comprehensive evaluation is irrigation works, followed by flood control works, and the smallest are other works. In Qinghai and Ningxia, the main expenditure directions of water conservancy investment are flood control and irrigation, and the smallest are other works. The most influential indicators in the comprehensive evaluation of Xinjiang province are irrigation works, followed by institutional capacity building, and the least influential is other works.

Table 2. Water investment structure weights and positive and negative ideal solutions (by use).

Province	Index	Central Investment	Local Investment	Domestic Loan	Corporate and Private Investment	Others	Unfinished Investment
Shaanxi	ω_j	0.1075	0.1149	0.0946	0.1111	0.0960	0.0894
	X^*	0.1075	0.1149	0.0946	0.1111	0.0960	0.0894
	X^-	0.0000	0.0000	0.0000	0.0000	0.0000	0.0000
Gansu	ω_j	0.1114	0.1158	0.1046	0.1009	0.1016	0.1008
	X^*	0.1114	0.1158	0.1046	0.1009	0.1016	0.1008
	X^-	0.0000	0.0000	0.0000	0.0000	0.0000	0.0000
Qinghai	ω_j	0.1232	0.1120	0.0930	0.0953	0.1086	0.0987
	X^*	0.1232	0.1120	0.0930	0.0953	0.1086	0.0987
	X^-	0.0000	0.0000	0.0000	0.0000	0.0000	0.0000
Ningxia	ω_j	0.1233	0.1160	0.0892	0.0475	0.1042	0.1056
	X^*	0.1233	0.1160	0.0892	0.0475	0.1042	0.1056
	X^-	0.0000	0.0000	0.0000	0.0000	0.0000	0.0000
Xinjiang	ω_j	0.1095	0.1106	0.0999	0.1060	0.0999	0.1120
	X^*	0.1095	0.1106	0.0999	0.1060	0.0999	0.1120
	X^-	0.0000	0.0000	0.0000	0.0000	0.0000	0.0000

4.2 Analysis of the reasonableness of the water conservancy investment structure.

Figure 2 shows the spatial distribution of Euclidean distance of the source structure of water conservancy investment in Shaanxi, Gansu, Qinghai, Ningxia, and Xinjiang. Figure 3 shows the spatial distribution of the Euclidean distance of the source of water conservancy investment in each province. The results of the calculation of the structure rationality of the source of water conservancy investment funds in each province from 2009 to 2019 are shown in Table 3, and the results of the calculation of the structure rationality of the use of water conservancy investment funds are shown in Table 4.

4.2.1 Analysis of the rationality of the source of water investment structure

From Figure 2, it can be seen that the Euclidean distances of positive and negative ideal solutions of water investment source structure are distributed between (0.16, 0.39) and (0.04, 0.34) in Shaanxi, between (0.14, 0.37) and (0.05, 0.33) in Gansu, between (0.20, 0.35) and (0.16, 0.30) in Qinghai, between (0.24, 0.35) and (0.09, 0.29) in Ningxia, and between (0.17, 0.35), (0.07, 0.30) in Xinjiang, respectively. In cross-sectional comparison, the distribution of the distance between the integrated vector of evaluation indicators and the positive ideal solution for the source structure of Shaanxi, Gansu, and Xinjiang are all relatively concentrated. The distribution of the distance between the negative ideal solution and the negative ideal solution is relatively scattered in the two-dimensional data space, reflecting that the rationality of each source of water conservancy investment varies in a wide range.

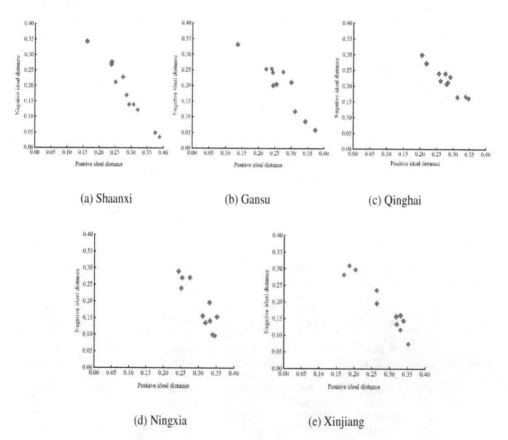

Figure 2. Spatial distribution of positive and negative ideal distance two-dimensional data of water investment structure (by source).

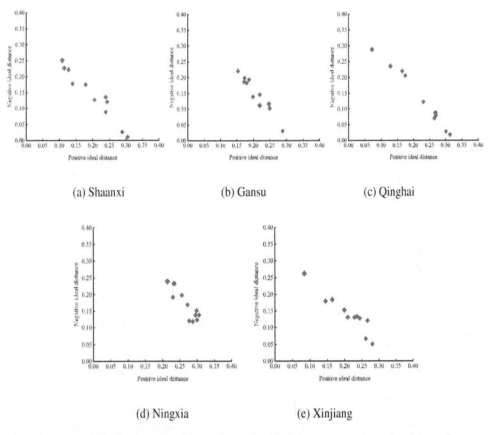

Figure 3. Spatial distribution of positive and negative ideal distance two-dimensional data of water investment structure (by use).

The Euclidean distances in Qinghai and Ningxia provinces have minor variations, and the sources of water conservancy investment do not change much.

The results in Table 3 show that, from the overall trend, the structural rationality of water conservancy investment sources in Shaanxi, Gansu, Qinghai, and Xinjiang have all

Table 3. The water investment structure in five provinces during 2009-2019 (by source).

Year	Distance of the evaluation scheme from the optimal ideal reference point					Reasonableness of water investment structure				
	Shaanxi	Gansu	Qinghai	Ningxia	Xinjiang	Shaanxi	Gansu	Qinghai	Ningxia	Xinjiang
2009	0.3821	0.3615	0.1978	0.0851	0.2351	8.34%	13.66%	31.81%	40.07%	30.01%
2010	0.3624	0.2767	0.1867	0.0736	0.2301	11.74%	27.42%	33.23%	41.65%	29.58%
2011	0.2718	0.3209	0.0777	0.1091	0.2970	27.64%	20.12%	48.46%	36.65%	17.50%
2012	0.2113	0.2033	0.0985	0.1625	0.2497	37.56%	41.34%	45.55%	28.26%	26.21%
2013	0.2424	0.1736	0.1183	0.1766	0.1183	32.24%	44.61%	42.60%	25.93%	47.25%
2014	0.2500	0.1697	0.1714	0.0903	0.2181	31.28%	45.17%	34.67%	39.65%	32.77%
2015	0.1608	0.0000	0.1163	0.0866	0.1464	45.39%	70.95%	42.96%	40.43%	42.69%
2016	0.1582	0.1643	0.1104	0.0000	0.2119	45.85%	47.07%	44.16%	51.29%	33.10%
2017	0.1057	0.1397	0.0296	0.0104	0.0344	53.11%	49.92%	55.57%	50.24%	59.44%
2018	0.1015	0.1172	0.0000	0.0853	0.0153	53.67%	53.19%	59.37%	40.71%	62.46%
2019	0.0000	0.1290	0.0919	0.1191	0.0276	67.90%	51.57%	46.61%	35.39%	62.28%

gradually improved and diversified their funding sources. At the same time, the structural reasonableness of the water conservancy investment sources shows a fluctuating growth trend over time. According to the magnitude of fluctuations in the structural rationality of water conservancy investment sources in these four provinces, the changes in their rationality from 2009 to 2019 are classified as a period of small fluctuations, a period of sharp fluctuations, and a period of structural stability.

Shaanxi's investment structure for 2009-2010 is a period of small fluctuations, low structural reasonableness, and relatively small fluctuations. In 2011-2016 fluctuations were more intense, from 2017 onwards into a stable growth phase, and higher reasonableness in 2019 reached a maximum of 67.9%. The reasonableness of Gansu, Qinghai, and Xinjiang fluctuated slightly from 2009 to 2012. It fluctuated sharply between 2013 and 2017, with the highest reasonableness of water conservancy investment sources in Gansu at 70.95% in 2015 and the reasonableness of Qinghai and Xinjiang reaching 55.57% and 59.44% in 2017, respectively. The reasonableness degree is high and stable, above 50%, while the reasonableness degree of Qinghai experienced a slight fluctuation during this period, and the reasonableness degree dropped to 46.61%.

Ningxia's water conservancy investment source rationality is in a state of substantial fluctuations. In 2019, rationality was 35.39%, less than 40.07%; in 2009, it is not difficult to find that the proportion of uncompleted investment in Ningxia province is large, and the structure of the source of funds is not reasonable. This is similar to the pattern reflected by the distribution of the distance between the comprehensive vector of evaluation indicators and the positive and negative ideal solutions in the two-dimensional data space of the water conservancy investment structure.

4.2.2 *Analysis of the rationality of the use of water investment structure*

As seen from Figure 3, the Euclidean distances between the positive and negative ideal solutions of the water conservancy investment use structure in Shaanxi Province, Qinghai, and Xinjiang are all more varied, indicating that the distribution of feasible solutions is more dispersed. The reasonableness of the water conservancy use structure involved in the evaluation varies widely. In Gansu and Ningxia, the distribution of the distance between the integrated vector of evaluation indicators and the positive and negative ideal solutions are all relatively concentrated, and the range of variation of the water conservancy investment use structure is small and prone to unreasonable changes.

As seen from Table 4, overall, the rational degree of use structure of the five provinces is on the rise. The difference lies in the magnitude of the trend fluctuations vary greatly. Shaanxi's water investment use structure has grown steadily, achieving a breakthrough from within 10% to above 60%. The reasonableness of the use of funds in Qinghai and Xinjiang reached a maximum of 80.12% and 75.75% in 2018 and 2017, respectively. There is a high reasonableness after stabilizing at more than 55% in 2019. Gansu and Ningxia provinces have a fluctuating trend and frequent change cycles in the fund use structure, and the final reasonableness fell below 50% in 2019, with low reasonableness.

Comparing the difference in the flow of funds between Gansu and Ningxia in various years, we found that the investment in irrigation, flood control, and other infrastructure projects is large. In contrast, the investment in soil and water conservation and ecological management, and other aspects of water conservancy projects is less. The proportion of unfinished investment is also higher, indicating the shifting funds from traditional irrigation and flood control to ecology, water conservation, and other directions. Paying attention to scientific regulation of recharge, balancing conservancy allocation, and strengthening watershed ecological protection is a meaningful way to promote the continued rationalization of the structure of water conservancy investment use.

Table 4. The water investment structure in five provinces during 2009-2019 (by use).

Year	Distance of the evaluation scheme from the optimal ideal reference point Shaanxi	Gansu	Reasonableness of water investment structure Qinghai	Ningxia	Xinjiang	Shaanxi	Gansu	Qinghai	Ningxia	Xinjiang
2009	0.3821	0.3615	0.1978	0.0851	0.2351	8.34%	13.66%	31.81%	40.07%	30.01%
2010	0.3624	0.2767	0.1867	0.0736	0.2301	11.74%	27.42%	33.23%	41.65%	29.58%
2011	0.2718	0.3209	0.0777	0.1091	0.2970	27.64%	20.12%	48.46%	36.65%	17.50%
2012	0.2113	0.2033	0.0985	0.1625	0.2497	37.56%	41.34%	45.55%	28.26%	26.21%
2013	0.2424	0.1736	0.1183	0.1766	0.1183	32.24%	44.61%	42.60%	25.93%	47.25%
2014	0.2500	0.1697	0.1714	0.0903	0.2181	31.28%	45.17%	34.67%	39.65%	32.77%
2015	0.1608	0.0000	0.1163	0.0866	0.1464	45.39%	70.95%	42.96%	40.43%	42.69%
2016	0.1582	0.1643	0.1104	0.0000	0.2119	45.85%	47.07%	44.16%	51.29%	33.10%
2017	0.1057	0.1397	0.0296	0.0104	0.0344	53.11%	49.92%	55.57%	50.24%	59.44%
2018	0.1015	0.1172	0.0000	0.0853	0.0153	53.67%	53.19%	59.37%	40.71%	62.46%
2019	0.0000	0.1290	0.0919	0.1191	0.0276	67.90%	51.57%	46.61%	35.39%	62.28%

5 CONCLUSION

This paper takes five provinces of Northwest China, including Shaanxi, Gansu, Qinghai, Ningxia, and Xinjiang, as the research object. It uses the modified TOPSIS method to analyze and compare the rationality of the water conservancy investment structure in five provinces by source. It uses classification and obtains several conclusions as follows:

1) Looking at the water conservancy investment by source, the source structure rationality shows a fluctuating growth trend over time. Government investment, including central and local government investment at all levels, corporate and private investment, and other funds involved in the construction of water conservancy projects, accounted for a relatively small percentage of total investment in the five northwestern provinces.
2) In terms of water investment uses, although the overall structure of water investment uses in the five northwestern provinces is increasing, there is still a problem of uneven investment in different water projects. Traditional flood control, irrigation, water supply, and other infrastructure projects are large investments, but there is less investment in water and soil conservation and ecological management of water projects. Looking at the water conservancy investment by source, the source structure rationality shows a fluctuating growth trend over time. Government investments account for the majority of funding for water conservation projects in the five provinces of western China. Bank loans, corporate and private investments, and other funds used in the construction of water conservation projects account for a small portion of total funding.
3) As a whole, the reasonableness of the change in the structure of water conservancy investment can be a good evaluation of the change in the structure of water conservancy investment. The size of the reasonableness of the change can be measured quantitatively. Both by source and by use, Shaanxi's water investment structure is more reasonable. In contrast, Ningxia's water investment structure is less reasonable and needs to be adjusted to make it more reasonable.

ACKNOWLEDGMENTS

This work was supported by the Shaanxi Provincial Department of Education Key Scientific Research Project (number 22JT031).

REFERENCES

Fang G.H and Huang X.F (2011) *Multi-Objective Decision Theory, Methods and their Applications*, Science Press, Beijing.

Fang G.H., Geng J.Q. and Mao G.N. (2010) Jiangsu Province in 2005-2007 Water Investment Structure Analysis. *J. China Rural Water and Hydropower*, (04):119–121.

Fang G.H., Zhuang JH. and Tan W.X. (2012) Rationality Analysis of Water Conservancy Investment of Jiangsu Province in the Eleventh Five-years Planning. *J. Water Resources and Power*, 30(08):119–121.

Han N., Chen K. and Liu X.W. (2004) Shenzhen "Ninth Five-Year Plan" Water Infrastructure Investment Structure Optimization. *J. China Rural Water and Hydropower*, (09):123–124.

Hwang C.L. and Yoon K. (1981) *Multiple Attribute Decision Marking: Methods and Applications*. Springer-Verlag, Berlin.

Jing X.N. (2021). *Research on the Change of Investment Structure of Water Conservancy in the East and West of China*. Xi'an University of Technology.

Li S.H. (2017) *Research on the Influence of Investment Structure on the Performance of Farmland Water Conservancy Investment*. South China Agricultural University.

Liu D.Y., Liang ZM., Wang SY. and Yi Z.Z. (2011). Application of TOPSIS Objective Weighting Method to Drought Comprehensive Assessment. *J. South-to-North Water Resources and Power*, 29(06):8–10+92.

Liu M.Y., Hua L., Wang SY. and Liu C. (2007) Comprehensive Evaluation on Water Environment Quality in the Wenyu River by Improved TOPSIS. *J. South-to-North Water Transfers and Water Science & Technology*, (03):57–60.

Pan J., Fang G.H. and Gao Y.Q. (2011) Jiangsu Province from 2005 to 2009 Water Investment Structure Analysis. *J. Water Resources and Power*, 29(06):143–144+67.

Ren L.F., Wang Y.R., Zhang YQ. and Sun Z.Q. (2017) Improvement of TOPSIS Method and Comparative Study. *Chinese Journal of Health Statistics*, (01):64–66.

Wang C. (2018) Countermeasures and Measures to Deeply Promote Structural Reform on the Supply Side of Water Resources. *J. Water Economy*, 36(01):20–23+89.

Zhang W.S. (2004) Establishment and Application of an Analytical Model for Optimization of Water Resources Investment Structure. *J. Yangtze River*, (09):50–51.

Zhao Y.H., Chen CQ., and Dong ZX. (2019). Estimation and Adjustment of Investment in Water Project Construction During the 13th Five-Year Plan period in Jiangsu and Suggestions for Fund Rising. *J. Zhi Hui*, (04):79–81.

Zhuang J.H., Fang GH., Pan J. and Bai Y.F. (2012) Correlation Analysis Between Types and Rationality of Water Conservancy Investment Structure in Jiangsu Province. *J. Water Resources and Power*, 30(09):119–121+140.

Civil Engineering and Energy-Environment – Gao & Duan (Eds)
© 2023 the Author(s), ISBN 978-1-032-56059-5

A numerical simulation of the impact of hydropower development on regional air temperature in Canyon district

Hailong Wang
Huaneng Yarlung Zangbo River Lower Reaches Development Leading Group Office, China

Bei Zhu*, Chang Liu, Shiyan Wang, Shilin Zhao, Xu Ma, Yiqian Tan & Xing Yang
Department of Water Ecology and Environment Research, China Institute of Water Resources and Hydropower Research, Beijing, China

Huazhang Sun
Huaneng Tibet Yarlung Zangbo River Hydropower Development Investment Co., Ltd, China

ABSTRACT: The change of subsurface caused by the change of watershed area after the construction of hydroelectric projects in the canyon area changes the heat transport and moisture circulation between the ground and air systems. This will probably affect the climate of the surrounding river valley to some extent. This paper is based on the CFD refinement model (ANSYS Fluent). It realizes the water-heat transport between water bodies and the atmosphere by compiling a custom module UDF to establish a three-dimensional water-air heat and moisture exchange model for a typical section of the river in a reservoir area and downstream water reduction section. It analyzes and predicts the effect on the river valley temperature produced by the change of subsurface caused by hydropower development in the canyon area. It shows that the temperature effect of the water body is related to the topography of the canyon, the width of the water surface, the distance from the water surface, and the temperature difference between the water body and the two banks. The temperature in the river valley is influenced by the combined effect of water surface and bank temperatures. It is more influenced by the river temperature effect at the near-water surface. The vertically influenced range of daytime cooling in the reservoir area before and after the reservoir construction is 123 m and 211 m, respectively. The range of daytime cooling in the downstream river is 16 m and 14 m, respectively. The larger the surface area, the higher the cooling effect of the water body. The temperature effect before and after the reservoir construction is higher than that of the river.

1 INTRODUCTION

Thanks to its rich water resources and great development potential, rivers in mountainous areas are of strategic importance in water resources and water energy allocation in China (Chang *et al.* 2010). The construction of large-scale water conservation projects greatly benefits regional water and water control, hydropower generation, shipping, etc. They greatly contribute to the social environment and economic construction of the region and the whole basin (Poff & Hart 2002). However, they inevitably have negative effects on local ecological characteristics and processes, being known or unclear, long-term or short-term (Liu *et al.* 2015). After the construction of the hydropower project in the canyon region,

---

*Corresponding Author: zhubei@iwhr.com

DOI: 10.1201/9781003433651-63

changes in the underlying surface caused by changes in the water area have changed the thermal conveyance and water cycle in the earth-atmosphere system (Sempreviva 2020). It will be possible to impact the atmosphere of the surrounding river canyon. Some studies show that local weather factors, such as wind, temperature, and humidity, will respond to water level changes in case the reservoir is finalized (Sui & Yang 2005; Wu *et al.* 2006). Therefore, the study is made on changes in the hydropower situation and local atmosphere effects after the construction and operation of hydropower projects. It is of great importance to the rational development and utilization of watershed climate resources.

The development of hydropower in mountainous areas affects the hydrological situation of the river, which results in a large rise and fall of the river level and a significant change in the river width during the year. Different River widths cause different evaporation on the water level. Variations in the local underlying surface affect the exchange of heat-inductive, potential heating, water, and momentum between the earth and atmosphere (Liu *et al.* 2006), which will affect the local atmosphere to a certain extent. Zhirnova *et al.* (2021) found from the climate data and annual ring chronology of Siberian larch and European red pine that the reservoir construction changes the local climate and tree growth, which depends on the specific sensitivity of species to climate change, the distance from the reservoir, geographical region, and other factors. Chen (1995) studied the influence of Dongting Lake waters on climate and found that the lake can regulate temperature and moderate temperature extremes, i.e., in case the lake area increases, precipitation increases in winter, decreases in summer, and air humidity increases. Adams and Dove (2012) showed that a 35 m wide river could reduce the surrounding temperature by 1~1. 5°C. Zhang's research (2011) shows that lakes in Qionghai have a moderating effect on the surrounding climate and can change the temperature and precipitation of the surrounding environment. Saburo *et al.* (1991) found that rivers can lower the surrounding temperature by more than 5°C in sunny summer, and a 260 m wide river can spread the cooling range to about 400 m. Yu *et al.* (1991) studied the climatic effects of water body changes in the Caohai in Weining, i.e., the average winter temperature decreased, relative humidity decreased, and precipitation decreased in case the water area of the Caohai decreased after drainage. Wang and Fu (1991) found via numerical simulations that the effect of water bodies on temperature diminished with decreasing water surface width, and the magnitude of the effect of water bodies with widths of 5 km and 1 km was 90%-95% and 50% of that of 10 km, respectively.

In terms of the research object, existing studies on local climate effects are mostly focused on the reservoir with open water. However, studies on the possible climate effects caused by the water body changes in the reduced section of the river with small water bodies are few and not systematic. Existing studies of local climate effects are mostly at large and generally use mesoscale meteorological models such as WRF and MM5. The model has good applicability for large areas of water, such as reservoirs and lakes. While the change in water surface width caused by hydropower development is less than 100 m, especially in the downstream river, the change may be only a few dozen meters. Therefore, mesoscale numerical models such as WRF and MM5 cannot be accurately simulated. For the complex geometry of the subsurface, the refined numerical simulation technology represented by CFD (Computational Fluid Dynamics) becomes an effective method for the fine-grained simulation of the local climate of canyon rivers. However, model building often requires fine mesh size, which requires high computational performance and modeling level. In addition, the existing refined CFD models are less comprehensive in considering the combined effects of water vapor evaporation diffusion, airflow field, and thermal environment. Thus, there are few fine-scale simulation studies on the energy transfer between canyon rivers and the atmosphere.

This study is based on a refined CFD model (ANSYS Fluent) that considers the sensible and latent heat exchange between the water body and the atmosphere as well as the influence of solar radiation. It realizes the moisture transfer between water bodies and the atmosphere by compiling a custom module UDF (User-Defined Function). It establishes a three-dimensional water-air heat and moisture exchange model for a typical section of the river in a reservoir and downstream water reduction section. This study analyzes and predicts the impact of the change of the substrate surface caused by hydropower development in the canyon area on the atmospheric temperature of the river valley. It provides a scientific basis for a comprehensive assessment of the environmental impact of hydropower projects in the high mountain canyon area and guidance for the construction and operation of hydropower projects.

2 CALCULATION MODEL

2.1 *Simulation object & calculation conditions*

In this study, a proposed hydropower station in a highland canyon is taken as an example. It selects a typical section of the reservoir (3579 m to the dam site) and a typical section of the downstream water reduction channel (4337 m to 8382 m below the dam), respectively. The role of solar radiation, subsurface properties, and temperature in the thermal environment of the river valley localities are fully considered based on local summer daytime midday meteorological conditions. The calculated working conditions are shown in Table 1.

Table 1. Conditions setting for summer daytime noon and nighttime calculation.

Working conditions			Initial temperature (°C)	Air pressure (hpa)	Relative humidity (%)	Specific humidity (g/kg)	Water temperature (°C)	Saturated specific humidity (g/kg)	Water surface width (m)
1	Before	Reservoir	24	714.2	50	13.10	15.2	15.18	143.5~356.2
2	reservoir construction	Downstream river	24	714.2	50	13.10	15.2	15.18	120.7~156.4
3	After the	Reservoir	24	714.2	50	13.10	15.1	15.08	353.0~561.1
4	reservoir construction	Downstream river	24	714.2	50	13.10	15.1	15.08	47.1.77.2.

The reservoir's water level rose, and its surface width increased significantly after it was built. The water surface width of the section at 3579 m before the dam increased from 356 m to 561 m, and the water surface width of the section at the dam site increased from 143 m to 353 m. The water level in the lower reaches of the river decreases slightly, and the water surface width decreases. The water surface width of the section at 4337 m below the YS dam decreases from 156 m to 77 m. The water surface width of the section at 8382 m below the YS dam was reduced from 121 m to 47 m. See Figures 1–2 for more details.

The RNG k-ε turbulence model and the component transport model are used to open the energy equation. This paper simulates the transport and diffusion of water vapor (H_2O) in the air produced by water evaporation. No chemical reaction occurs during the process. Therefore, the component transport model is chosen from the finite velocity model for the transport of non-reactive substances. The pressure and velocity coupling are solved by the SIMPLIC algorithm. The continuity equation, momentum equation, energy equation, and water vapor evaporation are in second-order windward differential format. The turbulent kinetic energy and turbulent dissipation rate are in first-order windward differential format. The continuity and momentum calculations converge to an accuracy of 10^{-3}. The convergence accuracy of water vapor mass fraction, temperature, turbulent kinetic energy, and turbulent dissipation rate is 10^{-6}.

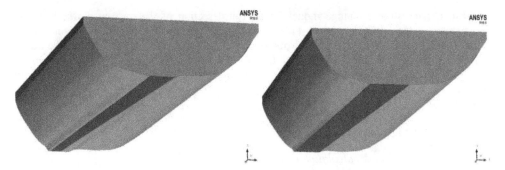

Figure 1. Geometric model of local climate before and after reservoir construction in a typical section of the reservoir area.

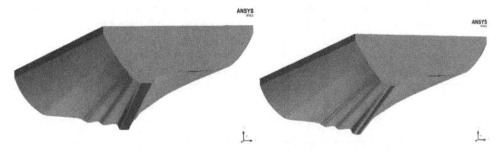

Figure 2. Geometric model of local climate before and after reservoir construction in a typical section of the lower river.

2.2 Water surface boundary setting

The water-air transport and heat exchange between the river and the atmosphere are the main considerations for studying the local climate of the river in the surrounding area. The water surface boundary is defined as the mass inflow boundary. The water-air-heat exchange is determined by the heat balance at the river surface. The mass flow rate of water-air transfer is determined by calculating the evaporation from the surface of the water body.

Solar short radiation received at the surface and atmospheric long radiation is balanced with sensible and latent heat transport from the surface to the atmosphere, long-wave radiation to the atmosphere, and heat transfer from the surface downward (Huang & Yan 1989).

$$K \downarrow + L \downarrow = L \uparrow + H + LE_0 + C_g H_g \qquad (1)$$

Where $K\downarrow$ is the downward solar short-wave radiation. $L\downarrow$ is the downward atmospheric long-wave radiation. $L\uparrow$ is the long-wave radiation radiated to the atmosphere from the surface. H and LE_0 are the sensible and latent heat fluxes delivered from the surface to the atmosphere. C_g is the surface-specific heat. H_g is the surface heat flux.

It is assumed that the river has constant runoff and almost no fluctuation of water temperature in a typical river section. Based on Equation (1), the surface water temperature of the water body is given by the energy balance equation as

$$\frac{d\theta_g}{dt} = \frac{1}{C_g}(K \downarrow + L \downarrow - L \uparrow - H_0 - LE_0) - H_g \qquad (2)$$

The above equation is defined as the temperature at the water surface boundary, which is loaded at the water surface boundary by UDF compilation o for a simplified model, K↓ is taken as 400 Wm^{-2}, and Cg is 4.186×106 Jm$^{-2\text{K}-1}$.

In the simulation, the moisture transfer and heat exchange must be linked to the wind speed, temperature, and humidity of the whole calculation domain to perform a coupled calculation and analysis of each field. It cannot be set directly in ANSYS Fluent but requires the use of the component transport module in Fluent. By preparing a custom module UDF, the defined calculation equations are imported into the model to realize the dynamic coupling of moisture and heat on the surface of the water body with the whole airflow field.

3 SIMULATION RESULTS AND ANALYSIS

3.1 *Temperature analysis of a typical section of the reservoir*

The middle section of the typical section of the reservoir (section at 1769 m before the dam) was selected. The temperature distribution before and after the reservoir construction is shown in Figure 3. A summer daytime at noon is characterized by a greater amount of solar radiation, while the thermal capacitance of both sides of the canyon is smaller. It is significantly exposed to solar radiation, which makes the temperature of the lower bedding surface higher. The water body is thermally capacious, and the river surface is cooler, so it warms up slower under the same solar radiation. The water temperature is only 16.2°C, showing a cooling effect on the area around the water body. There is a clear temperature stratification above the water body, and the cooling effect

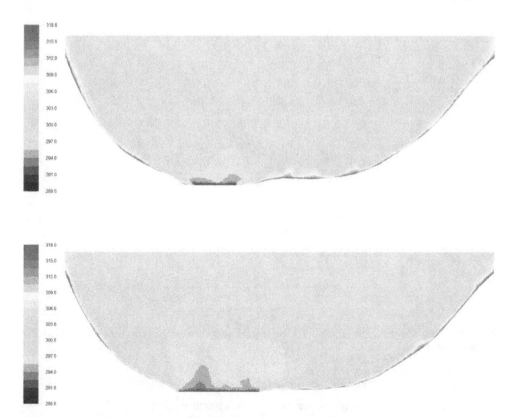

Figure 3. Temperature distribution profile before (top one) and after (lower one) the reservoir construction in a typical Section of the reservoir area.

is influenced by the topography of the canyon, the width of the water surface, and the distance from the water surface. The temperature in the river valley is affected by the combination of low temperature at the water surface and high temperature on both banks. It is influenced by the low temperature of the river near its surface. As a result of the reservoir construction, the average water surface width of the typical section of the reservoir increased from 249.5 m to 457.0 m, an increase of 83.2%. The range of water surface cooling influence is obviously increased.

The water surface vertical temperature distribution is analyzed according to the temperature values of different elevations on the water in the typical section surface (Figure. 4). The cooling effect generally decreases with increasing height. The cooling effect was significant within 40 m-70 m above the water surface. As the influence range of the water surface cooling effect is more than 0.5°C, the vertical influence range of water surface cooling in the reservoir before and after the reservoir construction is 123 m and 211 m, respectively. The vertical cooling range in the reservoir after the reservoir construction is 72% higher than that before the construction. The temperature above the water surface after the reservoir construction is lower than that before the construction. The temperature difference between before and after the reservoir construction, at the same distance above the water surface, can reach 1.3°C. As the distance from the water surface increases, the temperature difference before and after reservoir construction becomes smaller. For example, about 50 m above the water surface, the temperature difference before and after reservoir construction is 1.5°C; about 160 m above the water surface, the temperature difference before and after reservoir construction is 1°C; about 300 m above the water surface, the temperature difference before and after reservoir construction is 0.5°C.

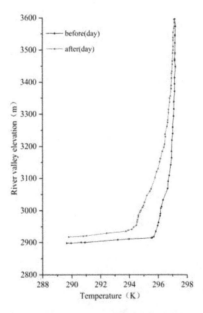

Figure 4. Temperature vertical distribution before and after the reservoir construction in a typical section of the reservoir area.

3.2 *Temperature analysis of the typical section of the downstream river*

The middle section (6360 m below the dam) of the typical section of the downstream river was selected. The temperature distribution before and after the reservoir construction is shown in Figure 5. The average water surface width of the typical downstream section was

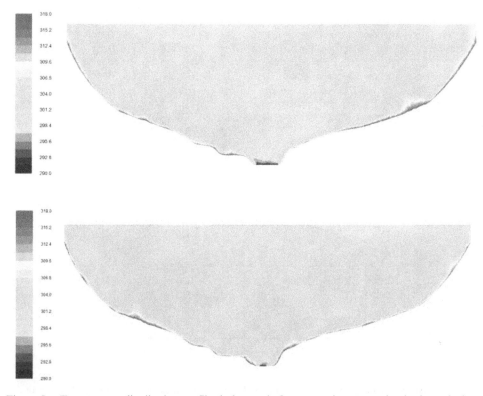

Figure 5. Temperature distribution profiles before and after reservoir construction in the typical section of the lower river channel (the lower picture: before reservoir construction; the upper picture: after reservoir construction).

reduced from 138.0 m to 62.0 m after the reservoir construction, a reduction of 55.1%. The influence range of the water surface cooling effect was significantly reduced. The water surface width of the downstream reduced section of the river before and after the reservoir construction was smaller than that of the reservoir section. Therefore, the scope of the cooling effect of the river is smaller compared with that of the reservoir.

The water surface vertical temperature distribution was analyzed according to the temperature values at different elevations on the water in the typical section surface (Figure 6). The temperature decreased over 0.5°C is taken as the influence range of the water surface cooling effect. Then the vertical influence range of water surface cooling in the downstream river before and after the reservoir construction is 16 m and 14 m, respectively. The vertical cooling range of the downstream channel before and after the reservoir construction is small. It lies in the fact that the width of the water surface before and after the reservoir construction is small, and the cooling effect is not obvious. At the same time, the two banks have a warming effect on the water and air, especially in the canyon area. The bottom center of the river valley is closer to the two banks, which are more warming. Therefore, the actual influence range of water surface cooling will be significantly larger than the calculated value. Overall, the temperature in the canyon is slightly higher after the reservoir construction. This is because the width of the water surface becomes narrower after the reservoir construction than before. However, the width of the water surface before and after the reservoir construction is not high, so the cooling effect of the water surface on the canyon environment is relatively limited.

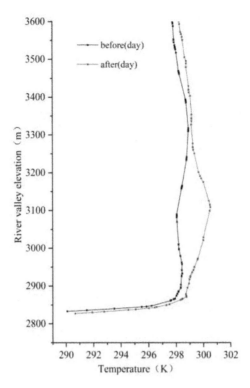

Figure 6. Distribution of temperature before and after reservoir construction in a typical section of the lower river.

In particular, the temperature vertical distribution characteristics were different before and after the construction due to the influence of the water surface width and topography. Immediately following the reservoir construction, within a height range of 50 m near the water surface, the temperature is mainly influenced by the low temperature of the water surface, i.e., the closer to the water surface, the lower the temperature. On the contrary, the temperature increases rapidly. During a range of 50 m to 280 m above the surface of the water, the temperature is affected by the combination of high temperature on both sides of the bank and low temperature on the water surface. The warming effect on both sides of the bank is greater than the cooling effect on the water surface. Therefore, the temperature in this range gradually increases with height, with the highest value reaching 27.4°C. However, as the elevation of the river valley continues to increase, the distance between the two banks and the center of the water surface also gradually increases. Hence, the warming effect decreases, and the temperature gradually drops to 25°C. Before the reservoir construction, the temperature was still mainly influenced by the low temperature of the water surface within the height range of 50 m near the water surface. Within the range of 50 m - 260 m on the water, the temperature was influenced by the high temperature on both banks and the low temperature on the water surface. Within the range of 50 m - 125 m, the temperature increased slightly with the elevation; within the range of 125 m - 260 m, the temperature decreased slightly with the elevation, and the variation was within 0.3°C; within the range of 260 m - 475 m on the water, the temperature increased with the elevation, the temperature decreased with the elevation, and the variation was within 0.3°C. The cooling effect of the water surface decreases with increasing elevation, and the temperature gradually increases to 25.8°C. It has been shown that the temperature of the area above 475 m above the water is predominantly influenced by high temperatures on both sides of the river, i.e., as the elevation increases, the cooling effect on both sides of the river decreases, and the temperature gradually decreases to 24.6°C.

4 CONCLUSIONS

In this study, we analyzed the effect of the change of water cover on the river valley temperature after the construction of a hydropower project through CFD 3D numerical simulation. We explored the possible temperature effect after constructing a power station in the high mountain canyon. The following conclusions can be drawn.

In summer daytime at noon, the water has a cooling effect on the atmospheric temperature. The temperature effect of the water is related to the topography of the canyon, the width of the water surface, the distance from the water surface, and the temperature difference between the water and the two banks. The temperature in the river valley is influenced by the combined effect of water surface and bank temperatures, and near the water surface, by the river temperature effect. It is 123 m and 211 m for daytime cooling of the water surface in the reservoir area before and after the reservoir construction, and 16 m and 14 m for daytime cooling of the water surface in the downstream river, respectively. The larger the water surface area, the higher the cooling effect of the water body. The temperature effect before and after the reservoir construction is higher than that of the river.

A three-dimensional water-air-heat exchange model is constructed. A scientific method of local climate refinement is proposed to analyze and predict the effects of changes in water cover caused by hydropower development in the canyon area on the temperature of the river valley to quantify and analyze the characteristics and extent of the effects of water temperature. However, the influence of the local climate is extremely complex. To some extent, this paper also generalizes the process of local climate change. In the future refined simulation of local climate, the conditions such as incoming wind speed, temperature, and solar radiation should be improved as much as possible to make CFD simulation more in line with the actual situation and provide a more accurate reference for the construction of hydropower projects in valley areas.

ACKNOWLEDGMENTS

This work is supported by the State Key Laboratory of Simulation and Regulation of Water Cycle in River Basin (WE1203B082022), Major Science and Technology projects of the Ministry of Water Resources (SKR-2022039).

REFERENCES

Adams L.W., Dove L.E. *Wildlife Reserves and Corridors in the Urban Environment: A Guide to Ecological Landscape Planning and Resource Conservation.* Wildlife Reserves and Corridors in the Urban Environment: A Guide to Ecological Landscape Planning and Resource Conservation, 2012.

Chang X., Liu X., Zhou W. et al. Hydropower in China at Present and its Further Development. *Energy*, 2010, 35(11): 4400–4406.

Chen B.L. *The Effects Of Water Body On Climate Over Dongting Lake Basin And Possible Climate Effects Of Sanxia Reservoir.* 1995, (03): 70–76.

Huang R.H, Yan B.L. A Numerical Model to Describe the Variability of River-Land Winds and its Numerical Experiment. *Chinese Journal of Atmospheric Sciences.* 1989, (1): 11–21.

Liu H.B, Zhang D.L, Wang B. Recent Advances in Regional Climate Modeling and Applications. *Climatic and Environmental Research.* 2006, (05): 649–668.

Liu S.L., An N.N., Dong S.K et al. The Effects of Hydropower Stations Construction on Vegetation Dynamics Based on NDVI: A Case Study of Cascade Hydropower Stations of Lancang River. *Mountain Research*, 2015, 33(1): 48–57.

Murakawa S., Sekine T., Narita K.I. et al. Study of the Effects of a River on the Thermal Environment in an Urban Area. 1991, 16(3-4), 993–1001.

Poff N.L., Hart D.D. How Dams Vary and Why it Matters for the Emerging Science of Dam Removal. *BioScience*, 2002, 52(8): 659–668.

Sempreviva A.M. Study of the Vertical Structure of the Coastal Boundary Layer Integrating Surface Measurements and Ground-Based Remote Sensing. *Sensors*, 2020, 20(22): 6516–6535.

Sui X. and Yang Z.F. Climatic Effects of Longyangxia Reservoir in Qinghai-Tibet Plateau and Trends with Time. *Mountain Research.* 2005, (3): 280–287.

Wang H. and Fu B.P. The Effects of Water Body on Temperature. *Scientia Meteorologica Sinica.* 1991, (3): 233–243.

Wu L., Zhang Q. and Jiang Z. Three Gorges Dam Affects Regional Precipitation. *Geophysical Research Letters*, 2006, 33(13):1–4.

Yu J.W Zhao G.Z and Tian Y. The Climatic Effects of Variation of the Water Body in Caohai, *Weining. Guizhou Science.* 1991, (1): 40–47.

Zhang D.P. Effect of the Changes of Qionghai's Water to Xichang Climate. *Journal of Xichang University (Natural Science Edition).* 2011, 25(2): 38–40.

Zhirnova D.F., Belokopytova L.V, Meko D.M et al. Climate Change and Tree Growth in the Khakass-Minusinsk Depression (South Siberia) Impacted by Large Water Reservoirs. *Nature Publishing Group*, 2021(11):14266–14278.

Civil Engineering and Energy-Environment – Gao & Duan (Eds)
© 2023 the Author(s), ISBN 978-1-032-56059-5

An empirical research on the ecological and economic effects of CCER trading in Chinese rural areas: Using synthetic control method

Qianhui Ma*, Yixin Lv & Guangyang Yu
School of Business, Sichuan University, Sichuan Province, Chengdu, China

Yuqi Liu & Yuhui Fan
School of Economics, Sichuan University, Sichuan Province, Chengdu, China

ABSTRACT: Facing the pressure of environmental protection and the accompanying economic development problems, this paper adopted the synthetic control method to study the ecological and economic effects of CCER trading. Calculation models of carbon sequestration and the index system were constructed to measure the effects. Considering the actual practices, the pilot province of CCER trading, Hubei province, was selected as the treated group. It was shown that the implementation of CCER trading saw an apparent increase in carbon sink and income in the rural area of the pilot province, that is, positive ecological and economic effects. The policy design of Hubei province is applicable and replicable nationwide, which will contribute to environmental protection and economic development.

1 INTRODUCTION

China was the largest carbon emitter in the world from 2005 to 2020. Given this grim situation, the Chinese government has committed itself to achieve the goal of carbon peak and carbon neutrality. Since 2013, the carbon trading policy has been launched in eight provinces and municipalities.

The Chinese Certified Emission Reduction (CCER) is the certified emission reduction of greenhouse gases produced by renewable energy and carbon sink projects that are mainly carried out in Chinese rural areas and usually provide farmers with employment opportunities. These CCERs can be purchased by third-party enterprises in the CCER market, which is an essential component of carbon trading. This may encourage enterprises to reduce emissions while also improving the rural environment and increasing farmers' income. In this case, CCER trading can greatly contribute to the goals of building rural areas with a thriving business, a pleasant living environment, and prosperity. However, considering the uncertain impact on the carbon sink and economic development, the government has suspended the approval of new CCER projects since 2017. There is still intense discussion on whether to resume CCER trading.

At present, the research on the ecological and economic effects of CCER trading mostly stays at the theoretical level, lacking empirical research. Thus, we established calculation models and an index system to measure them and adopted the synthetic control method to study the impact of CCER trading on rural carbon sinks and income increase (Abadie &

---

*Corresponding Author: qianhui_ma@stu.scu.edu.cn

DOI: 10.1201/9781003433651-64

Gardeazabal 2003). The results demonstrated that CCER trading had promoted the increase of carbon sink and farmers' income and showed positive ecological and economic effects. This can enrich the research on CCER trading related to agriculture and farmers and, concurrently, make up for the deficiency of existing research. This paper can also provide a reference for the government to improve the system and mechanism of CCER trading.

2   INDEX SYSTEM TO MEASURE THE ECOLOGICAL AND ECONOMIC EFFECTS

To cope with climate change, countries all over the world are trying to mitigate the aggravation of the greenhouse effect by removing CO_2 from the atmosphere and storing it in the land, forest and water by using the carbon sequestration capacity of the ecosystem.

Carbon sequestration was selected as the outcome variable of ecological effects in this paper. According to the main types of CCER projects, it was further divided into four types. Per capita disposable income and per capita wage income of rural residents were selected as the result variables of economic effects.

2.1   *Calculation of carbon sequestration in forest projects*

Using the forest volume method, the calculation formula is shown below.

$$C = A \cdot V \cdot BEFs \cdot D \cdot (1 + R) \cdot CF \tag{1}$$

In Equation (1), A×V represents forest volume; BEFs×D = BCEFs indicates biomass conversion and expansion coefficient; R is determined according to the biomass and local climate above the ground; CF indicates the carbon content of dry matter (He 2005; Guide for country reporting for FRA 2015; Xue *et al.* 2016).

2.2   *Calculation of carbon sequestration in biogas projects*

1) Manure management and energy substitution effects

$$ER = BE - PE \tag{2}$$

In Equation (2), ER represents annual emission reductions; BE represents annual greenhouse gas emissions generated by farmers in the benchmarking context; PE represents emissions generated by farmers each year in the project situation.
2) Carbon sequestration saved from deforestation

$$GZ = 1.63 \times 0.2727 \times S \times B \tag{3}$$

In Equation (3), GZ is the annual carbon sequestration of vegetation, 1.63 is the coefficient, 0.2727 is the mass fraction of carbon in CO_2, S is the stand area, and B is the annual net productivity of the stand (Mai *et al.* 2018; Fan 2016)

2.3   *Calculation of carbon sequestration in wetland projects*

The calculation can be conducted according to the following classifications (Ma *et al.* 2021):

1) Carbon sequestration in lake wetlands:

$$\begin{aligned} &\text{Annual carbon sequestration of Lake Wetland} \\ &= \text{Corresponding carbon sequestration rate in provinces} \\ &\times 365 \times \text{Lake wetland area in the year} \end{aligned} \tag{4}$$

2) Carbon sequestration in marsh wetlands:

$$\text{Annual carbon sequestration of mar}s h \text{ wetlands}$$
$$= \text{Carbon sequestration rate corresponding to mar}s h \text{ wetland types} \qquad (5)$$
$$\times 365 \times \text{Area of mars } h \text{ and wetlands in the year}$$

3) Carbon sequestration in constructed wetlands:

$$\text{Annual carbon sequestration of constructed Wetland}$$
$$= \text{Carbon sequestration potential} \qquad (6)$$
$$\times 365 \times \text{Constructed wetland area in that year}$$

2.4 *Calculation of carbon sequestration in photovoltaic projects*

$$C = \sum_{m=1}^{n} Sm \times Ie \qquad (7)$$

$$Sm = (Qt + v) \times T \qquad (8)$$

In Equations (7) and (8), Sm is the annual grid-connected electricity of photovoltaic power stations (kWh); Ie is the carbon emission coefficient of electricity ($kgCO_2/kWh$), which may vary by year and region; Qt is the total grid-connected photovoltaic power of each province in the previous year; v is the newly added grid-connected photovoltaic power of each province in the current year; T is the time coefficient reflecting the local effective sunshine hours (Chen *et al.* 2017; Huang *et al.* 2014; Sun *et al.* 2014)

2.5 *Methods of measuring economic effects*

Since the income obtained by farmers in the CCER projects will be directly reflected in their income, this study selected the per capita disposable income and per capita wage income of rural residents as the result variables of economic effects.

2.6 *Sample selection and data sources*

Hubei province, as one of China's carbon trading pilots, started carbon trading, including CCER trading, in April 2014. Hubei has the most complete CCER trading system and the most active practices. Additionally, the level of economic development and ecological protection in Hubei province is comparable to the national average. Therefore, Hubei province was chosen as the treated group. After excluding the remaining seven carbon trading pilot regions, Hong Kong, Macao, and Taiwan, 24 provinces were selected as the control group.

The data mainly comes from the statistical databases of the National Bureau of Statistics, the National Energy Administration, and the statistical yearbooks of local provinces ranging from 2004 to 2020.

3 EMPIRICAL RESEARCH

3.1 *Synthetic control method*

In this paper, the influence of CCER trading policy on rural economic and ecological effects in Hubei province was studied by using the synthetic control method. The policy implementation point was set at the beginning of the trading in 2014.

The ecologically and economically related control variable sets A1 and A2. The result variable sets B1 and B2 were collected from 2004 to 2020. The control variables and result variables related to ecological and economic effects in the control group from 2004 to 2013 (before the implementation of the policy) were matched with the optimal weight. The weight vector of structural synthesis control is as follows.

$$w \equiv \left(w_2, \ldots, w_{j+1}\right)^T \tag{9}$$

Calculating the evolution process of the explained variables in the synthetic control group during the sample period, we got the sequence of the explained variables.

$$y^* = Y_0 \cdot w \tag{10}$$

Where Y_0 is the matrix of the values of the interpreted variables in the control group during the sample period, y^* is the synthetic Hubei explained evolution series of the variables.

$$\alpha_t = y_t - y_t^* \tag{11}$$

Among them, y_t is the explanatory variable of Hubei province after the implementation of policy; y_t^* is the explanatory variable of Hubei province without policy intervention, and the difference α_t is the intervention effect of policy on the economy.

3.2 Benchmark regression analysis

3.2.1 Ecological effects

Table 1 shows that the synthetic control group was constructed as a convex combination of donors (i.e., photovoltaic project) by minimizing a weighted sum of squared deviations for the matching variables. Among them, Sichuan province accounted for the largest weight, reaching 0.63, while 19 provinces did not contribute to the synthesis.

Table 1. Optimal weight combination (photovoltaic project).

Provinces	Weight	Provinces	Weight	Provinces	Weight	Provinces	Weight
Hebei	0	Qinghai	0	Hainan	0.	Zhejiang	0.236
Shanxi	0	Inner Mongolia	0	Sichuan	0.63	Anhui	0
Liaoning	0	Guangxi	0.017	Guizhou	0	Jiangxi	0
Jilin	0.117	Tibet	0	Yunnan	0.024	Shandong	0
Heilongjiang	0	Ningxia	0	Shaanxi	0	Henan	0
Jiangsu	0	Xinjiang	0	Gansu	0	Hunan	0

The assessment results of ecological effects are shown in Figures 1–8. The real line showed the ecological effects of the treated group, and the dotted line showed the ecological effects of the synthetic control group. The difference between the actual and synthetic values of these variables was close to each other, which indicated that the model was well-fitting.

Our research showed that the implementation of CCER had produced the most significant ecological effects on the biogas project in Hubei province: the carbon sequestration in 2014-2017 was 18.19% higher than that of the synthetic group. The carbon sequestration of photostatic, wetland, and forestry in Hubei was 3.41%, 2%, and 0.57% higher than synthetic Hubei on average. Forestry carbon sequestration had a relatively large base, so an increase of 0.57% might also indicate a remarkable effect.

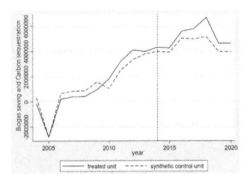
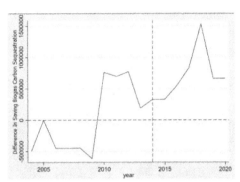

Figure 1. Comparison of biogas saving and carbon sequestration between Hubei and control groups.

Figure 2. Difference in saving biogas carbon sequestration.

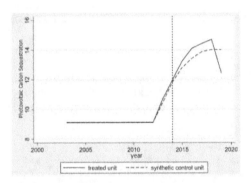
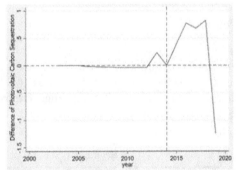

Figure 3. Comparison of photovoltaic carbon sequestration between Hubei and synthetic Hubei.

Figure 4. Difference in Photovoltaic carbon sequestration.

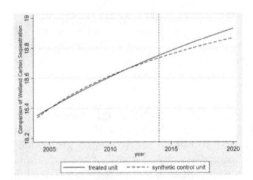
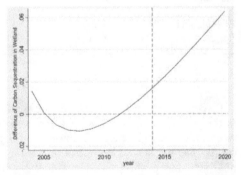

Figure 5. Comparison of wetland carbon sequestration between Hubei and control groups.

Figure 6. Difference in carbon sequestration in wetland.

To sum up, the CCER trading policy did increase the amount of carbon sequestration. That is to say, it had a positive impact on the carbon sink, and the order of its impact was biogas, photovoltaic, wetland and forestry. This conclusion may provide a reference for optimizing the CCER project layout. Resources should be put into the improvement of

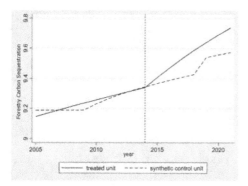 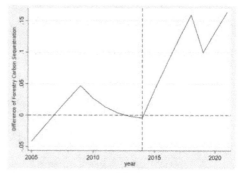

Figure 7. Comparison of forestry carbon sequestration between Hubei and control groups.

Figure 8. Difference in forestry carbon sequestration.

biogas facilities first to obtain greater carbon sequestration efficiency. Then, photovoltaic, wetland and forestry projects will be improved to maximize their carbon sequestration potential.

3.2.2 *Economic effects*
In terms of economic effects, Table 2 shows the composition of the synthetic control group, in which Hunan occupied the largest weight, followed by Shaanxi, Zhejiang, Xinjiang, Jiangsu, Hainan, and Heilongjiang. The other 17 provinces did not contribute to the synthesis of Hubei.

Table 2. Optimal weight combination constructed by the synthetic control method.

Provinces	Weight	Provinces	Weight	Provinces	Weight	Provinces	Weight
Hebei	0	Qinghai	0	Hainan	0.044	Zhejiang	0.169
Shanxi	0	Inner Mongolia	0	Sichuan	0	Anhui	0
Liaoning	0	Guangxi	0	Guizhou	0	Jiangxi	0
Jilin	0	Tibet	0	Yunnan	0	Shandong	0
Heilongjiang	0.021	Ningxia	0	Shaanxi	0.261	Henan	0
Jiangsu	0.047	Xinjiang	0.075	Gansu	0	Hunan	0.383

By comparing the data of Hubei and synthetic Hubei each year before the implementation of the policy, we found that the difference between the treated group and the control group was small. The difference was lower than the actual mean value of the control group, indicating that the synthetic control group had replicated the growth path of per capita disposable income of farmers in Hubei before the implementation of the CCER trading policy well.

Based on the above results, we further discovered that before the implementation of the CCER trading policy, the real line and the dotted line coincided with each other, indicating that the model was well-fitting. After the implementation, the per capita disposable income of real rural residents and the per capita wage income of rural households in Hubei province were significantly higher than the control group, which indicated that the rural CCER trading had increased the income of rural residents in Hubei province. The above is shown in Figure 9–12.

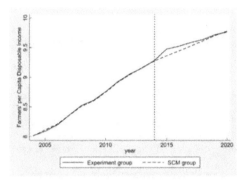
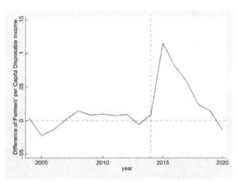

Figure 9. Farmers' per capita disposable income.

Figure 10. Difference in farmers' per capita disposable income.

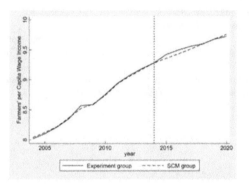
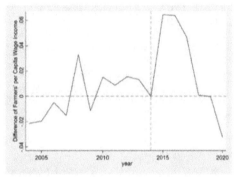

Figure 11. Farmers' per capita wage income.

Figure 12. Difference in farmers' per capita wage income.

3.3 Robustness test

To ensure the validity of the above empirical conclusions, we adopted the placebo test method to test the robustness of the results. The results showed that the MSPE ratio of Hubei was the highest, reaching 20.05, and the MSPE ratios of other provinces were lower than that of Hubei. This demonstrated that the probability of obtaining the same MSPE ratio as Hubei in the randomized arrangement was 1/24. It also indicated that the hypothesis that the CCER trading policy may positively impact Hubei could be accepted at the 96% significance level and was not caused by chance factors. Therefore, the results led to a certain degree of robustness.

4 CONCLUSION

Based on China's carbon peak and carbon neutrality targets, we empirically studied the ecological and economic effects of CCER trading by adopting the synthetic control method. The results showed that CCER trading could greatly increase regional carbon sequestration and rural residents' income, bringing positive ecological and economic effects to Hubei province. For this reason, resuming CCER trading may be what we can consider doing next.

REFERENCES

Abadie A. and Gardeazabal J. (2003), The Economic Costs of Conflict: A Case Study of the Basque Country, *American Economic Review*, 93(1):112–132.

Chen Y.W., Wang X.Y. and Gao H.Y. (2017), Development and Trading of Carbon Emission Reduction for Photovoltaic Power Generation Projects,. *China Energy*, 39(12):40–44.

Fan M. (2016) *Research on the Construction and Pricing Strategy of Rural Biogas Supply Chain based on Carbon Emission Reduction*, Nanchang University.

Guide for Country Reporting for FRA 2015 (2015). https://www.fao.org/forest-resources-assessment/past-assessments/fra-2015/zh/

He Y., (2005), A Review of Forest Carbon Sequestration Estimation Methods,. *World Forestry Research*, (01):22–27.

Huang J., Wang Y. and Wang Y.X. (2014), Discussion on the Calculation Method of the Environmental Effect of Energy Saving and Emission Reduction of Photovoltaic Power Generation, *North China Power Technology*, (10):67–70.

Ma Q.F., Yan H. and Li W. (2021), Analysis of Carbon Sequestration and Oxygen Release Service Functions of Wetland Ecosystems in Jilin Province, *Journal of Ecology and Environment*, 30(12):2351–2359.

Mai W.J., Li L.L. and Luan J.S. (2018), *Evaluation of Carbon Emission Reduction Effect of Rural Household Biogas Digested, Arid Zone Resources and Environment*, 32(02):75–80.

Sun Y.Y., Hou J.L. and He G.Q. (2014), A Method for Calculating Carbon Emission Reductions from Photovoltaic Power Generation Considering the Impact of Standby, *Power System Automation*, (17): 177–182.

Xue L.F., Luo X.F. and Wu X.R., (2016), Carbon Sequestration Efficiency in Four Major Forest Areas in China: Measurement,. *Drivers and Convergence, Natural Resources Report* 31(08):1351–1363.

Civil Engineering and Energy-Environment – Gao & Duan (Eds)
© 2023 the Author(s), ISBN 978-1-032-56059-5

Analysis of the siltation trend of beach on the south bank of Hangzhou Bay based on remote sensing images in China

Bohu Zhang[*] & Taoxiao Chen
Zhejiang Institute of Hydraulics & Estuary (Zhejiang Institute of Marine Planning and Design), Hangzhou, Zhejiang, China
Key Laboratory of Estuarine and Coastal of Zhejiang Province, Hangzhou, Zhejiang, China

ABSTRACT: Based on the Landsat remote sensing images of the beach on the South Bank of Hangzhou Bay in the four periods of low tide level in 1984, 1995, 2004, and 2015, the beach shape in each period is extracted by the water edge method, and the change characteristics of the beach and its response to the sharp decrease of sediment source are analyzed. The results show that during the 31 years from 1984 to 2015, the beach on the South Bank of Hangzhou Bay showed a stable siltation trend, and its evolution trend has no obvious response to the reduction of sediment from the Yangtze River.

1 INTRODUCTION

Beach is an important part of the coastal zone and a necessary condition for the maintenance of the estuarine and coastal ecosystem. Mastering the change process of the beach is of great significance for formulating reasonable beach utilization and protection strategies (Zhang 2012). Due to the high dynamics of tidal flats under the action of periodic tides, the conventional on-site workload is large, the cost is high, and risks exist (Zhang & Chen 2010). The remote sensing technology of extracting tidal flat information through the water edge line has the advantages of a wide observation range, a large amount of information, periodic period, etc. It is an effective means to study the evolution of tidal flats. In recent decades, remote sensing has been widely used in beach wetlands and other related research (Adam *et al.* 2010; Fan *et al.* 2017; Klemas, 2011; Liu *et al.* 2013). Taking Landsat remote sensing images as the data source, this paper dynamically monitors the beach on the South Bank of Hangzhou Bay based on the water edge method. It analyzes the response of the temporal and spatial evolution of the beach to the sharp reduction of sediment into the sea of the Yangtze River.

2 OVERVIEW OF THE STUDY AREA

Hangzhou Bay, located in the south wing of the Yangtze River Delta, is a world-famous strong tidal estuary. The sand from the Yangtze River has shaped the vast tidal flat wetland on the South Bank of Hangzhou Bay under the interaction of land and sea. It is mainly located between the West Third sea and Huangshan Mountain. It is arc-shaped and prominent in shape. The tidal flat area within the -5 m contour line is nearly 300 km^2. The location of the study area is shown in Figure 1. The tide in the study area is semi diurnal tide,

---

[*]Corresponding Author: zhangbohu@aliyun.com

DOI: 10.1201/9781003433651-65

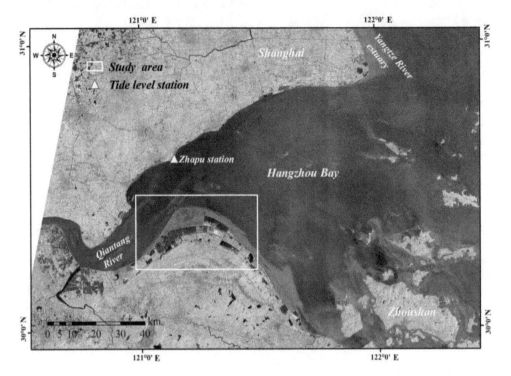

Figure 1. Study area.

the wave action is relatively weak, and the beach sediments are mainly silt. In recent 30 years, due to the implementation of large-scale siltation promotion and reclamation projects, the coastline has expanded significantly to Hangzhou Bay. Since 1984, it has extended an average of 4.4 km to the sea, with a cumulative reclamation area of about 230 km^2 (Zhang & Zhao 2016).

3 REMOTE SENSING DATA AND RESEARCH METHODS

3.1 *Remote sensing data*

Considering the long-term changes, the Landsat remote sensing data of the United States is adopted. Since its launch in 1972, the remote sensing data has accumulated more than 40 years of earth observation archives data, which is conducive to the analysis of long-term changes. The spatial resolution is 30 m, and the data resolution can meet the needs of this study. Four groups of Landsat TM / OLI images (Table 1 and Figure 2) in 1984, 1995, 2004,

Table 1. List of remote sensing image data.

Satellite	Sensor	Imaging date	Water level of Zhapu station at imaging time
LANDSAT5	TM	1984/09/14	−2.5 m
LANDSAT5	TM	1995/08/12	−2.6 m
LANDSAT5	TM	2004/08/20	−2.4 m
LANDSAT8	OLI	2015/08/03	−2.5 m

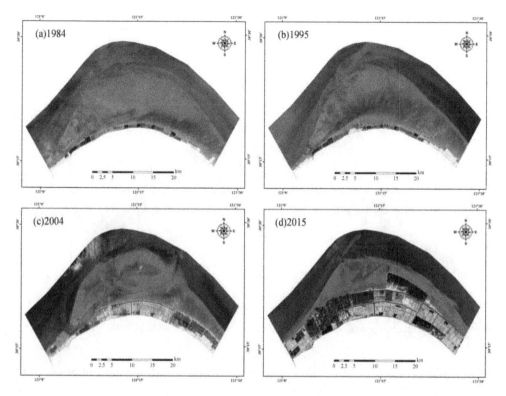

Figure 2. Remote sensing beach morphology from 1984 to 2015.

and 2015 were selected. The imaging time is different, and the product level is slightly different. The data is ll t (level 1t) product after accurate geometric correction and terrain correction. Before extracting the water edge line of the beach, the image data is atmospherically corrected. At the same time, in order to better reduce the impact of geometric distortion on the water edge line extraction results, the images in other years are subject to geometric fine correction based on the Landsat OLI image on August 3, 2015. Finally, the images after geometric fine correction are subject to edge enhancement.

The data selection not only considers the time interval of about 10 years but also takes into account the important time nodes of the change of water and sediment entering the sea of the Yangtze River. At the imaging time of phase IV remote sensing data, the Zhapu station near the long-term tide level station in the sea area is the low water level in spring tide (see Figure 1 for the location of the tide level station) and the instantaneous water level is about -2.5 m, which can support spatio-temporal comparison.

3.2 *Water edge extraction*

The beach water edge line is the dividing line between the seawater and the exposed beach, which presents different hue and texture characteristics in the remote sensing image. The supervised classification based on a support vector machine (SVM) has a good effect and high efficiency. This method is used for image binarization. After removing the noise, the Sobel algorithm is used to enhance the edge information of the waterline and finally combined with visual interpretation to modify and extract the final waterline (Su *et al.* 2018). Figure 3 shows the overall change of beach shape on the South Bank of Hangzhou Bay from 1984 to 2015, extracted from the remote sensing water edge. The beach area each year is

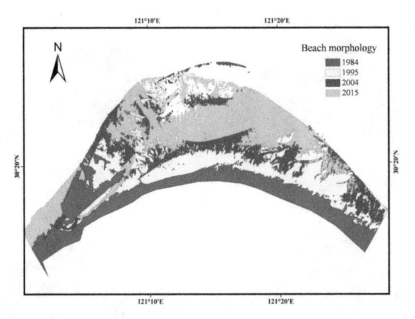

Figure 3. Overall change map of beach area (Measured by the coastline in 1984).

measured based on the 1984 coastline to reflect the comprehensive impact of beach reclamation and natural change on the beach.

4 RESULTS AND DISCUSSION

As can be seen from Figure 3, the beach on the South Bank of Hangzhou Bay shows a silting trend as a whole. According to the data, the beach areas in 1984, 1995, 2004, and 2015 were 180 km^2, 288 km^2, 317 km^2, and 430 km^2, respectively. The sediment entering the sea from the Yangtze River is the main sediment source in Hangzhou Bay (Zhang *et al.* 2022). Figure 4 shows the relationship between the beach change on the South Bank of Hangzhou Bay and the average annual sediment discharge into the sea of the Yangtze River. It can be seen from the figure that from 1984 to 2015, the amount of sediment entering the sea of the Yangtze River decreased from 480 million tons to 120 million tons per year. However, the temporal and spatial change of the beach did not respond significantly to the sharp decrease of sediment entering the sea of the Yangtze River.

The sustained silting and rising trend of the south bank beach should be closely related to the hydrodynamic conditions, reclamation of land from the sea, siltation promoting guide dike, and other human activities in the area. The overall sediment transport path of Hangzhou Bay is from north to south, and the sediment is mainly suspended sediment. The sediment entering the sea of the Yangtze River has decreased sharply. However, the sediment reservoir deposited in the sea area outside the Yangtze River estuary for thousands of years is gradually becoming the main sediment source of Hangzhou Bay. The sedimentation of sediment carried by the ebb tide on the side beach leads to continuous siltation of the coating surface. At the same time, the beach reclamation project generally adopts the method of promoting siltation first and then enclosing the embankment. By building the embankment perpendicular to the coastline, the sediment is deposited in the hidden areas of the bank sections on both sides and under relatively weak hydrodynamic conditions, which further promotes the continuous siltation of the beach surface.

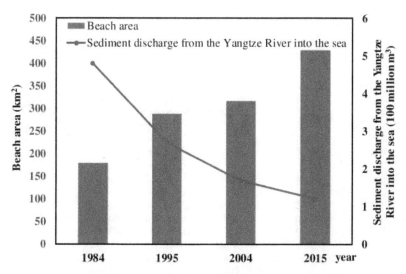

Figure 4. Relationship between the change of beach area on the South Bank of Hangzhou Bay and the amount of sediment entering the sea of the Yangtze River.

5 CONCLUSION

Based on the Landsat remote sensing images of the beach on the South Bank of Hangzhou Bay in the four periods of low tide level in 1984, 1995, 2004, and 2015, the beach area on the South Bank of Hangzhou Bay was dynamically monitored. The beach area was 180 km², 288 km², 317 km², and 430 km² in 1984, 1995, 2004, and 2015, respectively, showing an overall siltation trend. From 1984 to 2015, the sediment discharge of the Yangtze River into the sea decreased from 480 million tons/year to 120 million tons/year, but the response of the temporal and spatial change of the beach to the sharp decrease of sediment discharge into the sea is not apparent. Remote sensing data can improve the understanding of the evolution process of the tidal flat in Hangzhou Bay from the perspective of surface working mode. It can provide a data basis for dynamically mastering the spatial distribution pattern and siltation characteristics of the tidal flat in Hangzhou Bay under the new situation.

The medium and high-resolution satellite remote sensing data represented by Landsat data can provide an effective auxiliary means for the ground observation method of the spatio-temporal evolution of ground objects. The remote sensing observation results can improve the understanding of the evolution process of Hangzhou Bay tidal flat from the perspective of the area work model. It can also provide a data basis for dynamically grasping the spatial distribution pattern and the erosion and deposition law of tidal flat wetlands under the new situation.

ACKNOWLEDGMENTS

This work was financially supported by the National Natural Science Foundation of China (No. 41876095) and the Zhejiang Water Conservancy Science and Technology Project (RB2103).

REFERENCES

Adam E., Mutanga O. and Rugege D. (2010) Multispectral and Hyperspectral Remote Sensing for Identification and Mapping of Wetland Vegetation: A Review. *J. Wetlands Ecology and Management*, 18 (3): 281–296.

Fan Y.S., Chen S.L., Sun Y.G., *et al.* (2017) Modeling Formation and Evolution of Yuanyang Island in the Liaohe Estuary Using Waterline Approach. *J. Advances in Marine Science*, 35 (4): 579–592.

Klemas V. (2011) Remote Sensing Techniques for Studying Coastal Ecosystems: An Overview. *Journal of Coastal Research*, 27 (1): 2–17.

Liu Y.X., Li M.C., Zhou M.X., *et al.* (2013) Quantitative Analysis of the Waterline Method for Topographical Mapping of Tidal Flats: A Case Study in the Dong-sha Sandbank, *China. J. Remote Sensing*, 5 (11): 6138–6158.

Su G.B., Chen S.L., Xu C.L., *et al.* (2018) Quantitative Retrival of Tidal Flat Elevation with GF-1 Images in the Yellow River Mouth. *J. Marine Geology Frontiers*, 34(11): 1–9.

Zhang B.H., Pan C.H., Pan D.Z., *et al.* (2022) Study on Variation Trend of Surface Suspended Sediment Based on Remote Sensing in the Hangzhou Bay. *Journal of Sediment Research*, 47(3): 56–63.

Zhang B.H. and Zhao L.L. (2016) Reclamation Process Research on Hangzhou Bay Andong Beach Based on Landsat Remote Sensing Data. *J. Zhejiang Hydrotechnics*, (2): 16–19.

Zhang B.H. (2012) Analysis on the Characteristics and Influencing Factors of the Spatial-temporal Evolution of the Eastern Flat in the South Flank of Hangzhou Bay. *J. Zhejiang Hydrotechnics*, (6): 5–7.

Zhang Yang, Chen S.L. (2010) Super-resolution Mapping of Coastline with Remotely Sensed Data and Geostatistics. *Journal of Remote sensing*, 14 (1): 157–172.

Civil Engineering and Energy-Environment – Gao & Duan (Eds)
© 2023 the Author(s), ISBN 978-1-032-56059-5

Intelligent inspection of enterprise environment based on AR helmet

Yongliang Peng[*], Hongyu Zhao[*] & Ming Li
Department of Artificial Intelligence, Southwest Jiaotong University, Shanghai, China

ABSTRACT: With the vigorous development of China's manufacturing industry, hazardous waste inspection has become a necessary safeguard. However, the current hazardous waste inspection mainly relies on human experience and cannot efficiently and accurately identify violations. AR helmet is an integrated helmet with a variety of intelligent interactive functions. Based on the optical waveguide, 5 G, the Internet of Things, artificial intelligence, and other technologies, it provides a fast and convenient new method for enterprise inspection, making enterprises completely break away from traditional. In the inspection mode, the inspection work becomes more intelligent, which effectively improves the intrinsic safety management level of the enterprise. This paper specifically analyzes the product structure and product characteristics of AR helmets. It briefly describes the basic content of daily inspection of enterprises, the specific functions of the product, and the results of inspection and identification. The inspection work is changed from the traditional inspection method. The research is of great significance for enterprises to carry out intelligent and efficient inspection work.

1 INTRODUCTION

1.1 *Briefly*

The daily inspection of the enterprise environment is the daily work of the personnel of the EHS department of the enterprise. The traditional inspection work generally only records the inspection results. Paper documents or mobile APP software are used to monitor the operation of the environmental protection treatment and monitor equipment and hazardous waste. It is difficult to record the identification signs of the warehouse and the storage of hazardous wastes, and it is difficult to record the process of daily inspection. This research adopts the method of embedded inspection software in the AR helmet, which can plan the daily inspection route. The daily inspection process can be recorded and uploaded to the remote background in the format of pictures or videos, and more environmental perception sensors can be loaded. For example, infrared imaging sensors and PH detection sensors can break through the bottleneck of traditional daily inspections, liberate people's hands, establish an inspection mode based on the AR helmet model, realize automatic collection, recording, and analysis of inspection data, and discover hidden dangers and existing problems, which lay a solid foundation for the standardized operation of daily inspections.

1.2 *The necessity of daily inspection of the enterprise environment*

With the promulgation of the "Regulations on the Central Ecological Environmental Protection Supervision Work" issued by the General Office of the CPC Central Committee

---

[*]Corresponding Authors: 362584981@qq.com and hyzhao@swjtu.edu.cn

DOI: 10.1201/9781003433651-66

and the General Office of the State Council, it has become the norm for the national environmental protection department to conduct environmental protection supervision of enterprises, such as routine supervision, special supervision, and "look back" and other supervision work. The inspection contents include failure to publicize the environmental impact assessment, no environmental protection approval procedures, low-frequency noise or excessive noise, potential safety hazards in the operation of the motor unit, privately set up hidden pipes for sewage discharge, excessive COD in the filter tank, etc. This involves environmental protection procedures and water, sound, gas, solid environmental pollution, and other issues. In order to cope with the increasingly frequent environmental inspection work by the environmental protection department and also for the enterprise's own environmental protection management, the daily inspection of the enterprise environment is essential.

Mo Xiaoqing, in the article exploring the significance of boiler waste gas treatment in industrial enterprises, expounded that boiler waste gas inspection is necessary. This is because waste gas research can solve the problem of air pollution to a large extent and can also promote the pace of social environmental protection to continue to develop. Before Mo et al.'s work (2014), Yang (2021), in his discussion on the standardized management of hazardous waste warehouse managers, pointed out that the effective management of hazardous waste can not only ensure that the daily work is carried out according to the procedures, but also maintain the safety of production.

In the article "Management to Promote the Construction of a "Waste-Free City", it is listed in detail that since 2016, the country has successively issued many environmental protection policies, attaching great importance to the management of hazardous waste. China has made strict requirements in this regard, providing relevant decision-making for enterprises. Policy guidelines show the importance of hazardous waste inspection work (Zheng et al. 2019). In Zhou et al.'s study (2016), they explored the organic solid waste treatment strategies for offshore islands and reefs for coral sand improvement, described the benign nature of marine ecosystems related to the disposal of hazardous wastes on islands and reefs by introducing the strategy of "sea power" in the report of the 18th National Congress of the Communist Party of China. Recycling, especially the treatment of hazardous wastes on islands and reefs in the South China Sea, is of great concern to the country (Zhou et al. 2016). The inspection of marine hazardous wastes should not be underestimated (Zhou et al. 2016). Rathi et al. (2021) mentioned water pollution in the article Critical review on hazardous pollutants in the water environment: occurrence, monitoring, fate, removal technologies, and risk assessment. The presence and persistence of harmful pollutants such as dyes, pharmaceuticals, personal care products, heavy metals, fertilizers, and pesticides and their transformation products are serious environmental and health issues. Persistent hazards can be avoided only by addressing pollution at the source; therefore, inspection is necessary to protect the earth's ecology (Rathi et al. 2021). Mmereki et al. (2016) discussed the management of hazardous waste in developing countries in the article "The Management of Hazardous Waste in Developing Countries", with special emphasis on industrial hazardous waste. They concluded that in developing countries, the hazardous waste management system lacks a systematic approach to managing waste management programs and cannot effectively collect and manage waste and reduce the negative impact of these activities. The hazardous waste inspection activities lack effective management and implementation (Mmereki et al. 2016).

2 PRODUCT DESIGN

At present, most chemical enterprises still adopt the traditional manual record management method and paper forms for the characteristics of equipment inspection workload, wide scope, and complex entity attribute description. Inspection personnel tick the points without exception and describe the problem with exception. With the development of information technology, these inspection results will also be regularly reported to the enterprise's EHS information management system, which is convenient for later statistics, analysis, and visual display. In recent years, with the development of AR technology, the combination of AR and inspection has promoted the intelligence of daily inspection of enterprises. In order to solve the problems of irregularity and low efficiency of hazardous waste inspection, this study designs and manufactures smart helmets and helmets suitable for hazardous waste environmental inspections based on AR technology combined with text recognition, image recognition, speech recognition, and temperature detection methods. The overall architecture is shown in Figure 1:

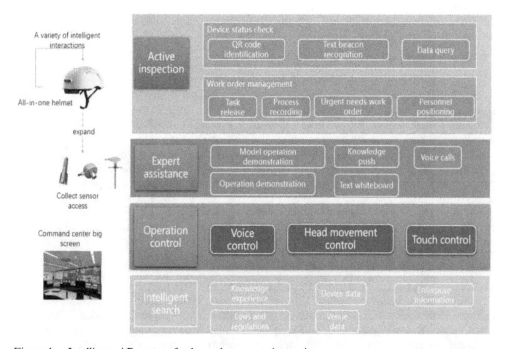

Figure 1. Intelligent AR system for hazardous waste inspection.

2.1 Introduction to AR helmet

2.1.1 Features

The AR helmet is an intelligent wearable device based on software and hardware technologies such as optical waveguide, 5G, the Internet of Things, artificial intelligence, etc. It combines high-performance cloud services so the staff can free their hands throughout the process and only need to control the device through intelligent voice. Equipment beacons can be recognized through images. Various sensors, such as thermal imaging and VOCs sensors, can be expanded to obtain on-site equipment information, the spatial distribution of environmental factors, etc. At the same time, through the camera of the device, what the on-site inspector sees can also be transmitted to the back end, realizing direct interaction and multi-level linkage between the command center and field personnel.

2.1.2 Product structure and future expansion

AR helmet is an all-in-one helmet with multiple intelligent interactive functions under 5G network coverage or offline environment. The product structure specifically covers four aspects: task distribution service, multimedia service, data information service, and expert support. Task distribution services mainly include on-site law enforcement and daily inspection services; multimedia services support audio and video calls, text exchanges, voice recognition, electronic whiteboards and frame synchronization services; data information services provide past knowledge and experience, audio and video materials, and inspection laws. Expert support means that you can see the live video scene on the expert interface and provide voice communication guidance, whiteboard guidance, model guidance, and text guidance to experts. The functions of the AR helmet are constantly improving. According to the specific situation of this product, the future expandable functions can be divided into three aspects, namely, switching the full offline operating environment, strengthening transmission security, and meter-level precision positioning. In the foreseeable future, the functions of AR helmets will become increasingly perfect. They will continue to develop into advanced, practical, reliable, and economical intelligent inspection equipment.

2.2 Contents of on-site inspection of enterprises

2.2.1 Daily inspection of waste gas treatment facilities
Spot inspection items for waste gas treatment facilities generally include pipelines, signs, activated carbon adsorption devices, fans, and the environment. For the pipeline, it is necessary to visually check whether it is blocked or damaged; for the identification plate, it is necessary to visually check whether it is dropped or damaged; for the activated carbon adsorption device, first, it is necessary to check and record the reading of the differential pressure gauge, and secondly, check whether the activated carbon is within the validity period, Finally, check whether the box is in good condition when the equipment is in use. In terms of the fan, it is necessary to visually check the fan belt and the oil level of the fan and check whether there is abnormal noise, vibration, and belt slip during its use. As for the environment, it is necessary to check whether there are combustibles around and whether the environment is clean. The inspection frequency of all items is once a day.

2.2.2 Daily inspection content of hazardous waste warehouse
The daily inspection of hazardous waste warehouses includes five aspects: hazardous waste storage facilities, hazardous waste storage containers, hazardous waste stacking, storage management, and storage facility safety protection. Among the common hidden dangers are unsealed warehouses, wrong or irregular hazardous waste warning signs, and untimely hazardous waste transfer. Intelligent inspection based on AR technology can effectively help inspectors identify risk points and propose countermeasures in a timely manner.

3 TEST RESULTS AND DISCUSSIONS

3.1 The reality of applying AR helmet inspection

3.1.1 AR helmet specific use function description
The functional design of AR helmets can be divided into five aspects, namely operation control, active patrol, expert assistance, intelligent data retrieval, and enhanced functions. Realize operation control functions include voice control, head movement control, and touch control; active patrol functions include equipment status inspection and emergency work order execution. Equipment status inspection is equipment QR code recognition, text beacon recognition, and equipment information consultation; expert auxiliary functions include a synchronous demonstration of model operation, voice calls, database entry browsing, and text whiteboard information synchronization. Intelligent data retrieval can retrieve knowledge, experience, laws, and inspection regulations. Enhanced functions include indoor and outdoor personnel positioning, heat source, and multivariate acquisition of noise.

3.1.2 AR inspection software function design

AR helmets can identify equipment beacons through images to obtain equipment asset information, operation information, etc. At the same time, through the camera of the device, what the on-site inspector sees can also be transmitted to the back end. Back-end experts can provide comprehensive expert support for personnel who go to on-site maintenance through the on-site assistance system, including two-way voice communication, data push, a content explanation based on cloud rendering (three-dimensional equipment model explanation, electronic whiteboard), text message prompts, and other multi-directional guidance functions. When it is found in the business system that there may be anomalies or hidden dangers at the site, business personnel often need to go to the site for investigation. The system sets up a "request list" function for such active inspection scenarios. The interface is shown in Figure 2. The platform initiates the designation of the content that needs to be surveyed, and the relevant staff brings mobile devices or AR devices to investigate and collect the situation.

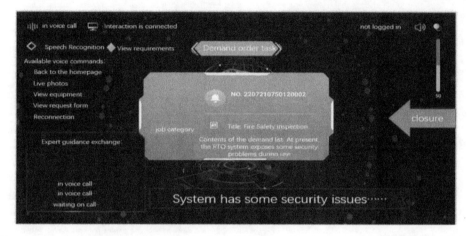

Figure 2. Send a request for a problem.

3.2 Analysis of inspection results

3.2.1 On-site identification and specification identification

There are two types of on-site signs: hazardous waste signs and hazardous waste storage facility signs. Hazardous waste labels need to indicate the main components of the hazardous waste, chemical name, hazard category, hazardous conditions and safety measures, the name, address, and contact number of the waste generating unit, contact person, and finally indicate the batch, quantity, and date of manufacture of the hazardous waste. Hazardous waste storage facility signs need to indicate the name of the affiliated company, responsible person and phone number, administrator and phone number, facility EIA approval, and facility building area. It should also include facility environmental pollution prevention and control measures, environmental emergency materials and equipment, and facility storage hazards waste list. During the identification process of on-site signs, if it is found that the characters of the signs are not clear and the items of the signs are incomplete, the inspector will fill in the inspection results of the signs in the AR software interface, record the abnormal situation and send it to the responsible person. The responsible person will carry out the inspection, as shown in Figure 3.

3.2.2 Environmental protection equipment operating status identification

The operating status of environmental protection equipment is mainly identified by thermal imaging. According to different types of environmental protection equipment, the specific

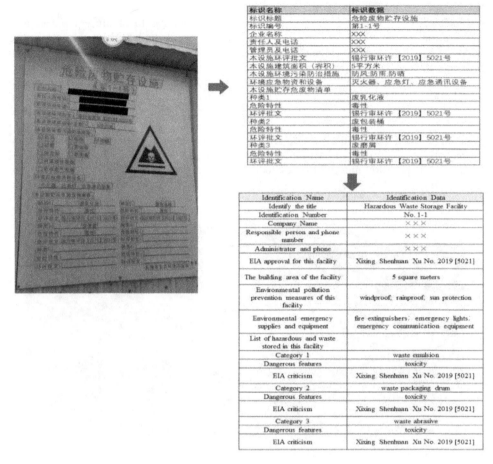

Figure 3. Schematic diagram of intelligent identification of hazardous waste identification.

identification content of environmental protection equipment is also different. However, it is necessary to identify whether the equipment can work normally when it is in normal use. If there is an abnormality, the inspection personnel need to fill in the specific fault of the equipment and the information about the spare parts that need to be replaced.

3.2.3 *Environmental recognition*

Environmental identification mainly identifies whether the hazardous waste storage warehouse and its surroundings are clean and tidy and contain combustibles. After visual inspection, the inspection personnel will score according to the standard. At the same time, they can note the problem to generate the environmental inspection result and send it to the relevant person in charge through the AR software.

4 CONCLUSION

The standardization and digitization of the daily inspection work of the enterprise environment is a difficult point for the EHS department of the enterprise. This study proposes that AR helmets assist enterprises in conducting intelligent inspections and provide information on intelligent diagnosis and intelligent decision-making for inspection work. The AR helmet can assist the inspectors in completing the daily inspection tasks of the enterprise with high quality. It can collect the image, sound, infrared thermal image, and other information of the inspection site in real-time, which is convenient for the query and reproduction of the fault problem. It improves inspection efficiency and reduces the maintenance cost that may be caused to the enterprise due to large-scale accidents.

REFERENCES

Mmereki D., Baldwin A., Hong et al. The Management of Hazardous Waste in Developing Countries. *Management of Hazardous Wastes*, 2016:39–51.DOI: 10.5772/63055.

Mo Xiaoqing, Lin Weiyi and Hu Yuanshun. Exploration on the Significance of Boiler Waste Gas Treatment in Industrial Enterprises. *Theoretical Research on Urban Construction*, 2014(13). DOI: 10.3969/j.issn.2095-2104.2014.13.1135.

Rathi B.S., Kumar PS., VO D.V.N. Critical Review on Hazardous Pollutants in Water Environment: Occurrence, Monitoring, Fate, Removal Technologies and Risk Assessment. *Science of the Total Environment*, 2021, 797. DOI: 10.1016 /j. scitotenv.2021.149134.

Yang Hangbo. Discussion on Standardized Management of Hazardous Waste Warehouse Managers. *Shandong Chemical Industry*, 2021, 50(3):145–146. DOI: 10.3969/j.issn.1008-021X.2021.03.055.

Zheng Yang, Jiang Wenbo, Li Ke et al. Standardizing Hazardous Waste Management and Promoting the Construction of a "no-waste City". *Environmental Protection*, 2019, 47(9):26–29.

Zhou Jihao, Shen Xiaodong, Zhang Ping et al. Study on the Organic Solid Waste Treatment Strategy of Offshore Islands and Reefs for the Purpose of Coral Sand Improvement. *Ecological Science*, 2016, 35 (6):176–181. DOI: 10.14108 /j. cnki.1008-8873.2016.06.025.

Civil Engineering and Energy-Environment – Gao & Duan (Eds)
© 2023 the Author(s), ISBN 978-1-032-56059-5

Study on mechanical properties of soil modified by ash in power equipment foundation under freeze-thaw cycle

Keyu Yue[*], Zhigang Wang, Yu Zheng & Hongdan Zhao
Liaoyuan Power Supply Company, State Grid Jilin Electric Power Co., Ltd, Liaoyuan, China

ABSTRACT: Carbonated soil is widely distributed in northeast China. Under the influence of salinization and the freeze-thaw cycles, the engineering performance of the soil body seriously affects the local regional construction and economic development. In order to ensure engineering safety and promote economic development, it is very necessary to conduct research on the saline soil improvement technology according to the regional characteristics of this region. Lime is an inorganic cementing material mainly comprises CaO and a small amount of MgO. After hydration, it forms a stable structure between the soil skeleton. The improvement of lime for carbonate soil has good results and good freezing and thawing properties. Therefore, this paper is based on China's general change and the relevant experience of good soil; after preliminary screening, the study finally determined to use lime and fly ash, two kinds of inorganic binder, to improve the saline soil test. By changing the amount of lime and fly ash for dry and wet cycle and freeze-thaw cycle test, and the improved saline soil in the case of different amendments content performance change, the study concluded that when the lime content of 8% and with a ratio of 1:2, the water stability and frost resistance is good, which provide a reference for the frozen area carbonate soil modification research.

1 INTRODUCTION

Salted soil is the soil formed under the action of various saline-alkali components (Komarov *et al.* 2016). At present, in the definition of saline soil at home and abroad, it is generally believed that when the soil in the salt content of the soil exceeds a particular value, it is defined as saline soil (Triffault-Bouchet *et al.* 2010). Salted soil mainly shows the engineering hazards such as subsidence, salt swelling, and corrosion, and according to the different types of site utilization, the limits given by each specification are also different. The main engineering hazards of saline soil include subsidence, salt swelling, erosion, and others (Xing *et al.* 2022; Xu *et al.* 2022). Decontamination refers to when the saline soil is rich in soluble salt in water, the soil body of soluble salt crystals dissolved in water, thus destroying the soil structure and strength, which can easily cause the uneven settlement of the foundation. The degree of dissolution is related to the type of soil, the type of soluble salt, the soluble salt content, and the distribution of soluble salt in the soil layer. The problem of salt swelling is mainly related to sulfate saline soil. Under the crystallization action of $NaSO_4$, the resulting $NaSO_4 \cdot 10H_2O$ crystal expansion is several times larger than the original volume, which makes the soil expand. The foundation damage caused by saline soil and salt swelling generally only occurs in shallow foundation-buried buildings. The damage caused by corrosion generally refers to the physical erosion of saline soil and saline groundwater into the material

---

[*]Corresponding Author: yang20220618@outlook.com

516   DOI: 10.1201/9781003433651-67

pores, such as foundations and pipelines, resulting in physical erosion and chemical corrosion of materials.

To study saline soil, Yang & Wang (2015), through the improvement of natural salt using saline soil related field research summary, combined with the practice of saline-alkali improvement in Xinjiang, put forward the combination of the natural salt drainage and water conservancy improvement comprehensive mode and realized the new mode of the sustainable development of irrigation agriculture. People choose quicklime, #325 ordinary Portland cement materials, as the improvement of saline soil curing material. The study found that the new polymer material SH mixed with quicklime. Cement-lime combination curing agents can effectively improve the water stability coefficient of saline soil for saline soil subgrade in the rainy season construction and extend the service life of the highway, which has essential application value. People used different curing agents to cure sulfate soil and chloride salt soil to study its dry shrinkage and wet swelling properties. It was found that replacing some cement parameters with a fly ash and slag mixture can optimize the hydration process of cement. At the same time, NaCl and Na_2SO_4 can weaken the chemical reaction degree of the hydration process. Researchers studied the influence of the physical and chemical properties of the freeze-thaw cycle on the improved saline soil of lime through cascade analysis and cation exchange amount. They analyzed the mechanism of carbonate soil after the freeze-thaw cycle through CT scanning, electron microscope scanning, and unlimited compressive strength (Chatterjee & Debnath 2020). Zhang *et al.* (2022) took the saline soil in the Songnen Plain in the northeast of China and conducted the freezing resistance test, conductivity test, and boundary moisture content test by setting the freeze-thaw cycles, lime mixing amount, and silica ash mixing amount. The results of the freezing resistance and conductivity test show that the mass loss of the modified soil is low after seven freeze-thaw cycles. After the maximum incorporation of 6% L, the conductivity is 3.56 times that of plain soil, but it is decreased by 35.3% with the increase of freeze-thaw cycles. With the increase of ash mixing amount, the liquid limit and plasticity index decrease, but the plastic limit is directly proportional to the amount of lime mixing. It is inversely proportional to the number of freeze-thaw cycles and the amount of silica ash mixing. The study found that incorporating lime will change the fine grading of saline soil, and adding lime will enhance the freezing resistance of saline soil.

Currently, most of the research on saline soil improvement in China is conducted from the agricultural perspective, while there is little research on saline soil improvement for engineering. The modified agents usually selected are cement, lime, fly ash, polymer materials, and multi-material mixture. Based on the relevant experience of generally improved soil in China, two inorganic combinations of lime and fly ash were used to improve the saline soil after preliminary screening. The durability performance of the modified saline soil was studied, and the modified saline soil subgrade packing performance under different amendments was also analyzed. The research in this paper provides the necessary scientific basis for using saline soil, which is instructive for solving other saline soil areas.

2 DESIGN OF SALINE SOIL IMPROVEMENT TEST SCHEME

Soft soil foundation has the characteristics of large initial uneven settlement deformation, low bearing capacity, poor water stability, and high compressibility. The soft soil layer has large natural water content, large plastic index, large natural pore ratio, small gravity, and small permeability, which is a fluid plastic state. The shear strength of soft soil is very low and poor, which cannot be directly used for engineering construction, foundation treatment, etc. Foundation treatment is needed before it can be used for engineering construction.

In recent years, the preloading method is also common for dealing with soft soil foundation. Soft soil contains a large amount of pore water, and the water content is large, so the soil strength is very low. Therefore, when the load acts on the soft soil foundation, it is very likely to form excess pore pressure. Under the action of pressure, pore water will be discharged to the place with low pressure. As more pore water is discharged, the soil becomes more compact, so the strength of the filling geology will slowly increase. The preloading method is used to consolidate the soft soil foundation by drainage. The specific operation method is as follows. Before the construction begins, we prepressure the soft soil foundation with stones, the building itself, or soil material (the requirement is equivalent to the design load) to generate excess pore water pressure inside. Then, we promote the acceleration of pore water discharge and the consolidation rate of soft soil. At the same time, attention should be paid to adding drainage channels in the foundation, which are generally vertical and can be plastic drainage plates or sand wells. When the strength of the foundation meets the design requirements, the load accumulated in advance can be removed, and formal construction can begin.

The composite foundation method is also very common when dealing with soft soil foundations. The deep mixing pile method is a representative one in the composite foundation. The deep mixing pile method uses machinery into the deep layer of soft soil, and cement is injected and forced to stir fully in the soil. The soft soil reacts with the cement and reduces the moisture. The soil can then condense, increase its strength, and has a certain carrying capacity. Among them, cement acts as a curing agent that solidifies the soil.

In the process of preloading, different from the preloading method, the rate of preloading is not strictly required, and the operation is relatively easy. The preloaded load can be loaded quickly at one time, and the soil consolidation speed is faster, which greatly shortens the construction period. The soil will not produce lateral deformation. There will be no noise or disturbance in the construction process. We do not need a large amount of stacking materials to save transportation costs. However, in the construction process, vacuum pumps should be used for air extraction, and the power supply connection should be maintained. Thus, the requirement for power supply is very high, and the cost of electricity consumption is also large. Furthermore, in the process of consolidation of soft soil using this method, at the intersection of the consolidation area and the unconsolidated area, the soil in the unconsolidated area will lead to the consolidation area. The horizontal direction is likely to occur cracks, and this horizontal cracking may have a certain impact on the surrounding buildings. The type and dosage of the curing agent are shown in Table 1.

Table 1. Type and dosage design of modified agent.

Modifier type	Dine design	Lime admixture/%	Lime and fly ash	Ratio number
Lime	The amount of lime mixed increases by 2%	2		A-1
		4		A-2
		6		A-3
		8		A-4
Lime and fly ash	The ratio of lime to fly ash is 1 : 1 and 1 : 2	2	1:1	B-1
		4	1:1	B-2
		6	1:1	B-3
		8	1:1	B-4
		2	1:2	B-5
		4	1:2	B-6
		6	1:2	B-7
		8	1:2	B-8

Two inorganic combinations of lime and fly ash are used to improve the subgrade filling of saline soil. Lime is tested on the index of lime after self-digestion. Two basic improvement methods of lime, lime, and fly ash were used, respectively, and the amount of lime mixed in the lime scheme was increased by 2%.

3 TEST METHOD FOR PERFORMANCE EVALUATION OF IMPROVED SALINE SOIL WITH INORGANIC COMBINATION MATERIALS

3.1 Dry and wet cycle test

The improved saline soil's dry and wet cycle resistance ability refers to its ability to resist the damage caused by water changes in the natural environment. It is one of the most critical indicators to improve the durability of the improved saline soil. Because it is generally located above the groundwater level, the groundwater is inhaled under the capillary action. It loses water through transpiration and evaporation, resulting in constant changes in the water content, and its strength also changes. The dry and wet cycle test is to simulate the water absorption-water loss cycle of the field road and land and to analyze the influence of the dry and wet cycle on the mechanical properties of the specimen. First, the specimen is put in the sink, and water is added about 2 cm away from the bottom of the specimen. After soaking for 24 hours, the surface is dried and put into a 20°C oven for one day, which is a cycle. The specimen strength was measured after six times.

3.2 Frozen-thaw cycle

This paper uses the slow freezing method according to the concrete frost resistance test standard. The test piece is formed according to the unlimited compressive strength molding method and maintained in the standard curing box until the specified age period. The test piece is weighed one day before the test age period, and then the test piece is soaked in $22 \pm 3°C$ water for 8 h, and the water depth is 2 cm higher than the top surface of the test piece. The dry surface of the test piece is put into the test box and then into the refrigerator. It is then frozen for 4 h at $-17\sim -20°C$. After the specimen is frozen, it is melted together with the specimen box into the water at $20 \pm 3°C$ for 4 h. This is a freeze-thaw cycle and is repeated. When five cycles are reached, the freeze-thaw test piece is taken out to weigh its mass and calculate the mass loss rate.

4 STUDY ON IMPROVED PERFORMANCE OF SALINE SOIL

4.1 Dry and wet cycle test

The standard health 6-month test piece was soaked in 20°C of water for one day and then dried in a 20°C oven for one day, which is a dry and wet cycle. After six dry and wet cycles, the unlimited compressive strength test was conducted to obtain the influence of the dry and wet cycle on the improved saline soil.

The test samples of A1, A4, B5, and B8 were selected for the dry and wet cycle tests. The test results are shown in Table 2 and Figure 1.

Table 2. Strength before and after six dry and wet cycles with different ratios.

Ratio number	Normal maintenance is mild / MPa	Intensity after six cycles	slip /%
A1	0.87	0.61	30
A4	1.28	1.03	20
B5	1.32	1.06	20
B8	1.97	1.69	14

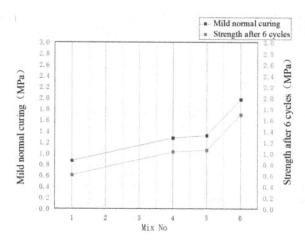

Figure 1. Comparison of dry and wet cycle strength in different ratios.

According to the analysis of the test results in Table 2 and Figure 1, the strength of the lime-modified saline soil decreased significantly after six dry and wet cycles. The strength reduction rate of the improved saline soil with ratio A1 reached 30%, with the most significant reduction rate. The strength reduction rate of the ratio of A4 improved saline soil by 20%, indicating poor water stability of the improved lime. The ratio of B8 lime and fly ash improved saline soil is reduced to 14% after six dry and wet cycles, which shows that the incorporation of fly ash improves the water stability of the improved saline soil and the water stability of the improved saline soil is better.

4.2 *Freeze and melt cycle*

The freeze-thaw cycle test uses five cycles, with quality loss as the evaluation index. The ratio of A1, A4, B5, and B8 was selected for the freeze-thaw cycle test, and the test results are shown in Table 3 and Figure 2.

Table 3. Results of the freeze-thaw cycle test.

Ratio number	Water absorption rate/ % for one day	Mass loss rate after freeze-thaw cycle/%
A1	25.24	58,1
A4	23.42	35.2
B5	29.89	11.8
B8	33.21	6.2

Table 3 and Figure 2 show that lime-improved saline soil has a low water absorption rate, significant mass loss rate, and poor freezing resistance after the frost-thaw cycle. For the modified saline soil of lime, the mass loss of ratio A1 is greater than that of ratio A4, indicating that the mixing amount of lime is directly proportional to its freezing resistance. Lime and fly ash have better frost resistance. With the increase of the coal ash mixing amount, the freezing resistance of the improved saline soil improves, which shows that the fly ash can increase its freezing resistance.

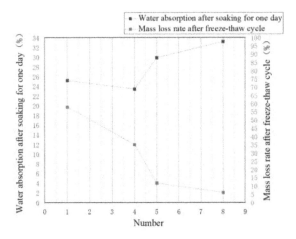

Figure 2. Freeze-thaw cycle test.

5 CONCLUSION

In this paper, we propose a fault segment location method based on CEEDMAN-energy relative entropy to address the difficulty of ground fault location in a small current grounding system. Through simulation, we verify that this method has obvious advantages over traditional methods, and the accuracy of location results is high under different fault initial phase angles and different transition resistances, which can achieve accurate fault segment location. Based on the relevant experience of generally improved soil in China, two materials, lime and fly ash, are finally determined to improve the saline soil after preliminary screening. The dry and wet cycle test shows that the strength loss of lime-modified saline soil is significantly more significant than that of lime and fly ash. The strength loss decreases with increased lime mixing in lime-improved saline soil. In the improved saline soil of lime and fly ash, with the increase of fly ash incorporation, the dry and wet resistance circulation capacity is significantly improved, and the strength loss is also reduced accordingly. The melting cycle test shows that the lime-modified saline soil has the most quality loss and poor freezing resistance. With the increase of lime mixing amount, the freezing resistance of lime-improved saline soil is also improved accordingly.

ACKNOWLEDGMENTS

This work was supported by the Science and Technology Project of State Grid Jilin Electric Power Co., Ltd. (Grant NO. 2022-17).

REFERENCES

Biswapriya Chatterjee and Sudipta Debnath. Cross-Correlation Aided Fuzzy Based Relaying Scheme for Fault Classification in Transmission Lines, *Engineering Science and Technology, an International Journal*, Volume 23, Issue 3,2020, Pages 534–543, ISSN 2215-0986.

Komarov I.A., Anan'ev V.V. and Bek D.D. (2016) Using the Cooling Capacity of Liquefied Natural Gas to Freeze Saline Soil that Contains Cryopegs. *Soil Mechanics and Foundation Engineering*, 53 (2): 132–138. doi:10.1007/s11204-016-9376-6.

Morteau B., Galvez R., Leroueil S et al. (2010) *Treatment of Salted Road Runoffs Using, and A Comparison of Their Salt Removal Potential*. doi:10.1520/STP48911S.

Xing S., Wu T., Li Y. and Miyamoto Y. (2022). Shaking Table Test and Numerical Simulation of Shallow Foundation Structures in Seasonal Frozen Soil Regions. *Soil Dynamics and Earthquake Engineering*, 159, 107339. https://doi.org/10.1016/j.soildyn.2022.107339

Xu S., Liu D., Li T., Fu Q., Liu D., Hou R., Meng F., Li M. and Li Q. (2022). Spatiotemporal Evolution of the Maximum Freezing Depth of Seasonally Frozen Ground and Permafrost Continuity in Historical and Future Periods in Heilongjiang Province, China. *Atmospheric Research*, 106195. https://doi.org/10.1016/j.atmosres.2022.106195

Yang and Wang. (2015). Current Situation and Improvement of Saline Soil Resources in China. *Shandong Agricultural Sciences*, 4, 125–130.

Zhang Yuan, Meng Fanxiang, Fu Qiyang, Wang Zongliang. Study on Improved Saline Soil Boundary Water Content of Lime Silicon Ash Under Freeze-Melting Cycle. *Jilin Water Conservancy*, 2022 (09): 9–14.

Zhang Yuan, Meng Fanxiang, Fu Qiyang, Wang Zongliang. Study on the Chemical Characteristics of Lime Silicon Ash Under Freeze-Thaw Cycle. *Henan Science and Technology*, 2022,41 (18): 98–102.

Civil Engineering and Energy-Environment – Gao & Duan (Eds)
© 2023 the Author(s), ISBN 978-1-032-56059-5

Improvement in the extraction and classification accuracy of vegetation

Xuefei Zhang*
Strasbourg Cedex France

ABSTRACT: Semi-natural factors are generally between urban areas and forests and play an important role in protecting the environment. The current vegetation extraction methods are relatively traditional, such as supervised classification, unsupervised classification, and object-oriented, and the overall extraction does not further subdivide the vegetation. The purpose of this paper is to automatically extract semi-natural factors, which has achieved the purpose of improving classification accuracy and reducing running time. This paper mainly uses two different algorithms Random Forest (RF) and Support Vector Machine (SVM) based on supervised classification and uses the height information of Digital Surface Model (DSM) data to improve the classification results. In this paper, classified images and precision statistics based on two different algorithms and filtering windows are obtained. The classification results prove that the combination of DSM and satellite image data improves the accuracy of traditional vegetation classification, and the classification accuracy is improved from 69% to 79%. In pixel-based vegetation extraction, RF provides better vegetation extraction results than SVM, with classification accuracy increased from 67% to 78%, and RF has a large advantage in runtime. In terms of spatial properties, the classification of semi-natural factors has achieved good results.

1 INTRODUCTION

1.1 *The important role of semi-natural factors in the ecological environment*

The erosion of biodiversity is partly due to the fragmentation of habitats of animal and plant species. In France, the regulatory framework of the "Trame verte et bleue (TVB)" was established by the Grenelle 2 law to maintain or reconstruct the terrestrial and freshwater ecological continuities and to form a coherent ecological network for species movement. (Alavoine-Mornas *2018*). Theoretically, this object is based on the different registers of landscape ecology, planning, or politics, and prefigures all the needs to preserve biodiversity. TVB is based on the principle that every species needs space to live, feed and reproduce. They also need to move between these spaces and rely on an ecological network of connectivity. The purpose of this regulation is, therefore, to protect vegetated areas and water resources, whatever they may be, by limiting the current trend towards fragmentation of territories through urbanization. Identified as one of the major threats to biodiversity, fragmentation leads to a reduction of the areas and numbers of habitats available for the species and their isolation (Fahrig 2003).

A solution to combat this fragmentation is then to restore a connection between habitats (Fahrig & Merriam 1985). Peri-urban outskirts areas are probably the areas of major interest in planning issues. The fragmentation of urban suburbs includes many semi-natural factors, such as hedges, isolated trees, antlers, forests, etc. They promote functional linkages between

---

*Corresponding Author: 3180100260@caa.edu.cn

DOI: 10.1201/9781003433651-68

habitats of various species, thus allowing the dispersal and migration of these species, but there are also significant contradictions between urbanization and cropland. The management of semi-natural factors is particularly important. Indeed, they are subject to regulations aimed at achieving economic management of space (Loi relative à la Solidarité et au Renouvellement Urbains (SRU) of 13 December 2000 supported by the law Grenelle II). Using "instruments de planification urbaine (SCoT)" and "Plan Local d'Urbanisme (PLU)" are incentives to reduce the consumption of natural and agricultural areas by encouraging densification of already urbanized or peripheral areas.

In this context, local and regional authorities need data and methods to identify the interest zone in local areas. In 2015, the project Alsace-Aval-Sentine (A^2S) began to reorganize. In 2016, The National Centre for Space Studies (CNES) launched the "Satellite Data Application" program, and Kalideos is the main data source platform for this program. This is to offer methods and tools for analyzing landscape dynamics for Urban/land use (Artificialization-Urbanization, agricultural cycle, deforestation, *et al.*). It needs to have two characteristics. One is to meet users' needs in a short time, and the other is to function complementary to existing products.

1.2 State of the art

1.2.1 Classification with multiple source information
The combination of multi-source information is mainly in two aspects: a combination of multi-source remote sensing images and that of auxiliary information and remote sensing images. For example, in a combination of multi-spectral image and DSM, DSM is a derivative of radar data, so it has a more accurate position and height information, but the object has less geometric information, and high-resolution images can provide very detailed object information, such as spectral features, textures, shapes, etc. But the edge of the DSM is usually very rough, so it does not involve the segmentation step. Therefore, DSM can only be used for classification.

Due to a large number of foreign materials with the same spectrum but different materials, as well as the disturbance factors of topography and soil moisture, the classification of forest vegetation is limited, but the application of auxiliary information such as elevation, slope, and aspect can improve the accuracy of vegetation classification (You & Wang 2003).

Huo *et al.* (2001) studied the influence degree of auxiliary information such as elevation, aspect, soil water content, and soil roughness on the accuracy of classification, and used a fuzzy algorithm to quantify the relationship between slope direction, elevation, and vegetation, through supervision and non-supervision. The fuzzy matrix is established by classification, and the results show that the three-dimensional information of the features can effectively improve classification accuracy.

1.2.2 Fuzzy classification
A fuzzy classification approach is usually useful in mixed-class areas and was investigated for the classification of suburban land cover from remote sensing imagery (Zhang & Foody 1998), the study of medium-to-long-term (10–50 years) vegetation changes (Okeke & Karnieli 2006) and the biotic-based grassland classification (Sha *et al.* 2008).

The fuzzy algorithm is generally applicable to mixed-pixel regions. Fuzzy classification is a kind of probability classification. In theory, probability-based classification is more suitable for mixed pixel regions. Because these units are interlaced, they cannot be simply divided into one class.

1.3 Aim of this paper

Different vegetation types may possess similar spectra, which makes it very hard to obtain accurate classification results either using the traditional unsupervised classification or supervised classification. Creating improved classification methods is always a hot research topic. At present, most of the research is directly extracting vegetation as a whole, lacking further subdivision of vegetation. With the application of ultra-high resolution remote sensing images, the vegetation community can be well distinguished, and the combination with the auxiliary data can improve the extraction accuracy. A DSM is usually acquired by LASER scanning or from stereoscopic pairs of aerial or satellite images. By contrast, a direct DTM acquisition traditionally requires more effort on the terrain or for scene object classification and the automatic derivation of a DTM from a DSM still faces difficulties. This paper combines ultra-high-resolution images with DSM data to improve extraction and classification accuracy. By extracting DTM data directly from DSM data to form normalized digital surface model (nDSM) data and adding nDSM data, we can improve extraction accuracy. And then by combining them with ultra-high resolution images, we can extract more subtle vegetation information, such as low-level semi-natural factors. Shrubs, single trees, or fences can also be better classified by using nDSM data.

2 STUDY AREA AND DATASETS

2.1 *Area study*

A mixed vegetation area in the northwest of Strasbourg, namely Zone 1 is selected as the research objective. The study area is part of the urban suburbs, with a range of 1 km by 1 km. Zone 1 has different types of semi-natural factors, and some of them exist in the city, so it is more conducive to the comparison of extraction accuracy. The landscape elements that we seek in this study to extract and classify are plant groups marked by great diversity in terms of their morphological and textural characteristics.

2.2 *Datasets*

Stands of isolated trees in Alsace are mostly orchards. For this type of fruit tree, it is estimated that the smallest crown in full turgescence vegetation is about 3 meters wide (less than 10 m^2 surfaces). Therefore, Très haute résolution spatiale (THRS) images with less than 2.5 m spatial resolution must be used. In a satellite image, an object is represented by a grouping of several pixels (Mass 2013). The minimum unit of representation of these tree elements will be directly influenced by the chosen image. The choice of the chosen image is also limited by other constraints such as the entire recovery of our study area, the acquisition date respecting an adequate seasonality, a low rate of engagement, or presenting the least possible shadow effects possible, we finally chose two images multi-spectral images (MS) and panchromatic image (PAN) SPOT 7 dating from 30 September 2015 covering our entire study area. Its specific feature attributes are shown in Table 1.

Table 1. Characteristics of SPOT 7(Source: Kalideos, Airbus).

					Angle		
Zone géographique	Capteur	Date	Heure	Mode	d'incidence	Caractéristiques spectrales	Résolution spectrale
			10:06:29	Pan	12.3222	0,450-0,745µm Bleu:0.450 - 0.520µm	1,25 m
Strasbourg	SPOT 7	30.09.2015	10:14:50	MS	16.0527	Vert:0.530 - 0.590µm Rouge:0.625 - 0.695µm Infrarouge proche:0.760 - 0.890µm	9,48 m

Their resolutions are 1.25 (Panchromatic, PAN) meters and 9.48 (Multi-Spectral, MS) meters, and their combined image resolution is 1.25 meters. The DSM data is provided by Group Atelier 3D du CNES, which is generated by the Pléiades satellite imagery. The pixel size is 1 and the time is September 22, 2016. The pixel value of the DSM is matched with that of the fused spot7 image, and the image accuracy is not generated due to high and low resolution. We also have the data of BD TOPO OF IGN (2011), which has very good geometric quality IRC 50 cm.

3 METHODOLOGY

The proposed scheme includes five parts:

3.1 *Data pre-processing*

The data pre-processing stage consists of two major steps: (1) processing and correction of no data portion in original DSM data and (2) nDSM generation from DSM.DSM contains some errors (No data) due to incorrect image matching or inefficient morphological filtering of high objects. (Sohel 2005), Consequently, we need to mark the No-data area in DSM in spot7. The purpose is to prevent the 2D data from overlapping the pixels of No-data in the DSM, resulting in errors in the extraction results. First, the No-data mask will be extracted from the original DSM, and it will be fused with the original spot7 to obtain the complete mask data. At the same time, the no-data mask and the final mask data need to be consistent to facilitate the formation of nDSM and the creation of the final classification image.

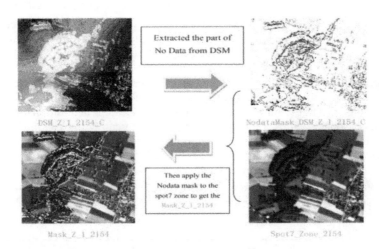

Figure 1. No Data mask processing.

3.2 *Establishment of nDSM*

It is very important to extract DTM directly from DSM, otherwise, there will be a height difference, that is the minimum value of DSM minus DTM is not equal to zero. This means that there is an error in the DSM or the DTM. Maybe these results are from a problem in the DSM geoid, or it is caused by different satellite sensors. To compensate for the error, it is especially important to extract DTM directly from DSM. Different from directly acquiring DSM data from LiDAR, there are currently few studies that automatically extract DTM from DSM. Firstly, close gaps (which is based on the nearest neighbor) are mainly to find and close No-data, then get DSM rempli, and then calculate the slope intermediate value to provide a threshold for the separation of the ground and the ground below; DTM Filter is to separate the surface and the ground, thus getting DTM, use close_gaps_with_spline again to make the image smoother, get the final DTM, and then use the raster calculator to get nDSM.

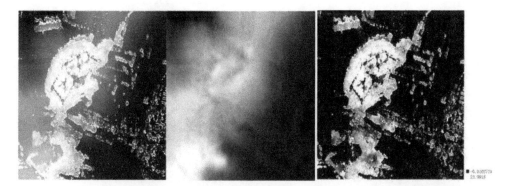

Figure 2. Result of extracting DTM from DSM:1) DSM, 2) DTM, 3) nDSM (left to right).

Finally, the nDSM data and the original spot7 data are fused to obtain the final classified image (Full feature image) with vegetation height, as shown in Figure 3.

Figure 3. Full feature image (nDSM & spot7).

3.3 *Extraction*

To compare the accuracy, the experiments were divided into two groups: (1) direct extraction; (2) adding nDSM for extraction. Whether it is direct extraction or nDSM extraction, the process is the same. The only difference is that the source data is different. Spot 7 was used in direct extraction, nDSM extraction is the full feature that is superimposed on No-data and nDSM. Figure 4 shows the overall extraction step.

3.3.1 *NDVI*
NDVI (Normalized Difference Vegetation Index) can turn multi-spectral data into a single image band for displaying vegetation distribution. Higher NDVI indicates more green vegetation.

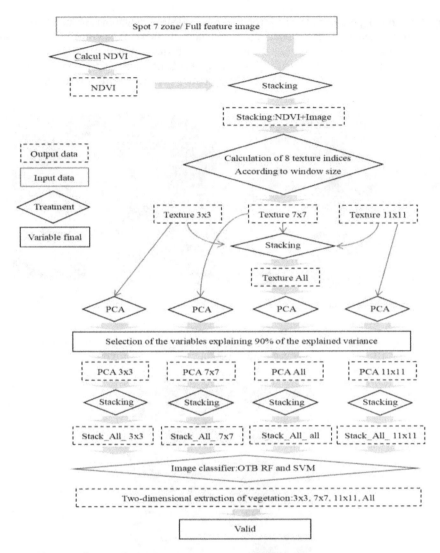

Figure 4. The overall extraction steps.

3.3.2 *Calculation of texture indices (co-occurrence measures)*

A geographic object on a digital image is represented by a group of pixels and/or mixes. This object is then characterized by both a specific spectral value and its spatial arrangement (texture). Numerous studies have demonstrated the contribution of texture to the discrimination of the tree layer. Therefore, the creation of texture variables seems relevant. For this, a set of 8 texture indices implemented will be calculated according to three different window sizes: average, variance, homogeneity, contrast, dissimilarity, entropy, second moment, and correlation. Texture algorithms work like spatial filtering. Indices are calculated pixel by pixel according to the variability of the numerical values of the surrounding pixels according to a defined convolution window (Natural Resources Canada 2013). In our case, we choose 3 different window sizes: 3×3, 7×7, and 11×11. An "ALL" variable was created by summing all the indices for all window sizes.

3.3.3 *PCA*

The various bands of a multi-spectral image are often highly correlated, and their DN values and displayed visual effects are often similar. The principal component analysis removes more information between bands and compresses multi-band image information into a few conversion bands that are more efficient than the original band. In general, the first principal component (PC1) contains 80% of the variance information in all bands, and the first three principal components contain 95% of the information in all bands. Due to the irrelevance between the individual bands, the main component band can generate more colors and better-saturated color composite images.

3.3.4 *Presentation of the classifiers*

Two classifiers are used to distinguish the tree from the non-tree by SVM and RF. The results and performances of these two classifiers will be compared once the classifications have been validated.

3.4 *Classification*

For the classification of semi-natural factors, we have obtained two classification results, one is the classification without high information, and the other is the classification based on nDSM with high information. This method does not need to set the threshold in advance, only needs to set the relevant spatial attributes related to the vegetation. The spatial attribute is calculated according to the object (polygon) of the image segmentation and the combined result, so the band information is not needed, so it is unsupervised. Classification and each spatial attribute have their calculation formula and method, do not interfere with each other, and is a fully automatic classification. The limitation is that because it is an unsupervised classification, it has higher accuracy requirements for the results of previous segmentation and merging, which will directly affect the subsequent classification results, and cannot take into account the spectral characteristics of vegetation, and limit the classification conditions and thus the image classification results.

Table 2. Expression for spatial properties of vegetation.

Elongation	$area / (pi () * (($perimeter / (2* pi ()))))
Compactness	$perimeter / 2*sqrt (pi () * $area)
Morton	$area / (pi () * 0.5 * $perimeter)
Convexity	(bounds_height($geometry) / bounds_width($geometry)) * ($area / $perimeter)

4 RESULTS AND DISCUSSION

4.1 *Choice of classifier*

In the two-dimensional vegetation extraction results in Figures 5 and 6, RF has better detail processing than SVM, SVM has an excessive classification, and RF can handle the details well, such as grassland and forest in forest and the roads in the forest can be divided clearly. the SVM tends to encompass the contours of objects while the RF can sometimes give a "pepper and salt" appearance.

In the vegetation extraction map with nDSM, as shown in Figures 7 and 8, RF has been over-classified, especially in the vegetation in the city, there are many salts and pepper phenomena, and the SVM better clusters the vegetation information. This phenomenon mainly involves the separation of vegetation or it was caused by spectral confusion.

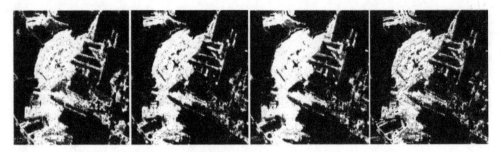

Figure 5. Classifications RF 2D of the vegetation(white)/non-vegetation(black) layer for Zone according to the different variables (left to right:3×3;7×7;11×11; ALL).

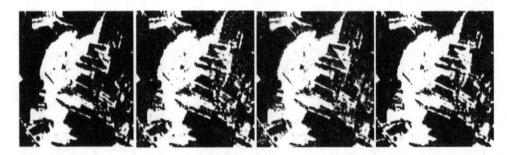

Figure 6. Classifications SVM 2D of the vegetation(white)/non-vegetation(black) layer for Zone according to the different variables (left to right:3×3;7×7;11×11; ALL).

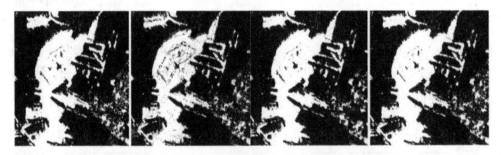

Figure 7. Classifications RF nDSM of the vegetation(white) / non-vegetation(black) layer for Zone according to the different variables (left to right:3×3;7×7;11×11; ALL).

4.2 *Choice of the variable*

It can be seen from the statistics obtained from the confusion matrix that RF is generally better than SVM, whether it is a general two-dimensional vegetation extraction or nDSM extraction.

As shown in Figures 9 and 8, for the RF algorithm, the 11×11 window shows the best classification accuracy, and in the time operation, the 11×11 window is the least time-consuming, because the RF algorithm is determined by the mode in the region. Therefore, RF analyzes the largest number of pixels of the same type in the study area and then classifies them into one class. The larger the region we choose, the more pixel values, the more the number of samples obtained, and the higher the classification accuracy. It is more

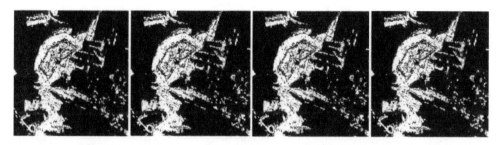

Figure 8. Classifications SVM nDSM of the vegetation(white) / non-vegetation(black) layer for Zone according to the different variables (left to right:3×3;7×7;11×11; ALL).

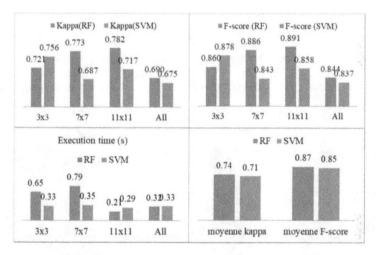

Figure 9. Matrix of confusion and validation of the vegetation layer (2D).

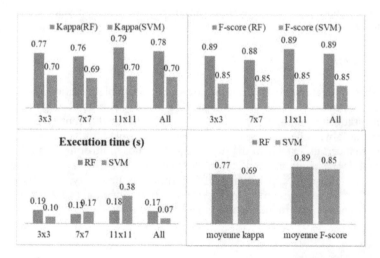

Figure 10. Matrix of confusion and validation of the vegetation layer (nDSM).

suitable for pixel-based extraction; SVM is a linear classification, which needs to find the best classification interval, which can be either spatial or planar, but when encountering a more complex mixed pixel area, or separating between two objects When the sex is poor, the classification of the SVM becomes weaker. The size of the different windows has little effect on the accuracy of SVM classification results, and the 3×3 window is the best.

In general, the algorithm and window size affect the classification results to some extent, but we can't rule out the artificial factors. In the sample selection, we still manually select the training samples and the verification samples. The sample selection of different qualities will directly affect the final classification results and accuracy.

4.3 *The distinction between semi-natural elements*

Figure 11 is a comparison of the whole classification result with validation, Figure 11–1 is a classification without height information, and Figure 11–2 is a classification with height information. From the perspective of classification, the biggest difference between the two images is the vegetation in the building. In Figure 11–1, almost all the vegetation in the building is classified as a fence, but in Figure 11–2, after adding the height information, part of the timber is distinguished. Due to the classification accuracy of the original image, we found the individual vegetation is classified as a small polygon produced by the phenomenon of "salt and salt" and does not match the vegetation of a single plant in validation.

Figure 11. The distinction of semi-natural elements (left to right 2Dclassification; nDSM classification; Validation: Digitalization of the tree).

5 CONCLUSION

RF in pixel-based vegetation classification shows a more reliable classification accuracy than SVM; the combination of DSM data and satellite imagery is a good way to improve classification accuracy and requires more advanced research; for different types of vegetation, classification has always been a research direction that people have neglected. This research fills this gap to a certain extent. In the future, the fusion of multi-source data is a direction, such as time series or a combination of satellite imagery and DSM data; machine learning is the future direction, and deep learning of algorithms is necessary. The classification of vegetation is not limited to monitoring and solving environmental problems. In the future, smart cities will need more refined management of vegetation.

REFERENCES

Alavoine-Mornas F. and Girard S. 2017. Green Belts in the Hands and Minds of Farmers: a Socioagronomic Approach to Farmers' Practices. *Journal of Rural Studies, Elsevier*, pp.30–38. ff10.1016/j.jrurstud.2017.09.005ff. ffhal-01899071f.

Fahrig L. 2003. Effects of Habitat Fragmentation on Biodiversity. *Annu Rev Ecol Evol Syst* 34: 487–515, DOI:10.1146/annurev.ecolsys.34.011802.132419.

Fahrig L. and Merriam G. 1985. Habitat Patch Connectivity and Population Survival. *Ecology* 66(6), 1762–1768.

Huo Hongtao, Wang Renhua et al. 2001. Research on Improving the Accuracy of Image Classification by Three-Dimensional and Related Auxiliary Information. *Journal of Beijing Forestry University*, 23 (2): 28–31.

Okeke F. and Karnieli A. 2006. Methods for Fuzzy Classification and Accuracy Assessment of Historical Aerial Photographs for Vegetation Change Analyses. *Part I: algorithm Development, Int J Remote Sens*, vol. 27 (pg. 153–76).

Ressources Naturelles Canada, 2013, *Classification et Analyse des Images, Consulté en Janvier* 2018 et Disponible en Ligne [http://urlz.fr/6vqy].

Sha Z., Bai Y., Xie Y. et al. 2008, Using a Hybrid Fuzzy Classifier (HFC) to Map Typical Grassland Vegetation in Xilinhe River Basin, Inner Mongolia, China. International Journal of Remote Sensing.

Sohel Syed, & Paul Dare. 2005. Fusion of Remotely Sensed Multi-spectral Imagery and Lidar Data for Forest Structure Assessment at the Tree Level. *Commission VII, WG VII/6*, p.1089–1094.

Sheeren D., Mass A., Ducrot D., Fauvel M., Collard F., Ma S. 2012. la télédétection Pour la Cartographie de la Trame Verte en Milieu Agricole. Revue Internationale de Géomatique, n° 4/2012, 539-563 L. Breiman, 2001, *Random Forests, Machine Learning*, vol. 45, p. 5–32.

Yu Xiaobin & Wang Lei 2003. Research on the Application of Auxiliary Information to Improve Forest Classification and Forest Zoning Capacity. *Journal of Beijing Forestry University*, 25(special issue): 41–42.

Zhang J. and Foody G.M. 1998. A Fuzzy Classification of Sub-Urban Land Cover from Remotely Sensed Imagery, *Int J Remote Sens*, vol. 19 (pg. 2721–38).

Civil Engineering and Energy-Environment – Gao & Duan (Eds)
© 2023 the Author(s), ISBN 978-1-032-56059-5

Retail package design in the context of full-scale commercial and industrial entry

QingChun Li, Nan Liu* & Ye Zhang
Liaoning Electric Power Trading Center Co., Ltd, Ltd, Shenyang, Liaoning, China

Yang Qi
Economic and Technical Research Institute of State Grid Liaoning Electric Power Co., Ltd, Shenyang, Liaoning, China

QianQiao Zhao
State Grid Liaoning Marketing Service Center, Shenyang, Liaoning, China

ABSTRACT: Under the background of comprehensive industry and commerce entering the market, in the fully competitive electricity market, how to establish a scientific and efficient retail product package to meet the differentiated needs of users has become the key exploration goal of electricity-selling enterprises. This paper will discuss the basic theory and mechanism of retail package design based on in-depth research on the industry's current situation and future development. Moreover, it talks about commerce entering the market and designing retail packages that meet users' needs to maximize the utility of electricity and maximize the utility of electricity in their profits.

1 INTRODUCTION

In the current tight supply of electricity and coal, the power system has ushered in a new round of reform. Commercial and industrial customers have been fully liberalized to enter the power market to participate in market transactions, the number of participants in the retail power market has increased, and opportunities and risks exist in the power sales market. In the context of full commercial and industrial market entry, it has become a key objective for power sales companies to explore how to establish scientific and efficient retail packages to meet the differentiated needs of customers in a fully competitive power sales market. In this paper, we discuss the basic theory and mechanism of retail package design based on an in-depth study of the current situation and future development of commercial and industrial market entry. We design a retail package that maximizes customers' utility and profit.

2 ANALYSIS OF THE CURRENT SITUATION OF FULL-SCALE COMMERCIAL AND INDUSTRIAL ENTRY

2.1 Analysis of policies of full-scale commercial and industrial entry

At this stage of tight power supply and scarcity of coal resources, to improve energy utilization efficiency, alleviate tension, and guarantee people's livelihood security. The National

---

*Corresponding Author: 27011150@qq.com

Development and Reform Commission issued the "Notice on Further Deepening the Market Reform of Feed-in Tariff for Coal-fired Power Generation" (NDRC Price [2021] No. 1439) in October 2021 (in the future referred to as "No. 1439"). No. 1439 pointed out the full liberalization of the feed-in tariff for coal-fired power generation, the abolition of the industrial and commercial catalog sales tariff, and the promotion of all industrial and commercial users to enter the market (Weng 2022). This reform was in 2015, the second round of power system reform for six years after exploring another important turning point in the history of power reform.

The reform is large in scale, wide in scope, and strong in execution. Many new users have changed from receiving electricity prices to participating in market transactions, with many commercial and industrial users not having the conditions to enter market transactions. To solve this problem, the Development and Reform Commission issued a supporting document, "National Development and Reform Commission General Office on the organization of power grid enterprises agent to purchase electricity notice" (Development and Reform Office price [2021] No. 809) (in the future referred to as "No. 809"). No. 809 proposed power grid enterprises agent to purchase electricity mechanism, pointed out that the power grid Enterprises can use agents to help commercial and industrial users without the ability to purchase electricity in the electricity market to conduct electricity purchase transactions.

2.2 Analysis of the current situation of the development of the electricity sales side under the full-scale commercial and industrial entry

Coal-fired power generation feed-in tariffs and commercial and industrial users are fully liberalized, making the number of market users grow substantially and the scale of market-based transactions expand rapidly. The commodity attributes of electricity become obvious, the types of market transactions become rich, and the frequency of transactions gradually increases. Retail users with diverse electricity characteristics and very different service needs enter the market. These changes have put higher requirements on the trading ability and business level of electricity sales users.

As the number of participants in the electricity retail market increases, opportunities and risks exist in the electricity sales market. Suppose electricity sales companies want to survive and thrive in the electricity market and provide customers with better prices and services. In that case, they must continue to develop and innovate, adjust their business models and improve their service levels. Therefore, in the new context of full-scale commercial and industrial entry into the market, there may be limitations to the retail packages previously offered by power sales companies. For power sales companies, the influx of competitors has increased the cost of selling electricity, and retail package pricing needs to be adjusted. Customers with more retail packages to choose from will need a more comprehensive understanding of their electricity needs and the rules of the retail market to select a reasonable and efficient retail package from the many options available.

3 RETAIL PACKAGE DESIGN IN THE CONTEXT OF FULL-SCALE COMMERCIAL AND INDUSTRIAL ENTRY

The highly competitive electricity sales market has both opportunities and challenges. Power sales companies face a complex and volatile wholesale market. They also need to provide lower prices and better service to retail customers to gain a foothold in the retail market. They need to design a tariff product that is highly adaptable and profitable, and the design of retail packages is a key decision point for them. As a more reasonable and fair market-based competitive tool, retail packages have become a key exploration target for power sales companies.

3.1 Retail package basic concept

Electricity retail packages present electricity in clear detail with parameters such as quality, energy level, power, and tariff, and help the electricity sales company to provide services to electricity customers as a presentation method. Retail packages turn invisible electricity into tradable commodities, meeting customers' rich and diverse needs. Scientific and reasonable retail packages play an important role in maintaining the reliability of the power system and the efficient operation of the power market, as well as ensuring the adequacy and cost-effectiveness of future power system investments (Wu *et al.* 2019).

3.2 Retail package pricing theory

Before designing a retail package, the underlying theory of package tariff pricing needs to be analyzed so the package can be priced more scientifically and rationally. This section introduces competitive pricing theory and demand-oriented pricing theory.

3.2.1 Competitive pricing theory

With the liberalization of China's electricity sales market, there are more and more electricity sales companies, and the competition on the electricity sales side is intensifying. Therefore, in participating in the electricity sales business, the electricity sales companies should study their competitors' service status and price level. They set the corresponding retail electricity prices based on their strength concerning the cost and supply and demand conditions so that customers can enjoy higher electricity utility than the packages provided by the same industry companies. This pricing method is the competitive pricing method. When setting the initial retail package tariff, the basic requirement is to ensure the cost recovery of the electricity sales company, and the goal is to maximize the customer's utility and profit. When a competitor emerges, the company needs to adjust its marketing strategy according to the competitor's retail tariff (He 2021).

3.2.2 Demand-driven pricing theory

All production operations should be centered on consumer demand and fully reflected in all production operations. The method of determining prices based on consumer demand is called demand-driven pricing. The demand-driven pricing method mainly includes price discounts and demand differentiation pricing methods. The price discount pricing method attracts customers to encourage customers to pay on time. The commonly used discount pricing strategy mainly includes cash discounts, function discounts, and time discounts to provide customers with a certain discount on the original retail price. For example, when the electricity demand is low, the electricity sales company offers a low valley tariff or night tariff. Demand differential pricing is a method of setting multiple prices for the same commodity, where the price difference is not caused by a difference in the cost of the commodity but by a difference in consumer demand. For example, electricity sales companies offer green power packages based on customers' environmental needs to help customers get the most out of their electricity while meeting individual needs.

The two pricing theories can make the electricity sales company's pricing conform to the maximum market demand and promote retail package sales, which is conducive to the electricity sales company to obtain the best economic benefits and customers to obtain the maximum utility of electricity.

3.3 Retail package design in the context of full-scale commercial and industrial entry

In the new competitive retail environment, many customers with different electricity consumption characteristics and service needs have entered the market. Electric sales companies represent many customers and offer various retail packages for customers to choose from. This paper classifies electricity users into large commercial and industrial users and small commercial and industrial users and residents according to their electricity consumption characteristics. Large-scale commercial and industrial customers often purchase large quantities so that the electricity sales company can design packages based on historical data. In contrast, for small-scale commercial and industrial customers and residents, the number of such customers is large, the number of single purchases is small, the electricity consumption varies greatly, and the purchase situation is complex. Hence, the electricity sales company often needs to provide additional personalized services in addition to the generally applicable retail packages. Therefore, we design separate retail packages for these customers in this paper.

3.3.1 *Standard retail package design*
Electricity-selling companies design standard retail packages based on the characteristics of large industrial and commercial customers' electricity demand, electricity consumption habits, and tariff sensitivity, which is conducive to the supervision of the retail electricity market and the protection of electricity customers' rights and interests by the relevant authorities.

3.3.1.1 *Minimum spending package*
Considering the unstable electricity demand of some large commercial and industrial customers, the power sales company provides minimum consumption packages for customers with fluctuating electricity consumption to protect their interests and minimize the risk of electricity purchase.

The minimum consumption package, when the customer's electricity consumption is below a certain threshold, is paid following the minimum consumption amount, and the portion exceeding the threshold is billed according to the established tariff of the package contract. The formula for calculating the package price paid by the customer every month is shown in the following equation.

$$P = \begin{cases} P_0, & Q < Q_0 \\ P_0 + P_F \times (Q - Q_0), & Q \geq Q_0 \end{cases} \tag{1}$$

In the formula, P is the package price; P_0 is the minimum consumption amount; Q is the actual electricity consumption of the customer in the month; Q_0 is the electricity threshold; P_F is the established tariff of the retail package contract.

3.3.1.2 *Stepped tariff package*
Most large commercial and industrial customers show an increasing trend of electricity consumption with the stable development of the industry, so the electricity sales companies provide step tariff packages. The use of step tariff packages is, on the one hand, conducive to promoting energy conservation and emission reduction, guiding users with high electricity consumption to adjust their electricity consumption behavior and promoting electricity conservation; on the other hand. It is conducive to the retail electricity market, considering users' price affordability in different industries and establishing a fair electricity pricing mechanism. Table 1 shows the electricity value standards and the corresponding tariff standards for different gradients.

Table 1. Stepped tariff package gradient division.

Gradient	Amount of electricity	Electricity tariff
1st gradient	Q_1	P_1
2nd gradient	Q_2	P_2
...	...	...
nth gradient	Q_n	P_n

The formula for calculating the package price paid by the user monthly is shown in the following equation.

$$P = P_1 \times Q_1 + P_2 \times Q_2 + \cdots + P_n \times Q_n \tag{2}$$

The formula P is the package price Q_n, the amount of electricity in the nth gradient P_n, and the tariff of the nth gradient.

3.3.1.3 *Peak and valley tariff package*

Power sales companies offer peak and valley tariff packages for customers with obvious peak and valley characteristics and high flexibility in adjusting peak and valley power. Peak and valley tariff packages can effectively avoid problems such as huge losses and fuel wastage caused by the shutdown and restart of generating units and help reduce customer production and operation costs.

The formula for calculating the package price paid by the user monthly is shown in the following equation.

$$P = P_U \times Q_U + P_S \times Q_S + P_D \times Q_D \tag{3}$$

In the formula, P is the package price; P_U is the peak hour tariff; Q_U is the number of peak hour electricity consumption; P_S is the tariff when it is flat; Q_S is The number of electricity consumption during the smoothing period; P_D is the tariff in the low valley; Q_D is the number of electricity consumption during low hours.

3.3.2 *"Basic + Green power" package design*

Retail electricity packages can be divided into basic and customized packages. Among them, the basic package refers to the package issued by the electricity sales company to the customer, the electricity sales company can entrust the trading center for information dissemination, and the customer passively selects. The customized package refers to the retail package issued by the retail customer to invite demand. The electricity sales company actively accepts and customizes the development of the retail package according to the invitation's content (Hua 2021).

3.3.2.1 *Basic retail package*

The basic package tariff is the retail package purchase price for end users and is calculated as shown in the following equation.

$$P_B = P_P + P_C + P_E \tag{4}$$

In the formula, P_B is the retail package price charged by the electricity sales company to the end user; P_P is the power purchase price for power sales companies in the electricity market; P_C is the projected price for power sales companies to avoid market risk; P_E is the price of government funds and surcharges. The purchase price is adapted to customers' needs with seasonal production characteristics and is divided into quarterly tariffs according to four seasons: spring, summer, autumn, and winter.

The formula for calculating the monthly purchase price of electricity in the electricity market for power sales companies is shown in the following equation.

$$P_E = P_i \times Q_i \tag{5}$$

In the formula, P_i is the tariff for the month; Q_i is the number of electricity consumption for the month. The electricity price is P_S for March, April, and May; P_M for June, July and August; P_F for September, October and November; and P_W for December, January and February.

3.3.2.2 *Green power package*

With the global low-carbon development strategy, new energy sources such as wind and light, and renewable energy sources such as hydropower, which are zero-emission power sources, are taking up an increasingly high proportion of electricity. Thus, power sales companies are offering green power packages. Green power packages mainly use wind, photovoltaic, and other green power generation to meet customers' electricity needs. The higher the percentage of green power purchased by the customer, the greater the discount on the power purchase. Due to the randomness and volatility of new energy generation, the green power package in this paper is currently only available to small commercial and industrial customers and residents. The discount rate varies according to the percentage of green power, as shown in Table 2.

Table 2. Green power package tariff discount rate.

Percentage of green power	Discount rate
<20%	a
20%–60%	b
>60%	c

The unit tariff discount rate is when green power generation accounts for less than 20% of the total electricity. The unit tariff discount rate is b when green power generation accounts for 20%–60% of the total electricity; the unit tariff discount rate is c when green power generation accounts for more than 60% of the total electricity.

The formula for calculating the package price paid by the user monthly is shown in the following equation.

$$P = a \times P_L \times Q_G + P_L \times (Q - Q_G) \tag{6}$$

In the formula, P is the package price; a is the discount rate; According to the percentage of green power into a, b, c three levels; P_L is the electricity generation tariff for green power energy; Q_G is the amount of electricity generated for green power; Q is the number of customers using electricity in the month.

4 CONCLUSION

The full-scale commercial and industrial market entry have made the retail electricity market more competitive. Electricity sales companies must design well-matched electricity retail packages and provide personalized services according to customers' needs (Lu *et al.* 2020). In this paper, we analyze the current situation of the electricity sales side of the retail electricity market against the background of full-scale industrial and commercial market entry. We Propose various electricity retail package models: minimum consumption packages, tiered tariff packages, and green electricity packages to maximize customers' utility and profit to provide a reference for electricity sales companies to design retail packages. Although some research progress has been made in this paper, a further in-depth research is still needed. In the future, electricity sales companies can consider the pricing of retail packages from the perspective of maximizing the optimal allocation of resources to improve the efficiency of resource utilization better.

ACKNOWLEDGMENTS

On the completion of this paper, I would like to express my heartfelt thanks and sincere respect to all the leaders and colleagues of the research on the main problems facing the construction of Liaoning's electricity market and its strategies (Item number: SGLNJY00ZLJS2200058). Thank you for your patient guidance and for providing me with many learning materials. The successful completion of this paper cannot be separated from your help.

REFERENCES

He Qing. *Research the Retail Price Package Pricing of Electricity Selling Companies in the Spot Market*. North China Electric Power University (Beijing), 2021. DOI: 10.27140/d.cnki.ghbu.2021.000474.

Hua Jingwen, *Research on User-oriented Energy Value-added Services and Retail Package Pricing Mechanism*. North China Electric Power University (Beijing), 2021. DOI: 10.27140/d.cnki.ghbu.2021.000764.

Lu En, Bei pei, Wang Haohao, Chen Qing, Li Wenxuan and Gu Weizhen. Pricing Strategy Design for Retail Electricity Packages Considering Customer's Autonomous Selectivity. *Power System Automation*, 2020,44 (19):177–184.

Weng Shuang. New Power Sales Bureau. *Guangxi Electric Power*, 2022 (04): 58–62.

Wu Jinghui, Zhang Jie, Pan Shuyan, Wu Feng, Li Kaixin, Yao Xing'an, Yang Liu, Zhan Weixu and Hanghang E. Standard Retail Package Design in the Spot Market of Electric Power – Analysis Based on User Clustering. *Price Theory and Practice*, 2019 (12): 132–136. DOI: 10.19851/j.cnki.CN11-1010/F.2019.12.039.

Civil Engineering and Energy-Environment – Gao & Duan (Eds)
© 2023 the Editor(s), ISBN 978-1-032-56059-5

Author index

An, X. 58

Cai, M. 234
Cao, Q. 112
Cao, Y. 241
Chang, G. 350
Chen, H. 112, 382
Chen, L. 186
Chen, T. 503
Chen, W. 9
Chen, Y. 234
Cheng, L. 226
Cui, T. 58

Deng, Q. 72
Deng, W. 64
Dong, B. 410
Dong, L. 241
Dong, S. 58
Du, Y. 192, 342

Fan, B. 396
Fan, S. 263
Fan, Y. 495
Fan, Y.L. 457
Fang, H. 361, 375
Fang, Y. 72
Feng, S. 64
Feng, Z. 410
Ma, X. 423
Fu, Q. 335
Fu, S. 361, 375

Gong, L. 78
Guo, S. 199

He, H. 39
He, J. 478

He, Z. 263
Hong, B. 119
Hong, Y. 72
Hu, L. 277
Hu, Z. 254
Huang, M.T. 335
Huang, Q. 139
Huang, R. 78
Huang, X. 23, 129
Huang, Y. 316

Jia, D. 241
Jian, W. 316, 423
Jiang, Y. 174, 451
Jie, Y. 443

Kang, L. 335
Kang, Q. 126
Ke, Y. 271
Kou, W.W. 457

Lei, Y. 207
Lei, Y.L. 303
Li, D. 226, 436
Li, F. 199
Li, G. 234
Li, H. 277
Li, J. 249
Li, L. 293
Li, M. 509
Li, Q.C. 534
Li, S. 263, 361, 375
Li, T. 156
Li, W. 129, 285
Liang, D. 249
Liang, J. 350
Liang, T. 78
Liang, W. 285

Liang, Z. 91
Liao, S. 254
Lin, L. 241
Lin, X. 361, 375
Liu, B. 162
Liu, C. 58, 485
Liu, H. 119, 216
Liu, J. 326
Liu, N. 534
Liu, P. 350
Liu, Y. 119, 199,
   207, 495
Lu, P. 112
Lu, W. 402
Lu, Z. 207, 234
Luo, W. 293
Lv, X. 14
Lv, Y. 495

Ma, L. 309
Ma, Q. 495
Ma, R. 9
Ma, X. 174, 425, 485
Ma, Y. 112, 382
Ma, Z. 78, 199
Meng, F. 100

Pan, X. 241
Peng, Y. 509

Qi, J. 199, 371
Qi, Y. 254, 534
Qian, X. 174
Qin, Z. 199
Qiu, H. 100
Qu, X. 309

Ren, Z. 402
Ruan, W. 222

Shao, Z. 199
Shen, Q. 72
Shi, W. 9, 382
Shu, S. 263
Shuai, B. 309
Sun, H. 485
Sun, Z. 199

Tan, H. 335
Tan, C. 342
Tan, W. 91
Tan, Y. 485
Tan, Z. 342
Tang, Y. 156
Tao, H. 91
Tao, Y. 382
Tian, N. 309

Wan, L. 216
Wang, C. 23, 119, 129
Wang, F. 174, 249
Wang, H. 309, 342, 485
Wang, J. 39, 249, 350, 361, 375, 396
Wang, K. 199
Wang, L. 350
Wang, P. 285, 382
Wang, R. 156, 390
Wang, S. 350, 467, 485
Wang, T. 249
Wang, X. 1
Wang, Y. 78, 271, 390
Wang, Z. 162, 516
Wei, L. 199
Wu, J. 156
Wu, Q. 342

Wu, W. 186, 199
Wu, Y. 72

Xi, J. 78
Xia, B. 303
Xia, N. 408
Xiao, C. 309
Xiao, F. 271
Xiao, L. 126
Xie, G. 112
Xie, J. 149
Xiong, Y. 467
Xu, G. 263
Xu, J. 390
Xu, W. 309
Xu, X. 408

Ya, W. 443
Yan, X. 58, 199
Yan, Z. 396
Yang, B. 14
Yang, J. 119, 174
Yang, K. 293
Yang, L. 326
Yang, S. 139
Yang, X. 485
Yu, G. 495
Yu, M. 309
Yuan, B. 78
Yuan, W. 467
Yuchi, X. 168
Yue, K. 516

Zeng, C. 271
Zhai, C. 263

Zhang, A. 174
Zhang, B. 503
Zhang, J. 408, 474
Zhang, K. 369
Zhang, L. 14, 342
Zhang, N. 1
Zhang, Q. 100
Zhang, R. 64
Zhang, S. 241, 390
Zhang, X. 523
Zhang, X.M. 457
Zhang, Y. 241, 534
Zhang, Z. 14, 156, 249
Zhao, G. 85, 415
Zhao, G.C. 48
Zhao, H. 474, 509, 516
Zhao, Q.Q. 534
Zhao, S. 162, 485
Zhao, W. 402
Zhao, X. 91
Zheng, X. 126
Zheng, Y. 516
Zhou, B. 58
Zhou, J. 402
Zhou, K. 126
Zhou, T. 91
Zhou, X. 369
Zhu, B. 485
Zhu, J. 293, 474
Zhu, X. 234, 271
Zhuang, C. 174
Zou, C. 293
Zou, Q. 156